# Achievements and Prospects of Advanced Pavement Materials Technologies

# Achievements and Prospects of Advanced Pavement Materials Technologies

Editors

**Zhanping You**
**Jian-long Zheng**
**Hainian Wang**

MDPI • Basel • Beijing • Wuhan • Barcelona • Belgrade • Manchester • Tokyo • Cluj • Tianjin

*Editors*
Zhanping You
Department of Civil, Environmental, and Geospatial Engineering
Michigan Technological University
Houghton
United States

Jian-long Zheng
School of Traffic and Transportation Engineering
Changsha University of Science and Technology
Changsha
China

Hainian Wang
College of Transport Infrastructure
Chang'an University
Xi'an
China

*Editorial Office*
MDPI
St. Alban-Anlage 66
4052 Basel, Switzerland

This is a reprint of articles from the Special Issue published online in the open access journal *Applied Sciences* (ISSN 2076-3417) (available at: www.mdpi.com/journal/applsci/special_issues/Pavement_Materials_Technologies).

For citation purposes, cite each article independently as indicated on the article page online and as indicated below:

| LastName, A.A.; LastName, B.B.; LastName, C.C. Article Title. *Journal Name* **Year**, *Volume Number*, Page Range. |
|---|

**ISBN 978-3-0365-5566-9 (Hbk)**
**ISBN 978-3-0365-5565-2 (PDF)**

# Contents

# About the Editors

**Zhanping You**

Dr. Zhanping You earned his Ph.D. in Civil Engineering from the University of Illinois at Urbana - Champaign in 2003. Dr. You is a Distinguished Professor at Michigan Technological University. Dr. You has completed research projects related to road materials, pavement engineering, and sustainable building materials. His contribution to pavement and materials research has led to journal articles, book chapters, and advances in engineering practice. He has led research projects from pavement science to engineering practices with funding from federal, state, and local agencies. His contribution to pavement and materials research has been featured in newspapers, magazines, and other media. Dr. You has published over 400 papers in peer reviewed journals and conference proceedings. Dr. You has received numerous recognitions. He was awarded the U.S. Department of Transportation's Dwight David Eisenhower Transportation Fellowship in 2001. In both 2004 and 2005, he was awarded the Dwight David Eisenhower Transportation Faculty Fellowship. He earned the prestigious Michigan Tech Research Award in 2019 and University Distinguished Professorship. He was named a Fellow of ASCE and a Fellow of the EMI. Dr. You served as the Chair of ASCE Engineering Mechanics Institute's Pavements Committee and CI's Bituminous Materials Committee. He is also the guest editor for other journals.

**Jian-long Zheng**

Dr. Zheng is a member of the Chinese Academy of Engineering. He is a highway engineering expert, professor, and doctoral supervisor of Changsha University of Science and Technology, director of the National Engineering Reesearch Center of Highway Maintenance Technology, director of the Key Laboratory of Highway Engineering of the Ministry of Education, the vice director of Teaching Guiding Committee for Civil Engineering of the Ministry of Education, the vice chairman of China Highway & Transportation Society and the executive director of the International Road Federation. Dr. Zheng has been long engaged in developing and practicing the technology of highway and geotechnical engineering, with over 260 articles and 7 academic books published. He holds 60 patents of national in inventions. He led a well know project on the Technology of Highway Construction in Expansive Soil Area, and he received the First Prize of National Science and Technology Progress Award in 2009 due to this achievement. His effort on the Technology of Asphalt Pavement State Design Method and Structural Performance Improvement Technology led to him receiving the Second Prize of National Science and Technology Progress Award in 2012. Meanwhile, he won a number of special awards. He won a Second Prize of National Teaching Achievement and two First Provincial Level Prizes. He was also awarded the Outstanding Achievement Award in Transportation Technology from the Ministry of Transport among his numerous key awards.

**Hainian Wang**

Dr. Hainian Wang is the Chairman of the College of Transport Infrastructure and Dean of Chang'an Dublin International College of Transportation at Chang'an University. He has engaged in teaching and scientific research work on transportation infrastructure for nearly two decades. His research interests include pavement structure design and regional environment analysis, airport runway design and rehabilitation, sustainable pavement materials development of evaluation, microstructure characterization, modeling, etc. He has supervised over 34 MS and 6 Ph.D. thesis research studies. Dr. Hainian Wang is serving as the leading scientist of the National Key Research and Development Program of China. He has published more than 130 papers as a lead author or co-author, including more than 60 SCI papers, cited 1941 times, h impact factor of 27. He was selected as the 2020 and 2021 Most Cited Chinese Researchers by ELSEVIER, and is on the world's top 2% scientists list by Standford University. He has hosted five international academic conferences, served as the guest editor and published five Special Issues of international journals, and been invited to give more than 20 international academic presentations outside the United States, Germany, the United Kingdom, South Africa and other countries.

applied sciences

MDPI

Editorial

# Achievements and Prospects of Advanced Pavement Materials Technologies

**Zhanping You [1,*], Jian-long Zheng [2] and Hainian Wang [3]**

1 Department of Civil and Environmental Engineering, Michigan Technological University, 1400 Townsend Drive, Houghton, MI 49931, USA
2 School of Traffic and Transportation Engineering, Changsha University of Science and Technology (CSUST), Changsha 410004, China; zjl@csust.edu.cn
3 College of Transport Infrastructure, Chang'an University, South Erhuan Middle Section, Xi'an 710064, China; wanghn@chd.edu.cn
* Correspondence: zyou@mtu.edu

Received: 8 September 2020; Accepted: 8 September 2020; Published: 2 November 2020

Road transportation is a basic need for mobility and daily life. Currently, there are a number of challenges in dealing with distressed pavement and seeking new materials with sustainability in mind in pavement systems. Therefore, it is important for us to conduct further research in the following areas: (1) pavement structure, materials, and design; (2) pavement models as better solutions for pavement constructions; (3) pavement mechanics for improved understanding and mechanism analyses; (4) utilization of recycled materials for environmentally friendly solutions; (5) maintenance and rehabilitation for an extended life span of pavement; (6) intelligent construction for project management, energy conservation, and future constructions; and (7) innovative approaches to test and evaluate the performance of pavement materials. The purpose of this Special Issue "Achievements and Prospects of Advanced Pavement Materials Technologies" is to explore new research ideas for pavement materials as described above.

The Special Issue "Achievements and Prospects of Advanced Pavement Materials Technologies" is part of the journal *Applied Sciences* (ISSN 2076-3417). This Special Issue belongs to the section "Materials." This Special Issue contains 25 technical articles [1–25]. All of the 25 papers have been peer reviewed under the journal's rigorous review criteria. The collection includes invited papers from experts in international communities and articles that have been selected from the 2019 World Transport Convention (WTC) held in June 2019 in Beijing, the 4th International Conference on Transportation Infrastructure and Materials in Jinan, Shandong, China in 2019, and the 5th Chinese-European Workshop on Functional Pavements in 2019 in Changsha, China.

Modified asphalt binders and mixtures are the most important construction materials for pavement, and therefore they always attract research interest all over the world. Nearly half of the papers collected in this Special Issue are related to these topics. M. Hasan, Z. You, M. Satar, M. Warid, N. Kamaruddin, D. Ge, and R. Zhang evaluated the effects of Titanate coupling agent on the engineering properties of asphalt binders and mixtures incorporating LLDPE-$CaCO_3$ pellets [8]. Another paper presents an investigation on a damage model of an eco-friendly basalt fiber-modified asphalt mixture under freeze–thaw cycles [18], authored by W. Wang, Y. Cheng, G. Ma, G. Tan, X. Sun, and S. Yang. Additionally, C. Zhang, H. Wang, Z. You, J. Gao, and M. Irfan investigated the performance of styrene-butadiene-styrene (SBS)-modified asphalt based on the different evaluation methods [22]. The micromechanism of the dispersion behavior of polymer-modified rejuvenators in aged asphalt materials is presented by M. Zhao, F. Shen, and Q. Ding. Studies on manufacturing modifiers are also included, such as preparation of polyacrylate hollow microspheres [4] and design of SBS-modified bitumen stabilizer powder based on the vulcanization mechanism [23].

Nanotechnology has been adopted by researchers to reinforce properties of asphalt binders and mixtures. J. Rafi, M. Kamal, N. Ahmad, M. Hafeez, M. Faizan ul Haq, S. Aamara Asif, F. Shabbir, and S. Bilal Ahmed Zaidi used carbon black nanoparticles to reinforce an asphalt mixture [15]. In another study, C. Li, Z. Fan, S. Wu, Y. Li, Y. Gan, and A. Zhang also evaluated the effects of carbon black nanoparticles from the pyrolysis of discarded tires on the performance of asphalt and its mixtures [11]. Yet another study presents the dispersion homogeneousness and performance enhancement of a carbon nanotubes (CNTs)-modified asphalt binder [7], authored by M. Haq, N. Ahmad, M. Nasir, Jamal; M. Hafeez, J. Rafi, S. Zaidi, and W. Haroon. In addition, the durability and the properties of an asphalt concrete with nano hydrophobic silane silica were evaluated considering the spring-thawing effect [5], in an article authored by W. Guo, X. Guo, M. Sun, and W. Dai.

Different types of waste oil have been utilized by investigators to improve virgin binder properties or rejuvenate aged binders. T. Shoukat and P. Yoo present a study on the rheology of an asphalt binder modified with 5w30-viscosity grade waste engine oil [16]. Meanwhile, J. Gao, H. Wang, Z. You, M. Mohd Hasan, Y. Lei, and M. Irfan used wood-derived bio-oil to modify an asphalt binder and evaluate its rheological behavior and sensitivity [5]. On the other hand, researchers investigated the possibility of using waste vegetable oil to rejuvenate aged asphalt binders [19]. Reclaimed asphalt pavement (RAP) is another hot research topic in recycling technology. In this Special Issue, two papers have been collected on this topic. The first one is an evaluation of the fatigue life of asphalt concrete mixtures with RAP [1], authored by W. Bańkowski. The second paper presents a method for evaluating the particle clustering phenomenon in RAP [20], authored by G. Xu, T. Ma, Z. Fang, X. Huang, and W. Zhang.

The mechanical properties of paving materials are always included within the scope of research interests. In this Special Issue, J.-C. Carret, H. Di Benedetto, and C. Sauzéat characterized asphalt mixes' behavior from dynamic tests and made a comparison with conventional cyclic tension–compression tests [2]. S. Lv, X. Fan, C. Xia, J. Zheng, D. Chen, and L. You characterized the moduli decay of an asphalt mixture under different loading conditions [14]; T. Huang, S. Qi, M. Yang, S. Lv, H. Liu, and J. Zheng presented a strength criterion of asphalt mixtures in three-dimensional stress states under freeze–thaw conditions [9]; X. Li, X. Lv, X. Liu, and J. Ye analyzed an indirect tensile fatigue test of an asphalt mixture using the discrete element method [12]; J. Chen, C. Yao, H. Wang, W. Huang, X. Ma, and J. Qian investigated the interface shear performance between a porous polyurethane mixture and asphalt sublayer [3]. Additionally, the deformation and damping of cement treated and expanded with polystyrene mixed lightweight subgrade fill were characterized under cyclic load [13], by W. Lu, L. Miao, J. Zhang, Y. Zhang, and J. Li.

Portland cement concrete is another widely used material for pavement construction. Two studies on this topic are included in this Special Issue. A case study in China was conducted by L. Yu, X. Yang, X. Yan, X. Zhang, T. Zhao, C. Duan, and J. Mills-Beale, presenting the design and construction of oblique prestressed concrete pavement [21]. M. Zheng, Y. Tian, X. Wang, and P. Peng investigated the durability of the anti-skid performance of grooved concrete pavement [25].

Pavement management and maintenance are essential for keeping transportation infrastructure in good service condition. Y. Tian, B. Ma, K. Tian, N. Li, and X. Zhou determined the components of light screening preventive maintenance agent (LS pre-maintenance agent) for strong ultraviolet (UV) radiation areas and evaluated its performance [17]. C. Kou, A. Kang, P. Xiao, P. Mikhailenko, H. Baaj, L. Sun, and Z. Wu presented a source pollution control measure based on the spatial-temporal distribution characteristic of the runoff pollutants at urban pavement sites [10].

**Acknowledgments:** This Special Issue would not be possible without the contributions of the above authors, hundreds of dedicated volunteer reviewers, and the editorial team of *Applied Sciences*.

**Conflicts of Interest:** The authors declare no conflict of interest.

## References


1. Bańkowski, W. Evaluation of Fatigue Life of Asphalt Concrete Mixtures with Reclaimed Asphalt Pavement. *Appl. Sci.* **2018**, *8*, 469. [CrossRef]
2. Carret, J.-C.; Di Benedetto, H.; Sauzéat, C. Characterization of Asphalt Mixes Behaviour from Dynamic Tests and Comparison with Conventional Cyclic Tension–Compression Tests. *Appl. Sci.* **2018**, *8*, 2117. [CrossRef]
3. Chen, J.; Yao, C.; Wang, H.; Huang, W.; Ma, X.; Qian, J. Interface Shear Performance between Porous Polyurethane Mixture and Asphalt Sublayer. *Appl. Sci.* **2018**, *8*, 623. [CrossRef]
4. Chen, P.; Chen, P.; Ye, N.; He, C.; Tang, L.; Li, S.; Sun, L.; Li, Y. Preparation of Polyacrylate Hollow Microspheres via Facile Spray Drying. *Appl. Sci.* **2019**, *9*, 228. [CrossRef]
5. Gao, J.; Wang, H.; You, Z.; Mohd Hasan, M.R.; Lei, Y.; Irfan, M. Rheological Behavior and Sensitivity of Wood-Derived Bio-Oil Modified Asphalt Binders. *Appl. Sci.* **2018**, *8*, 919. [CrossRef]
6. Guo, W.; Guo, X.; Sun, M.; Dai, W. Evaluation of the Durability and the Property of an Asphalt Concrete with Nano Hydrophobic Silane Silica in Spring-Thawing Season. *Appl. Sci.* **2018**, *8*, 1475. [CrossRef]
7. Ahmad, N.; Nasir, M.A.; Hafeez, M.; Rafi, J.; Zaidi, S.B.A.; Haroon, W. Carbon Nanotubes (CNTs) in Asphalt Binder: Homogeneous Dispersion and Performance Enhancement. *Appl. Sci.* **2018**, *8*, 2651.
8. Hasan, M.R.M.; You, Z.; Satar, M.K.I.M.; Warid, M.N.M.; Kamaruddin, N.H.M.; Ge, D.; Zhang, R. Effects of Titanate Coupling Agent on Engineering Properties of Asphalt Binders and Mixtures Incorporating LLDPE-$CaCO_3$ Pellet. *Appl. Sci.* **2018**, *8*, 1029. [CrossRef]
9. Huang, T.; Qi, S.; Yang, M.; Lv, S.; Liu, H.; Zheng, J. Strength Criterion of Asphalt Mixtures in Three-Dimensional Stress States under Freeze-Thaw Conditions. *Appl. Sci.* **2018**, *8*, 1302. [CrossRef]
10. Kou, C.; Kang, A.; Xiao, P.; Mikhailenko, P.; Baaj, H.; Sun, L.; Wu, Z. A Source Pollution Control Measure Based on Spatial-Temporal Distribution Characteristic of the Runoff Pollutants at Urban Pavement Sites. *Appl. Sci.* **2018**, *8*, 1802. [CrossRef]
11. Li, C.; Fan, Z.; Wu, S.; Li, Y.; Gan, Y.; Zhang, A. Effect of Carbon Black Nanoparticles from the Pyrolysis of Discarded Tires on the Performance of Asphalt and its Mixtures. *Appl. Sci.* **2018**, *8*, 624. [CrossRef]
12. Li, X.; Lv, X.; Liu, X.; Ye, J. Discrete Element Analysis of Indirect Tensile Fatigue Test of Asphalt Mixture. *Appl. Sci.* **2019**, *9*, 327. [CrossRef]
13. Lu, W.; Miao, L.; Zhang, J.; Zhang, Y.; Li, J. Characteristics of Deformation and Damping of Cement Treated and Expanded Polystyrene Mixed Lightweight Subgrade Fill under Cyclic Load. *Appl. Sci.* **2019**, *9*, 167. [CrossRef]
14. Lv, S.; Fan, X.; Xia, C.; Zheng, J.; Chen, D.; You, L. Characteristics of Moduli Decay for the Asphalt Mixture under Different Loading Conditions. *Appl. Sci.* **2018**, *8*, 840. [CrossRef]
15. Rafi, J.; Kamal, M.A.; Ahmad, N.; Hafeez, M.; Faizan ul Haq, M.; Aamara Asif, S.; Shabbir, F.; Bilal Ahmed Zaidi, S. Performance Evaluation of Carbon Black Nano-Particle Reinforced Asphalt Mixture. *Appl. Sci.* **2018**, *8*, 1114. [CrossRef]
16. Shoukat, T.; Yoo, P.J. Rheology of Asphalt Binder Modified with 5W30 Viscosity Grade Waste Engine Oil. *Appl. Sci.* **2018**, *8*, 1194. [CrossRef]
17. Tian, Y.; Ma, B.; Tian, K.; Li, N.; Zhou, X. Study on Components Determination and Performance Evaluation of LS Pre-Maintenance Agent. *Appl. Sci.* **2018**, *8*, 889. [CrossRef]
18. Wang, W.; Cheng, Y.; Ma, G.; Tan, G.; Sun, X.; Yang, S. Further Investigation on Damage Model of Eco-Friendly Basalt Fiber Modified Asphalt Mixture under Freeze-Thaw Cycles. *Appl. Sci.* **2019**, *9*, 60. [CrossRef]
19. Cao, X.; Cao, X.; Tang, B.; Wang, Y.; Li, X. Investigation on Possibility of Waste Vegetable Oil Rejuvenating Aged Asphalt. *Appl. Sci.* **2018**, *8*, 765.
20. Xu, G.; Ma, T.; Fang, Z.; Huang, X.; Zhang, W. The Evaluation Method of Particle Clustering Phenomena in RAP. *Appl. Sci.* **2019**, *9*, 424. [CrossRef]
21. Yu, L.; Yang, X.; Yan, X.; Zhang, X.; Zhao, T.; Duan, C.; Mills-Beale, J. Design and Construction of Oblique Prestressed Concrete Pavement: A Case Study in China. *Appl. Sci.* **2018**, *8*, 607. [CrossRef]
22. Zhang, C.; Wang, H.; You, Z.; Gao, J.; Irfan, M. Performance Test on Styrene-Butadiene-Styrene (SBS) Modified Asphalt Based on the Different Evaluation Methods. *Appl. Sci.* **2019**, *9*, 467. [CrossRef]
23. Zhang, W.; Ding, L.; Jia, Z. Design of SBS-Modified Bitumen Stabilizer Powder Based on the Vulcanization Mechanism. *Appl. Sci.* **2018**, *8*, 457. [CrossRef]

24. Zhao, M.; Shen, F.; Ding, Q. Micromechanism of the Dispersion Behavior of Polymer-Modified Rejuvenators in Aged Asphalt Material. *Appl. Sci.* **2018**, *8*, 1591. [CrossRef]
25. Zheng, M.; Tian, Y.; Wang, X.; Peng, P. Research on Grooved Concrete Pavement Based on the Durability of Its Anti-Skid Performance. *Appl. Sci.* **2018**, *8*, 891. [CrossRef]

**Publisher's Note:** MDPI stays neutral with regard to jurisdictional claims in published maps and institutional affiliations.

applied sciences

MDPI

Article

# Performance Test on Styrene-Butadiene-Styrene (SBS) Modified Asphalt Based on the Different Evaluation Methods

**Chen Zhang [1,2], Hainian Wang [3,*], Zhanping You [4], Junfeng Gao [3,4] and Muhammad Irfan [5]**

1 NO. 259 West Erhuan, Xi'an Aeronautical University, School of Energy and Architecture, Xi'an 710077, Shaanxi, China; zlw19861108@aliyun.com
2 Key Laboratory of Special Environment Road Engineering of Hunan Province, Changsha University of Science & Technology, Changsha 410004, Hunan, China
3 South Erhuan Middle Section, Key Laboratory of Transport Industry of Road Structure and Materials (Xi'an), Chang'an University, Xi'an 710064, Shaanxi, China; gaojunfeng@chd.edu.cn
4 Department of Civil and Environmental Engineering, Michigan Technological University, 1400 Townsend Drive, Houghton, MI 49931, USA; zyou@mtu.edu
5 College of Civil Engineering, National University of Sciences and Technology (NUST), NUST Campus, Risalpur 24080, Pakistan; mirfan@mce.nust.edu.pk
* Correspondence: wanghn@chd.edu.cn

Received: 8 December 2018; Accepted: 22 January 2019; Published: 30 January 2019

**Abstract:** To uniform the evaluation indicators of Styrene-Butadiene-Styrene (SBS) modified asphalt, the SK70# and SK90# matrix asphalt were modified by different SBS modifier dosage in this study. The test methods in China and Superpave were used to test the performance of each SBS-modified asphalt respectively, from which the appropriate evaluation index of SBS-modified asphalt was determined. The results showed that the addition of SBS modifier improved the high temperature performance and lowered the temperature sensitivity of asphalt binder, while it increased the viscosity of asphalt binder in high temperatures. Due to the variability that appeared in the results of the penetration test by the swelling of SBS-modified asphalt, the penetration test was not recommended to evaluate the performances of SBS-modified asphalt. The softening point of SBS-modified asphalt with the modifier dosages of 4.5%, 5%, 5.5% and 6% increased 5.7%, 12.8%, 22.5% and 26.4% respectively compared to the matrix asphalt for SK70# matrix asphalt, and increased 21.2%, 26.3%, 33.6% and 46.6% respectively compared to the matrix asphalt for SK90# matrix asphalt. The effect of SBS-modifier on the softening point of SK90# matrix asphalt is significantly better than that of SK70# matrix asphalt. The improvement effect of SBS modifier on low temperature performance of matrix asphalt decreased with a decrease in test temperature. When studying the influence of the SBS modifier on the low temperature performance of asphalt binder, it was recommended to use the bending beam rheometer (BBR) test to evaluate the low temperature performance of SBS-modified asphalt.

**Keywords:** highway engineering; SBS modified asphalt; laboratory test; evaluation index

## 1. Introduction

Styrene-butadiene-styrene (SBS) is a common modifier with high molecular polymer, which could make the asphalt binder modified by miscible with asphalt binder [1]. The SBS-modified asphalt could improve the high temperature rutting resistance, low-temperature crack resistance and anti-fatigue performance of asphalt pavement [2]. The SBS-modified asphalt has been applied widely in many high-grade pavement in China at present to satisfy the increasing traffic. SBS modified asphalt has wide scope of application [3]. In recent years, many scholars in China and other countries research much about SBS-modified asphalt. Khodaii (2009) conducted dynamic creep test on unmodified and SBS-modified samples, and the creep behavior of the samples was estimated by the three-stage creep model. The result showed that dense-graded mixtures had higher permanent deformation susceptibility than coarse graded mixtures, and lower stress levels in dynamic creep test could not show the actual behavior of asphalt mixtures and particularly the modified mixtures [4]. Forough (2014) used the creep curves derived from the dynamic creep tests to investigate the effects of loading frequency and temperature on the moisture sensitivity of dense-graded polymer-modified asphalt mixtures. The results showed that both the variables of loading frequency and temperature had significant effects on the permanent strains of both the dry and wet asphalt mixtures [5]. Huang (2015) used multiple stress creep recovery (MSCR) test to investigate the effect of cross-linking agent and SBS content on SBS-modified asphalt. The result showed the effect of increasing SBS content was more prominent for binders at lower SBS content. MSCR test failed to distinguish 5.0% and 5.5% SBS-modified asphalt in the study [6]. Wang (2017) conducted three points bending test and deformations test to evaluate the low-temperature performance and fatigue resistance of recycled SBS-modified asphalt mixture. The results showed that fatigue resistance of modified recycling of asphalt mixture with different RAPs did not vary much under low-temperature while displaying an obvious difference under higher temperature [7]. However, the test index used to evaluate the performance of SBS-modified asphalt is diversified, and lack of unification. This study based on the Chinese test methods of penetration test, softening point test, ductility test, elastic recovery test and Superpave test methods of dynamic shear rheometer test (DSR), Brookfield rotary viscosity test, bending beam rheometer (BBR) test, to analyze the influence of SBS modifier on asphalt performance. By comparing each performance index, it provides theoretical support for unifying the performance evaluation index of SBS modified asphalt.

## 2. Material

### 2.1. Matrix Asphalt

The matrix asphalt used in this study was SK70# and SK90#, and the technical indicators are shown in Table 1.

**Table 1.** The technical indicators of matrix asphalt.

| Technical Indicators | Units | SK90# | | SK70# | |
|---|---|---|---|---|---|
| | | Test Result | Technical Requirement | Test Result | Technical Requirement |
| Penetration (15 °C, 100 g, 5 s) | 0.1 mm | 31 | / | 26 | / |
| Penetration (25 °C, 100 g, 5 s) | 0.1 mm | 91 | 80–100 | 73 | 60–80 |
| Penetration (30 °C, 100 g, 5 s) | 0.1 mm | 162 | / | 129 | / |
| PI | | −1.24 | −1.5–+1.0 | −1.39 | −1.5–+1.0 |
| Softening point | °C | 47 | ≥45 | 49 | ≥46 |
| Ductility (5 °C, 5 cm/min) | cm | 105.4 | ≥45 | 79.8 | ≥20 |
| Rotary viscosity (135 °C) | Pa·s | 0.756 | ≤3 | 0.723 | ≤3 |
| Density (15 °C) | g/cm$^3$ | 0.913 | / | 0.974 | / |

### 2.2. SBS Modifier

The line type SBS modifier was used in this study, which is development by a materials company and widely used in Shaanxi province, China. According to the manufacturer's suggestion, the SBS modifier dosages were 4.5%, 5%, 5.5% and 6%. Technical indicators of this SBS modifier are shown in Table 2.

**Table 2.** Technical indicators of line type SBS modifier.

| Technical Parameters | S/B Mass Ratio | Volatile (%) | Tensile Strength (MPa) | Ash (%) | 300% Stress at Definite Elongation (MPa) |
|---|---|---|---|---|---|
| Test Results | 30/70 | ≤0.7 | 18.5 | ≤0.2 | 2.4 |

## 3. Results and Discussion

### *3.1. Performance Test of SBS Modified Asphalt with Chinese Evaluation Methods*

#### 3.1.1. Penetration

The test temperatures were set as 15 °C, 20 °C, 25 °C and 30 °C in this study. The penetration of matrix asphalt and SBS-modified asphalt for different test temperatures is shown in Figure 1. For SK70# and SK90# matrix asphalt, the SBS modifier could lower the penetration of them, which lead to the asphalt binder being hardened. Take the test temperature of 25°C and SK90# matrix asphalt as example, the penetration of SBS modified asphalt with SBS dosages of 4.5%, 5%, 5.5%, and 6% decreased by 10.3%, 12.9%, 14.4% and 15.5% respectively, compared to the matrix asphalt.

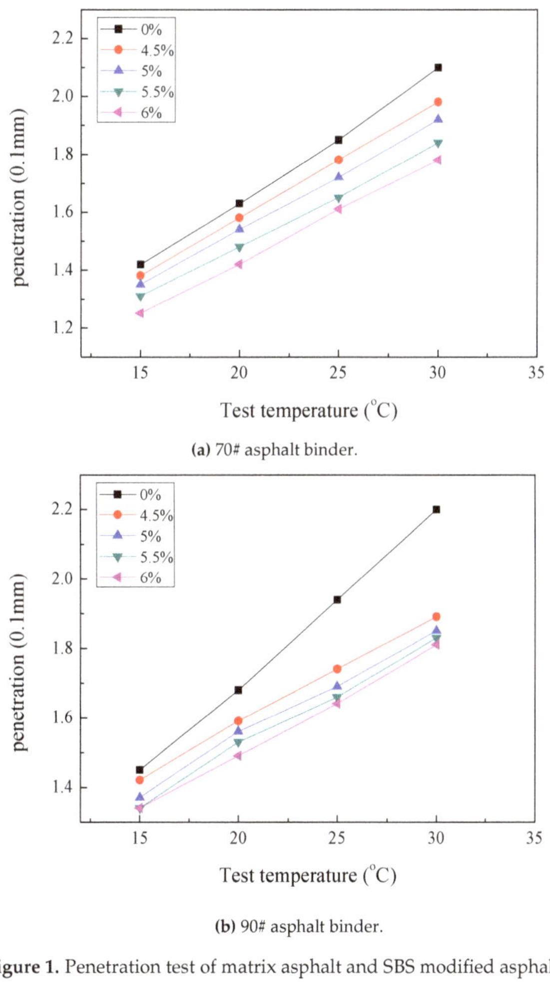


**(a)** 70# asphalt binder.

**(b)** 90# asphalt binder.

**Figure 1.** Penetration test of matrix asphalt and SBS modified asphalt.

For studying the influence of SBS modifier dosage on penetration of asphalt binder, the variance analysis of temperature and SBS modifier dosage on penetration of SK70# and SK90# matrix asphalt were conducted, as shown in Table 3.

**Table 3.** Variance analysis of penetration for SBS modified asphalt.

| Source | | df | *F*-Value | *p*-Value |
|---|---|---|---|---|
| SBS modifier dosage | 70# matrix asphalt | 4 | 7.65 | 0.064 |
| | 90# matrix asphalt | 4 | 5.89 | 0.076 |
| Temperature | 70# matrix asphalt | 2 | 37.64 | 0.001 |
| | 90# matrix asphalt | 2 | 16.88 | 0.007 |

In Table 3, the significance level is 0.05, when the *p*-value less than 0.05, that means SBS modifier dosage or temperature has a significant impact on penetration in significance level of 0.05. The smaller the *p*-value is, the more significant the effect is. The results showed that, the temperature had a significant effect on penetration of SBS modified asphalt, while the SBS modifier dosage had no considerable influence on it [8,9]. In addition, the aromatic hydrocarbon and resin in matrix asphalt could be absorbed by SBS modifier, which lead to the swelling behavior. This caused the great variability in the test result of penetration.

### 3.1.2. Softening Point

Softening point is one of the indicators used to characterize the high temperature performance of asphalt binder. It is the critical temperature which the physical state of asphalt shifts from viscid-plastic to viscous flow. The higher the softening point, the better the high temperature performance of asphalt binder [10]. The softening point of the asphalt binder with different SBS modifier dosages is as shown in Figure 2.

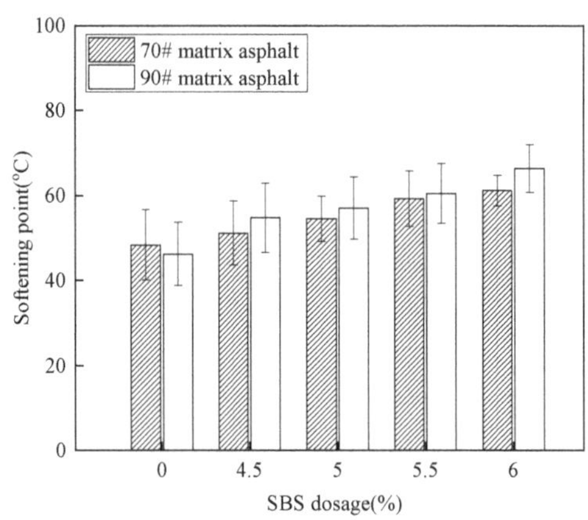


**Figure 2.** Softening point of matrix asphalt and SBS modified asphalt.

From Figure 2, for SK70# and SK90# matrix asphalt, the addition of SBS modifier increased the softening point of the asphalt binder. The softening point of SBS modified asphalt with the modifier dosage of 4.5%, 5%, 5.5% and 6% increased 5.7%, 12.8%, 22.5% and 26.4% respectively compared to the SK70# matrix asphalt. The increase was 21.2%, 26.3%, 33.6% and 46.6% respectively, compared to the SK90# matrix asphalt [11]. The single factor variance analysis was conducted with different asphalt binder types and different SBS modifier dosages for softening point of the asphalt binder. The results are as shown in Table 4.

From Table 4, the SBS modifier had significant influence on the softening point in both SK70# and SK90# matrix asphalt, which meant the addition of SBS modifier could improve the temperature performance of asphalt binder [12]. The F-value of two kinds of asphalt binder in the variance analysis was compared and the results showed that the F-value of SK90# matrix asphalt was bigger than that of SK70# matrix asphalt. Therefore, the improving effects of SBS modifier on the softening point of SK90# matrix asphalt were significantly more than that of SK70# matrix asphalt, which meant the SBS modifier had the better improving effects on fluxed bitumen.

**Table 4.** Variance analysis of softening point for SBS modified asphalt.

| Source | | df | *F*-Value | *p*-Value |
|---|---|---|---|---|
| SBS modifier dosage | 70# matrix asphalt | 4 | 268.47 | 0.25 |
| SBS modifier dosage | 90# matrix asphalt | 4 | 8641.65 | 0.02 |

### 3.1.3. Ductility and Elastic Recovery

Ductility is a test index used to characterize the low temperature performance of asphalt binder. The higher the ductility, the better the low-temperature crack resistance of asphalt binder. The elastic recovery of asphalt binder is a mechanical index, which could reflect the elasticity capacity of asphalt binder from stress to recovery [13]. The elastic recovery index directly reflects the high temperature, low temperature, fatigue and durability performance. The temperature of ductility test was 5 °C, the stretching velocity was 5cm/min, and the temperature of the elastic recovery test was 25 °C. The results of ductility test and elastic recovery test of matrix asphalt and SBS modified asphalt are shown in Figure 3.

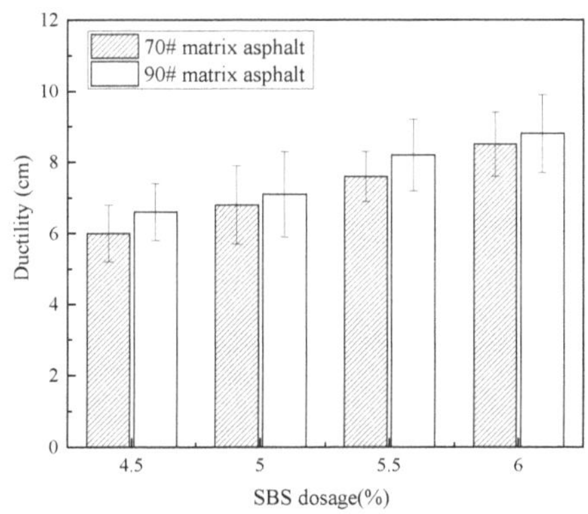


(a) Ductility.

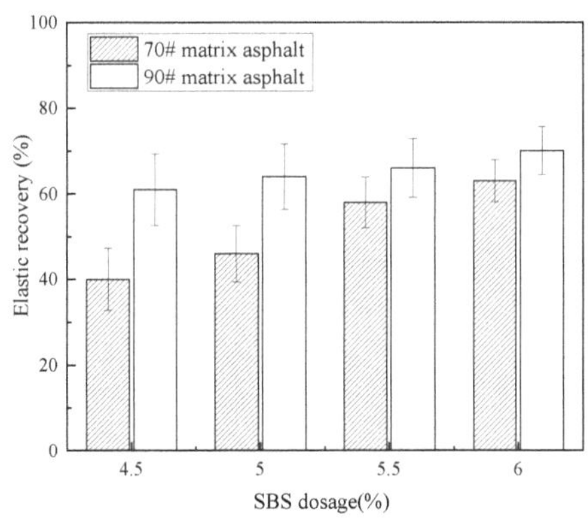


(b) Elastic recovery.

**Figure 3.** Ductility and elastic recovery result of matrix asphalt and SBS modified asphalt.

From Figure 3, the ductility and elastic recovery of 90# matrix asphalt was higher than that of 70# matrix asphalt. The ductility and elastic recovery of asphalt binder increases with the increase of SBS modifier dosage, which meant the addition of SBS modifier could improve the low temperature performance and elastic recovery performance of asphalt binder [14,15]. The variance analysis for the influence of SBS modifier on the ductility and elastic recovery of asphalt binder is shown in Table 5.

**Table 5.** Variance analysis result of ductility and elastic recovery for SBS modified asphalt.

| Source | | df | F | Sig. |
|---|---|---|---|---|
| Ductility | 70# matrix asphalt | 3 | 72.58 | 0.001 |
| | 90# matrix asphalt | 3 | 36.47 | 0.002 |
| Elastic recovery | 70# matrix asphalt | 3 | 874.62 | 0.001 |
| | 90# matrix asphalt | 3 | 153.87 | 0.002 |

It can be seen that SBS modifier had a significant influence on the ductility and elastic recovery of asphalt binder. For the ductility and elastic recovery of asphalt binder, the F-value of 70# matrix asphalt was bigger than that of 90# matrix asphalt, which meant the effects of SBS modifier on the ductility and elastic recovery of SK70# matrix asphalt was significantly greater than that of SK90# matrix asphalt.

### 3.2. *Test Result of Superpave Method for SBS Modified Asphalt*

#### 3.2.1. Dynamic Shear Rheological (DSR) Test

The anti-rutting factor $G^*/\sin\delta$ of SBS modified asphalt was tested to study the anti-rutting performance of SBS modified asphalt at the temperature of 45 °C, 50 °C, 55 °C, 60 °C, 65 °C and 70 °C respectively. The larger the $G^*/\sin\delta$, the better the high-temperature performance of asphalt, which means the high-temperature rutting resistance is better [16]. The modifier dosages were 4.5%, 5%, 5.5% and 6%, and the matrix asphalt was SK90# asphalt binder. The test results are shown in Figure 4.

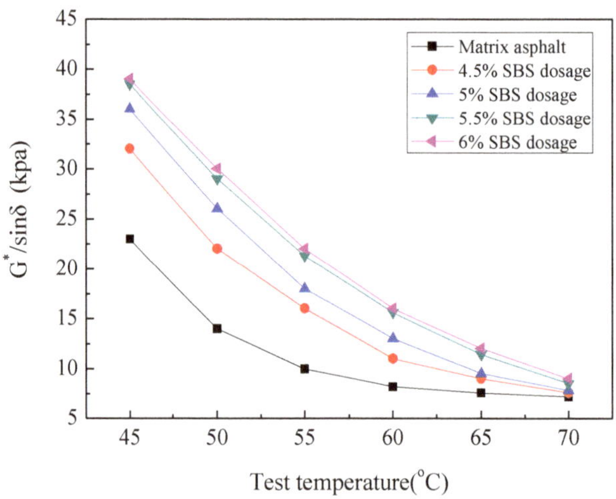


**Figure 4.** Rutting resistance factor of matrix asphalt and SBS modified asphalt at different temperatures.

From Figure 4, for the matrix asphalt and SBS modified asphalt, the $G^*/\sin\delta$ decreased with the increase of temperature, and the addition of SBS modifier can significantly improve the $G^*/\sin\delta$ of asphalt binder, and enhance the high-temperature rutting resistance of asphalt [17].

The difference analysis of G*/sinδ was conducted in different SBS modifier dosage, and the analysis results are shown in Table 6. The results showed that when the SBS modifier dosages increase from 0% to 5.5%, the high-temperature rutting resistance of SBS-modified asphalt increases continuously, and the difference between each pair of them was significant. When the SBS modifier dosages increased from 5.5% to 6%, there were no obvious difference between each pair of them, which meant the increase of high-temperature rutting resistance of asphalt was not significant when the modifier dosages increased from 5.5% to 6%.

**Table 6.** Difference analysis of G*/sin δ at different SBS modifier dosages.

| | 0% | 4.5% | 5% | 5.5% | 6% |
|---|---|---|---|---|---|
| 0% | – | S | S | S | S |
| 4.5% | | – | S | S | S |
| 5% | | | – | S | S |
| 5.5% | | | | – | N |
| 6% | | | | | – |

Note: S = Significant; N = Non-significant.

3.2.2. Brookfield Rotary Viscosity Test

Another index to measure the performance of asphalt binder is the viscosity. The smaller the viscosity of asphalt at high temperature, the better the asphalt mixture can be mixed and compacted [18]. Therefore, the SHRP method requires the rotary viscosity at 135 °C shall not exceed 3 Pa.s. The Brookfield rotary viscometer was adopted in this paper to determine the rotational viscosity of SBS modified asphalt. In the test, the rotor speed was 20 rpm/min, the rotor of SBS modified asphalt was 27#, and the mass was 10.5 g, the mass of 90# matrix asphalt sample was 8.5 g using 21# rotor. The test temperature were 135 °C, 140 °C, 150 °C, 160 °C, 170 °C, 177 °C and 190 °C respectively, and the dosages of SBS modifier were 0%, 4.5%, 5%, 5.5% and 6%. The test results are shown in Figure 5.

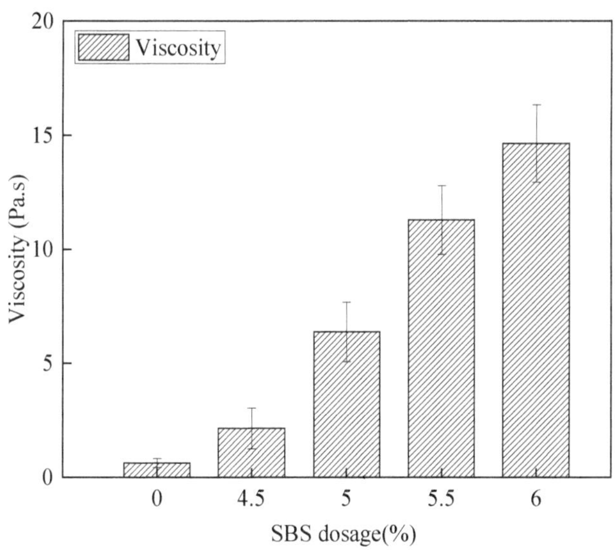


**(a)** Viscosity of SBS modified asphalt with different modifier dosages

**Figure 5.** *Cont.*

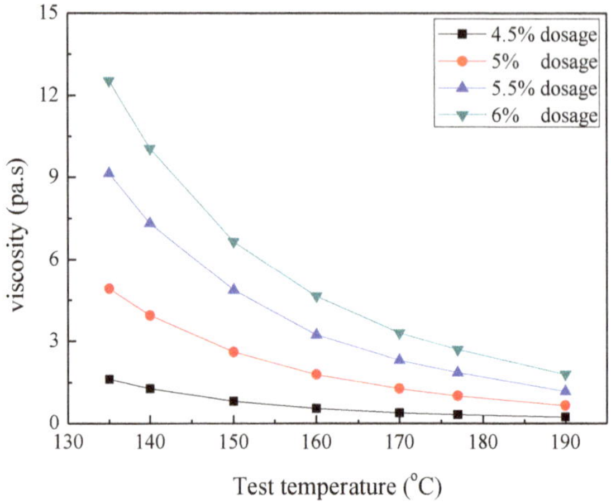


**(b)** Viscosity of SBS modified asphalt with different temperature

**Figure 5.** Rotary viscosity of asphalt binder.

From Figure 5, the addition of SBS modifier could significantly improve the rotary viscosity of asphalt binder. Compared to the rotary viscosity at 135°C of matrix asphalt, when the SBS modifier dosages were 4.5%, 5%, 5.5% and 6% respectively, the rotary viscosity of asphalt binder increased by 2.42 times, 9.14 times, 16.96 times and 22.31 times respectively. And with the increase of SBS modifier dosage, the rotary viscosity of asphalt binder increased significantly. In general, the viscosity-temperature curve was used to characterize the relations of viscosity and temperature of SBS modified asphalt, as shown in Equation (1) [19].

$$\log\log(\eta) = n - VTS\log(T) \tag{1}$$

Which, $\eta$ is the asphalt viscosity (cPa·s); $T$ is the test temperature; $n$ is the regression coefficient; VTS is represents the temperature sensitivity of asphalt binder.

The viscosity of SBS modified asphalt with different SBS modifier dosage was fitted by Equation (1), and the relations of viscosity and temperature was determined. The results are shown in Table 7.

**Table 7.** Fitting result of viscosity-temperature curve for SBS modified asphalt.

| SBS Modifier Dosage | N | VTS | $R^2$ |
|---|---|---|---|
| 4.5% | 3.426 | 0.869 | 0.997 |
| 5% | 3.351 | 0.743 | 0.997 |
| 5.5% | 3.254 | 0.715 | 0.995 |
| 6% | 3.139 | 0.637 | 0.993 |

From Table 7, the fitting coefficient $R^2$ of viscosity-temperature curve of SBS modified asphalt were all greater than 0.99, which meant that the viscosity-temperature relationship of SBS modified asphalt could be characterized better by Refutas curve. The VTS represent the temperature sensitivity of SBS modified asphalt, and the bigger value of VTS, the more distinct temperature sensitivity [20]. From Table 7, the VTS value decreased with the increase of SBS modifier dosage, which meant that the SBS modifier could significant lower the temperature sensitivity of asphalt binder.

3.2.3. Bending Beam Rheometer (BBR) Test

The bending beam rheometer test (BBR) is used to analyze the stiffness of SBS modified asphalt in SHRP test methods, by which the low temperature performance of SBS modified asphalt could be characterized [21]. The SBS modified asphalt sample is conducted short term aging with RTFOT method first, and then conducted stress aging with pressure aging vessel (PAV).

The test is loaded 240 s at the stress of 100 g (980 mN), and the creep stiffness S(t) at t = 60 s is used as one of the evaluation index of low temperature performance of SBS modified asphalt. The smaller the S(t), the better low temperature performance of SBS modified asphalt. The m value is the other index of BBR test, which characterize the change of S(t) over time. The S(t) value is required less than 300 MPa, and the m value is required more than 0.3 in SHRP method. The test temperature of −12 °C and −18 °C was used in this study, and three parallel-samples were prepared for the test. The test results of S(t) of SK90# matrix asphalt and SBS modified asphalt are shown in Figure 6.

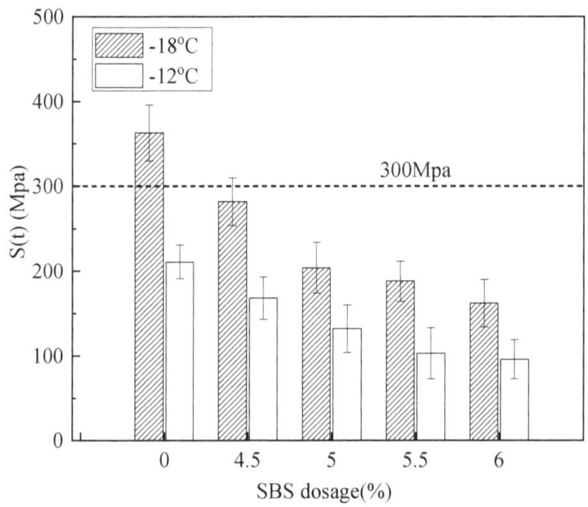


**Figure 6.** Creep stiffness S(t) of SK90# matrix asphalt and SBS modified asphalt.

From Figure 6, the S(t) values at −18 °C was bigger than that at −12 °C, and the S(t) values of both decreased with the increase of SBS modifier dosage at the temperature of −12 °C and−18 °C, which meant the low temperature performance of SBS modified asphalt had been improved. To analyze the influence of SBS modifier dosage on S(t), the single factor variance analysis for S(t) at different temperature was conducted. The results are shown in Table 8.

**Table 8.** Variance analysis for S(t) of SBS modified asphalt.

| Source | | df | F | Sig. |
|---|---|---|---|---|
| SBS modifier dosage | −18 °C | 4 | 198.68 | 0.001 |
| | −12 °C | 4 | 376.57 | 0.001 |

Form Table 8, the *p*-value was both 0.001, which meant the SBS modifier dosage had significant influence on S(t) of asphalt binder. Through the compared of two *F*-value at −12 °C and −18 °C, the influencing effect of SBS modifier dosage on stiffness S(t) at −18 °C was less than that at −12 °C. Therefore, the improvement effect of SBS modifier dosage on the low temperature performance of asphalt binder reduced with the decrease of the test temperature.

## 4. Conclusions

In China, many researchers are confronted with two sets of indicators in evaluating SBS-modified asphalt, Chinese test methods and Superpave test methods, which are random in practical application. In order to make the evaluation of SBS-modified asphalt performance more reliable and targeted, this paper evaluates the adaptability of each indicator through the quantitative analysis of laboratory tests. The research results provide theoretical support for unifying the performance evaluation index of SBS-modified asphalt, and provide suitable evaluation index for the selected SBS modifier which is development by a materials company and widely used in Shaanxi province, China. Some conclusions are as follows:

(1) According to the performance evaluation tests conducted in China and other countries, the addition of SBS modifier can significantly improve the high-temperature performance of asphalt, and can also significantly reduce the temperature sensitivity of asphalt. In addition, it increased the viscosity of asphalt at high temperatures, and increased the difficulty of mixing and compaction of the asphalt mixture.

(2) For SK70# and SK90# matrix asphalt, the addition of the SBS modifier can reduce the penetration of asphalt binder in a certain extent, but the effect was not obvious, and the penetration index of asphalt binder had no obvious trends between different modifier contents. In addition, the swelling effect of SBS modified asphalt might lead to great variability of penetration test results, so the penetration test was not recommended to evaluate the performance of SBS modified asphalt.

(3) The addition of SBS modifier can significantly improve the softening point of asphalt binder, and this conclusion was in line with the dynamic shear rheological (DSR) test results in Superpave. Therefore, the softening point can be taken as an evaluation index for the high-temperature performance of SBS modified asphalt. The variance analysis shows the effects of SBS modifier on the softening point of SK90# matrix asphalt were significantly better than that of SK70# matrix asphalt.

(4) SBS modifier can considerable enhance the ductility and elastic recovery of asphalt binder, which was in line with the bending beam rheometer (BBR) test results in Superpave. The analysis of variance showed that the influence of SBS modifier on the ductility and elasticity recovery of 70# asphalt was greater than that of 90# asphalt, and the influence of SBS modifier on stiffness was greater than that of the ductility. Therefore, in the study of the influence of SBS modifier on the low temperature performance of asphalt binder, the effect of BBR test was more obvious than the low temperature ductility test. In addition, the low temperature ductility value of SBS modified asphalt was small. Therefore, the possibility of error was greater during the test. Thus, it was recommended to using the BBR test to evaluate the low-temperature performance of SBS modified asphalt.

**Author Contributions:** C.Z. and H.W. conceived and designed the experiments; C.Z. performed the experiments; C.Z. and H.W. analyzed the data; H.W. and Z.Y. contributed reagents/materials; C.Z. wrote the paper; J.G. and M.I. reviewed and edited the paper.

**Funding:** This research was funded by the research project of the National Natural Science Foundation of China (NSFC) (no. 51578075, 51878063), the Open Fund of Key Laboratory of Special Environment Road Engineering of Hunan Province (Changsha University of Science & Technology) (KFJ180503), and the Scientific Research Foundation of Xi'an Aeronautical University (2018KY0212).

**Conflicts of Interest:** The authors declare no conflict of interest.

## References

1. Wen, G.; Zhang, Y.; Zhang, Y.; Sun, K.; Fan, Y. Rheological characterization of storage-stable SBS-modified asphalts. *Polym. Test.* **2002**, *21*, 295–302. [CrossRef]
2. Park, T.S.; Cho, H.K. Evaluation of Mechanical Characteristics of SBS Polymer Modified Asphalt Mixtures upon Variation of Filler Contents. *Neurochem. Res.* **2002**, *33*, 1979–1989.
3. Zhao, X.; Wang, S.; Wang, Q.; Yao, H. Rheological and structural evolution of SBS modified asphalts under natural weathering. *Fuel* **2016**, *184*, 242–247. [CrossRef]

4. Khodaii, A.; Mehrara, A. Evaluation of permanent deformation of unmodified and SBS modified asphalt mixtures using dynamic creep test. *Constr. Build. Mater.* **2009**, *23*, 2586–2592. [CrossRef]
5. Forough, S.A. Investigating the effects of loading frequency and temperature on moisture sensitivity of SBS-modified asphalt mixtures. *J. Mater. Civ. Eng.* **2014**, *26*, 897–903.
6. Huang, W.; Tang, N. Characterizing SBS modified asphalt with sulfur using multiple stress creep recovery test. *Constr. Build. Mater.* **2015**, *93*, 514–521. [CrossRef]
7. Wang, Y.; Sun, L.; Zhou, J. Pavement performance evaluation of recycled styrene–butadiene–styrene-modified asphalt mixture. *Int. J. Pavement Eng.* **2017**, *18*, 404–413. [CrossRef]
8. Fu, H.; Xie, L.; Dou, D.; Li, L.; Yu, M.; Yao, S. Storage stability and compatibility of asphalt binder modified by SBS graft copolymer. *Constr. Build. Mater.* **2007**, *21*, 1528–1533. [CrossRef]
9. Zhao, Y.L.; Fan, G.U.; Huang, X.M. Analysis on SBS Modified Asphalt Aging Characterization Based on Fourier Transform Infrared Spectroscopy. *J. Build. Mater.* **2011**, *14*, 620–623.
10. Becker, M.Y.; Müller, A.J.; Rodriguez, Y. Use of rheological compatibility criteria to study SBS modified asphalts. *J. Appl. Polym. Sci.* **2003**, *90*, 1772–1782.
11. Kim, T.W.; Baek, J.; Lee, H.J.; Choi, J.Y. Fatigue performance evaluation of SBS modified mastic asphalt mixtures. *Constr. Build. Mater.* **2013**, *48*, 908–916. [CrossRef]
12. Zhang, R.; Wang, H.; Gao, J.; Yang, X.; You, Z. Comprehensive performance evaluation and cost analysis of SBS-modified bioasphalt binders and mixtures. *J. Mater. Civil. Eng.* **2017**, *29*, 04017232. [CrossRef]
13. Gao, J.; Wang, H.; You, Z.; Mohd Hasan, M.; Lei, Y.; Irfan, M. Rheological behavior and sensitivity of wood-derived bio-oil modified asphalt binders. *Appl. Sci.* **2018**, *8*, 919. [CrossRef]
14. Morrison, G.R.; Lee, J.K.; Hesp, S.A. Chlorinated polyolefins for asphalt binder modification. *J. Appl. Polym. Sci.* **2010**, *54*, 231–240. [CrossRef]
15. Jasso, M.; Bakos, D.; Stastna, J.; Zanzotto, L. Conventional asphalt modified by physical mixtures of linear SBS and montmorillonite. *Appl. Clay Sci.* **2012**, *70*, 37–44. [CrossRef]
16. Background, A.; Objective, B.; Materials, C. *Understanding The Performance of Modified Asphalt Binders in Mixtures: High-Temperature Characterization*; Federal Highway Administration: Washington, DC, USA, 2002.
17. Galooyak, S.S.; Dabir, B.; Nazarbeygi, A.E.; Moeini, A. Rheological properties and storage stability of bitumen/SBS/montmorillonite composites. *Constr. Build. Mater.* **2010**, *24*, 300–307. [CrossRef]
18. Zhang, L.; Greenfield, M.L. Relaxation time, diffusion, and viscosity analysis of model asphalt systems using molecular simulation. *J. Chem. Phys.* **2007**, *127*, 194502. [CrossRef] [PubMed]
19. Ji, X.P.; Zheng, N.X.; Yang, D.Q.; Hou, Y.Q. Research on Mixing Temperature of Hot-recycled Asphalt Mixture Based on Composite Curve of Viscosity-temperature. *China J. Highw. Transp.* **2010**, *23*, 16–21.
20. Zhang, R.; Wang, H.; Gao, J.; You, Z.; Yang, X. High temperature performance of SBS modified bio-asphalt. *Constr. Build. Mater.* **2017**, *144*, 99–105. [CrossRef]
21. Ma, T.; Huang, X.; Zhao, Y.; Yuan, H. Aging Behaviour and Mechanism of SBS-Modified Asphalt. *J. Test. Eval.* **2012**, *40*, 1186–1191. [CrossRef]

*applied sciences*

MDPI

*Article*

# The Evaluation Method of Particle Clustering Phenomena in RAP

**Guangji Xu, Tao Ma *, Zhanyong Fang, Xiaoming Huang and Weiguang Zhang**

School of Transportation, Southeast University, 2 Sipailou, Nanjing 210096, Jiangsu, China; guangji_xu@seu.edu.cn (G.X.); 220162666@seu.edu.cn (Z.F.); huangxm@seu.edu.cn (X.H.); wgzhang@seu.edu.cn (W.Z.)

* Correspondence: matao@seu.edu.cn; Tel.: +86-1580-516-0021

Received: 31 December 2018; Accepted: 24 January 2019; Published: 27 January 2019

**Abstract:** The particle clustering phenomena in reclaimed asphalt pavement (RAP) particles is one of the most important factors to affect the efficient recycling of asphalt concrete. In this study, the particle composition, clustering degree, crushing properties and clustering stability of RAP were studied by extraction test and cantabro-crushing test. It was found that the particles above 4.75 mm were composed mainly of small particles with a large degree of cluster and poor stability. The coarse particles (>4.75 mm) had a great influence on the variation of 4.75 mm sieve. Quantitative indexes of Percentage Loss rate (*PL*) and Stability Index (*w*) were proposed to evaluate the clustering degree and the stability of RAP. It provided a meaningful reference for the comparison of different RAP and the different crushing processes. In addition, the RAP could be divided into three kinds of structures, including weak cluster structure, strong cluster structure and old aggregate. The process of crushing the RAP was divided into three stages, which are weak structure-dominated, strong structure-dominated and the broken old aggregate. The weak structure had the largest degree of cluster and worst stability, resulting in a large variability of RAP, and it should be avoided in the crushing process.

**Keywords:** Green pavement; Clustering index; Extraction test and Cantabro-crushing test; RAP; Breakage behavior; Cluster phenomenon

---

## 1. Introduction

The asphalt RAP mixture has an enhanced ductility and a higher strain capacity when compared to the control asphalt mixture. These improved properties are useful in the construction of the road bases and sub-bases [1,2]. The average RAP content added in the new asphalt mixture has steadily increased in recent years [3]. From an environmental perspective, higher percentages of RAP (e.g., >50%) are being utilized to reduce costs and save natural resources. The demolished old asphalt pavement and waste asphalt mixture will not only occupy a large amount of land resources, but also cause environmental pollution and ecological damage [4]. Currently, the asphalt pavement recycling technology provides an effective way for the disposal of waste asphalt mixture [5–7]. According to the construction site and temperature, asphalt pavement regeneration technology generally could be divided into four categories: in-plant hot recycling technology, hot in-place recycling technology, in-plant cold recycling technology and cold in-place recycling technology [8–10]. However, these recycling methods are faced with a common problem. The utilization rate of RAP is low, the performance variability of recycled materials is large, and the road performance thus cannot be fully guaranteed. So as to improve the utilization rate of old materials, a large number of studies have been carried out. Tang et al. used two methods to evaluate asphalt mixtures with high RAP contents based on the performance of fatigue-cracking resistance. It was found that faster fatigue degradation was observed for the 40% RAP binder and RAP mixture when subjected to repeated loading [11]. Fakhri studied the glass fiber modified warm mix asphalt mixtures with high RAP content, and improved performance of the WMA mixture was shown due to the glass fiber and higher RAP percent [12]. Stimilli et al., proposed that an accurate mix design and the selection of adequate binder could overcome the potential drawbacks related to the use of high RAP percentage, given the possibility to produce suitable recycled mixtures [13]. The performance variability of recycled materials cannot be solved fully, and the reason is because there is a great variability of RAP [14–17]. The particle cluster phenomenon in RAP is one of the most important factors to affect the gradation variability and road performance of regenerated mixtures. Many studies had been carried out to investigate the clustering phenomena in regenerated asphalt mixtures. Bressi (2015) contended that the rheology, and in particular the complex modulus, could be used to study the cluster phenomenon in mixtures containing RAP [18]. In Bressi's more recent research, he concluded that the quantity and the quality of virgin aggregates play a significant role in the clusters' formation [19].

According to related studies, the cluster level of different sources of RAP materials was quite different [20–22]. It could be affected by the pavement aging degree, material composition, gradation characteristics, milling equipment, milling temperature, pavement humidity, pavement layer and many other factors [23–25]. If the old asphalt mixtures were not broken fully, it would seriously affect the performance of the regenerated mixture and reduce the utilization rate of the RAP [26–28]. Therefore, to fully reduce the clustering degree of old asphalt mixtures and improve the utilization rate of old materials, it is necessary to study the clustering status, the fragmentation behavior and the disposal measures of RAP mixtures.

The objectives of this study are to study the cluster phenomenon in RAP mixtures, including the clustering degree, particle composition in RAP, law of particle breakage, and the stability of the cluster. These quantitative evaluation indexes are suggested with a purpose to put forward some efforts to reduce the cluster phenomenon in RAP materials.

## 2. Materials

In this study the following two RAP materials were used:

(1) The RAP 1 was obtained during the process of pavement milling, in which the asphalt binder content was 5.16%. The RAP 1 material was first dried to constant weight at 60 °C and collected for the following experiment.

(2) The RAP 2 was collected from crushing material of plant, and it contained 5.0% of asphalt binder. Similar with RAP 1, RAP 2 samples were also dried to constant weight at 60 °C, and then were sampled for the following experiments.

The gradation and particles distribution of RAP 1 and RAP 2 materials are important material properties, and it can be obtained from the sieving experiment. Therefore, both RAP mixtures were first subjected to sieving analysis, before and after extraction. The gradation results of RAP 1 and RAP 2 materials are shown in Figure 1.

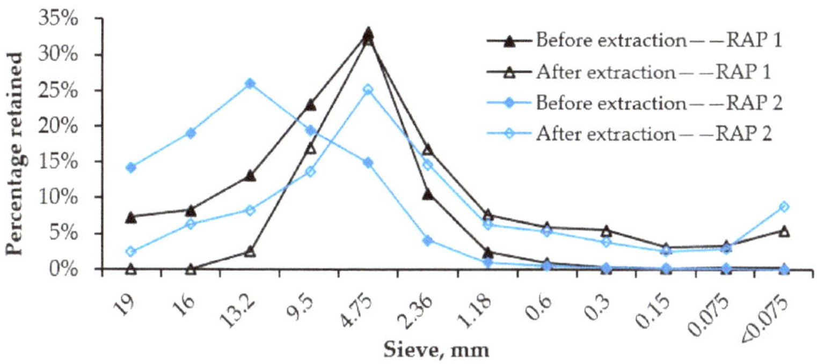


**Figure 1.** Percentage retained of reclaimed asphalt pavement (RAP) 1 and RAP 2 before and after extraction.

## 3. Methodology

This study aims to investigate the cluster phenomena in RAP materials. The extraction test and Cantabro-crushing test were used to analyze the clustering phenomena in RAP materials of AC-13 asphalt mixture. Validation tests were also performed on the RAP materials of Stone Mastic Asphalt (SMA) and Open Graded Friction Course (OGFC) asphalt mixtures. Besides, aggregate image measurement system (AIMS) was used to calculate the angularity and sphericity of RAP materials. A detailed description is given below.

### 3.1. Extraction Test

According to the Chinese specification JTG E20-2011 T0722, which is "Standard Test Methods of Asphalt and Asphalt Mixtures for Highway Engineering", the centrifugal separation method is used to extract the RAP. This extraction test is often used to determine the aggregate gradation or binder content of asphalt mixtures. But different from the traditional method of extraction test, in this study, the extraction test was aimed at the size of each of the RAP particles, including 2.36 mm, 4.75 mm, 9.5 mm, 13.2 mm, 16 mm and 19 mm. Then the extraction test was carried out separately. The main process of the experiment included the following steps:

a. Sieving the RAP mixtures and weighing 1 kg to 1.5 kg of RAP materials of each size which is above 2.36 mm;
b. Carrying out the extraction test for each particle size.
c. Drying and weighing the materials after extraction test, sieving, and calculating the percentage retained of each size.
d. Calculating the Percentage-Loss index by Equation (1):

$$PL_i = (1 - a_i) \times 100\% \tag{1}$$

where $PL_i$ = The loss percentage of the $i$th particle size; $a_i$ = The percentage retained of the $i$th sieve after extraction.

### 3.2. Cantabro-Crushing Test

To simulate the break-up process of each size of RAP materials, the Cantabro-Crushing test is used here. The Cantabro-Crushing test was designed to study the breaking law and the stability of the cluster comprising of each RAP particle size. In this test the Los Angeles abrasion tester was used, as is shown in Figure 2. The experiment process was described as follows:

a. Sieving the RAP materials and weighing 2 kg of RAP material of each size which is above 4.75 mm;

b. Carrying out the crushing test at room temperature, and the rotation number of Los Angeles abrasion tester is set to 50 r, 100 r, 200 r, 300 r.
c. After the test, taking out the samples and sieving, weighing the mass of the remaining RAP materials on the sieve;
d. Calculating the index of Crushing-Loss rate following Equation (2):

$$CL_i = \frac{m_{1i} - m_{2i}}{m_{1i}} \times 100\% \quad (2)$$

where $CL_i$ = Crushing loss rate of the $i$th particle size; $m_{1i}$ = The mass of the $i$th particle before test, kg; $m_{2i}$ = The mass of the remaining RAP particles on the $i$th sieve, kg.

(a) (b) (c)

**Figure 2.** Samples before and after the Cantabro-crushing test, (**a**) Cantabro-crushing test apparatus; (**b**) Samples before the test; (**c**) Samples after the test.

### 3.3. Aggregate Image Measurement System

The aggregate image measurement system (AIMS) is an integrated system, consisting of image acquisition hardware and computers for system operation and data analysis. The image acquisition hardware uses cameras, microscopes, aggregate trays, backlight and overhead lighting systems. The computer software, including the hardware and the user interface, were designed for the purpose of data analysis. It can also combine the software, such as Excel or Matlab, to output data in the form of chart or figure. In this paper, the AIMS was used to capture the shapes of RAP materials and to analyze clustering degree with respect to angularity and sphericity [29,30]. RAP material particles with sizes of 13.2 mm to 16 mm were subjected to the test, and a total of five kinds of mixtures were used. They include new materials, secondary crushing old materials, pavement milling materials, flat secondary crushing old materials, and flat pavement milling materials, respectively. The angularity (GA) and sphericity index (SP) were calculated.

Angularity (GA) is defined to character the shape of the particle boundary variation, which is an efficient way to measure the change in high gradient value along the boundary of polygonal particles. The calculation of GA is shown in Equation (3).

$$\text{GA} = \frac{1}{\frac{n}{3} - 1} \sum_{i=1}^{n-3} |\theta_i - \theta_{i+3}| \quad (3)$$

where: $\theta$-Azimuth at an edge point; $n$-The total number of points; $i$-The $i$th point on the edge.

Sphericity (SP) is applicable for coarse aggregate size and describes the overall three-dimensional shape of particles, as is defined in Equation (4).

$$\text{SP} = \sqrt[3]{\frac{d_s d_I}{d_L^2}} \quad (4)$$

where $d_S$: Minimum particle size; $d_I$: Intermediate particle size; $d_L$: Intermediate particle size.

### 3.4. Evaluation Methods of Particle Clustering Phenomena

To quantitatively analyze the overall stability of RAP material, this study combines the results of crushing test and extraction test to define the index of stability, which is w, as is shown in Equation (5). The larger the index of stability ($w$), the higher the clustering stability of the coarse particle:

$$w = \frac{1}{n}\sum_{i=1}^{n}\left(\frac{S_{c,i} - S_i}{S_{c,i}}\right) \times 100\% \quad (5)$$

where $w$ = Index of stability; $n$ = The number of the particle size; $S_{c,i}$ = The loss rate of the $i$th particle size after extraction test; $S_i$ = The loss rate of the $i$th particle size after crushing test.

According to the results analysis, a flow chat is illustrated here to evaluate RAP materials clustering degree. As is shown in Figure 3, old materials can be divided into three structural types which are respectively strong clustering structure, weak clustering structure and aggregates after needle-like particles tests. Then the extraction test and the breakage test can be conducted to calculate extraction loss rate and breakage loss rate, which are used to calculate the stability index. On the basis of the stability index, the clustering degree of old materials could be determined, and the clustering characteristics of recycled asphalt pavement can be comprehensively evaluated.

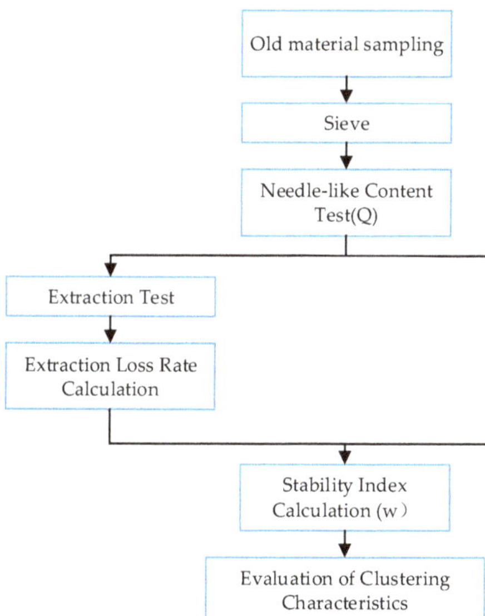


**Figure 3.** A flow chart of the evaluation method.

## 4. Results and Discussion

### 4.1. Particle Composition Analysis of Each Particle Size

Figure 4 shows the result of the percentage retained on each size of RAP materials after the extraction test. It can be found that one particle size of the RAP was mainly composed of the next particle size. Further, the cluster of coarse aggregate (>4.75 mm) was mainly composed of aggregates with its next particle size and 4.75 mm size. It indicates that the cluster phenomenon of coarse aggregates in RAP materials would have a great impact on the variability of its next particle size and the particle size of 4.75 mm. In addition, for RAP 1 materials, which is milled from the old asphalt pavement directly, it can be found that the particle size of 16mm and 19 mm were mainly composed of aggregates of 9.5 mm and 4.75 mm, and this is different from other sizes. This is probably because the raw materials were from AC13 type asphalt mixture, thus the content of particles of 13.2 mm size was less, and the RAP did not break sufficiently during the milling procedure.

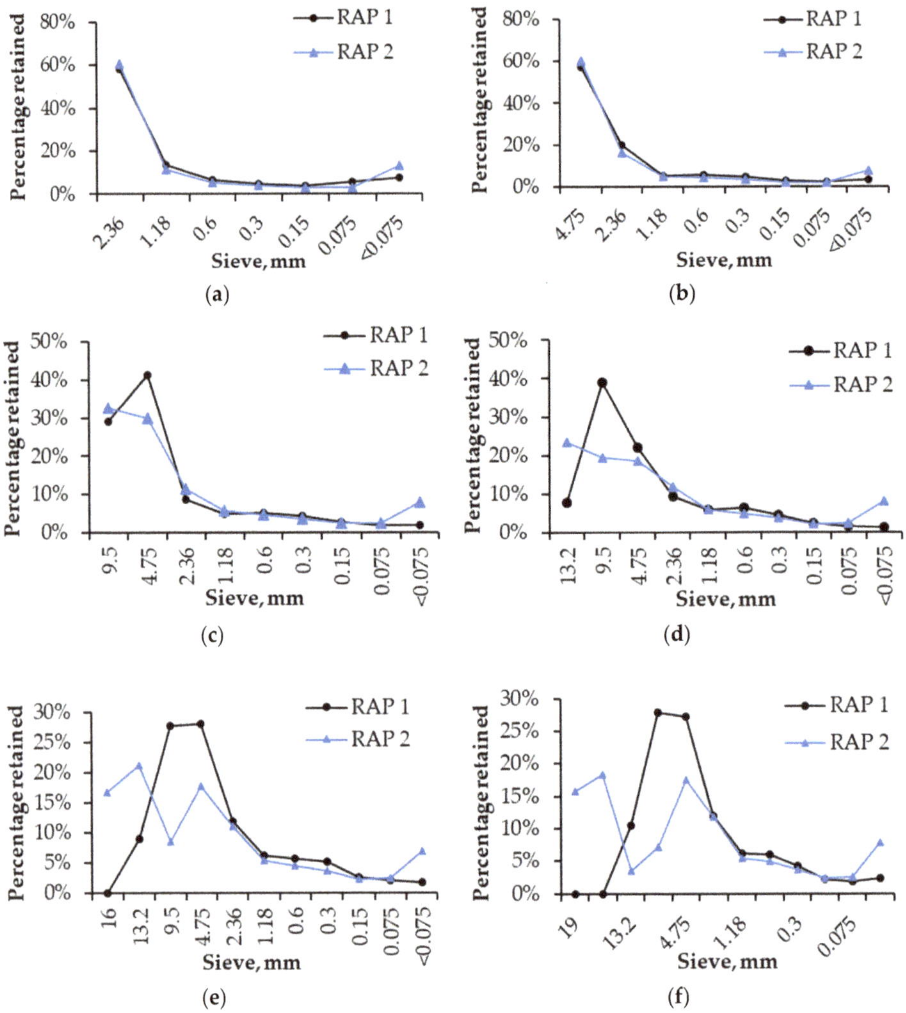


**Figure 4.** The Percentage Retained of Each Sieve after the Extraction Test, (**a**) 2.36 mm sieve; (**b**) 4.75 mm sieve; (**c**) 9.5 mm sieve; (**d**) 13.2mm sieve; (**e**) 16 mm sieve; (**f**) 19 mm sieve.

Therefore, from the analysis of the composition of each particle size, it can be found that the content of its next particle size was higher, and the cluster of coarse aggregate (>4.75 mm) was mainly composed of its next particle size and 4.75 mm particle size aggregate. In terms of fines (< 4.75 mm), the 2.36mm particle was mainly composed of its next size aggregate and mineral powder. It also has to be noted that the particle composition of fines needs further study.

### 4.2. Clustering Degree Analysis of Each Particle Size

To further analyze the clustering degree of RAP materials with different particle sizes, the Percentage-Loss (*PL*) rate after extraction was used as the evaluation index. The greater the *PL*, the more serious the clustering degree. Results are shown in Figure 5.

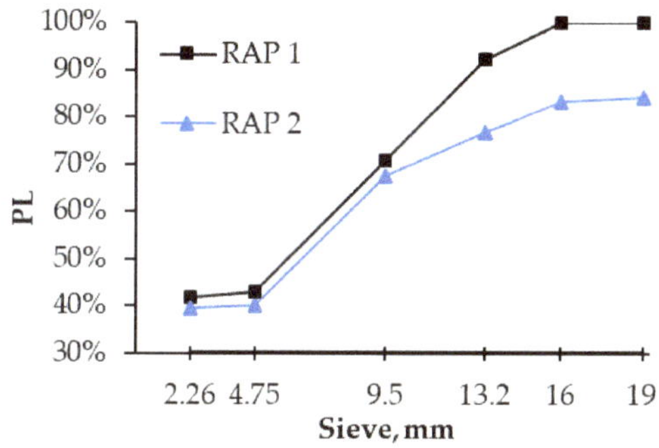


**Figure 5.** Percentage Loss (*PL*) Rate of Each Particle Size after Extraction Test.

As is shown in Figure 5, with the increase of particle size, the loss rate of the two kinds of RAP materials showed a general increasing trend. It indicates that the larger the size of the RAP particle, the more severe the clustering degree. Besides, there were two change phases, and 4.75 mm sieving holes can be seen as the cut-off point. The loss rate of the particle size less than 4.75 mm changes smoothly and the clustering degree was smaller; while the particle size which are above 4.75 mm, the loss rate increased rapidly, and the clustering degree was more serious.

Therefore, the particle size larger than 4.75 mm would seriously affect the variability of RAP and as a result, required further crushing treatment. Furthermore, when compared the RAP1 with RAP2, the clustering degree of RAP1 was greater. To finely crush the RAP materials, it is necessary to reduce the clustering degree of the coarse aggregate (above 4.75 mm), which is of great significance in reducing the variability of the RAP mixtures and to improve the performance of the recycled mixtures.

### 4.3. Analysis of the Breakage Behavior of RAP

The extraction test found that the particle clustering phenomena occurred mainly in the coarse aggregate (>4.75 mm). To further analyze the crushing law and clustering stability of RAP, in this paper, the Cantabro-Crushing test was used to crush the particle (>4.75 mm) respectively. The cluster stability of each particle size was evaluated by the index of the Crushing loss rate (*CL*) for RAP 1 and RAP 2, and results are shown in Figures 6 and 7, respectively.

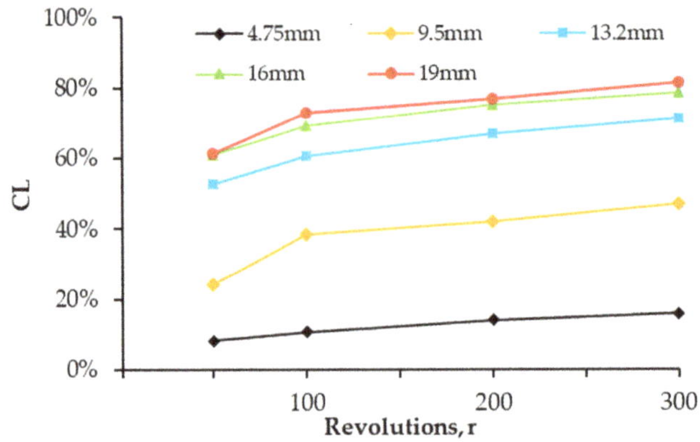


**Figure 6.** Effect of Revolutions on the Crushing Loss Rate of RAP 1.

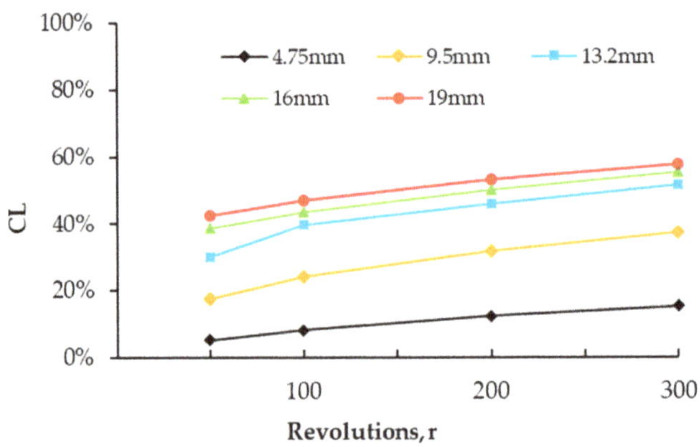


**Figure 7.** Effect of Revolutions on the Crushing Loss Rate of RAP 2.

Figures 6 and 7 illustrate that the loss rates of the two kinds of RAP materials gradually increase with the increase of the revolutions. As for the same RAP material, with the increase of the particle size, the loss rate of the particles gradually increased at the same revolutions. It indicated that the clustering stability of the RAP materials decreases with the increase of the particle size or revolutions. Compared with these two kinds of RAP, it can be found that the crushing loss rate of RAP 1 was obviously greater than that of RAP 2, indicating that the clustering stability of RAP 1 was lower than that of RAP 2.

To describe the broken law, the RAP materials can be divided into three types of structures firstly, as is shown in Figure 8. The first one is weak cluster structure. It mainly refers to some flat, needle-shaped particles, and it mainly is composed of small particles aggregates wrapped with asphalt binder. The second one is strong cluster structure. This structure mainly refers to some particles with tight clusters and better grain shape, and it is generally formed by small particles wrapping around the large particle aggregates. The last one is old aggregate, and which is composed of independent stone particles.

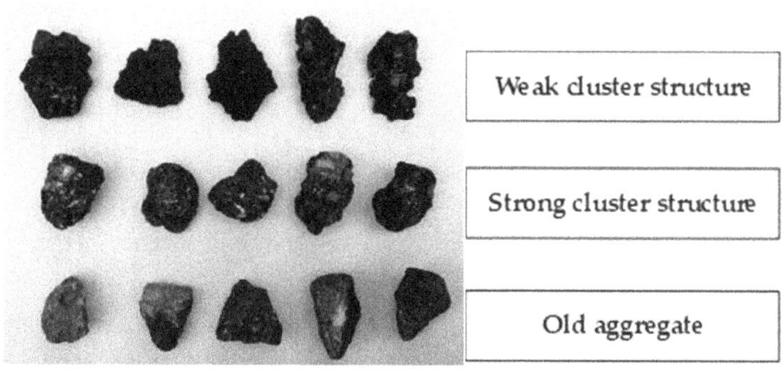


**Figure 8.** Three Different Cluster Structures of RAP.

Therefore, according to the results of Figures 6 and 7, the process of RAP crushing can be further divided into three stages, qualitatively. The first breaking stage is dominated by weak cluster structure, which is the crushing of the flat, needle-shaped particles. Followed by the second breaking stage, it is dominated by strong cluster structure which was the crushing of blocky-shaped particles. The last stage is the crushing of aggregates. This was consistent with a process in which the trend of RAP loss rate is from rapid to steady.

## 4.4. Quantitative Analysis of Clustering Stability of RAP

Table 1 summarizes the stability index results of two types of RAP materials. As seen in Table 1, as the number of revolutions increases, the stability index of coarse particles gradually decreased. The larger the number of revolutions, the greater the impact on the cluster stability of the coarse particle. Compared the RAP 1 with the RAP 2, it can be found that the stability index of RAP 2 was higher than RAP 1, indicating that RAP 2 material has a higher stability. Therefore, the stability index ($w$) can be used to quantitatively describe and compare the clustering stability of different RAP mixtures, as wells the crushing effect of different processes. It could provide a valuable reference for the comparison of the advantages and disadvantages of different RAP mixtures and different crushing processes.

**Table 1.** The Stability Index ($w$) of Two Kinds of RAP Materials.

| | Revolutions | $w$ |
|---|---|---|
| **RAP 1** | 50 r | 53.59% |
| | 100 r | 42.54% |
| | 200 r | 36.64% |
| | 300 r | 31.85% |
| **RAP 2** | 50 r | 65.09% |
| | 100 r | 57.01% |
| | 200 r | 47.98% |
| | 300 r | 41.23% |

## 4.5. Comparison and Analysis of Weak Cluster Structure, Strong Cluster Structure and Old Aggregate

In this paper, the particle size of 16–19 mm in RAP 2 was used to further analyze the three kinds of structures in RAP, including the proportion, the clustering stability and the clustering degree. The proportion of three kinds of structures is shown in Table 2. The results of the crushing test and extraction test are shown in Figures 9 and 10 respectively.

**Table 2.** The Proportion of Three Kinds of Structures (16–19 mm).

| Number | 1 | 2 | 3 | 4 | 5 | Average |
|---|---|---|---|---|---|---|
| Weak cluster structure | 36.06% | 36.61% | 33.10% | 38.13% | 34.03% | 35.58% |
| Strong cluster structure | 45.29% | 42.41% | 48.53% | 41.43% | 43.66% | 44.27% |
| Old aggregate | 18.65% | 20.98% | 18.37% | 20.44% | 22.32% | 20.15% |

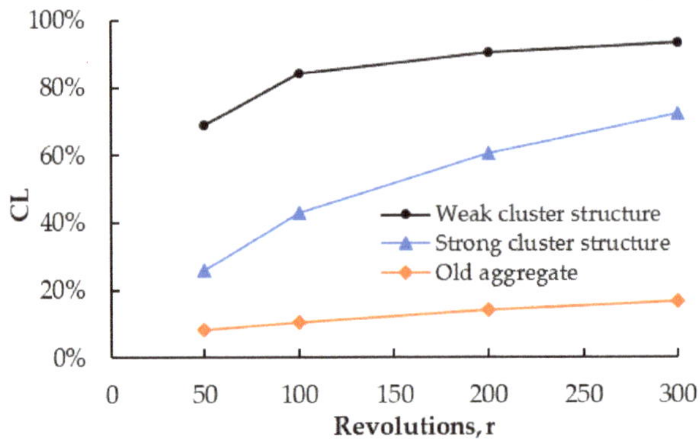


**Figure 9.** The Crushing Loss Rate of the Three Kinds of Structures.

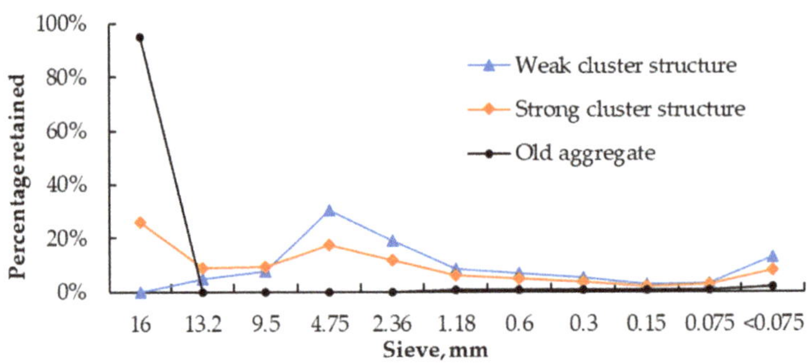


**Figure 10.** The Percentage Retained of Three Kinds of Structures after the Extraction Test.

As shown in Table 2, it can be found that the content of strong cluster structure was the highest, followed by the weak cluster, and the content of old stones was the least. Figure 9 shows that the loss rate of the weak cluster structure increases fastest, and the loss rate was the largest; followed by the strong cluster structure; the loss rate of the old aggregate particles was least. This indicates that the stability of the weak cluster structure was worst, easily broken. In contrast, the strong cluster structure was relatively stable, but would experience a greater loss under a full effect of external forces. The old aggregate particles were most stable, mainly corner damage could be expected.

Figure 10 shows that when the sieve size was above 16mm, the percentage retained of weak cluster structures was 0. The weak cluster structure was mainly composed of 4.75 mm and below fines, and the content of mineral powder was high. This indicated that the clustering degree of the weak cluster structure was the largest, and this is likely to have an adverse effect on gradation variation. According to the result of extraction test, the loss rate of the strong cluster structure was larger, and it mainly affected the content of coarse particles. For the old aggregate, there is no loss and it thus had little effect on gradation variation.

Therefore, from the comparative analysis of these three kinds of structures, it can be found that the order of influence on the gradation variability of the RAP was: weak cluster structure > strong cluster structure > old aggregate. Therefore, to reduce the cluster phenomenon and reduce the variability of RAP, the formation of the weak cluster structure should be avoided as much as possible during crushing process at first. It is important to reduce the content of strong cluster structures, and to increase the content of old aggregates. It is suggested that the nonlinear viscoelastic response of RAP mixtures at high temperatures should be considered in the future investigations [31]. The particle packing ability before and after crushing for different types of RAP should be considered to provide a better understanding of the effect of particle clustering on the overall performance of the mixture in the future study [32–34].

### *4.6. Validation of Clustering Characteristics*

To verify the applicability of other mixture types, RAP materials were collected from OGFC and SMA asphalt pavement. Figure 11 shows the extraction loss rate before and after crushing with respect to OGFC and SMA RAP material. Three particles sizes were compared, which are 4.75 mm, 9.5 mm and 13.2 mm. As shown in Figure 11, the extraction loss rate increases with the increase in particle size no matter of OGFC or SMA mixture type, and each loss rate of size beyond 4.75 mm exceeds 30% for all sizes, indicating a large clustering degree and relatively poor stability. This is consistent with the cases in dense grade mixtures of AC type asphalt mixture. Additionally, materials of each particle size decrease in extraction loss rate after a second time crushing, it is understandable because the clustering materials went through crushing process tended to cluster fewer aggregates.

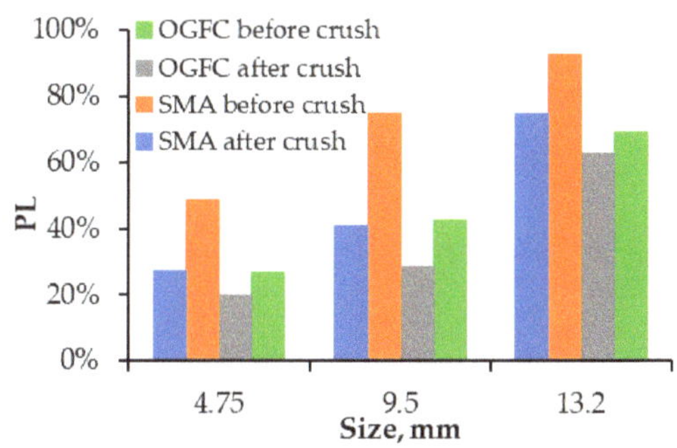


**Figure 11.** Percentage Loss (*PL*) Rate of RAP material.

Figure 12 shows the comparison of breakage loss rate between OGFC and SMA. On one hand, for both types, the loss rate increases as the particle size increases, especially for 13.2 mm aggregates which has a distinct increase in breakage loss compared with others. On the other hand, the increase rate of 4.75 mm size aggregates in both OGFC and SMA materials is very small. The breakage loss changed little as rotate number increases, indicating a stable status of 4.75 mm old clustering materials. For the other two size materials, the breakage rate increases as rotate number increases, and the rate of OGFC materials is higher than that of SMA materials. This is probably because OGFC is open grade mixture and SMA is semi-open grade mixture, and the air voids in OGFC materials is larger and thus OGFC materials are easier to be separated.

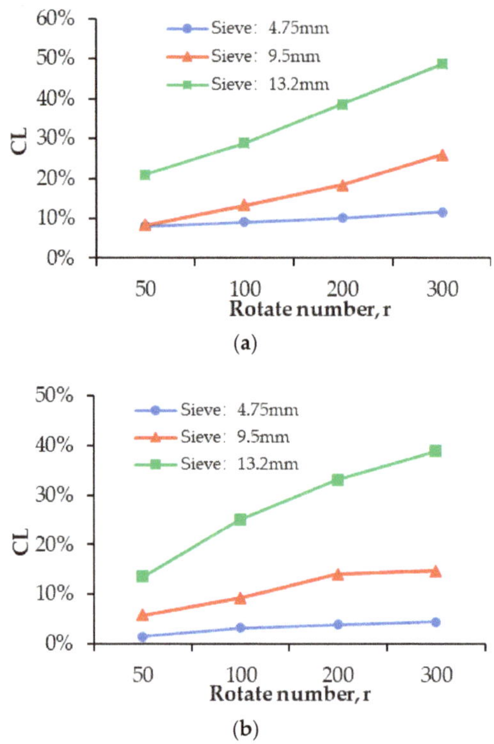


**Figure 12.** Crushing Loss rate of old (**a**) Open Graded Friction Course (OGFC) materials; (**b**) Stone Mastic Asphalt (SMA) materials.

From the breakage rate results, it was found that OGFC suffered from a greater loss in RAP materials. In Figure 13, the stability index between OGFC and SMA was compared. As shown in Figure 12, with the increase of rotate number, the stability index of both materials decreases. Although the decrease trends are similar, the OGFC materials presented a lower stability index than SMA materials at all rotate number. It could be concluded that old OGFC materials are more difficult to reuse due to their low stability and high clustering degree.

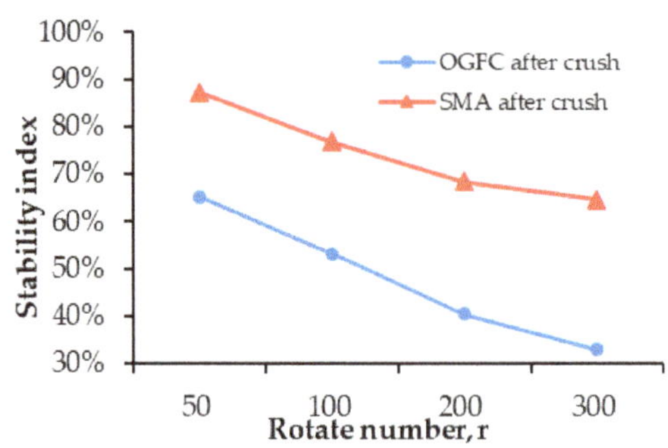


**Figure 13.** Stability Index of old OGFC and SMA materials.

### 4.7. Other Attempts to Evaluate the Old material Clustering Degree

The angularity results of all aggregates are shown in Figure 14. Compared with the new aggregates, the clustering of the old aggregates results in an increase of the edge angle, which makes the RAP aggregates more fragile. RAP aggregates subjected to a secondary crushing can reduce the edge angle to a certain extent, which also accounts for the relatively high clustering stability of the old aggregates. From the results of the angularity test, RAP material particles in a flat shape tend to have a higher edge angle and poorer cluster stability, and thus the edge angle could to some extent represent the clustering degree of RAP materials. The greater the edge angle, the larger the RAP material clustering degree. The worse stability of the cluster could cause gradation variation of rejuvenated RAP material.

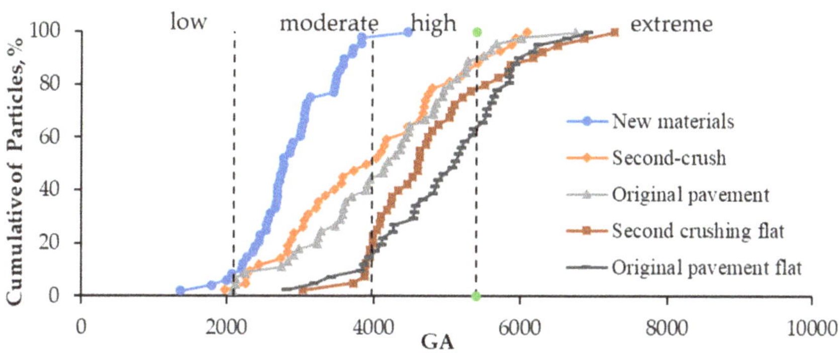


**Figure 14.** Comparison of edge angle between new and RAP materials.

The sphericity results of RAP materials are shown in Figure 15. As is shown in Figure 15, compared with the new materials and the flat shape of the old materials, the sphericity of flat old material is obviously smaller. Given that the old flat particles clustering degree is higher and the clustering stability is poorer, the sphericity index might be used to represent the old materials clustering degree. The smaller the sphericity value is, the higher the RAP material clustering degree and the worse the stability of the cluster will be.

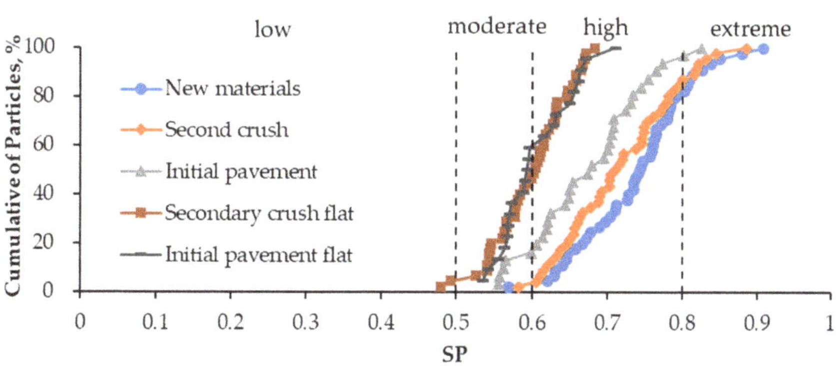


**Figure 15.** Comparison of Sphericity between new and RAP materials.

## 5. Conclusions

The particle clustering phenomena in RAP was one of the most important factors to limit its utilization rate in pavement engineering. In this paper, the clustering characteristics of RAP, such as the particle composition of each particle size, the clustering degree and the clustering stability, were investigated by the extraction test and crushing test. The following conclusions can be drawn:

(1) The cluster phenomenon in RAP occurred mainly in the coarse part (>4.75 mm), and the clustering degree was more serious as the particle size increased. In terms of the RAP particle composition, the cluster of coarse aggregate (>4.75 mm) was mainly composed of its next particle size and 4.75 mm particle size aggregate.
(2) The RAP materials can be divided into three types of structures, namely weak cluster structure, strong cluster structure and old aggregate. The crushing process of RAP can be divided into three stages. The first stage was dominated by weak cluster structure, which was the crushing of the flat, needle-shaped particles; the second stage was dominated by strong cluster structure, which was the crushing of blocky-shaped particles; and the last stage was the crushing of aggregates. It is important to reduce the content of the weak cluster structure, because the clustering degree of the weak cluster structure accounts for the largest part and is more likely to have an adverse effect on gradation variation.
(3) The quantitative indexes, loss rate (*PL* and *CL*) and stability index ($w$), were proved to evaluate the clustering degree, the crushing characteristics and the clustering stability of RAP materials. They provided meaningful references for the comparison of different RAP materials and different crushing processes.

Recommendations for reducing the particle clustering phenomena in RAP:

(1) Due to the high content of 4.75 mm particles, as well as the large clustering degree of particles above 4.75 mm, it is suggested that the coarse aggregate, especially particles of 4.75 mm key sieve size, be fully crushed and broken.
(2) The weak cluster structure in RAP is very fragile, and had a serious impact on the RAP variability. It is suggested that the formation of weak cluster structure should be avoided and the content of the old aggregate particles should be increased in the process of milling and planning. Separating the weak cluster structure, or subject to secondary breaking, could also reduce the variability of RAP.
(3) The method of measuring the proportion of weak cluster, strong cluster and old aggregate in RAP could be used to analyze the advantages and disadvantages of crushing processes. The stability index ($w$) can be used for quantitative comparison.

**Author Contributions:** Conceptualization, T.M.; methodology, G.X. and T.M.; validation, G.X. and Z.F.; formal analysis and investigation, T.M and G.X.; resources and data curation, G.X. and W.Z.; writing—original draft preparation, Z.F.; writing—review and editing, G.X.; supervision, T.M and X.H.; project administration, T.M.

**Funding:** This research was funded by the National Natural Science Foundation of China (grant number 51808116 and 51878164), Natural Science Foundation of Jiangsu Province with grant number BK20180404, and Fundamental Research Funds for the Central Universities, grant number 2242018K40003.

**Conflicts of Interest:** The authors declare no conflict of interest.

## References

1. Hassan, K.E.; Brooks, J.J.; Erdman, M. *The Use of Reclaimed Asphalt Pavement (RAP) Aggregates in Concrete*; Waste Management Series; Elsevier: Harrogate, UK, 2000; Volume 1, pp. 121–128.
2. Gu, F.; Ma, W.; West, R.C.; Taylor, A.J.; Zhang, Y. Structural performance and sustainability assessment of cold central-plant and in-place recycled asphalt pavements: A case study. *J. Clean. Prod.* **2019**, *208*, 1513–1523. [CrossRef]
3. Zaumanis, M.; Mallick, R.B.; Frank, R. 100% Hot Mix Asphalt Recycling: Challenges and Benefits. *Transp. Res. Procedia* **2016**, *14*, 3493–3502. [CrossRef]
4. Singh, D.; Chitragar, S.F.; Ashish, P.K. Comparison of moisture and fracture damage resistance of hot and warm asphalt mixes containing reclaimed pavement materials. *Constr. Build. Mater.* **2017**, *157*, 1145–1153. [CrossRef]

5. Xu, G.J.; Wang, H. Diffusion and interaction mechanism of rejuvenating agent with virgin and recycled asphalt binder: A molecular dynamics study. *Mol. Simul.* **2018**, *44*, 1433–1443. [CrossRef]
6. Xu, G.J.; Wang, H. Molecular dynamics study of oxidative aging effect on asphalt binder properties. *Fuel* **2017**, *188*, 1–10. [CrossRef]
7. Hou, S.; Chen, C.; Zhang, J.; Shen, H.; Gu, F. Thermal and mechanical evaluations of asphalt emulsions and mixtures for microsurfacing. *Constr. Build. Mater.* **2018**, *191*, 1221–1229. [CrossRef]
8. Sheng, X.H.; Wang, M.; Xu, T.; Chen, J. Preparation, properties and modification mechanism of polyurethane modified emulsified asphalt. *Constr. Build. Mater.* **2018**, *189*, 375–383. [CrossRef]
9. Niazi, Y.; Jalili, M. Effect of Portland cement and lime additives on properties of cold in-place recycled mixtures with asphalt emulsion. *Constr. Build. Mater.* **2009**, *23*, 1338–1343. [CrossRef]
10. Farooq, M.A.; Mir, M.S. Use of reclaimed asphalt pavement (RAP) in warm mix asphalt (WMA) pavements—A review. *Innov. Infrastruct. Solut.* **2017**, *2*, 10. [CrossRef]
11. Tang, S.; Williams, R.C.; Cascione, A. Reconsideration of the fatigue tests for asphalt mixtures and binders containing high percentage RAP. *Int. J. Pavement Eng.* **2017**, *18*, 443–449. [CrossRef]
12. Fakhri, M.; Hosseini, S.A. Laboratory evaluation of rutting and moisture damage resistance of glass fiber modified warm mix asphalt incorporating high RAP proportion. *Constr. Build. Mater.* **2017**, *134*, 626–640. [CrossRef]
13. Stimilli, A.; Virgili, A.; Giuliani, F.; Canestrari, F. Mix design validation through performance-related analysis of in plant asphalt mixtures containing high RAP content. *Int. J. Pavement Res. Technol.* **2017**, *10*, 23–37. [CrossRef]
14. Mučinis, D.; Sivilevičius, H.; Oginskas, R. Factors determining the inhomogeneity of reclaimed asphalt pavement and estimation of its components content variation parameters. *Baltic J. Road Bridge Eng.* **2009**, *4*, 69–79.
15. Sivilevičius, H.; Bražiūnas, J.; Prentkovskis, O. Technologies and principles of hot recycling and investigation of preheated reclaimed asphalt pavement batching process in an asphalt mixing plant. *Appl. Sci.* **2017**, *7*, 1104. [CrossRef]
16. Lu, D.X.; Saleh, M.; Nguyen, N.H.T. Effect of rejuvenator and mixing methods on behaviour of warm mix asphalt containing high RAP content. *Constr. Build. Mater.* **2019**, *197*, 792–802. [CrossRef]
17. Ding, X.H.; Ma, T.; Huang, X.M. Discrete-Element Contour-Filling Modeling Method for Micro-and Macro- Mechanical Analysis of Aggregate Skeleton of Asphalt Mixture. *J. Transp. Eng. Part B Pavements* **2019**, *145*, 04018056. [CrossRef]
18. Bressi, S.; Dumont, A.G.; Pittet, M. Cluster phenomenon and partial differential aging in RAP mixtures. *Constr. Build. Mater.* **2015**, *99*, 288–297. [CrossRef]
19. Bressi, S.; Pittet, M.; Dumont, A.G.; Partl, M.N. A framework for characterizing RAP clustering in asphalt concrete mixtures. *Constr. Build. Mater.* **2016**, *106*, 564–574. [CrossRef]
20. Zaumanis, M.; Oga, J.; Haritonovs, V. How to reduce reclaimed asphalt variability: A full-scale study. *Constr. Build. Mater.* **2018**, *188*, 546–554. [CrossRef]
21. Bressi, S.; Cavalli, M.C.; Partl, M.N.; Tebaldi, G.; Dumont, A.G.; Poulikakos, L.D. Particle clustering phenomena in hot asphalt mixtures with high content of reclaimed asphalt pavements. *Constr. Build. Mater.* **2015**, *100*, 207–217. [CrossRef]
22. Nazzal, M.D.; Mogawer, W.; Austerman, A.; Qtaish, L.A.; Kaya, S. Multi-scale evaluation of the effect of rejuvenators on the performance of high RAP content mixtures. *Constr. Build. Mater.* **2015**, *101*, 50–56. [CrossRef]
23. Kou, C.; Xiao, P.; Kang, A.; Mikhailenko, P.; Baaj, H.; Wu, Z. Methods to evaluate the aging grades of reclaimed asphalt binder. *Appl. Sci.* **2017**, *7*, 1209. [CrossRef]
24. Xu, T.; Wang, Y.; Xia, W.; Hu, Z. Effects of flame retardants on thermal decomposition of SARA fractions separated from asphalt binder. *Constr. Build. Mater.* **2018**, *173*, 209–219. [CrossRef]
25. Shen, W.; Shan, L.; Zhang, T.; Ma, H.; Cai, Z.; Shi, H. Preparation and Effectiveness of Composite Phase Change Material for Performance Improvement of Open Graded Friction Course. *J. Clean. Prod.* **2019**, *214*, 259–269.
26. Moghaddam, T.B.; Baaj, H. The use of rejuvenating agents in production of recycled hot mix asphalt: A systematic review. *Constr. Build. Mater.* **2016**, *114*, 805–816. [CrossRef]

27. Shannon, C.; Mokhtari, A.; Lee, H.D.; Tang, S.; Williams, C.; Schram, S. Effects of high reclaimed asphalt-pavement content on the binder grade, fatigue performance, fractionation process, and mix design. *J. Mater. Civ. Eng.* **2016**, *29*, 04016218. [CrossRef]
28. Pan, P.; Kuang, Y.; Hu, X.; Zhang, X. A comprehensive evaluation of rejuvenator on mechanical properties, durability, and dynamic characteristics of artificially aged asphalt mixture. *Materials* **2018**, *11*, 1554. [CrossRef]
29. Ding, X.H.; Ma, T.; Gao, W. Morphological characterization and mechanical analysis for coarse aggregate skeleton of asphalt mixture based on discrete-element modeling. *Constr. Build. Mater.* **2017**, *154*, 1048–1061. [CrossRef]
30. Bessa, I.S.; Castelo Branco, V.T.F.; Soares, J.B. Evaluation of polishing and degradation resistance of natural aggregates and steel slag using the aggregate image measurement system. *Road Mater. Pavement Des.* **2014**, *15*, 385–405. [CrossRef]
31. Bazzaz, M.; Darabi, M.K.; Little, D.N.; Garg, N. A straightforward procedure to characterize nonlinear viscoelastic response of asphalt concrete at high temperatures. *Transp. Res. Rec.* **2018**, *2672*, 481–492. [CrossRef]
32. Li, J.; Zhang, J.; Qian, G.; Zheng, J.; Zhang, Y. Three-Dimensional Simulation of Aggregate and Asphalt Mixture Using Parameterized Shape and Size Gradation. *J. Mater. Civ. Eng.* **2019**, *31*, 04019004. [CrossRef]
33. Pouranian, M.R.; Haddock, J.E. Determination of voids in the mineral aggregate and aggregate skeleton characteristics of asphalt mixtures using a linear-mixture packing model. *Constr. Build. Mater.* **2018**, *188*, 292–304. [CrossRef]
34. Salemi, M.; Wang, H. Image-aided random aggregate packing for computational modeling of asphalt concrete microstructure. *Constr. Build. Mater.* **2018**, *177*, 467–476. [CrossRef]

applied sciences

MDPI

*Article*

# Discrete Element Analysis of Indirect Tensile Fatigue Test of Asphalt Mixture

**Xuelian Li [1,2,*], Xinchao Lv [1], Xueying Liu [3] and Junhong Ye [1]**

1. School of Traffic and Transportation Engineering, Changsha University of Science and Technology, 960, 2nd Section, Wanjiali South RD, Changsha 410114, China; xinchao_lv@163.com (X.L.); yejunhong_csust@163.com (J.Y.)
2. State Engineering Laboratory of Highway Maintenance Technology, Changsha University of Science and Technology, 960, 2nd Section, Wanjiali South RD, Changsha 410114, China
3. China Railway Major Bridge Reconnaissance & Design Institute Co., Ltd., 34, Hanyang RD, Wuhan 430050, China; lxy19930319@126.com

* Correspondence: lixuelian@csust.edu.cn; Tel.: +86-131-8700-7683

Received: 3 December 2018; Accepted: 7 January 2019; Published: 17 January 2019

**Abstract:** In order to investigate the damage to microstructure and some other micromechanical responses during a fatigue test on asphalt mixture, Particle Flow Code (PFC) was used to reconstruct a two-dimensional discrete element model of asphalt mixture, based on computed tomography (CT) images and image-processing techniques. The indirect tensile fatigue test of asphalt mixture was simulated with this image-based microstructural model, and verified in the laboratory. It was found that there were four stages during the fatigue failure: no crack, crack initiation, crack developing, and interconnected crack. Cracks mainly developed between the aggregate and asphalt mortar, near the loading axis. The corresponding stages of failure, the developing trend and the distribution characteristics of the cracks matched well with those in the laboratory test. Furthermore, the trends of both the time-load curve and time-displacement curve from the simulation test were also consistent with those from the experimental test. In short, the distribution characteristics of cracks and internal forces of asphalt mixture show that it is feasible to simulate the fatigue performance of the asphalt mixture by a discrete element method (DEM).

**Keywords:** discrete element; asphalt mixture; digital image; fatigue; crack

---

## 1. Introduction

The fatigue failure is one of the most common distresses in asphalt pavements [1]. The fatigue properties of asphalt mixture could be evaluated by an indirect tensile fatigue test [2]. At present, the fatigue test of asphalt mixture is mainly carried out in the laboratory. However, the variability of asphalt mixture is very difficult to control during a laboratory test. Because of its heterogeneity, the fatigue properties are impossible to be fundamentally explored and evaluated by the laboratory test. Therefore, it is necessary to find a way to study the behavior and mechanism of the fatigue failure based on a more non-uniform microstructure [3].

Numerical simulation analysis is frequently used to study the micromechanical properties of the asphalt mixtures. Recently, some researchers have been trying to use a discrete element method (DEM) to analyze the fatigue properties of asphalt mixtures. Ma et al. analyzed the impacts of different parameters of air voids on the creep behavior of an asphalt mixture by the DEM [4,5]. Cao et al. conducted a two-point bending beam fatigue simulation test with the trapezoidal samples. The relationship between the fatigue properties of a French high modulus asphalt mixture and the Burger's parameters were investigated by the DEM [6]. Gao et al. analyzed the failure procedure of an asphalt mixture by the DEM [7]. A local discrete element model was established to discuss the damage and cracks caused by the cyclic fatigue load [8]. Yu analyzed the effect of the aggregate packing on the dynamic modulus of an asphalt mixture by the DEM [9]. These achievements mentioned above demonstrate that the DEM is applicable for the research on fatigue characteristics of pavement asphalt mixes. However, the discrete element models of asphalt mixtures are generated by computer algorithms, which are different from the actual microstructures. The distribution characteristics of cracks and internal forces of asphalt mixtures are also seldom described in the indirect tensile fatigue test.

Since microstructure images captured by X-ray computed tomography (X-ray CT) can show the internal distribution and composition of the material, CT and the DEM have been widely applied to asphalt mixture together recently. You et al. obtained the images of asphalt mixture samples by the optical scanning, and reconstructed two-dimensional discrete element models based on these optical images [10]. Based on many two-dimensional slices of the same sample, the three-dimensional discrete element model of an asphalt mixture was reconstructed [11,12]. Zelelew et al. described the microstructure of asphalt mixtures obtained by CT and established a two-dimensional discrete element model to predict the creep compliance of uniaxial loading [13]. Zhou investigated the dynamic deformation process of an asphalt mixture based on the image-based discrete element model [14]. Tan believed that image-based discrete element model could provide a potential detailed insight into the failure mechanism of the heterogeneous rocks at the microscopic level [15]. Generally speaking, the asphalt mixture is a non-uniform complex viscoelastic material. It can be simulated more accurately by the DEM with CT image-based microstructural modeling. Therefore, the CT image-based microstructural modeling is feasible to simulate the asphalt mixture.

Therefore, in this paper, the two-dimensional discrete element model was reconstructed by the Particle Flow Code (PFC) software, based on the CT images. The indirect tensile fatigue simulation test was carried out to analyze the fatigue properties of the asphalt mixtures. The distribution of internal forces, damage to the microstructure, and some other micromechanical responses in the model were investigated. Furthermore, the simulation test was verified by the laboratory fatigue test.

## 2. Materials and Methods

### 2.1. Materials

AC-13 asphalt mixture was designed according to JTG F40-2004 [16]. Figure 1 shows mix grading with upper and lower limits. The binders used the styrene-butadiene-styrene (SBS) modified asphalt, is the most common binders used in asphalt pavement [17,18]. Both the coarse and fine aggregates were basalt, and the filler was limestone. Experimental tests are carried out to evaluate the properties, they meet the requirement of the specification [19]. According to the Marshall method [20], the optimum asphalt content was 5.2%, and the filler content was 7%. Samples, with a diameter of 101.6 mm and height of 63.5 $\pm$ 1.3 mm, were compacted 75 times on each side according to JTG F40-2004 [16].

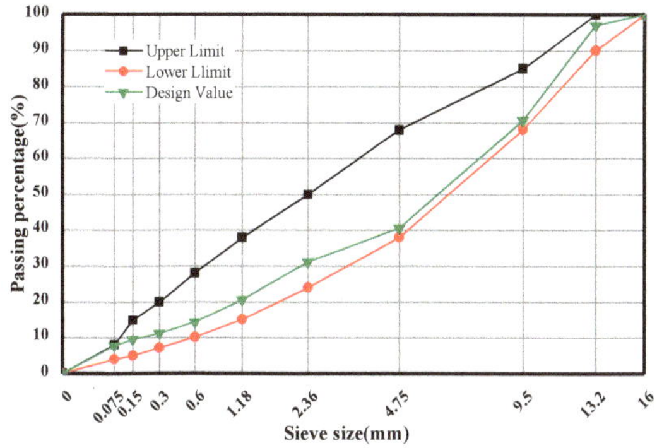


**Figure 1.** Gradation of aggregate for asphalt mixture AC-13.

### 2.2. Methods

An image-based model was reconstructed by the DEM, and the discrete element analysis of an indirect tensile fatigue test was carried out. The distribution of the cracks and other simulation test results of the asphalt mixtures were validated by the laboratory test. The steps of the methodology are shown in Figure 2.

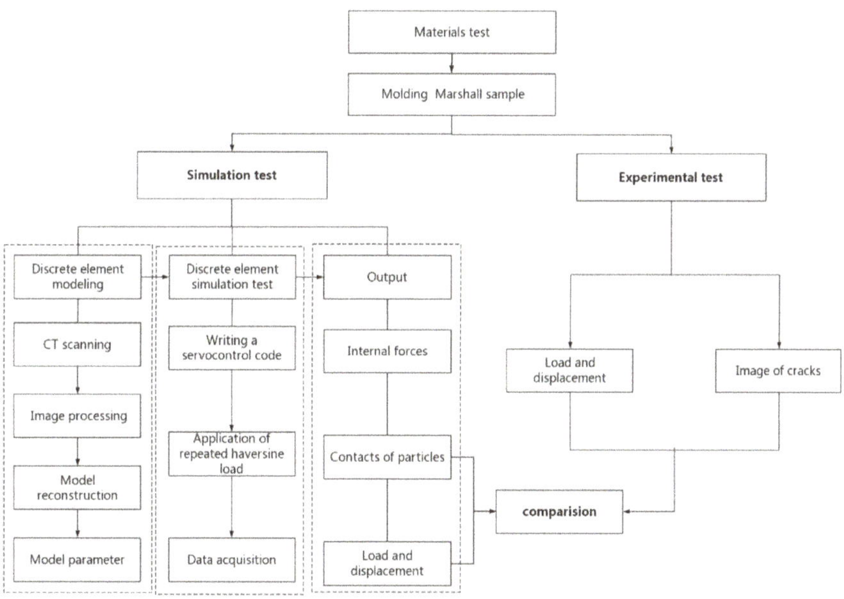


**Figure 2.** Flowchart of steps of the test.

#### 2.2.1. Discrete Element Analysis of the Indirect Tensile Fatigue Test

(1) Reconstruction based on the computed tomography (CT) images

Four Marshall samples were compacted in the laboratory. In order to distinguish among the aggregate particles, mortar and voids, an X-ray CT was used to obtain the images of the internal cross-sections of the samples. These images can be transferred to the model geometry in PFC2D. One of the sections at the height of 10–50 mm is present in Figure 3a. The aggregate, asphalt mortar and air voids in the CT image were separated by the image-processing technique. The image was converted to a binary format, in other words, the image was in the form of a digital matrix (from 0 to 255). Then image pixels became the particles in the PFC2D model. The position and geometric information of each phase were introduced into PFC in the form of coordinates, to reconstruct the model. Because of the accuracy and efficiency of calculation, the radius of the particle in the model was set at 0.75 mm, and the discrete element model is illustrated in Figure 3b.

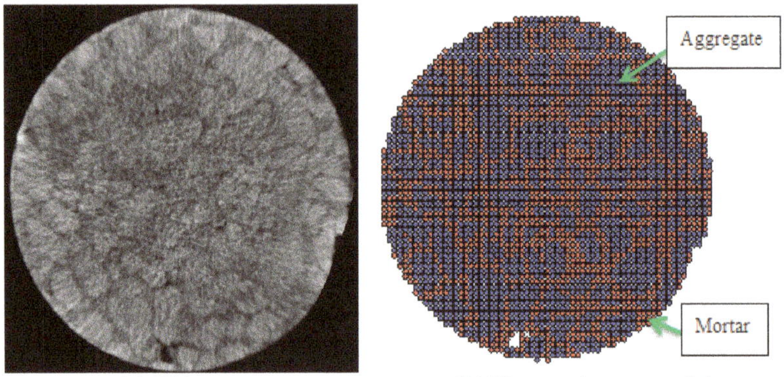


(**a**) CT image (**b**) Discrete element model

**Figure 3.** The computed tomography (CT) image and the corresponding discrete element model.

(2) Micromechanical parameters

In this paper, the Burgers' model was utilized to describe the properties of the contacts between the aggregate and asphalt mortar. Because the aggregates were all covered by the asphalt mortar, the contacts between different aggregates were ignored. The schematic diagrams of different contact models are shown in Figure 4. Different contact models between the different particles are present in Table 1. According to the micromechanical parameters in the paper of Feng [21] and Zelelew [13], the model parameters were set based on empirical values [22] and previous experimental results [5,23]. The parameters were also adjusted appropriately to match the experimental results of the indirect tensile test from the laboratory, i.e., peak force and stiffness. The parameters of the Burger's model are shown in Table 2, the normal stiffness kn and the shear stiffness ks of each phase are illustrated in Table 3. The damp and calculating time-step of the model was a default value.

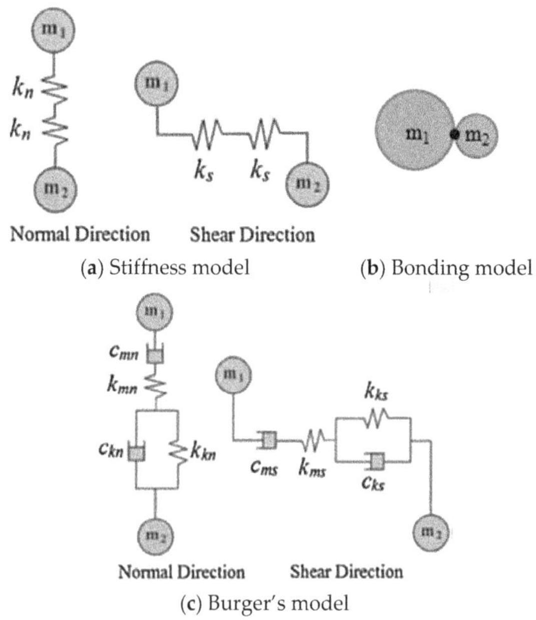


**Figure 4.** Schematic diagrams of different contact models.

**Table 1.** Contact models of the asphalt mixture.

| Contact Points | Contact Models |
| --- | --- |
| Between aggregates | Stiffness model + Bonding model |
| Within asphalt mortar | Stiffness model + Burger's model + Bonding model |
| Between aggregate and asphalt mortar | Stiffness model + Burger's model + Bonding model |

**Table 2.** Micromechanical parameters of different contacts.

| Micro Parameters | kn | ks |
| --- | --- | --- |
| Between aggregates | $4.2 \times 10^8$ | $3.0 \times 10^7$ |
| Within asphalt mortar | $6.3 \times 10^7$ | $0.6 \times 10^7$ |

**Table 3.** Micromechanical parameters of Burger's model.

| Parameters | Maxwell | | Kelvin | |
| --- | --- | --- | --- | --- |
| | Stiffness | Viscosity | Stiffness | Viscosity |
| Normal | $6.04 \times 10^5$ | $2.2 \times 10^7$ | $1.33 \times 10^5$ | $2.77 \times 10^6$ |
| Tangential | $2.42 \times 10^5$ | $8.81 \times 10^6$ | $5.3 \times 10^4$ | $1.11 \times 10^6$ |

(3) Fatigue load and boundary conditions

Two rigid walls were set at the top and bottom of the model (as shown in Figure 5). A stress-control mode was used in the simulation test, which was in accordance with the laboratory test. The rigid bottom wall of the model was fixed. In PFC2D, walls are assigned at constant or variable velocity. The variable speed of the top wall can be regulated by writing a servocontrol code. The displacement of the rigid top wall of the model was the vertical deformation of the model. The total stiffness of the particles was updated at each time-step. The multiplication of the total stiffness and the vertical deformation was the load on the model. In each cycle, the moving speed of the top wall could be back-calculated by the target load and stiffness. The speed of the top wall was controlled by the continuous sinusoidal load function. Its frequency was 10 Hz. The unbalanced force collected from the top wall was the fatigue load of the simulation test.

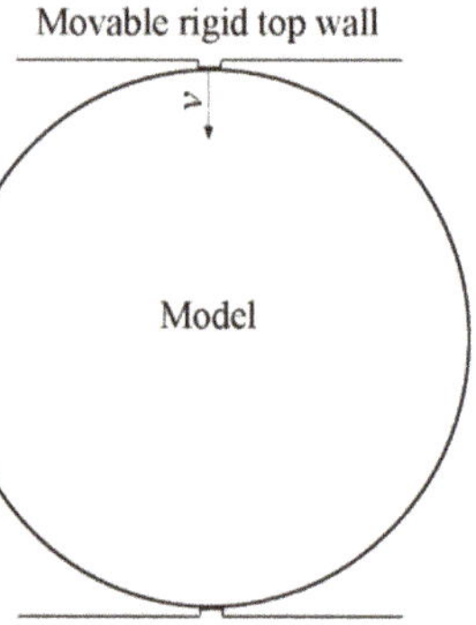


**Figure 5.** Boundary condition of the model.

2.2.2. Experimental Test Validation in the Laboratory

To compare and analyze the distribution and development characteristic of the cracks, the indirect tensile fatigue test at the stress level of 0.7 was carried out. Samples with a diameter of 101.6 mm and height of 63.5 ± 1.3 mm were compacted using the Marshall method.

The indirect tensile fatigue test was carried out with an electro-hydraulic closed-loop servo MTS810 (as shown in Figure 6) material test system. The fixture was designed for the indirect tensile test. The width of the loading bar was set for 12.7 mm. In the indirect tensile fatigue test, the stress control mode was adopted, the load was subjected to a continuous semi-positive wave load, and the load level was determined by 0.7 of the indirect tensile strength.

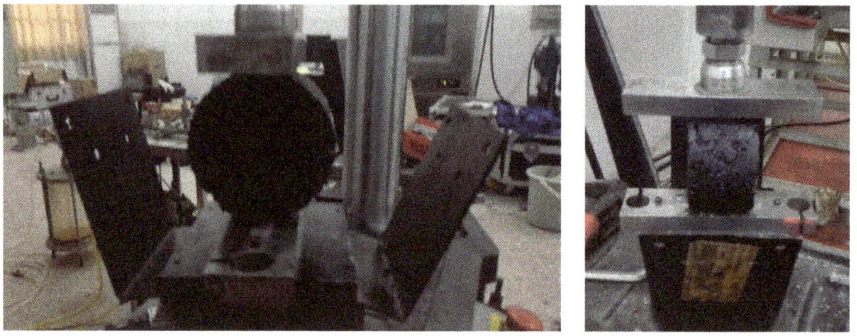

(**a**) Cross-section (**b**) Lateral section

**Figure 6.** Indirect tensile fatigue test in the laboratory.

## 3. Results and Discussions

### *3.1. Discrete Element Analysis of Indirect Tensile Fatigue Test*

The fatigue simulation test has a high requirement for computer performance. The lower the stress level of the fatigue test, the longer the fatigue life the sample will have. In this case, each simulation loading cycle will take about one hour. When the stress level is 0.7, the fatigue life of the sample is more than 300 cycles, which takes about 300 h to calculate on the computer. Therefore, the stress level was set as 0.9 in these simulation tests to reduce the calculating time. The internal forces distribution, contact condition and displacement of the particle during the fatigue simulation test would be analyzed. The simulation test was divided into four stages: no crack, crack initiation, crack development and interconnected crack.

#### 3.1.1. No Crack

At the early loading stage, the internal force distribution of the model under two different loads is shown in Figure 7a,b. The black chain and red chain in the model are the compression and tension, respectively. Contacts and displacement of particles at early loading stage are illustrated in Figure 8.

(1) From Figure 7, it can be found that internal forces mainly concentrate along the loading axial at the early loading stage. The forces on the bottom and top are bigger, but the forces in the center are smaller. After loading for a while, the force of the center increases, which means the forces are transferred from the top and bottom to the center of the sample.
(2) In Figure 8, although almost all the particles in the model move downward, the contacts among the particles are good, and no crack appears. Moreover, the displacement near the top wall is the biggest, while the displacement near the bottom wall is the smallest.

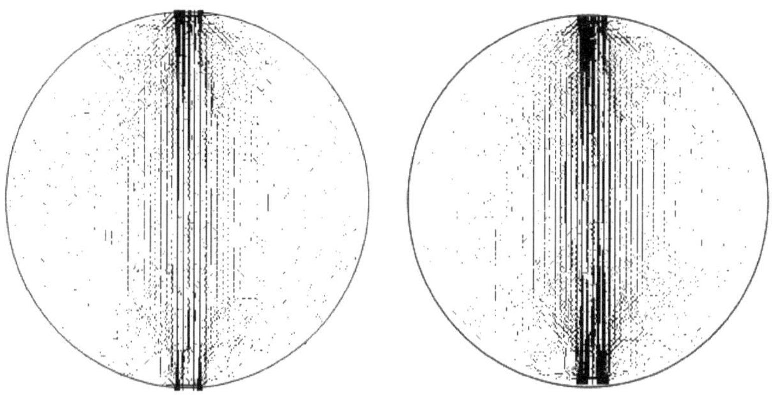

(**a**) Internal forces under the lower load (**b**) Internal forces under the higher load

**Figure 7.** Internal forces distribution at early loading stage.

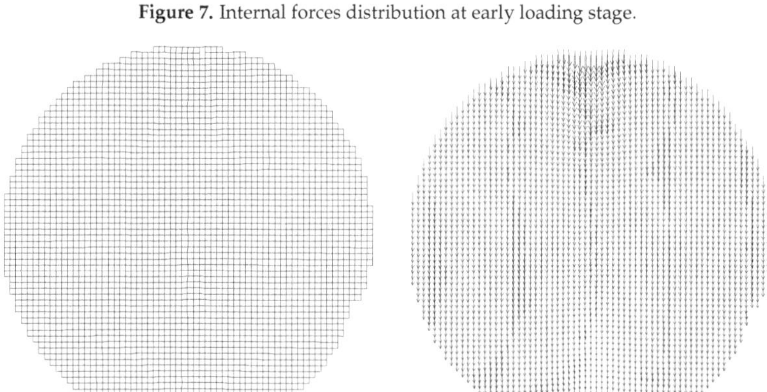

(**a**) Contact among the particles (**b**) Displacement of the particles

**Figure 8.** Model at early loading stage.

3.1.2. Crack Initiation

(1) According to Saint Venant's principle, the transverse tensile forces will come out when the internal forces spread from the top and the bottom of the model to the whole cross-section under the fatigue load [24]. The thicker red force chain appears near the top wall at this stage, as shown in Figure 9a. Those mean that there are more significant tensile forces at the top of the model.

(2) From Figure 9b, it can be found that the contacts among the particles break and the cracks initiate. The thicker black force chain also appears near the cracks, as shown in Figure 9a. The internal forces of the model redistribute, and the intensive forces come out near the cracks.

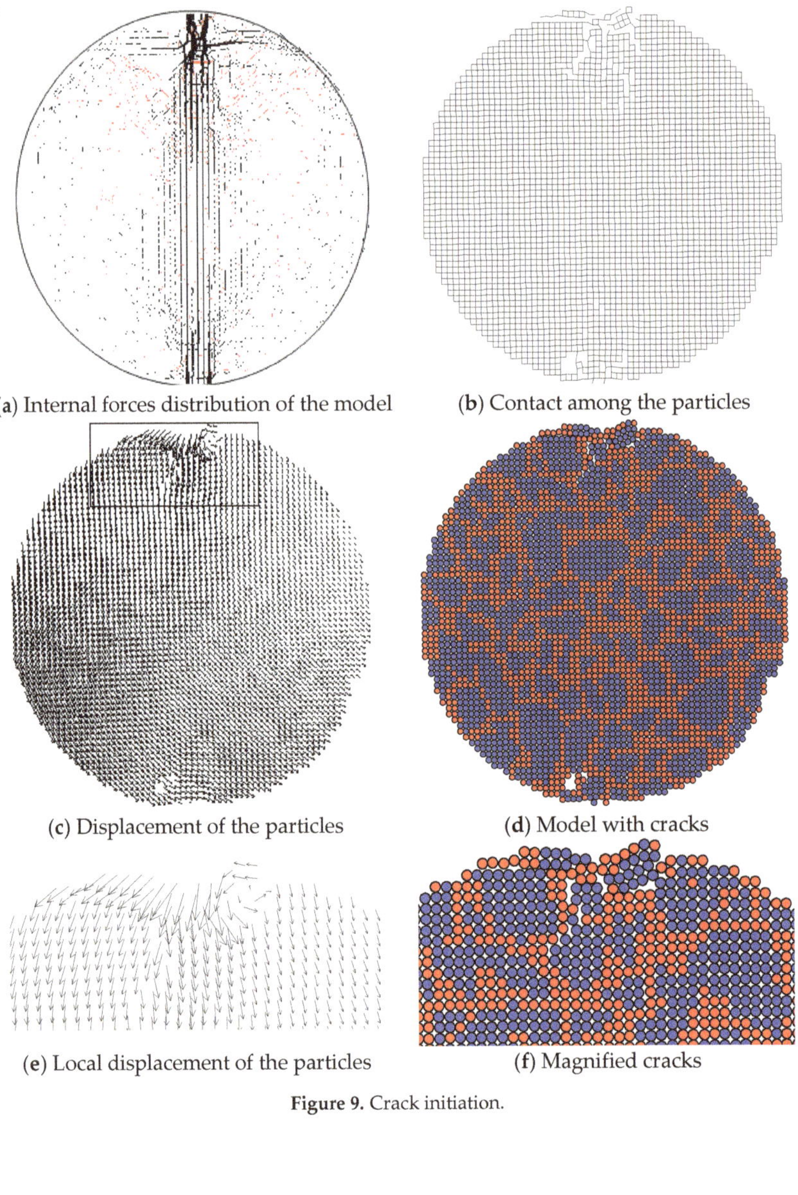

(**a**) Internal forces distribution of the model (**b**) Contact among the particles

(**c**) Displacement of the particles (**d**) Model with cracks

(**e**) Local displacement of the particles (**f**) Magnified cracks

**Figure 9.** Crack initiation.

(3) The displacement of the particles of the model is present in Figure 7c–d, and the local displacement is also shown in Figure 9e,f. The particles near the top wall tend to move towards both sides of the model. As compared with other parts, the displacement where the contact failed is bigger.

According to the analysis of the displacement and the forces distribution, it can be found that the crack is a kind of tensile force failure caused by the tensile stress; under the effect of the repeated transverse tensile forces, the contacts among the particles fail, and the cracks initiate.

#### 3.1.3. Crack Development

After loading for a while, the crack develops downwards continually. This is at the crack development stage.

(1) From the forces distribution in Figure 10a. The chains, especially the red force chains, also continue to move downward, the black force chains change from the linear force chains with a uniform distribution to the curve force chains with the non-uniform distribution.
(2) From the contact condition of the particles in Figure 10b, it can be found that the contact failure mainly distributes linearly near the top wall. This means that the cracks are mostly near the loading axis which connects the top wall.
(3) In Figure 10d, there is displacement in different directions. In addition, there is also plenty of displacement in the same direction and different magnitudes.

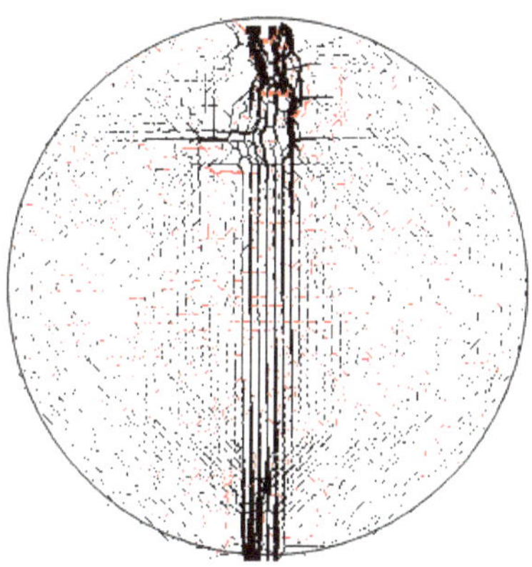
(**a**) Internal forces distribution of the model

(**b**) Contact among the particles

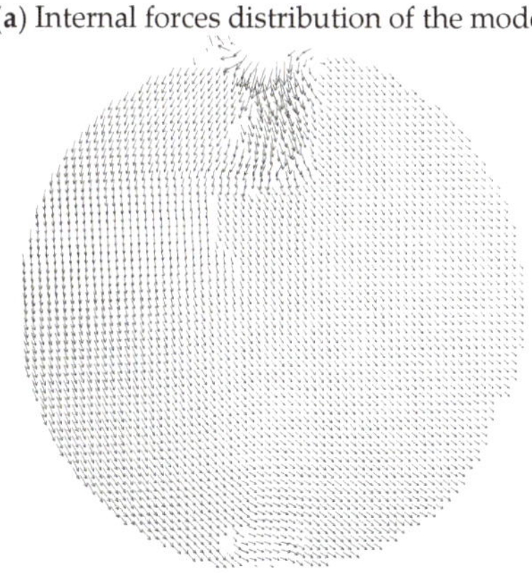
(**c**) Displacement of the particles

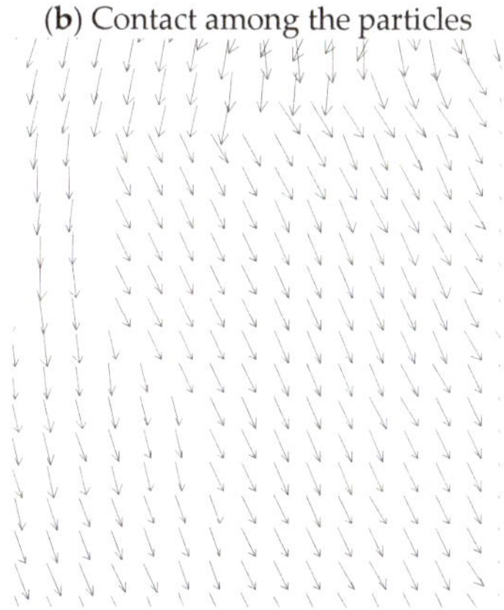
(**d**) Local displacement of the particles

**Figure 10.** Crack development.

Based on the internal forces, crack and displacement, it can be found that besides the failure of the tensile forces, there is also shear failure caused by inhomogeneous compressive forces in the model at this stage. Because of the high stiffness and strength of the aggregates, the cracks primarily come out at the interface between the aggregate and asphalt mortar [22].

#### 3.1.4. Interconnected Crack

(1) When the cracks are interconnected, the forces distribution is present in Figure 11a. The internal forces are mainly near the loading axis. There are more intensive tensile forces and compressive forces where the contact fails, and the tensile forces are primarily at the top of the model.

(2) The distribution of the cracks is given in Figure 11b; it is found that the cracks are also mainly near the loading axis. Only a few small cracks at the top of the model develop towards both sides of the model. The cracks primarily come out within the asphalt mortar or extend downwards along the interface between the aggregate and asphalt mortar.

(3) From the displacements of the particles in Figure 11c,d, it is found that the displacement at different directions and displacement in the same direction come out in the model at the same time.

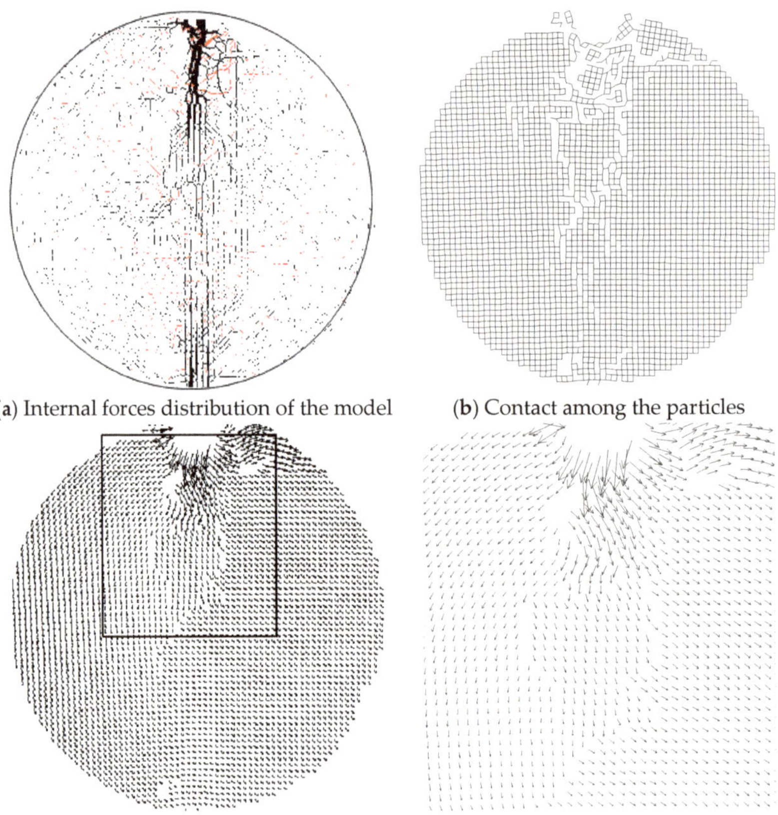

(**a**) Internal forces distribution of the model (**b**) Contact among the particles (**c**) Displacement of the particles (**d**) Local displacement of the particles

**Figure 11.** Interconnected crack.

According to the distribution of the force and displacement of the particles, it is found that the forces failure and shear failure happen simultaneously during the failure process. This failure is mainly caused by the shear failure during the latter process. The tensile failure occurs under the repeated transverse tensile forces, and the cracks come out. Under the comprehensive effect of the non-uniform compressive forces and intensive tensile forces, the cracks develop downwards along the interface between the aggregate and asphalt mortar.

### 3.2. *Experimental Test Validation in the Laboratory*

The indirect tensile fatigue test at a low-stress level will need too much time, while the samples at the high-stress levels will break quickly. Although the fatigue life is different at different stress levels, the internal stress-strain responses of the model and development process of the crack are similar. Therefore, to compare and analyze the distribution and development characteristic of the crack, the indirect tensile fatigue simulation test and laboratory test were carried out at the stress level of 0.9 and 0.7, respectively. The time-load curves were obtained by both the simulation test and the laboratory test at the stress level of 0.9.

#### 3.2.1. Characteristics of the Crack

The samples were loaded at several stages. At each stage, the vertical deformation was controlled at 1.5 mm. After each loading stage, the sample was taken out for CT scanning. During the laboratory test, the samples were broken after three loading stages. The CT images at each stage were scanned and analyzed. The distribution and development characteristics of the crack from both simulation test and laboratory test were compared. The contact condition of the model and CT images at these three loading stages: no crack, crack developing, and crack interconnected are shown in Figure 12, respectively. In Figure 12, the CT images of the sample, acquired during the loading stages, are not perfect circles. Some part of the image on the right was missing because one X-ray tube of the CT was damaged. Since the cracks mainly distribute in the middle part of the section, the deficiency of the image can be ignored. According to the development and distribution characteristics of the crack at the three loading stages, the analysis is as follows:

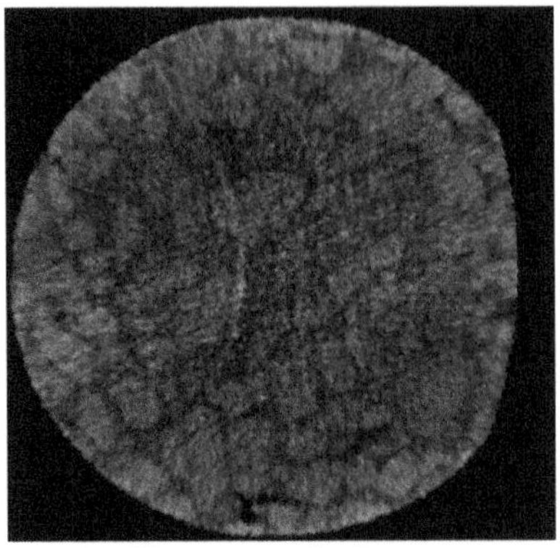
(a) CT images at the first stage

(b) Contact condition at the first stage

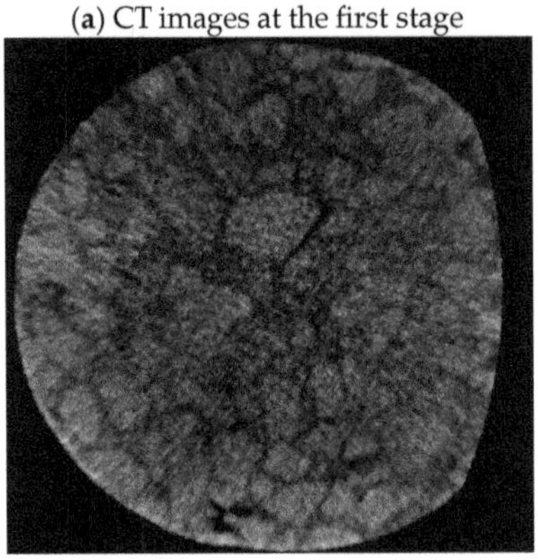
(c) CT images at the second stage

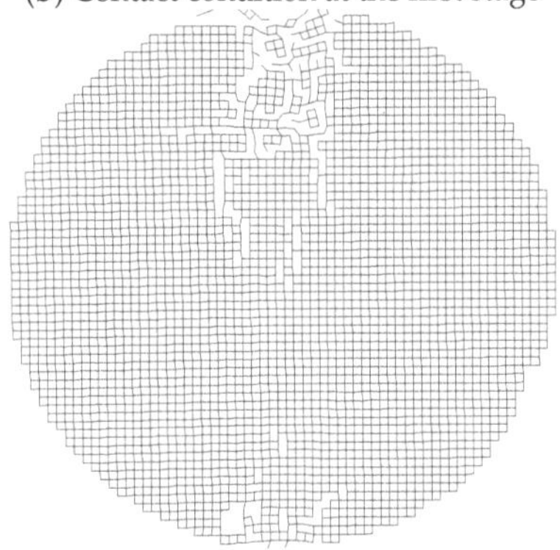
(d) Contact condition at the second stage

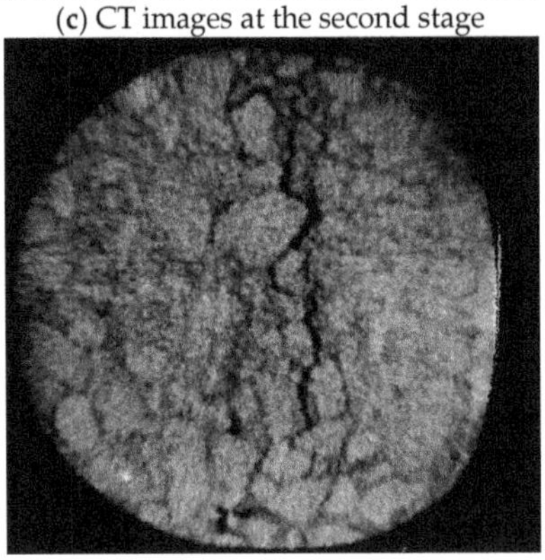
(e) CT images at the third stage

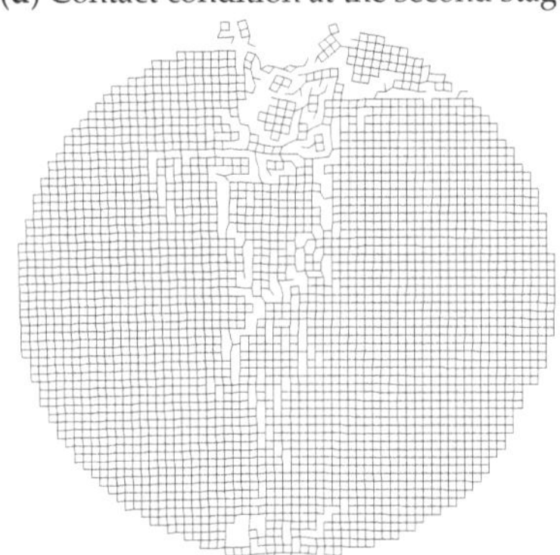
(f) Contact condition at the third stage

**Figure 12.** Cracks in the simulation test and laboratory test.

(1) At the first stage, when the vertical deformation of the sample is 1.5 mm, there is no crack in the CT image, as in Figure 12a. The stress is stable and there is no crack under the corresponding contact conditions in Figure 12b. It can be seen that the contacts among particles are intact. From the displacement of the particles in Figure 8c, it can be found that nearly all the particles move downwards under the load, which is in accordant with that in the laboratory test, for the sample is elastically compressed.

(2) The fatigue loading stop after the top wall has moved downwards at the second 1.5 mm. In Figure 12c, the cracks have extended to the middle of the section, and the length is about 50 mm. Most of the cracks distribute near the loading axis of the section. From Figure 12d, it can be found that the contacts within the asphalt mortar, which is near the top wall or the loading axis, are completely damaged. According to Figure 12c,d, the length and distribution of the cracks are consistent.

(3) After loading continually for a while, the cracks continue developing downwards until the main crack is interconnected. That is, the sample is damaged. In Figure 12e, as compared with the former stage, the sample breaks and has lots of small discontinuous cracks. It may result from that that there are only a few aggregates in the top of the model, and the model cannot bear the load after the cracks are interconnected. A few small cracks develop towards two sides of the model, which may be caused by the small cracks near the top wall, because the small cracks will develop outwards under the tensile stress. From Figure 12e,f, it can be found that the shape, length and distribution of the cracks from the simulation test are consistent with those from the laboratory test.

(4) From Figure 12a–f, it can be found that the development and distribution characteristics of the cracks of the sample in each loading stage are in accordance with those of the discrete element model. As far as the characteristics of the crack are concerned, the indirect tensile fatigue simulation test matches the indirect tensile fatigue laboratory test very well.

#### 3.2.2. The Time-Load Curve and Time-Displacement Curve

In the laboratory test, the load, loading time and vertical deformation of the sample were collected, and the time-load curve and time-displacement curve are plotted in Figure 13a. In the simulation test, the load is the unbalanced force received from the top wall. The loading time is the multiplication of the time-step length and the cycle times. The time-load curve and time-displacement curve of the simulation test are plotted in Figure 13b. Since the whole experiment would take too much time and produce lots of experimental data, only the first four cycles of the curve were compared between these two tests.

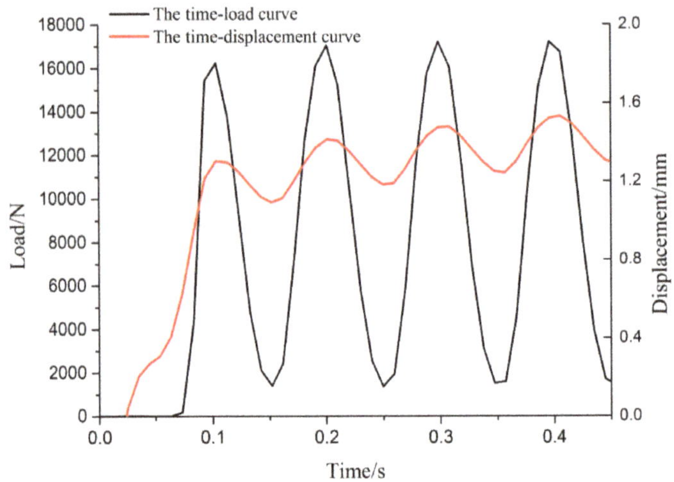


(**a**) Laboratory test

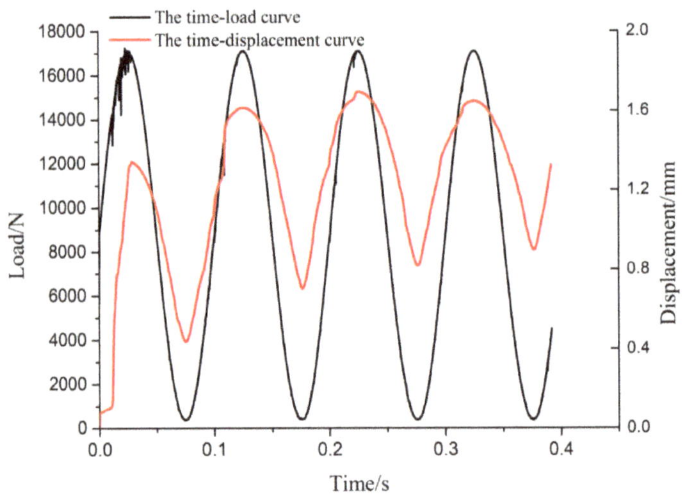


(**b**) Simulation test

**Figure 13.** Time-load curves and time-displacement curves.

The loads of these two tests are both continuous sinusoidal loads. From the time-displacement curve in Figure 13, it can be found that the tendency of the vertical deformation in both tests is consistent: (1) the elastic deformation comes out under the sinusoidal load. The elastic deformation follows the sinusoidal law, just like the load. (2) The plastic cumulative deformation also occurs at the same time. The time-displacement curve of the model is a rising sinusoidal curve. Firstly, the bottom value of the sinusoidal curve increases rapidly after loading. When the stress becomes stable, it tends to be relatively flat. After the model breaks, it grows again quickly. In the laboratory test, just like the simulation test, the vertical deformation of the sample also can be divided into three stages: growing rapidly, relatively flat, and growing rapidly, as illustrated in Figure 14. Therefore, the stress-strain response of the simulation test is consistent with that of the laboratory test.

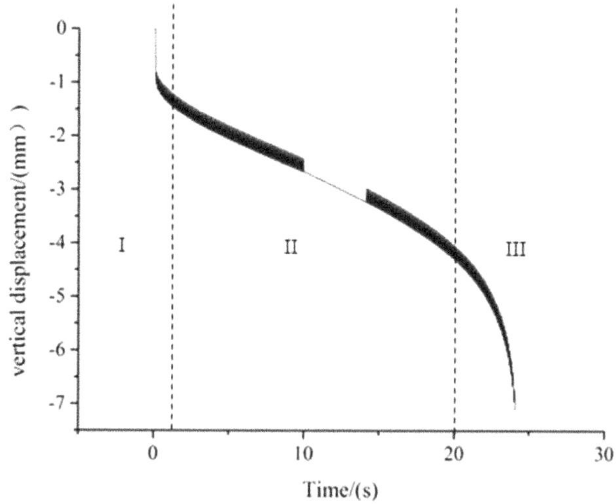


**Figure 14.** Time-displacement curve of the sample in the laboratory test.

## 4. Conclusions

In this paper, the discrete element model was reconstructed based on the CT image of an actual sample, and an indirect tensile fatigue simulation test was carried out. The distribution characteristics of the cracks and internal forces in the simulation test were verified in the laboratory. The following conclusions can be drawn:

(1) At the beginning of the loading, the internal forces of the model came down from the top of the model. The contacts among the particles were intact. The structure of the model was complete and there was not any crack. When the cracks came out, the contacts among the particles near the top wall were damaged first, and the particles with damaged contacts moved towards both sides of the crack. The failure here was caused by the tensile forces.
(2) The model was loaded continually after cracking, and the particles whose contacts were damaged moved either to the different directions or the same direction with different distances. It was found that the cracks developed from the shear failure under the non-uniform compressive forces and the tensile failure under the intensive tensile forces.
(3) When the model was damaged, the contact failure mainly happened near the loading axis of the model. The damaged contacts came out between the aggregates and asphalt mortar. The displacement at different directions and in the same direction was produced at both sides of the cracks. During the failure process, the shear failure and tensile failure co-occurred, but later development of the crack was mainly caused by the tensile failure.
(4) The corresponding stages of the fatigue failure, development and distribution of the cracks, time-load curve and time-displacement curve of the indirect tensile fatigue simulation test were consistent with those of the laboratory test. That is, the applicability and feasibility of the fatigue simulation test were verified by the experimental test.

**Author Contributions:** Conceptualization, X.L. (Xuelian Li) and X.L. (Xinchao Lv); Methodology, X.L (Xuelian Li); Software, X.L. (Xueying Liu); Resources, X.L. (Xuelian Li); Data Curation, X.L. (Xinchao Lv); Writing—Original Draft Preparation, X.L. (Xinchao Lv); Writing-Review & Editing, X.L. (Xuelian Li); Visualization, J.Y.; Project Administration, X.L. (Xuelian Li); Funding Acquisition, X.L. (Xuelian Li).

**Funding:** This work was funded by the National Natural Science Foundation of China [number 51308075], Department of Transport of Hainan Province (Number JT20160898009), and Open Fund of State Engineering Laboratory of Highway Maintenance Technology (Changsha University of Science & Technology) [number kfj160101, kfj140104].

**Acknowledgments:** Our special thanks go to all the subjects who participated in the data acquisition.

**Conflicts of Interest:** The authors declare no conflict of interest.

## References

1. Zhang, Y.; Luo, R.; Lytton, R.L. Mechanistic Modeling of Fracture in Asphalt Mixtures under Compressive Loading. *J. Mater. Civ. Eng.* **2012**, *25*, 1189–1197. [CrossRef]
2. Said, S.F.; Vieira, J.M.; Hakim, H.; Eriksson, O.; Nilsson, R.; Cocurullo, A. Interlaboratory experiment of asphalt concrete using indirect tensile fatigue test. In Proceedings of the 5th Eurasphalt & Eurobitume Congress, Istanbul, Turkey, 13–15 June 2012.
3. Chen, J.; Zhang, M.; Wang, H.; Li, L. Evaluation of thermal conductivity of asphalt concrete with heterogeneous microstructure. *Appl. Therm. Eng.* **2015**, *84*, 368–374. [CrossRef]
4. Ma, T.; Zhang, D.; Zhang, Y.; Zhao, Y.; Huang, X. Effect of air voids on the high-temperature creep behavior of asphalt mixture based on three-dimensional discrete element modeling. *Mater. Des.* **2016**, *89*, 304–313. [CrossRef]
5. Ma, T.; Zhang, Y.; Zhang, D.; Yan, J.; Ye, Q. Influences by air voids on fatigue life of asphalt mixture based on discrete element method. *Constr. Build. Mater.* **2016**, *126*, 785–799. [CrossRef]
6. Cao, Q.; Liu, X.; Hao, W.; Huang, X.; Transportation, S.O.; University, S. Simulation and analysis of two-point bending fatigue test of asphalt concrete based on discrete element model. *J. Southeast Univ.* **2017**, *33*, 286–292.
7. Gao, X.; Koval, G.; Chazallon, C. A discrete element model for damage and fracture of geomaterials under fatigue loading. In Proceedings of the 8th International Conference on Micromechanics on Granular Media, Montpellier, France, 3–7 July 2017.
8. Gao, X. Modelling of Nominal Strength Prediction for Quasi-Brittle Materials: Application to Discrete Element Modelling of Damage and Fracture of Asphalt Concrete under Fatigue Loading. Ph.D. Thesis, University of Strasbourg, Strasbourg, France, 2017.
9. Yu, H.; Shen, S. Impact of aggregate packing on dynamic modulus of hot mix asphalt mixtures using three-dimensional discrete element method. *Constr. Build. Mater.* **2012**, *26*, 302–309. [CrossRef]
10. You, Z. Development of a Micromechanical Modeling Approach to Predict Asphalt Mixture Stiffness Using the Discrete Element Method. Ph.D. Thesis, University of Illinois at Urbana, Champaign, IL, USA, 2003.
11. You, Z.; Adhikari, S.; Kutay, M.E. Dynamic modulus simulation of the asphalt concrete using the X-ray computed tomography images. *Mater. Struct.* **2009**, *42*, 617–630. [CrossRef]
12. You, Z.; Buttlar, W.G. Discrete Element Modeling to Predict the Modulus of Asphalt Concrete Mixtures. *J. Mater. Civ. Eng.* **2004**, *16*, 140–146. [CrossRef]
13. Habtamu, M.Z.; Athanassios, T.P. Micromechanical Modeling of Asphalt Concrete Uniaxial Creep Using the Discrete Element Method. *Road Mater. Pavement Des.* **2015**, *11*, 613–632.
14. Ji, Z.; Qiong, T.; Yong-qin, R.; Hong, Y. A DEM Geometric Modeling Method of HMA Based on Digital Imaging Processing. *J. Civ. Archit. Environ. Eng.* **2012**, *34*, 136–140.
15. Tan, X.; Konietzky, H.; Chen, W. Numerical Simulation of Heterogeneous Rock Using Discrete Element Model Based on Digital Image Processing. *Rock Mech. Rock Eng.* **2016**, *49*, 4957–4964. [CrossRef]
16. *Technical Specifications for Construction of Highway Asphalt Pavements*; JTG F40-2004; China Communications Press: Beijing, China, 2004.
17. Behnood, A.; Gharehveran, M.M. Morphology, rheology, and physical properties of polymer-modified asphalt binders. *Eur. Polym. J.* **2018**. [CrossRef]
18. Li, X.; Zhou, Z.; You, Z. Compaction temperatures of Sasobit produced warm mix asphalt mixtures modified with SBS. *Constr. Build. Mater.* **2016**, *123*, 357–364. [CrossRef]
19. *Test Methods Aggregate of Highway Engineering Aggregate*; JTG E42-2005; China Communications Press: Beijing, China, 2005.
20. *Application Handbook of Standard Test Methods of Bitumen and Bituminous Mixtures for Highway Enginerring*; JTG E20-2011; China Communications Press Co., Ltd.: Beijing, China, 2011.
21. Feng, H.; Pettinari, M.; Stang, H. Study of normal and shear material properties for viscoelastic model of asphalt mixture by discrete element method. *Constr. Build. Mater.* **2015**, *98*, 366–375. [CrossRef]
22. Chen, J.; Zhang, D. *Application of Particle Flow Code in Rode Engineering*; China Communications Press Co., Ltd.: Beijing, China, 2015.

23. Liu, X. The Research on Microscopic Mechanism of Top-Down Cracking in Asphalt Pavement. Master's Thesis, Changsha University of Science & Technology, Changsha, China, 2017.
24. He, Z.Q.; Liu, Z. Investigation of Bursting Forces in Anchorage Zones: Compression-Dispersion Models and Unified Design Equation. *J. Bridge Eng.* **2011**, *16*, 820–827. [CrossRef]

applied sciences

MDPI

Article

# Preparation of Polyacrylate Hollow Microspheres via Facile Spray Drying

**Pingxu Chen [1,*], Nanbiao Ye [1], Chaoxiong He [1], Lei Tang [1], Shuliang Li [2,3], Luyi Sun [3,*] and Yuntao Li [2]**

[1] National Engineering Laboratory of Plastics Modification and Processing, and Research and Development Center, Kingfa Science and Technology Company, Ltd., Guangzhou 510663, China; yenanbiao@kingfa.com.cn (N.Y.); hechaoxiong@kingfa.com.cn (C.H.); tanglei@kingfa.com.cn (L.T.)
[2] College of Materials Science and Engineering, Southwest Petroleum University, Chengdu 610050, China; shuliang.li@stu.swpu.edu.cn (S.L.); yuntaoli@swpu.edu.cn (Y.L.)
[3] Polymer Program, Institute of Materials Science and Department of Chemical & Biomolecular Engineering, University of Connecticut, Storrs, Connecticut, CT 06269, USA
* Correspondence: pinger@kingfa.com.cn (P.C.); luyi.sun@uconn.edu (L.S.)

Received: 20 November 2018; Accepted: 29 December 2018; Published: 10 January 2019

**Abstract:** Polyacrylate microspheres with a hollow structure were prepared by a facile spray drying method. The effects of spray drying process parameters, including inlet temperature, atomizer rotational speed, and feed speed, on the particle size, bulk density, and morphology of the resultant polyacrylate hollow microspheres were investigated and discussed. The mechanism for the formation of the polyacrylate hollow microspheres was proposed. This facile and scalable method for preparing hollow polymer microspheres is expected to be valuable to prepare various polymer hollow structures for widespread application.

**Keywords:** polyacrylate; hollow microspheres; spray drying

## 1. Introduction

Polymer hollow microspheres have drawn major attention because of their large specific surface area, relative low density, and high encapsulation capability [1–4]. As a result, they have found a wide range of applications, including drug delivery, catalysis, and coatings [5–7]. Specifically, hollow microspheres can be used as drug carriers, improving the flowability and packability when compared with the raw crystals of drugs [8]. Single-hole hollow polymer microspheres with specific high-capacity uptake of target species may also provide new opportunities in the capsulation of drugs [9]. Silica/polymer hollow functional polymer microspheres prepared by reversible addition–fragmentation chain transfer (RAFT) polymerization were applied as a reservoir for nitric oxide (NO) [10]. Moreover, conductive polymer hollow microspheres with high specific surface areas and outstanding electrical properties exhibited superb microwave absorption performance [11]. Therefore, it is meaningful to develop a facile and scalable method to prepare hollow microspheres for various applications.

Polymer hollow microspheres are mainly prepared via three approaches: self-assembly, template assisted synthesis, and emulsion polymerization [12–15]. Self-assembly and template assisted synthesis are demanding on reaction system and processing procedures, while emulsion polymerization is more industry-friendly and scalable. In addition, this emulsion polymerization method can effectively control sphere size, structure, and composition. Therefore, emulsion polymerization is an ideal method to prepare polymer hollow microspheres, especially at a large scale [16,17]. Kobayashi et al. prepared hollow polystyrene particles by seeded emulsion polymerization [18], which has many advantages, including being less dependent on the type of base polymer used, more friendly to environment, and more easy to scale up. However, further water absorption by the particles may occur during the polymerization, thus making it challenging to control the hollow structure.

In recent years, preparation of polymer microspheres by spray drying has attracted attention, in which an atomizer is employed to make small droplets with subsequent flash evaporation by hot air to obtain microspheres. Compared with other methods, spray drying has the advantages of a facile process, high controllability, and ease of commercialization [19,20]. However, preparing hollow microspheres with controllable size and morphology by spray drying remains a challenge. We aim to develop a facile and scalable method to prepare polymer hollow microspheres by combining emulsion polymerization and spray drying.

## 2. Materials and Methods

A high-speed centrifugal spray dryer (model LPG-8, Changzhou Lima Drying Engineering Co., Ltd., Changzhou, China) was used. A conventional polyacrylate emulsion was used as the model emulsion for this project to prepare polyacrylate hollow spheres, while other polymer emulsions should work similarly effectively. First, a polyacrylate emulsion with a solid content of 33 wt % and latex particle size of 30–50 nm was prepared by using methyl methacrylate acrylate (MMA)/butyl acrylate (BA)/acrylic acid (AA) (85/15/3 in mass ratio) as comonomers (MMA, BA, and AA were all industrial grade and obtained from Guangzhou Guanglin Chemical Co., Ltd., Guangzhou, China) and using ammonium sulfate allyloxy nonylphenoxy poly(ethyleneoxy)(10) ether (DNS-86, industrial grade, Guangzhou Shuangjian Trade Co., Ltd., Guangzhou, China) as emulsifier via semi-continuous emulsion polymerization [21]. The glass transition temperature ($T_g$) of the synthesized polyacrylate was characterized to be 83 °C by differential scanning calorimetry (DSC) at a heating rate of 10 °C/min.

The synthesized polyacrylate emulsion was filtered with a 200 mesh filter and transferred into a centrifugal spray drying tower (Figure 1) using a peristaltic pump (model BT300, Changzhou Purui Fluid Technology Co., Ltd., Changzhou, China). The emulsion was subsequently converted into small droplets by a high-speed atomizer (model DPG-5, Xiangsu Xinglun Electromechanical Equipment Co., Ltd., Taizhou, China) because of the atomization effect. The formed small droplets were in full contact with the hot air in the tower body, and were dried into microspheres after being heat-exchanged along their transportation path, and then separated by a cyclone separator. The microspheres were finally collected in a receiving tank and filtered through a 100-mesh filter and then sealed. The entire process is briefly shown in Figure 1. The resultant microspheres were characterized by scanning electron microscopy (SEM) and laser diffraction. For SEM, the polyacrylate microspheres were coated with a thin gold layer before being imaged using a HITACHI S-3400N scanning electron microscope (HITACHI, Hitachinaka, Japan) operated at a working voltage of 20 kV. For laser diffraction, the microspheres were dispersed in deionized water to prepare a dilute suspension (0.1–1.0 wt %) with the assistance of ultrasonication, and subsequently characterized on a Malvern Laser Particle Sizer (Mastersizer 3000, Malvern, UK). Three specimens of each sample were characterized, and the average results were recorded.

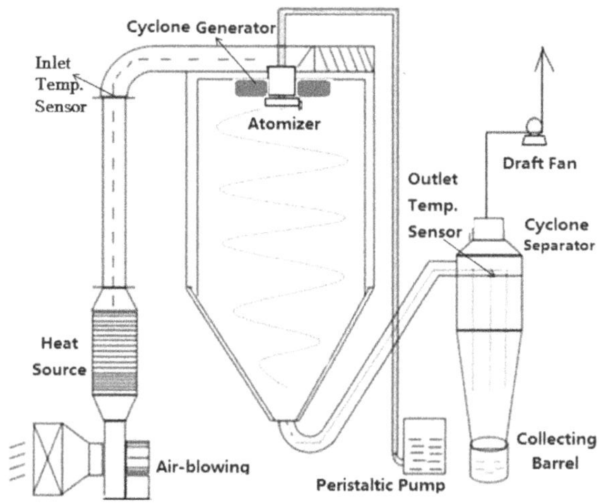


**Figure 1.** Schematic of a high-speed centrifugal spray dryer.

## 3. Results and Discussion

During the entire process, three key parameters affect the drying result and therefore the morphology of the formed polyacrylate hollow microspheres: (1) temperature, including inlet air temperature (i.e., the temperature of the hot air enters the tower) and outlet air temperature (i.e., the temperature of the cyclone separator above the receiver); (2) size of the droplets, which is mainly decided by the speed of the atomizer; and (3) feed rate of the emulsion.

Properly tuning the inlet and the outlet air temperatures is particularly critical to the formation of the polyacrylate hollow microspheres. These temperatures not only affect the water evaporation rate, moisture content, and drying efficiency, but also directly determine the microsphere nucleation mode, thereby affecting the microsphere size and morphology. The effect of inlet and outlet temperatures on the microsphere bulk density is illustrated in Table 1. At a 28,000 rpm atomizer rotational speed and an emulsion feed rate of 160 g/min, the bulk density of the hollow microspheres decreased with an increasing inlet/outlet temperature, especially when the inlet air temperature was at 200–220 °C. As shown in Figure 2, inlet air temperature has an effect on the size, size distribution, and morphology of the prepared polyacrylate microspheres. The microsphere size increased very marginally when the inlet temperature was increased from 140 to 200 °C. However, a slight size increase occurred in the range of 200–220 °C. For example, the D(50) of the microspheres increased from 34 to 43 μm when the inlet temperature was increased from 200 to 220 °C. The SEM images in Figure 2 show the morphology and surface structure of the microspheres prepared at various temperatures. One can observe that most of particles show a spherical morphology and a large portion of them are hollow, as shown in Figure 2a. Meanwhile, a small amount of broken fragments of microspheres are observed in Figure 2a–c, which is consistent with the literature [22].

**Table 1.** Effect of inlet/outlet temperature on the bulk density and dimensions of the microspheres.

| Inlet Temperature/°C | Outlet Temperature/°C | bulk Density/(g/cm$^3$) | D(10)/μm | D(50)/μm | D(90)/μm |
|---|---|---|---|---|---|
| 140 | 61 | 0.52 | 16 | 31 | 58 |
| 160 | 66 | 0.50 | 18 | 33 | 58 |
| 180 | 73 | 0.49 | 18 | 33 | 57 |
| 200 | 84 | 0.44 | 18 | 34 | 60 |
| 220 | 98 | 0.37 | 24 | 43 | 74 |

D(10), D(50), and D(90) particle sizes are the equivalent diameters of the largest particles with a cumulative volume distribution of 10%, 50%, and 90% in the distribution curve, respectively.

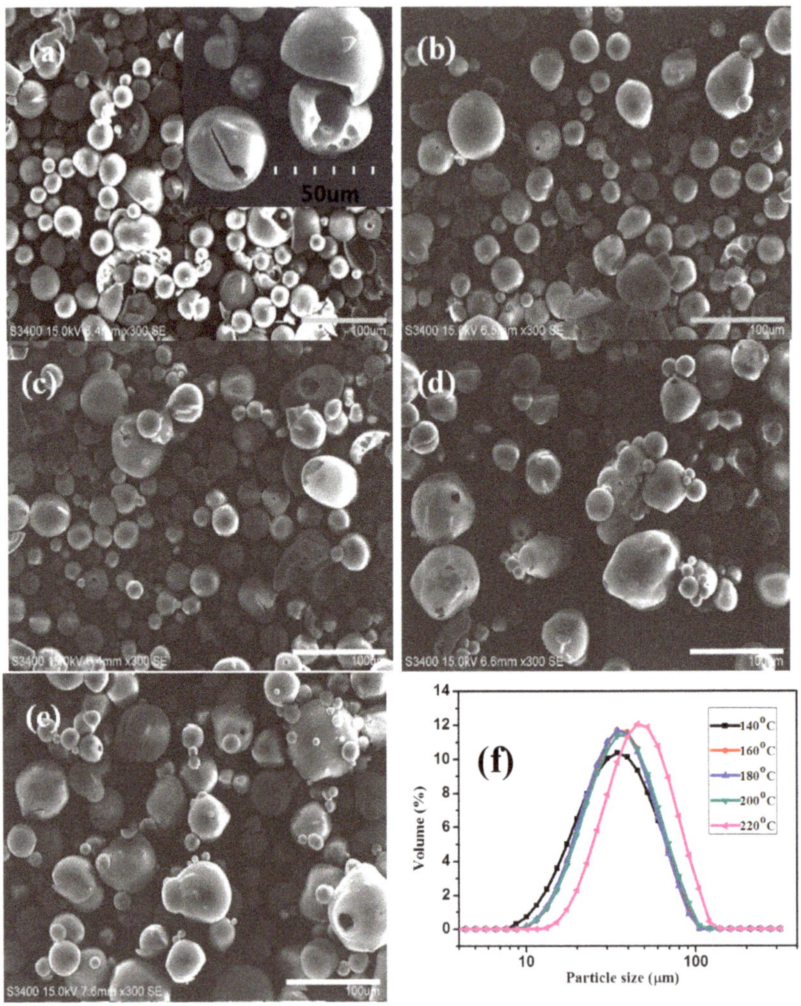


**Figure 2.** SEM images of the polyacrylate hollow microspheres prepared at different inlet temperatures: (**a**) 140 °C; (**b**) 160 °C, (**c**) 180 °C, (**d**) 200 °C, and (**e**) 220 °C; (**f**) Particle size distribution as a function of inlet temperature.

Based on the above results, the mechanism of microsphere formation is proposed and briefly illustrated in Figure 3. First, small droplets form when the emulsion goes through the atomizer. Due to the evaporation of water, the droplets shrink and the latex particles aggregate, leading to the initial formation of the wet shells of the microspheres. Wet shell microspheres are gradually dried and maintain the spherical morphology until the droplets completely solidify. The residual water within the spheres subsequently evaporates through the tiny pores and capillary channels between the latex particles. Finally, the hollow microspheres are generated. According to SEM characterization, this mechanism is applicable to the droplets with a size lower than ca. 20 μm. For larger droplets (>20 μm), there is a large amount of moisture within the wet shell. It evaporates quickly at elevated temperatures (ca. 140–180 °C), which is too much to volatilize through tiny pores or capillary channels, resulting in a gradual increase in internal pressure. When the internal pressure is higher than the mechanical strength of the shell, two situations occur: (a) if the inlet temperature is relatively low (i.e., 180 °C or lower), it creates a drying temperature lower than the glass transition temperature of polyacrylate shell, then the microspheres rapidly burst, generating debris and/or residues; (b) if the inlet temperature is very high (i.e., 200 °C or higher), which generates a surrounding temperature higher than $T_g$ of the shell, then bubble punching would happen due to viscoelastic failure of the shell, instead of breaking the shells.

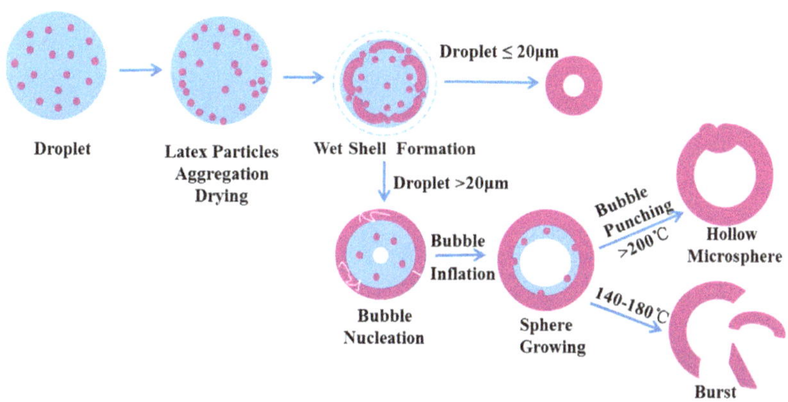


**Figure 3.** Formation mechanism of the polyacrylate hollow microspheres during centrifugal spray drying.

The speed of the atomizer disk directly determines the quality of the atomization effect and size of the atomized droplets, resulting in different microscopic morphologies. Figure 4 shows the effect of atomizer speed on the particle size and size distribution of the formed polyacrylate hollow microspheres at a 220 °C drying temperature and an emulsion feed rate of 160 g/min. With an increasing atomizer speed from 10,000 to 28,000 rpm, the average particle diameter D(50) of the microspheres gradually decreased from 76 to 35 μm. When the speed was further increased to 34,000 rpm, there was no obvious decrease in microsphere size. Overall, the higher the atomizer speed, the larger the centrifugal force and friction force subjected to the emulsion. Therefore, the emulsion is sheared and split into smaller droplets. With an improved atomization effect, the diameter of the dried microspheres is gradually reduced.

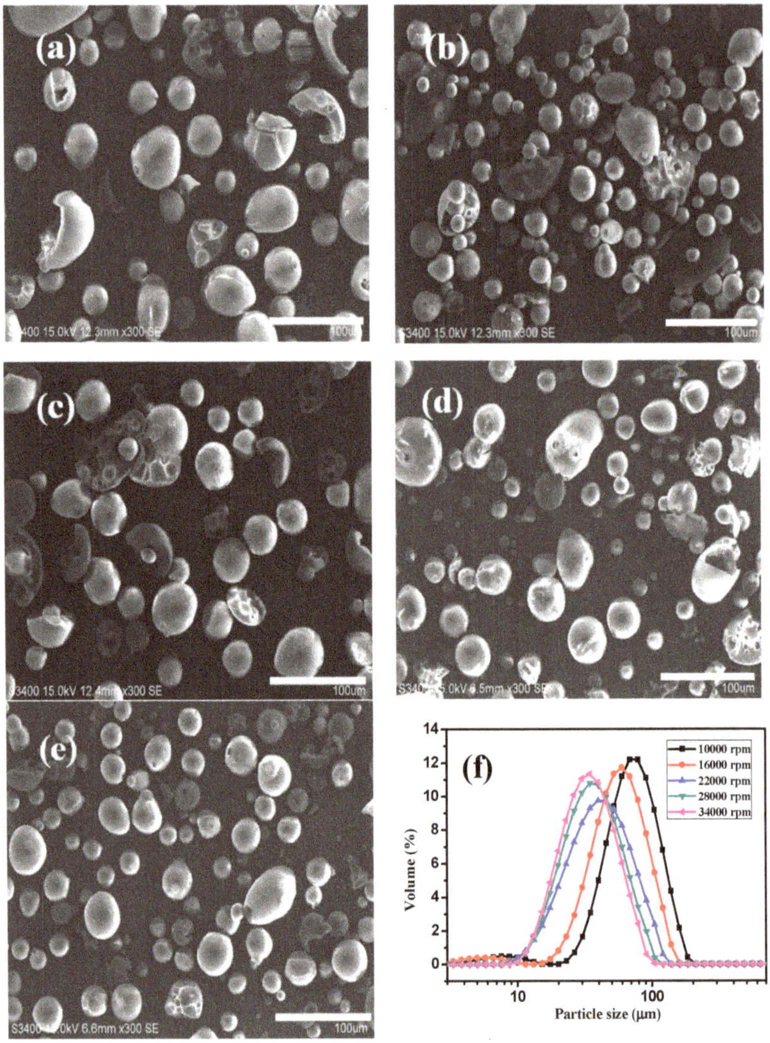


**Figure 4.** SEM images of the microspheres prepared at different atomizer rotational speeds: (**a**) 10,000 r/min; (**b**) 16,000 r/min; (**c**) 22,000 r/min; (**d**) 28,000 r/min, and (**e**) 34,000 r/min; (**f**) Effects of atomizer rotational speed on particle size and size distribution of the formed microspheres at 220 °C drying temperature.

When the inlet temperature of the spray drying (220 °C) and the speed of the atomizer (28,000 rpm) are determined, the drying effect and the size and morphology of the microspheres vary with the change of the feed rate of the emulsion. The effect of feed rate on particle size, size distribution, and morphology of the formed polyacrylate hollow microspheres are shown in Figure 5. At a sufficiently high drying temperature (220 °C), with the feed rate increased from 82 to 190 g/min, there are only marginal changes in the size and shape of the formed microspheres. The microspheres possess a smooth surface, clear interface, and high sphericity. However, some microspheres were broken. Increasing the feed rate can slightly increase the size of the microspheres, and hence the powder fluidity of the microspheres decreases because water evaporation becomes more difficult, but a high productivity can be obtained. The higher the feed rate, the higher the inlet air temperature required to dry the droplets into spheres. A similar research finding has been reported [23], and if both effectiveness and efficiency are taken into consideration, the optimized feed rate is 136 g/min.

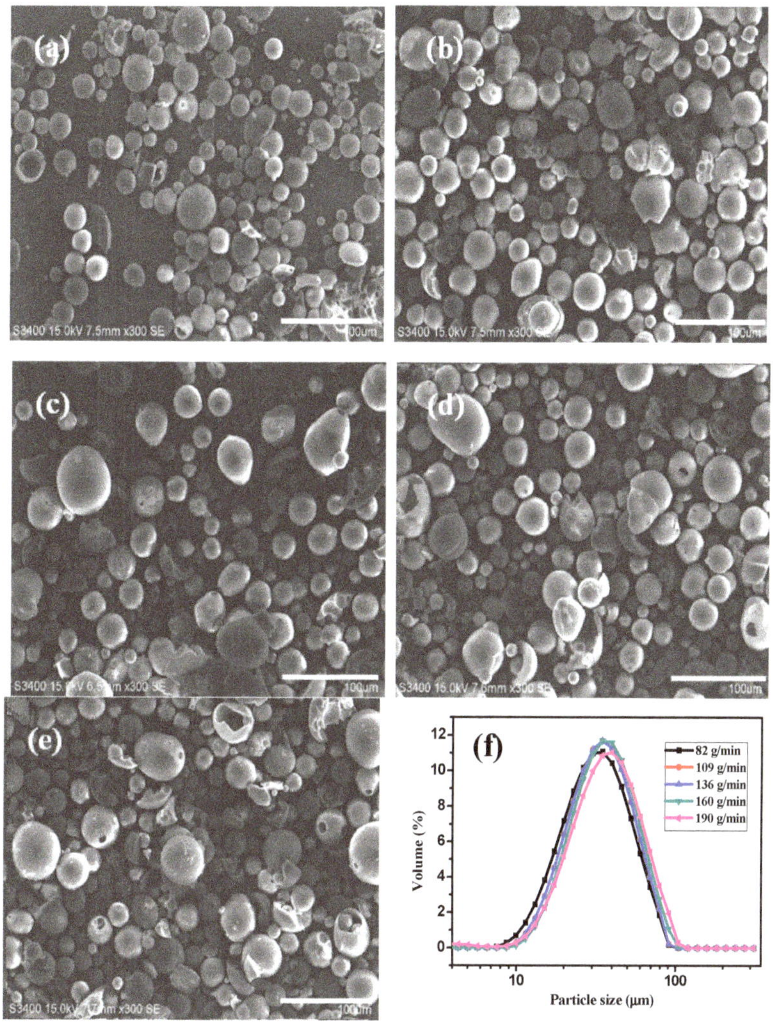


**Figure 5.** Effects of feed rate on particle size, size distribution, and morphology of the formed polyacrylate hollow microspheres: (**a**) 82 g/min; (**b**) 109 g/min; (**c**) 136 g/min; (**d**) 160 g/min, and (**e**) 190 g/min; (**f**) Particle size distribution as a function of feed rate.

## 4. Conclusions

Polyacrylate hollow microspheres were successfully prepared by a facile spray drying method. We found that when the temperature was increased from 140 to 220 °C, the average particle size increased, while the packing density of the microspheres decreased. With an increasing atomizer speed, the average particle size of the microspheres gradually decreased. Feed speed has a marginal effect on the average particle size at a sufficiently high drying capacity, but if the feed rate is too high, the drying process for microspheres will be much harder. We proved that spray drying of polyacrylate emulsion is a facile and scalable method to prepare polyacrylate hollow microspheres, and this method should be applicable to other systems to prepare various polymer hollow structures for widespread application.

**Author Contributions:** P.C. and N.Y. conceived and designed the experiments; L.S. and Y.L. optimized the process; P.C., N.Y., C.H., L.T., and S.L. performed the experiments; P.C., L.T., S.L., and L.S. wrote the first draft of the manuscript; all authors contributed to revise the manuscript.

**Funding:** This project is sponsored by the National Key Research and Development Program of China (2016YFB0302005), the Science and Technology Program of Guangzhou (201710010176), and the Guangzhou Development District (2017GH32).

**Conflicts of Interest:** The authors declare no conflict of interest.

## References

1. Han, J.; Liu, Y.; Guo, R. Reactive template method to synthesize gold nanoparticles with controllable size and morphology supported on shells of polymer hollow microspheres and their application for aerobic alcohol oxidation in water. *Adv. Funct. Mater.* **2009**, *19*, 1112–1117. [CrossRef]
2. Zhong, S.L.; Xu, R.; Zhang, L.F.; Qu, W.G.; Gao, G.Q.; Wu, X.L.; Xu, A.W. Terbium-based infinite coordination polymer hollow microspheres: Preparation and white-light emission. *J. Mater. Chem.* **2011**, *21*, 16574–16580. [CrossRef]
3. Wei, Z.; Wan, M. Hollow microspheres of polyaniline synthesized with an aniline emulsion template. *Adv. Mater.* **2002**, *14*, 1314–1317. [CrossRef]
4. Zhang, L.; Wan, M. Self-Assembly of Polyaniline—From Nanotubes to Hollow Microspheres. *Adv. Funct. Mater.* **2003**, *13*, 815–820. [CrossRef]
5. Han, J.; Song, G.; Guo, R. A facile solution route for polymeric hollow spheres with controllable size. *Adv. Mater.* **2006**, *18*, 3140–3144. [CrossRef]
6. Yang, X.; Chen, L.; Huang, B.; Bai, F.; Yang, X. Synthesis of pH-sensitive hollow polymer microspheres and their application as drug carriers. *Polymer* **2009**, *50*, 3556–3563. [CrossRef]
7. Wang, Z.; Qiu, T.; Guo, L.; Ye, J.; He, L.; Li, X. Polymerization induced shaping of Pickering emulsion droplets: From simple hollow microspheres to molecularly imprinted multicore microrattles. *Chem. Eng. J.* **2018**, *332*, 409–418. [CrossRef]
8. Arora, S.; Ali, J.; Ahuja, A.; Khar, R.K.; Baboota, S. Floating Drug Delivery Systems: A Review. *AAPS PharmSciTech* **2005**, *6*, E372–E390. [CrossRef]
9. Guan, G.; Zhang, Z.; Wang, Z.; Liu, B.; Gao, D.; Xie, C. Single-hole hollow polymer microspheres toward specific high-capacity uptake of target species. *Adv. Mater.* **2007**, *19*, 2370–2374. [CrossRef]
10. Liu, T.; Zhang, D.; Yang, X.; Li, C. Silica/polymer microspheres and hollow polymer microspheres as scaffolds for nitric oxide release in PBS buffer and bovine serum. *Polym. Chem.* **2015**, *6*, 1512–1520. [CrossRef]
11. Pang, R.; Hu, X.; Zhou, S.; Sun, C.; Yan, J.; Sun, X.; Xiao, S.; Chen, P. Preparation of multi-shelled conductive polymer hollow microspheres by using $Fe_3O_4$ hollow spheres as sacrificial templates. *Chem. Commun.* **2014**, *50*, 12493–12496. [CrossRef] [PubMed]
12. Okubo, M.; Kobayashi, H.; Huang, C.; Miyanaga, E.; Suzuki, T. Water Absorption Behavior of Polystyrene Particles Prepared by Emulsion Polymerization with Nonionic Emulsifiers and Innovative Easy Synthesis of Hollow Particles. *Langmuir* **2017**, *33*, 3468–3475. [CrossRef] [PubMed]
13. Zhao, D.; Wang, M.; Xu, Y.; Zhang, Z.; Ge, X. The fabrication and corrosion resistance of benzotriazole-loaded raspberry-like hollow polymeric microspheres. *Surf. Coat. Technol.* **2014**, *238*, 15–26. [CrossRef]
14. Tian, Q.; Yu, X.; Zhang, L.; Yu, D. Monodisperse raspberry-like multihollow polymer/Ag nanocomposite microspheres for rapid catalytic degradation of methylene blue. *J. Colloid Interface Sci.* **2017**, *491*, 294–304. [CrossRef] [PubMed]
15. Tang, C.; Liu, L.; Li, Y.; Bian, Z. Aerosol spray assisted assembly of $TiO_2$ mesocrystals into hierarchical hollow microspheres with enhanced photocatalytic performance. *Appl. Catal. B Environ.* **2017**, *201*, 41–47. [CrossRef]
16. Wichaita, W.; Polpanich, D.; Suteewong, T.; Tangboriboonrat, P. Hollow core-shell particles via NR latex seeded emulsion polymerization. *Polymer* **2016**, *99*, 324–331. [CrossRef]
17. Pan, J.; Yin, Y.; Gan, M.; Meng, M.; Dai, X.; Wu, R.; Shi, W.; Yan, Y. Fabrication and evaluation of molecularly imprinted multi-hollow microspheres adsorbents with tunable inner pore structures derived from templating Pickering double emulsions. *Chem. Eng. J.* **2015**, *266*, 299–308. [CrossRef]

18. Kobayashi, H.; Suzuki, T.; Moritaka, M.; Miyanaga, E.; Okubo, M. Preparation of multihollow polystyrene particles by seeded emulsion polymerization using seed particles with incorporated nonionic emulsifier: Effect of temperature. *Colloid Polym. Sci.* **2009**, *287*, 251–257. [CrossRef]
19. MacKová, H.; Horák, D.; Petrovský, E.; Kovářová, J. Magnetic hollow poly(*N*-isopropylacrylamide-*co*-*N*,*N*′-methylenebisacrylamide-*co*-glycidyl acrylate) particles prepared by inverse emulsion polymerization. *Colloid Polym. Sci.* **2013**, *291*, 205–213. [CrossRef]
20. He, X.D.; Ge, X.W.; Liu, H.R.; Deng, M.G.; Zhang, Z.C. Self-Assembly of pH-Responsive Acrylate Latex Particles at Emulsion Droplets Interface. *J. Appl. Polym. Sci.* **2007**, *105*, 1018–1024. [CrossRef]
21. Ramli, R.A.; Laftah, W.A.; Hashim, S. Core-shell polymers: A review. *RSC Adv.* **2013**, *3*, 15543–15565. [CrossRef]
22. Jiao, Y.; Lu, Y.P.; Xiao, G.Y.; Xu, W.H.; Zhu, R.F. Preparation and characterization of hollow hydroxyapatite microspheres by the centrifugal spray drying method. *Powder Technol.* **2012**, *217*, 581–584. [CrossRef]
23. Fang, X.; Shima, M.; Adachi, S. Effects of drying conditions on the oxidation of linoleic acid encapsulated with gum arabic by spray drying. *Food Sci. Technol. Res.* **2005**, *11*, 380–384. [CrossRef]

*applied sciences*

MDPI

*Article*

# Characteristics of Deformation and Damping of Cement Treated and Expanded Polystyrene Mixed Lightweight Subgrade Fill under Cyclic Load

**Weihua Lu [1], Linchang Miao [2], Junhui Zhang [3,*], Yongxing Zhang [1] and Jing Li [2]**

1 School of Civil Engineering, Nanjing Forestry University, Nanjing 210037, China; whlnjfu@njfu.edu.cn (W.L.); zhanguongxing81@aliyun.com (Y.Z.)

2 Institute of Geotechnical Engineering, Southeast University, Nanjing 210096, China; lc.miao@seu.edu.cn (L.M.); whl361@163.com (J.L.)

3 National Engineering Laboratory of Highway Maintenance Technology, Changsha University of Science & Technology, Changsha 410114, China

* Correspondence: zjhseu@csust.edu.cn; Tel.: +86-137-8716-2685

Received: 8 December 2018; Accepted: 27 December 2018; Published: 4 January 2019

**Abstract:** To investigate the deformation and damping characteristics of cement treated and expanded polystyrene (EPS) beads mixed lightweight soils, this study conducted a series of triaxial shear tests cyclic loading for different confining pressures, cement contents, and soil categories. Through repeated loading and unloading cycles, axial accumulative strain, resilient modulus, and damping ratio versus axial total strain were analyzed and the mechanical behavior was revealed and interpreted. Results show that the resilient modulus increases with increasing confining pressure and cement content. A decreasing power function can be used to fit the relationship between the resilient modulus and the axial total strain. Although sandy lightweight specimens usually own higher resilient modulus than silty clay lightweight specimens do, the opposite was also found when the axial total strain is larger than 8% with 50 kPa confining pressure and 14% cement content. For damping ratio the EPS beads mixed lightweight soil yields a weak growth trend with increasing axial total strain and a small reduction with higher confining pressure and cement content. For more cementations, the damping ratio of the sandy lightweight soil is always smaller than the silty clay lightweight soil. Nonetheless, the differences of damping ratios that were obtained under all of the test conditions are not significant.

**Keywords:** EPS lightweight soil; cyclic load; axial accumulative strain; resilient modulus; damping ratio

## 1. Introduction

With the rapid development of transportation in China, highways or high-speed railways would inevitably encounter the situation of soft soil foundations. How to deal with unacceptable settlements of embankments on these foundations has become one of the most challenging tasks for engineers [1,2]. Traditionally, the composite foundation and plastic drainage are always utilized to reinforce the soft soil foundation to diminish the potential settlement [3,4]. However, use of embankments with lightweight backfill material, such as expanded polystyrene (EPS) blocks (also called EPS geo-foam) or EPS beads mixed with soil and binder [5,6], can also reach this goal. Obviously, the latter is much more time saving, cost-effective, and even environmentally friendly.

Since Frydenlund [7] firstly reported that the EPS geo-foam was used as an embankment fill in Norway, more and more studies about this artificial material have been conducted and its good performances have been confirmed. For example, replacing the typical embankment fill material in highways [5], diminishing the maximum lateral earth pressure of a reinforced soil platform [8], and reducing the swelling pressure by expansion of soil behind the retaining wall [9], etc. At the same time, a lightweight fill consisting of dredged soil and air foam and cement (i.e., EPS beads) was adopted to reduce the embankment self-weight on soft foundations in Japan [10]. Miki [11] pointed out that the EPS beads could even reduce the weight of the fill to a great extent, even to be 50%, which is of great beneficial for the post-construction settlement control. Liu et al. [12] pointed out that the unconfined compressive strength of the lightweight fill, as well as the shear strength and stiffness, increases considerably if the cement-soil ratio of 10% to 15% is used. Moreover, Miao et al. [13] inspected the mechanical properties of the lightweight fill through a series of road performance tests, verifying that embankments with the lightweight backfill obtain an obviously smaller settlement over embankments with the conventional lime-stabilized fill.

As a subgrade filling replacement material, the lightweight soil should have good bearing capacity and deformation properties under static loading, but it also needs good mechanical performance under repeated traffic loading. As early as 2002, Minegishi et al. [14] has pointed out that the mechanical behaviors of lightweight soil under static and dynamic loads would be quite different from the natural soils. At present, the research on the dynamic characteristics of lightweight soils mainly focuses on the acquisition of basic dynamic characteristics parameters and the effects of various influencing factors. For example, Gao et al. [15] analyzed the characteristics of skeleton curve, dynamic shear modulus, and damping ratio of EPS composite soils based on nineteen combined axial-torsional tests on hollow cylinder specimens. Through resonant column and cyclic triaxial tests on sand-EPS bead mixtures, El-Sherbiny et al. [16] discovered that the material damping is relatively unaffected at small shear strains but it increases at larger strains, and the decrease in shear stiffness with increasing bead content occurs at all strain levels. Moreover, Alaie et al. [17] carried out a series of laboratory tests to evaluate the monotonic, cyclic, and post-cyclic behavior of the contact interface between lightweight soil and reinforced geogrid. Nonetheless, the road performance of the lightweight fill, especially the mechanical behavior under complex loading conditions as well as the difference with different natural soils is still not clear.

In this study, a series of triaxial shear tests, under repeated loading-unloading conditions, were conducted to investigate the characteristics of deformation and damping of EPS beads-mixed lightweight soils. The axial accumulative strain, the resilient modulus and the damping ratio versus the axial total strain under different confining pressures, cement contents, and soil categories were systematically analyzed. Subsequently, a further understanding of the mechanical properties of the EPS mixed lightweight soil under cyclic loading was obtained.

## 2. Materials and Methods

### 2.1. Materials

The silty clay and sand (as shown in Figure 1) that were adopted in this study were taken from the Beigu lake of Zhenjiang City, which belongs to the Yangtze River basin in the eastern China. Large particles that were more than 2 mm were removed from the material by a sieve. According to the Unified Soil Classification System (ASTM D2487-11) [18], the sand belongs to the poorly graded fine sand (SP). Its coefficient of uniformity (Cu) and coefficient of gradation (Cc) are 1.41 and 0.69, respectively. The grain size distributions of these two soils are illustrated in Figure 2. The main physical property indices of the silty clay are summarized in Table 1.

Figure 1. Test materials used in the tests: (a) Silty clay; (b) Sand; and, (c) expanded polystyrene (EPS) beads.

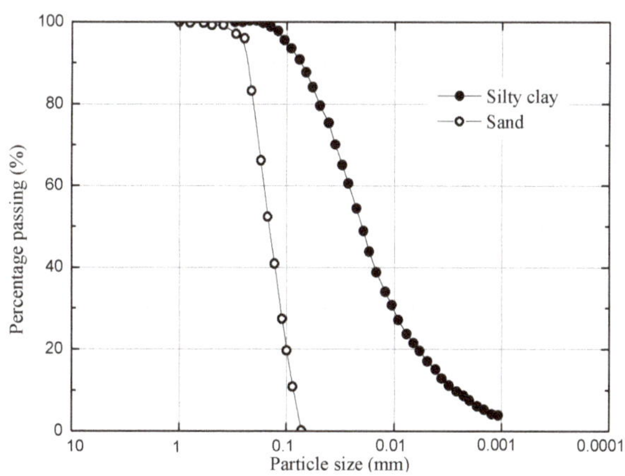


Figure 2. Particle size distribution of soils used in the tests.

Table 1. Properties of the test silty clay.

| Water Content (%) | Natural Density (g/cm$^3$) | Liquid Limit (%) | Plastic Limit (%) | Liquidity Index | Plasticity Index |
|---|---|---|---|---|---|
| 52.6 | 1.75 | 44.2 | 23.0 | 1.4 | 21.2 |

The EPS beads (as shown in Figure 1c) that were used in this study are manufactured by expandable polystyrene resin (from Suzhou Yizhan Purification Technology Co., Ltd. in China, 2017), containing microscopic cells that are foamed with pentanes or butanes. When the beads are formed after the blowing agent expands, volume of individual resin beads would increase by up to 40 to 50 times. The particle size of the round EPS beads ranges from 3 to 5 mm, with the bulk unit weight of 0.013 g/cm$^3$.

The Portland cement (P. O. 32.5) (Nanjing Pukou Youwei Cement Products Factory Co., Ltd. in China, 2018) was used as a binding material to bind the EPS beads with sand or silty clay, and water was used to carry out the hydration reaction and facilitate the mixing process. Subsequently, the lightweight mixture can be easier to be compacted to carry the required load after an appropriate curing time.

### 2.2. Specimen Preparation

To make standard triaxial specimens, the silty clay and sand were used as the raw material soil, and then cement and water were added by mass percent while the EPS beads were added by volume ratios. The cement content $a_w$ (or cement weight) was designed to be 14, 16, 18, and 20%, relative to the weight of silty clay or sand. The volume ratios of the sand/silty clay together with cement to the EPS beads were determined at 1:1. A machine mixer (Wuxi Chiba Mixing Equipment Co., Ltd. in China, 2014) was utilized to mix the mixture thoroughly, with the capacity bucket rotating at certain speeds of 102/204/388 r/min. The untreated soil was weighed to get the mass and firstly placed into the mixer, then the cement and the EPS beads were uniformly added into the soil and forcibly stirred for 5 min. At last, water was poured into the mixing bucket and stirred more than 5 min until the components were evenly distributed.

Once the lightweight soil was mixed thoroughly and stirred evenly, the weighed mixtures were put into a triple-piece mold and then compacted for 25 times in five layers, using a mini compaction hammer with a 295.8 g weight and a 12 cm drop distance. The size of the triple-piece mold was 3.91 cm in diameter and 8.0 cm in height. After compaction, the specimens were cured in a standard box with the temperature of 20 °C and the humidity of 100% for 24 h. Afterwards, the specimens were taken out and immediately put in plastic bags for a curing time of 14 days.

### 2.3. Test Procedure

To investigate the physical and mechanical properties of EPS composite soil, conventional testing methods, such as the unconfined compression test, the uniaxial compression test, the direct shear test, and the triaxial compression test, are always adopted. The current researches paid many attentions to interpret the mechanical behavior and its influence on strength, deformation, and failure modes under static loading [19–21]. However, as embankment fill subgrades, the EPS composite soil is likely to undertake cyclic loadings, such as the traffic loading or the seismic loading. Therefore, the dynamic properties of the EPS mixed lightweight fill are a concern for engineering [14]. In general, the conventional dynamic characteristic study on EPS composite soil is focused on the dynamic stress-state relationship, dynamic modulus and strength, damping ratio, and so on [15].

Although the conventional resonance column test and dynamic triaxial test would be the most commonly used methods to conduct the above research activities, the modification or new use of conventional test instruments can also fulfill some research tasks. Therefore, a conventional strain-controlled triaxial apparatus (Nanjing Soil Instrument Factory Co., Ltd. (type-ASPTTS) in China, 2012) was adopted to carry out the consolidated undrained (CU) test under confining pressures of 50, 100, and 150 kPa [22]. The CU test procedures in ASTM D4767-11 [23] were followed in the laboratory. In the traditional dynamic cyclic loading test, a cyclically changing direction dynamic load is applied to the specimen. However, the cyclic loading in this study is defined as a different repeated loading process. For example, under a confining pressure of 50 kPa, the specimen was firstly loaded by lifting the triaxial chamber with a rate of axial strain to a predetermined total strain value, and then the acted load was gradually removed by declining the chamber. Subsequently, the reloading and reunloading action would be developed repeatedly. In the scheme of this experimental study (see Figure 3), each specimen underwent nine repeated actions and the corresponding unloading strains were 2%, 3%, 4%, 6%, 8%, 10%, 12%, 14%, and 16%, respectively.

To reveal the dynamic properties of the EPS composite soil under repeated loading conditions, several affecting factors, such as the stress state, soil type, and cement content should be evaluated, as well as the effect variation on mechanical characteristics. The stress state and loading-unloading times were also checked for accessing the dynamic response of the lightweight material. Several mechanical indexes were evaluated, such as the axial cumulative strain, the resilient modulus, and the damping ratio.

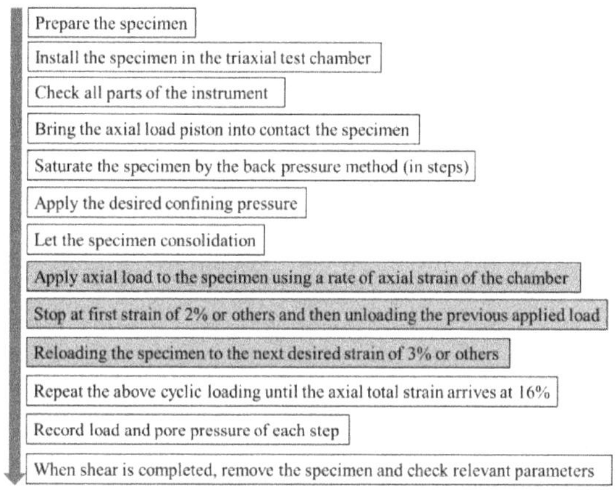


**Figure 3.** Test procedure for the consolidated undrained (CU) test in the laboratory under cyclic loading.

## 3. Results and Discussion

### 3.1. Axial Cumulative Strain

The axial cumulative strain is one of the important indexes to characterize the deformation ability of the lightweight soil. In this study, the axial cumulative strain is defined as the residual strain when the overlying load is removed. As shown in Figure 4 for sandy lightweight soil, the axial cumulative strain increases with the increasing cyclic loading times, as well as the axial total strain of the specimen. There exists a good linear relationship between the axial cumulative strain and the axial total strain (see Figure 5), and the ratio of axial cumulative strain to axial total strain remains constant. The same also applies to the tested silty clay lightweight soil specimens. However, the influence of confining pressure, cement content, and soil type on the ratio is very small through inspecting all of the tested specimens in this study (not shown here).

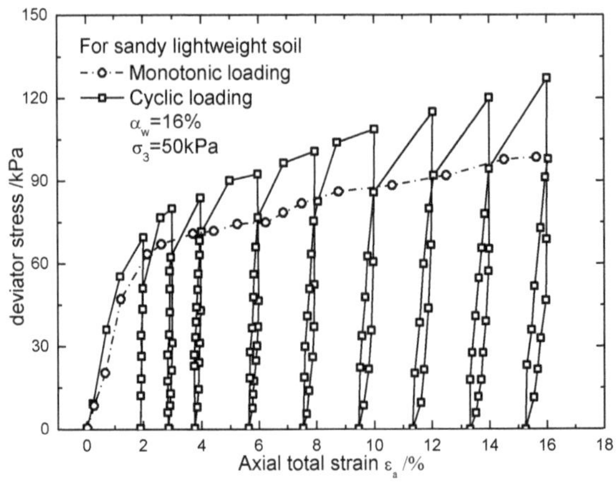


**Figure 4.** Typical stress-strain curve under monotonic loading and cyclic loading.

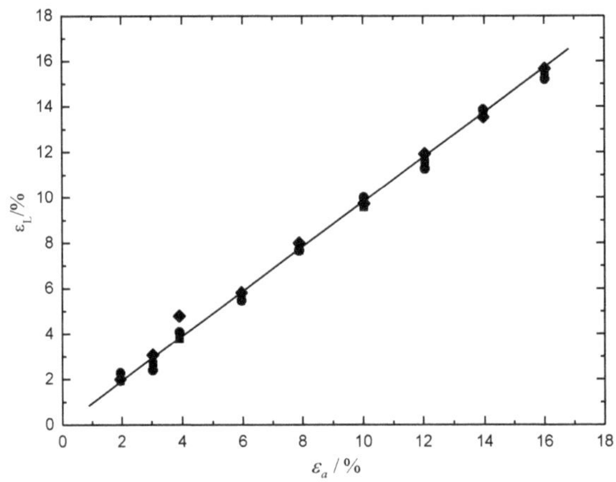


**Figure 5.** Relationship between axial accumulative strain and axial total strain.

Under lower confining pressures, the cementing structure (with relatively high cement content) of the mixed lightweight soil plays the main role in bearing the load capacity, and the damage of the specimen is very small. In this condition, the EPS particles are still constrained by the cementation force, indicating that the characteristics of larger elastic deformation cannot be brought into full activity. Although the elastic deformation of the lightweight soil would mainly come from the elastic deformation of the EPS particles, the elastic strain is always at a lower level when the stress state is relatively small. Therefore, the difference of the elastic strains under different conditions is not obvious.

### 3.2. Resilient Modulus

According to the linear relationship between the axial cumulative strain and the axial total strain, a formula can be expressed as:

$$\varepsilon_L = k\varepsilon_a \quad (1)$$

where $\varepsilon_L$ is the axial cumulative strain, $\varepsilon_a$ is the axial total strain, and $k$ is the ratio of the axial cumulative strain to the axial total strain.

Subjected to the monotonic loading, the relationship between the principal stress difference and the recoverable strain can be formulated as:

$$\sigma_1 - \sigma_3 = f(\varepsilon_a) = g(\varepsilon_h) = g((1-k)\varepsilon_a) \tag{2}$$

where $\sigma_1$ and $\sigma_3$ are the principal stresses and $\varepsilon_h$ is the recoverable strain.

The resilient modulus of the specimen under the monotonic loading can be defined as:

$$E_{ur} = \frac{\sigma_1 - \sigma_3}{\varepsilon_h} = \frac{\sigma_1 - \sigma_3}{(1-k)\varepsilon_a} \tag{3}$$

The resilient modulus here can also be defined as the average of the unloading modulus and the reloading modulus [24]. The unloading modulus is the ratio of the stress at the unloading point to the recoverable strain (i.e., elastic strain), and the reloading modulus is the ratio of the stress at which it is reloaded to the unloading point to the recoverable strain. Although the elastic strain that is mentioned above is very small, the variation will have a significant influence on the resilient modulus of the lightweight soil.

In Figures 6 and 7, the resilient modulus decreases with the increasing axial total strain, and it increases with the increasing confining pressure and cement content. Similar to the dynamic modulus [25], the resilient modulus will gradually approach a same critical value of 40 kPa, even under different confining pressures and cement contents in this study. Due to the increasing axial total strain, the cementation structure of the mixed lightweight soil is gradually damaged and its strength decreases step by step. Simultaneously, the constraint on the deformation of EPS particles is also reduced, and its elastic deformation is playing an increasingly important role afterwards. When the cementation structure is completely destroyed, the mixed lightweight soil tends to be loose, and whose resilient modulus arrives at a same level.

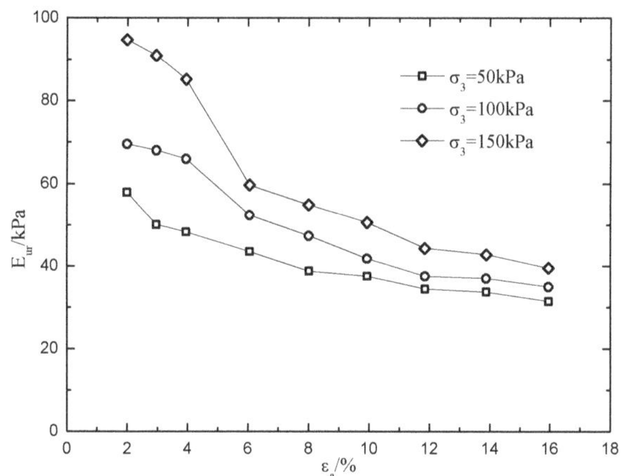


**Figure 6.** Resilient modulus versus axial total strain under different confining pressures.

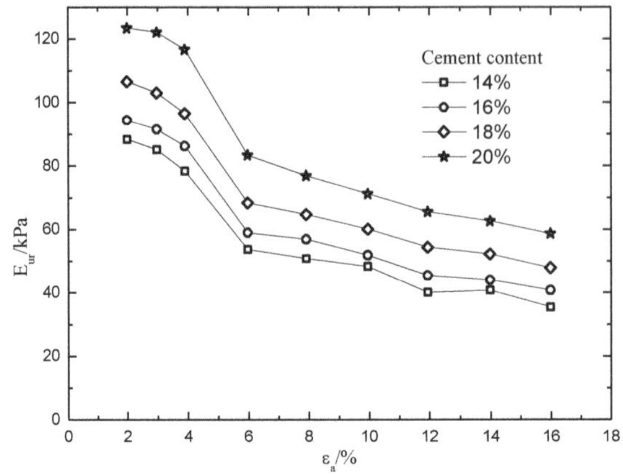


**Figure 7.** Resilient modulus versus axial total strain under different cement contents.

On the one hand, the increasing confining pressure enhances the restraint on the specimen and then on the EPS particles. On the other hand, the increasing confining pressure enhances the destruction of the cementation structure and thus weakens the restraint on the EPS particles. Therefore, the combination of the two actions increases the strength of the mixed lightweight soil increases, weakens the restraint, but increases the elastic deformation of the EPS particles. Obviously, the multiple of strength increase is larger than the multiple of elastic deformation increase, and this can be easily found in Figure 8.

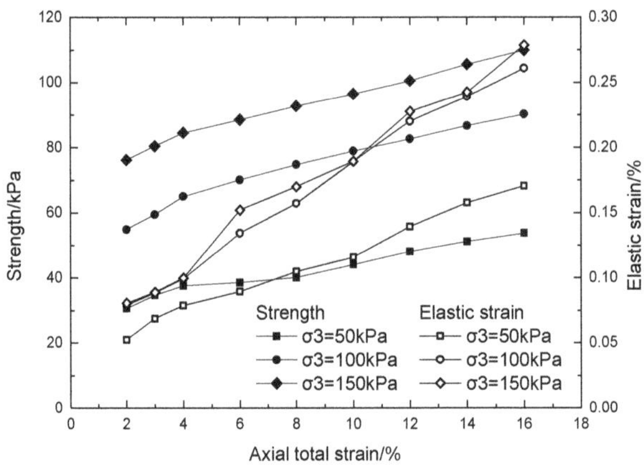


**Figure 8.** Strength and elastic strain corresponding to the resilient modulus.

With increasing cement content, the cementation becomes stronger, as well as the strength of the specimen. Deformation of the EPS particles is constrained, indicating that the elastic deformation would be reduced, which contributes to the increase of resilient modulus of the lightweight soil. Under the tested confining pressures and mixture ratios, the relationship between the resilient modulus and the axial total strain can be established as:

$$E_{ur} = a\varepsilon_a^b \quad (4)$$

where $a$ and $b$ are all the fitting parameters and they are related to the confining pressure and cement content. Subsequently, the fitting relation can be formulated by the following function:

$$\begin{cases} a = a_1 a_w^2 + b_1 a_w + c_1 \\ b = a_2 a_w^2 + b_2 a_w + c_2 \end{cases} \tag{5}$$

where $a_{1,2}$, $b_{1,2}$, and $c_{1,2}$ are all the fitting coefficients related to the confining pressure. Then, a general formula can be deduced as:

$$y = \alpha\sigma_3^2 + \beta\sigma_3 + \gamma \tag{6}$$

where $\sigma_3$ is the confining pressure, $\alpha$, $\beta$, and $\gamma$ are also the fitting coefficients, which can be deduced in Table 2.

**Table 2.** Deduced values of $\alpha$, $\beta$, and $\gamma$.

| Fitting Parameters | LSES | | | LCES | | |
|---|---|---|---|---|---|---|
| | $\alpha$ | $\beta$ | $\gamma$ | $\alpha$ | $\beta$ | $\gamma$ |
| $a_1$ | 4.082 | −871.570 | 47,389.000 | −0.887 | 130.350 | 4556.900 |
| $b_1$ | −1.372 | 295.900 | −15,571.000 | 0.194 | −23.908 | −1809.500 |
| $c_1$ | 0.118 | −25.067 | 1346.900 | −0.005 | 0.104 | 238.290 |
| $a_1$ | −0.032 | 7.820 | −446.140 | 0.003 | −0.125 | −45.564 |
| $b_2$ | 0.011 | −2.769 | 159.200 | −0.000 | −0.140 | 23.901 |
| $c_2$ | −0.001 | 0.243 | −14.364 | −0.000 | 0.030 | −3.080 |

Note: LSES is the sand mixed lightweight soil, LCES is the silty clay mixed lightweight soil.

As illustrated in Figure 9, the resilient modulus of sandy lightweight soil is higher than that of silty clay lightweight soil under confining pressures of 50 kPa and 150 kPa. The same condition also applies to the strength and recoverable elastic strain of the lightweight soil samples (as shown in Table 3). For the basic physical characteristics, the granular sand has smaller specific surface area and less activity than the silty clay. When cement is added, the granular sand and the cement have a rapid cementation speed with the help of water. The more cemented material is produced, the higher cementation strength would be generated to constrain the deformation of the EPS particles, resulting in smaller elastic deformation and higher resilient modulus. Although the silty clay is flattened with strong activity, the cement's hydrolysis and the hydration reactions mainly happen around a certain active medium. Therefore, the reaction is slow but needs a long time, and less cementing material is produced within a certain curing time. Moreover, the cemented silty clay has a lower strength to restrict the EPS particles, resulting in a lower strength than the sandy lightweight soil, as well as the resilient modulus. Due to the existence of a layer of bound water film around the fine-grained soil particle, silty clay has better deformation adaptability than sand, indicating that the silty clay lightweight soil has lower strength and resilient modulus than that of the sandy lightweight soil.

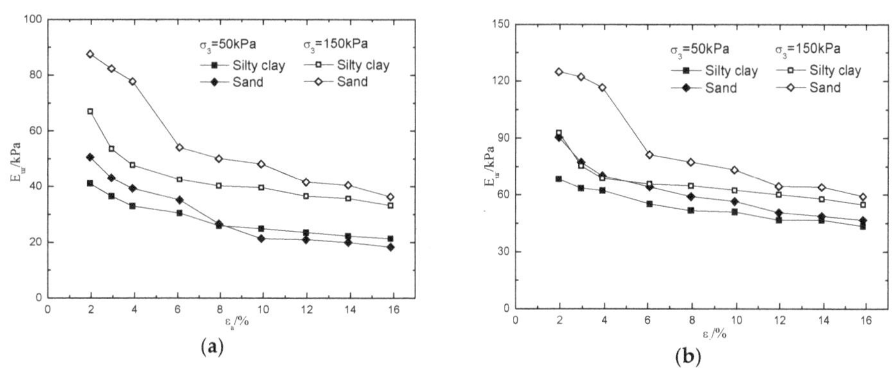


**Figure 9.** Resilient modulus versus axial total strain under different confining pressures: (**a**) aw = 14%; (**b**) aw = 20%.

As shown in Figure 9a, with a lower confining pressure of 50 kPa and a smaller cement content of 14%, when the axial total strain of the specimen exceeds 8%, the resilient modulus of the sandy lightweight soil is smaller than that of the silty clay lightweight soil. This is rather unusual in the overall condition of these tested samples. When the shear deformation of the specimen becomes larger (i.e., the axial total strain exceeds 8%), the degree of particle breakage increases, and the sandy lightweight specimen starts to loosen, resulting in a decrease in the cohesion, as well as the resilient modulus. Although the same thing happened to the silty clay lightweight soil, the above reduction in cohesion and resilient modulus under the same shear deformation are relatively smaller than the sandy soils. However, such a phenomenon does not exist when the confining pressure and cement content are higher, indicating that the constraint of higher confining pressure excels the loss of cohesion that is caused by shear deformation for sandy lightweight soil, and the higher cement content contributes to the increasing of cementing force in a short term. Within a limited curing time (7 days or 14 days), the resilient modulus of the sandy lightweight soil is usually larger that of silty clay type mixture (see Table 3).

**Table 3.** Strength and elastic strain corresponding to the resilient modulus in Figure 7.

| $\varepsilon_a$/% | $a_w$ = 14%, $\sigma_3$ = 50 kPa | | | | $a_w$ = 14%, $\sigma_3$ = 150 kPa | | | |
|---|---|---|---|---|---|---|---|---|
| | Strength/kPa | | $\varepsilon_h$/% | | Strength/kPa | | $\varepsilon_h$/% | |
| | LSES | LCES | LSES | LCES | LSES | LCES | LSES | LCES |
| 2 | 26.71 | 22.00 | 0.0528 | 0.0537 | 70.77 | 64.71 | 0.0809 | 0.0964 |
| 3 | 30.03 | 26.11 | 0.0696 | 0.0707 | 74.26 | 68.27 | 0.0893 | 0.1286 |
| 4 | 31.73 | 30.41 | 0.0801 | 0.0926 | 78.36 | 72.33 | 0.1005 | 0.1507 |
| 6 | 32.03 | 31.91 | 0.0911 | 0.0926 | 82.93 | 76.16 | 0.1526 | 0.1806 |
| 8 | 33.17 | 33.09 | 0.1254 | 0.1258 | 86.15 | 80.38 | 0.1718 | 0.2006 |
| 10 | 34.74 | 33.95 | 0.1600 | 0.1370 | 90.86 | 84.83 | 0.1925 | 0.2177 |
| 12 | 38.58 | 36.47 | 0.1845 | 0.1535 | 94.42 | 88.04 | 0.2319 | 0.2445 |
| 14 | 41.90 | 39.54 | 0.2116 | 0.1796 | 98.31 | 92.79 | 0.2432 | 0.2641 |
| 16 | 44.69 | 42.30 | 0.2431 | 0.1956 | 102.84 | 96.44 | 0.2792 | 0.2844 |
| $\varepsilon_a$/% | $a_w$ = 20%, $\sigma_3$ = 50 kPa | | | | $a_w$ = 20%, $\sigma_3$ = 150 kPa | | | |
| | Strength/kPa | | $\varepsilon_h$/% | | Strength/kPa | | $\varepsilon_h$/% | |
| | LSES | LCES | LSES | LCES | LSES | LCES | LSES | LCES |
| 2 | 45.90 | 35.41 | 0.0517 | 0.0521 | 100.23 | 83.75 | 0.0803 | 0.0893 |
| 3 | 50.29 | 42.30 | 0.0665 | 0.0670 | 107.24 | 92.63 | 0.0882 | 0.1223 |
| 4 | 53.49 | 49.19 | 0.0774 | 0.0794 | 114.00 | 100.53 | 0.0983 | 0.1469 |
| 6 | 56.06 | 56.01 | 0.0885 | 0.1018 | 121.73 | 105.70 | 0.1488 | 0.1616 |
| 8 | 59.61 | 59.56 | 0.1030 | 0.1176 | 128.95 | 113.99 | 0.1673 | 0.1797 |
| 10 | 63.81 | 62.91 | 0.1133 | 0.1268 | 135.03 | 121.57 | 0.1867 | 0.1989 |
| 12 | 69.23 | 68.69 | 0.1378 | 0.1501 | 142.43 | 130.90 | 0.2212 | 0.2214 |
| 14 | 73.90 | 72.90 | 0.1545 | 0.1578 | 149.64 | 139.36 | 0.2397 | 0.2440 |
| 16 | 77.21 | 77.20 | 0.1669 | 0.1790 | 156.20 | 151.32 | 0.2705 | 0.2710 |

Note: LSES is the sand mixed lightweight soil, LCES is the silty clay mixed lightweight soil.

### 3.3. Damping Ratio

According to the conventional definition [26], the damping ratio is related to the energy loss rate in a certain time, and the formula can be expressed as:

$$y = \alpha\sigma_3^2 + \beta\sigma_3 + \gamma \tag{7}$$

where $\lambda$ is the damping ratio, $\Delta W$ is the loss energy within a loading-unloading cycle, and $W$ is the total energy of a complete loading-unloading cycle.

In this study, the damping ratio is calculated by the energy loss rate in a loading-unloading cycle. The typical stress-strain relationship of the tested specimen can be illustrated in Figure 10.

In a complete loading-unloading cycle, the work that is done by extra loads can be defined as: at the beginning OA section, most of the work done by external load is converted into elastic potential energy, indicating that the specimen's deformation is mainly attributed to the elastic deformation. Subsequently, in the second AB section, plastic deformation is produced in the specimen and it becomes larger and larger. Although, the work done by external load on the specimen is mainly consumed by the plastic deformation and the viscous resistance, there will still be a small part stored as the elastic potential energy. Within the BC section, when the external load is removed, the elastic potential energy that is stored in sections of OA and AB will be gradually released and absolutely consumed by viscous resistance.

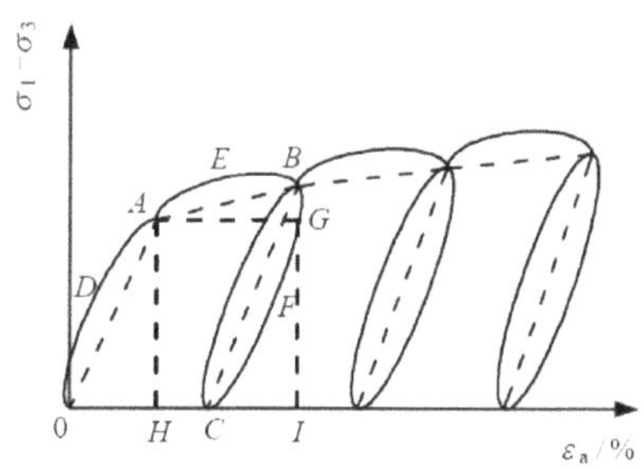


**Figure 10.** Simplified form of repeated loading and the stress-strain relationship.

According to the general form of stress-strain relationship and energy composition analyses, the EPS beads-mixed lightweight soil belongs to the viscoelastic plastic material. The $\Delta W$ and $W$ in Equation (7) can be calculated by the following expression:

$$\Delta W = S_{0DA} + S_{AEB} + S_{BFC} + S_{AGIH} \tag{8}$$

$$W = S_{0DAEBFC0} \tag{9}$$

where $S_{0DA}$, $S_{AEB}$, $S_{BFC}$, $S_{AGIH}$, and $S_{0DAEBFC0}$ are the areas of different regions in Figure 10. In Equation (8), the former three items are the energy consumed by damping, the last one is the energy consumed by plastic deformation. Afterwards, the damping ratio can be calculated out.

Figure 11 shows the variation of damping ratios versus axial total strains under different confining pressures, cement contents for different soil types. The damping ratio increases slowly with the increasing axial total strain, but its value ranges from 0.07 to 0.08. However, the damping ratios of the lightweight soil obtained by Gao et al. [15] have a larger variation range from 0.05 to 0.20 when the strain increases from 1% to 10% (see Figure 11d). This is mainly due to the different test methods and the corresponding formation of tested lightweight samples, with lower cement content and higher volume occupancy of EPS beads. Although the difference of damping ratios are relatively small under different conditions in this study, there are still some characteristics that can be detected. For example, the sandy lightweight soil has lower damping ratio than the silty lightweight soil, and larger confining pressure and higher cement content obtain smaller damping ratios.

There are two kinds of damping in soils, one is the dissipation damping, and the other is the material damping. The former is caused by the diffusing of energy that is accumulated in the soil to the outside world in forms of surface wave and body wave, while the latter is generated from the friction between particles and the viscosity of pore water and air. However, this study is not focusing on the dynamic problem under the action of high frequency. The wave is not considered and the dissipation damping can be assumed to be zero, then the damping can be deduced to be a completely material damping. Actually, the bonding strength of the specimen is relatively strong due to the adopted cement contents, so there is little difference between the dislocation and slip of particles under the confining pressures and the strain levels in this study.

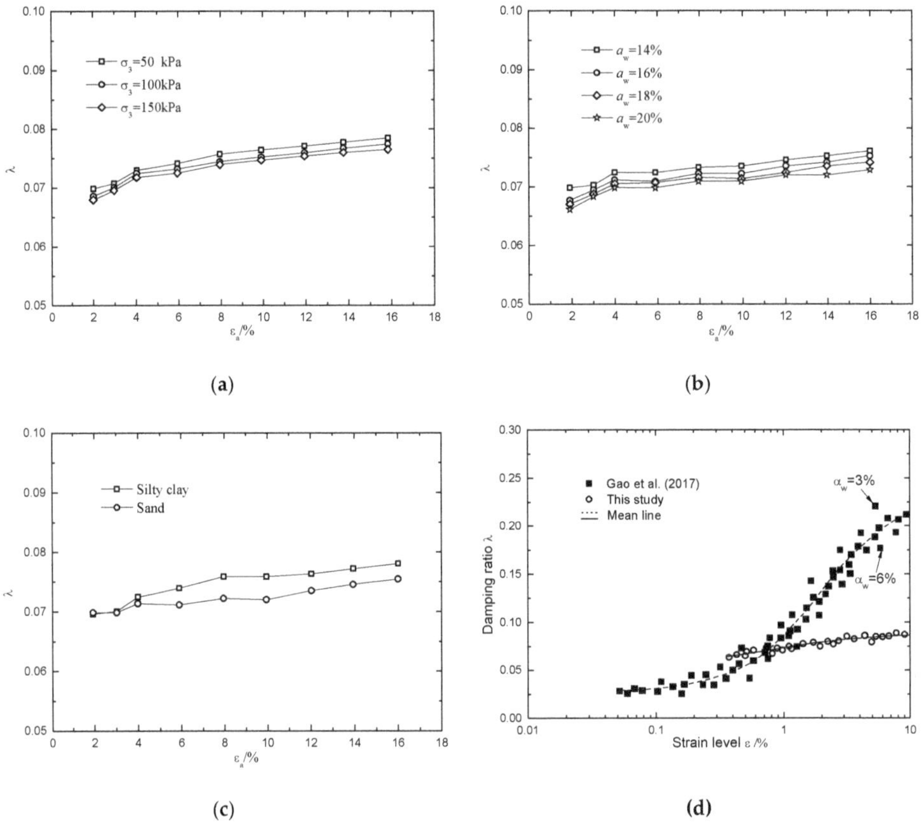


**Figure 11.** The damping ratio versus axial total strain under different condition: (**a**) Silty lightweight soil with $a_w$ = 14%; (**b**) Sandy lightweight soil with $\sigma_3$ = 100 kPa; (**c**) Under $a_w$ = 14% and $\sigma_3$ = 50 kPa; and, (**d**) Comparison with Gao et al.'s results.

Although little differences in hydration products, pore water, and pore gas have been detected, there are still some other points worth analyzing. The larger the strain is, the greater the damage of the specimen is. The greater the dislocation and slip between particles are, the greater the friction between particles is, and the greater the material damping ratio would be. With increasing confining pressure, the specimen is compressed denser, and then the particles' dislocation and slip are decreased with smaller friction, as well as the smaller damping ratio. The same situation is still applicable to the cement content. From the previous analyses, more cementing substances are produced in the sandy lightweight soil than in the silty clay lightweight soil for a certain curing time. Therefore, the interaction between grains of sandy lightweight soil is closer and the filling degree of the void is larger. Moreover, the dislocation, slip, and friction between grains are reduced, indicating that the discharge of pore water and pore gas in the lightweight soil is increased, and then a smaller material damping ratio is obtained.

## 4. Conclusions

The characteristics of deformation and damping of the sandy and silty clay EPS beads-mixed lightweight soil were investigated, regarding the axial accumulative strain, resilient modulus, and damping ratio under different confining pressures and cement contents. The main conclusions are as follows:

(1) Within a certain range of confining pressure, cement content, and strain level, the ratio of axial cumulative strain to axial total strain, which is less affected by soil category and loading factor, remains at a constant value. A linear relationship between the two can be deduced and the formula for resilient modulus can be derived indirectly.
(2) The springback deformation of EPS beads-mixed lightweight soil mainly comes from the springback deformation of the EPS particles. Due to the particularity of the lightweight soil, the confining pressure will cause certain damage to the cemented structure of the specimen, reducing the restriction on EPS particles but increasing the constraint on the specimen at the same time. However, the comprehensive effect depends on the strain level and the confining pressure.
(3) The resilient modulus of the EPS beads-mixed lightweight soil decreases with the increasing axial total strain and increases with the increasing confining pressure and cement content. When the axial total strain exceeds 8%, the resilient modulus of the sandy lightweight soil is smaller than the silty clay lightweight soil. However, when the axial total strain is smaller than 8%, the opposite results can be observed.
(4) The damping ratio of the EPS lightweight soil increases with the increasing axial total strain but decreases with the increasing confining pressure and cement content. Although the damping ratio of the sandy lightweight soil is smaller than that of the silty lightweight soil, the variation and difference are very small, and its value remains stable at around 0.07 to 0.08.

**Author Contributions:** Conceptualization, W.L., L.M. and J.Z.; methodology, W.L.; software, Y.Z.; validation, W.L. and L.M.; formal analysis, J.Z.; investigation, W.L.; resources and data curation, L.M.; writing—original draft preparation and writing—review and editing, W.L. and J.Z.; supervision, J.L.

**Funding:** This research was funded by the National Key Research and Development Program of China (Grant No. 2017YFC0805307), National Natural Science Foundation of China (Grant No. 51508279, 51578292, 51478054 and 51878078), Jiangsu Provincial Natural Science Fund Projects (Grant No. BK20150885), Excellent Youth Foundation of Natural Science Foundation of Hunan Province (Grant No. 2018JJ1026) Key Project of Education Department of Hunan Province (Grant No. 17A008). Open Research Fund of Science and Technology Innovation Platform of State Engineering Laboratory of Highway Maintenance Technology Changsha University of Science & Technology (Grant No. kfj150103), Ministry of Housing and Urban-Rural Development of China (2018-K4-008) and Open Fund of National Engineering Laboratory of Highway Maintenance Technology, Changsha University of Science & Technology (Grant No. kfj170101).

**Acknowledgments:** The authors still wish to thank the graduate students at who helped with specimen preparing and testing.

**Conflicts of Interest:** The authors declare no conflict of interest. The funders had no role in the design of the study; in the collection, analyses, or interpretation of data; in the writing of the manuscript, or in the decision to publish the results.

## References

1. Cao, W.; Shi, Q. Settlements of Existing and Widened Embankments over Soft Soils. *Adv. Mater. Res.* **2012**, *446–449*, 1167–1171. [CrossRef]
2. Zhang, J.; Peng, J.; Li, J.; Zheng, J.; Yao, Y. Characterization of stress and moisture-dependent resilient behaviour for compacted clays in south China. *Road Mater. Pavement* **2018**. [CrossRef]
3. Lu, W.; Miao, L. A simplified 2-D evaluation method of the arching effect for geosynthetic-reinforced and pile-supported embankments. *Comput. Geotech.* **2015**, *65*, 97–103. [CrossRef]
4. King, D.; Bouazza, A.; Gniel, J.; Rowe, R.; Bui, H. Geosynthetic reinforced column supported embankments and the role of ground improvement installation effects. *Can. Geotech. J.* **2018**, *55*, 792–809. [CrossRef]
5. Stark, T.; Horvath, J.; Leshchinsky, D. *Geo-Foam Application in the Design and Construction of Highway Embankments*; NCHRP Project No. 24-14; TRB: Washington, DC, USA, 2004.
6. Kim, Y.; Kim, H.; Lee, G. Mechanical behavior of lightweight soil reinforced with waste fishing net. *Geotext. Geomembr.* **2008**, *26*, 512–518. [CrossRef]
7. Frydenlund, T. *Expanded Polystyrene—A Lighter Way across Soft Ground*; International Report No. 1502; Norwegian Road Research Laboratory: Oslo, Norway, 1991.

8. Hatami, K.; Witthoeft, A.F. Reduction of backfill earth pressure in reinforced soil walls using geofoam. In Proceedings of the 86th TRB Annual Meeting, Transportation Research Board, Washington, DC, USA, 21–25 January 2007.
9. Ikizler, S.; Aytekin, M.; Nas, M. Laboratory study of expanded polystyrene (EPS) geofoam used with expansive soils. *Geotext. Geomembr.* **2008**, *26*, 189–195. [CrossRef]
10. Tsuchida, T. Super geo-material project in coastal zone. In Proceedings of the International Symposium on Ocean Space Utilization, COSU'95, Yokohama, 1995; pp. 22–31.
11. Miki, H. An overview of lightweight banking technology in Japan. In Proceedings of the International Symposium on EPS Construction Method (EPSTokyo'96), Tokyo, Japan, 2–4 October 1996; pp. 9–30.
12. Liu, H.; Deng, A.; Chu, J. Effect of different mixing ratios of polystyrene pre-puff beads and cement on the mechanical behavior of lightweight fill. *Geotext. Geomembr.* **2006**, *24*, 331–338. [CrossRef]
13. Miao, L.; Wang, F.; Han, J.; Lu, W.; Li, J. Properties and applications of cement-treated sand-expanded polystyrene bead lightweight fill. *J. Mater. Civ. Eng.* **2013**, *25*, 86–93. [CrossRef]
14. Minegishi, K.; Makiuchi, K.; Takahashi, R. Strength-deformation characteristics of EPS beads-mixed lightweight geomaterial subjected to cyclic loadings. In Proceedings of the International Workshop on Lightweight Geo-Materials, Tokyo, Japan, 26–27 March 2002; pp. 119–125.
15. Gao, H.; Bu, C.; Wang, Z.; Shen, Y.; Chen, G. Dynamic characteristics of expanded polystyrene composite soil under traffic loadings considering initial consolidation state. *Soil Dyn. Earthq. Eng.* **2017**, *102*, 86–98. [CrossRef]
16. El-Sherbiny, R.; Ramadan, S.; El-Khouly, M. Dynamic properties of sand-EPS bead mixtures. *Geosynth. Int.* **2018**, *25*, 456–470. [CrossRef]
17. Alaie, R.; Chenari, R. Cyclic and post-cyclic shear behavior of interface between geogrid and EPS beads-sand backfill. *KSCE J. Civ. Eng.* **2018**, *22*, 3340–3357. [CrossRef]
18. ASTM D2487-11. *Standard Practice for Classification of Soils for Engineering Purposes*; American Society for Testing and Materials: West Conshohocken, PA, USA, 2011.
19. Onishi, K.; Tsukamoto, Y.; Saito, R.; Chiyoda, T. Strength and small-strain modulus of lightweight geomaterials: Cement-stabilised sand mixed with compressible expanded polystyrene beads. *Geosynth. Int.* **2010**, *17*, 380–388. [CrossRef]
20. Gao, H.M.; Liu, J.Y.; Liu, H.L. Geotechnical properties of EPS composite soil. *Int. J. Geotech. Eng.* **2011**, *5*, 69–77. [CrossRef]
21. Edinçliler, A.; Özer, A.T. Effects of EPS bead inclusions on stress-strain behaviour of sand. *Geosynth. Int.* **2014**, *21*, 89–102. [CrossRef]
22. Zhang, J.; Li, J.; Yao, Y.S.; Zheng, J.L.; Gu, F. Geometric anisotropy modeling and shear behavior evaluation of graded crushed rocks. *Constr. Build. Mater.* **2018**, *183*, 346–355. [CrossRef]
23. ASTM D4767-11. *Standard Test Method for Consolidated Undrained Triaxial Compression Test for Cohesive Soils*; American Society for Testing and Materials: West Conshohocken, PA, USA, 2011.
24. Zhang, J.; Peng, J.; Li, J.; Zheng, J. Variation of resilient modulus with soil suction for cohesive soils in south China. *Int. J. Civ. Eng.* **2018**, *16*, 1655–1667. [CrossRef]
25. Gao, Y.F.; Wang, S.M.; Chen, C.B. A united deformation-strength framework for Lightweight Sand-EPS Beads Soil (LSES) under cyclic loading. *Soil Dyn. Earthq. Eng.* **2011**, *31*, 1144–1153. [CrossRef]
26. Rollins, K.M.; Evans, M.D.; Diehl, N.B.; Daily, W.D. Shear modulus and damping relationships for gravels. *J. Geotech. Geoenviron. Eng.* **1998**, *124*, 396–405. [CrossRef]

*applied sciences*

MDPI

*Article*

# Further Investigation on Damage Model of Eco-Friendly Basalt Fiber Modified Asphalt Mixture under Freeze-Thaw Cycles

**Wensheng Wang, Yongchun Cheng, Guirong Ma, Guojin Tan *, Xun Sun and Shuting Yang**

College of Transportation, Jilin University, Changchun 130025, China; wangws17@mails.jlu.edu.cn (W.W.); chengyc@jlu.edu.cn (Y.C.); magrjlu@163.com (G.M.); sunxunjlu@163.com (X.S.); yangstjlu@163.com (S.Y.)

* Correspondence: tgj@jlu.edu.cn; Tel.: +86-0431-8509-5446

Received: 21 November 2018; Accepted: 21 December 2018; Published: 25 December 2018

**Abstract:** The main distresses of asphalt pavements in seasonally frozen regions are due to the effects of water action, freeze-thaw cycles, and so on. Basalt fiber, as an eco-friendly mineral fiber with high mechanical performance, has been adopted to reinforce asphalt mixture in order to improve its mechanical properties. This study investigated the freeze-thaw damage characteristics of asphalt mixtures reinforced with eco-friendly basalt fiber by volume and mechanical properties—air voids, splitting tensile strength, and indirect tensile stiffness modulus tests. Test results indicated that asphalt mixtures reinforced with eco-friendly basalt fiber had better mechanical properties (i.e., splitting tensile strength and indirect tensile stiffness modulus) before and after freeze-thaw cycles. Furthermore, this study developed logistic damage models of asphalt mixtures in terms of the damage characteristics, and found that adding basalt fiber could significantly reduce the damage degree by about 25%, and slow down the damage grow rate by about 45% compared with control group without basalt fiber. Moreover, multi-variable grey models (GM) (1,N) were established for modelling the damage characteristics of asphalt mixtures under the effect of freeze-thaw cycles. GM (1,3) was proven as an effective prediction model to perform better in prediction accuracy compared to GM (1,2).

**Keywords:** asphalt mixture; basalt fiber; freeze-thaw cycle; logistic damage model; grey model

## 1. Introduction

Asphalt pavement has been widely used in flexible pavement constructions in a rapidly growing trend [1–4]. Asphalt mixtures are generally considered complex porous materials including asphalt, aggregates, and filler, as well as voids [5–7]. However, due to some environmental factors, there are many distresses in asphalt pavements such as spalling, crumbling, pavement potholes, etc., especially in seasonally frozen regions [8–10]. Therefore, researchers have been trying to modify asphalt mixture and explore its freeze-thaw damage.

Experiments about freeze-thaw cycle effects on asphalt mixtures have been investigated by many researchers recently [11–13]. Xu et al. [14] employed X-ray computed tomography technology (CT technology) to obtain and analyze internal images of asphalt mixture under different freeze-thaw cycles and investigated the influences of freeze-thaw cycles on the evolution of internal air voids. Moreover, Xu et al. [15] investigated the effects of freeze-thaw cycles on the thermodynamic characteristics of asphalt mixtures, based on the information entropy theory, CT, and digital image processing (DIP) technologies. Moreover, the influence of freeze-thaw cycles on permeability of asphalt mixtures were also evaluated by the means of flow state, as well as water conductivity, of asphalt mixtures [16]. Yan et al. [17] investigated stone matrix asphalt (SMA) mixtures under the effect of freeze-thaw cycles and evaluated the freeze-thaw resistance based on Marshall design indicators and water stability. Badeli et al. [18] conducted a rapid freeze-thaw cycle test for an asphalt mixture using thermomechanical tests.

Many researchers have also made efforts regarding freeze-thaw damage models of asphalt mixtures [19–21]. Tan et al. [22] investigated a freeze-thaw damage model and residual life prediction of asphalt mixture, based on damage theory, and they considered the compressive modulus as the index to evaluate freeze-thaw resistance and residual life prediction of an asphalt mixture under freeze-thaw cycles. Yi et al. [23] established generalized Maxwell and Drucker-Prager models to evaluate the viscoelastic-plastic damage under the condition of freeze-thaw cycles. A uniaxial compressive strength test was carried out to investigate the mechanisms of freeze-thaw failure in asphalt mixtures. Zhang [24] established a series of logistic damage models for different kinds of asphalt mixtures under the effects of water, temperature, and radiation to quantificationally analyze the damage degree of asphalt mixtures.

Fibers additives, such as cellulose fiber, polyester fiber, mineral fiber, etc., have been added into bitumen and proven as effective reinforcement materials for asphalt mixtures [25–28]. Imaninasab [29] investigated the effect of granular polymers on the rutting performance of stone matrix asphalt (SMA) compared with styrene-butadiene-styrene (SBS) modified and unmodified mixtures. Wang et al. [30] carried out many tests including indirect tensile strength, indirect tensile stiffness modulus, and dynamic shear rheological for polymer modified asphalt, in order to examine if the polymer modified asphalt has not been severely degraded. Hajikarimi et al. [31] used a dynamic shear rheometer to investigate the effect of polyphosphoric acid (PPA) and styrene-butadiene-styrene (SBS) modifications on the rheological and mechanical behavior of asphalt binders and asphalt mastics. Hajikarimi et al. [32] adopted a biphasic finite-element method to simulate an asphalt mastic as a heterogeneous medium consisting of aggregate particles, as inclusions within the asphalt binder, as the matrix for PPA and SBS. Basalt fiber, as a novel kind of eco-friendly mineral fiber, was produced from basalt rocks with high mechanical properties, low water absorption, and its by-products can be degraded directly in the environment without any harm [33]. Wang et al. [34,35] added basalt fiber into asphalt materials and evaluated their fatigue resistance by using direct tension, as well as fatigue tests. By means of X-ray tomography technology (i.e., CT technology) and the finite-element method, basalt fiber can release stress concentrations in critical areas and reduce fatigue damage. Gu et al. [36] compared and discussed basalt fiber and commonly used fibers and found that basalt fiber has a superior reinforcement effect on the high-temperature anti-rutting ability of bitumen mastic. Qin et al. [37] tested the reinforcement effects of basalt fibers with lengths of 3 mm, 6 mm, and 9 mm for asphalt mastics, with respect to lignin fiber and polyester fiber. Through leakage, penetration, strip-tensile, and dynamic shear rheometer (DSR) tests, basalt fiber, especially with length of 6 mm, has excellent comprehensive performances, due to a steady 3D networking structure in bitumen mastics. Zhang et al. [38] carried out repeated and multi-stress creep tests and used Abaqus software for analyzing the high-temperature performance of an asphalt mastic. Then Zhang et al. [39,40] conducted numerical simulations in Abaqus for compressive creep and bending creep tests, for the purpose of analyzing the distribution effect and reinforcement mechanism of basalt fiber. Wang et al. [41] explored the optimization design of SBS modified asphalt mixtures containing basalt fiber with the assistance of the central composite design method. Test results indicated that asphalt mixtures with a basalt fiber content of 0.34% and a length of 6 mm exhibited superior Marshall properties. Previous studies indicated that basalt fiber was effective in improving the mechanical properties of asphalt materials. Nevertheless, efforts undertaken for asphalt mixtures with basalt fiber under freeze-thaw cycles are still limited in this area.

In this present paper, freeze-thaw cycles were performed for asphalt mixtures reinforced with eco-friendly basalt fiber. The volume and mechanical properties (i.e., air voids, splitting tensile strength and indirect tensile stiffness modulus) of asphalt mixtures were adopted for freeze-thaw damage analysis. Furthermore, this study developed the logistic damage models of asphalt mixtures. in terms of the damage characteristics of the volume and mechanical properties. in order to quantificationally analyze the damage degree and damage grow rate. Subsequently, multi-variable grey models (GM) (1,N) were established for modelling the damage characteristics of the asphalt mixtures under the action of freeze-thaw cycles.

## 2. Raw Materials, Experimental Methods and Theory Background

### 2.1. Raw Materials and Specimen Preparation

#### 2.1.1. Raw Materials

In this study, asphalt of AH-90 produced from PetroChina Liaohe Petrochemical Company (Panjin, China) in Liaoning Province was selected, and the corresponding basic physical performances are presented in Table 1. Then, andesite mineral aggregates from a local quarry in Changchun, Jilin Province and limestone powder were chosen for the asphalt mixture. Additionally, basalt fiber was obtained from Jiuxin Basalt Industry Co., Ltd. (Changchun, China). The detailed physical properties of the aggregates and basalt fiber have been given in the previous study [5].

**Table 1.** Physical properties of asphalt AH-90 in this study.

| Properties | Measurement | Technical Criterion |
|---|---|---|
| Penetration @ 25 °C, 100 g, 5 s (0.1 mm) | 88 | 80~100 |
| Softening point (°C) | 47 | ≥44 |
| Ductility @ 10 °C, 5 cm/min (cm) | 43.5 | ≥30 |
| Solubility (trichloroethylene, %) | 99.8 | ≥99.5 |
| Density @ 15 °C (g/cm$^3$) | 1.05 | – |
| RTFOT | | |
| Mass loss (%) | 0.22 | ±0.8 |
| Penetration ratio @ 25 °C (%) | 66 | ≥57 |
| Ductility @ 10 °C, 5 cm/min (cm) | 28 | ≥8 |

#### 2.1.2. Specimen Preparation

Traditional dense-graded asphalt mixture is a well-graded asphalt mixture and is applied extensively in asphalt pavement construction in China [42]. In this study, the median gradation of asphalt mixture (AC-13) was selected, as shown in Figure 1, and the upper and lower limits of AC-13 are also illustrated. Asphalt mixtures reinforced with eco-friendly basalt fiber were made by a standard Marshall design method [43,44] approximately 0.4% basalt fiber with a length of 6 mm was added into asphalt mixtures. In accordance to JTG E20-2011 [45], the detailed preparation procedures are presented as follows:

- Step (1): the pre-heated aggregates and basalt fiber were mixed together in a mixing pot for 90 s in order to make basalt fiber uniformly dispersed in aggregates.
- Step (2): the pre-heated asphalt AH-90 was weighted and poured into the mixing pot and the mixture was blended for 90 s.
- Step (3): the pre-weighted limestone powder was added into the mixing pot and then blended for 90 s.
- Step (4): Marshall specimens of AC-13 with a diameter of 101.6 mm and a height of 63.5 mm were prepared by compacting 75 blows on each side.

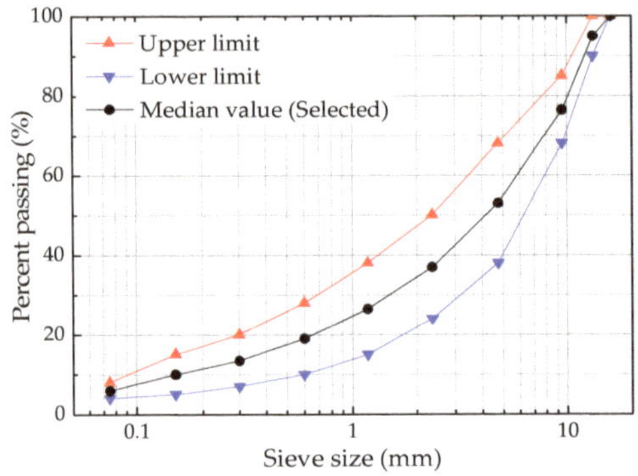


**Figure 1.** Gradation of asphalt mixture used in this study [5].

### 2.2. Freeze-Thaw Cycle Procedure

Figure 2 gives a detailed description of the freeze-thaw cycles in this study, in which the real simulation is shown in Figure 2a, and the one-time freeze-thaw cycle in the laboratory is shown in Figure 2b. Before the freeze-thaw cycle test, specimens were immersed into water and under vacuum (98.0 kPa) for 15 min, and soaked under atmospheric pressure for 30 min. Then, a one-time freeze-thaw cycle was carried out on specimens, in which the freezing condition of the refrigerator was set as −18 °C for 16 h, and the thaw condition of the thermostatic waterbath was in water at 60 °C for 8 h. A number of 84 specimens were treated for the freeze-thaw cycle test, in which each freeze-thaw cycle had 3 replicates. After 0, 1, 3, 6, 9, 12, and 15 freeze-thaw cycles, the experimental tests were carried out for asphalt mixtures.

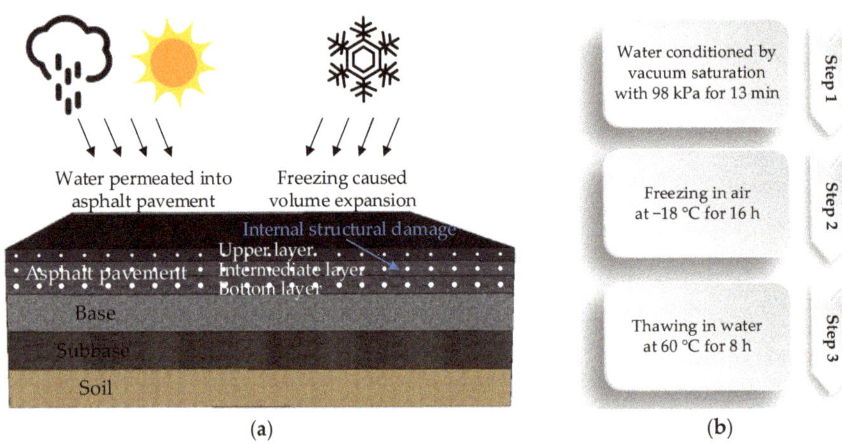


**Figure 2.** Freeze-thaw cycle in this study (**a**) schematic diagram; (**b**) experimental procedure.

### 2.3. Damage Characteristics Indicators

#### 2.3.1. Air Voids

Air voids are one of the key indicators for characterizing the quality of asphalt pavement. Almost all performances of asphalt pavement, like cracking resistance, anti-rutting, fatigue resistance, etc., are closely related to air voids. Asphalt pavement with higher air voids would incur many distresses, such as spalling, crumbling, pumping, and so on. However, lower air voids may lead to rutting and extensive oil, etc. Thus, an air voids indicator was selected to characterize the freeze-thaw damage in this study. In accordance to T0709 JTG E20-2011 [45], air voids (*VA*) can be determined through weighting asphalt mixture specimens in air and water at room temperature, following the Equations (1) and (2):

$$\gamma_f = m_a/(m_f - m_w), \quad (1)$$

$$VA = [1 - \gamma_f/\gamma_{TMD}] \times 100, \quad (2)$$

where $m_a$, $m_w$ and $m_f$ represent the mass of specimens in air, water, and the saturated surface dry mass, respectively. $\gamma_f$ is the bulk specific gravity and $\gamma_{TMD}$ is the theoretical maximum specific density which can be measured by vacuum sealing method.

### 2.3.2. Splitting Tensile Strength

A splitting test reflects mechanical performance at the moment of splitting, failure under specific temperature, and loading rate. Splitting tensile strength (*STS*) is considered as an effective indicator for asphalt mixture, and it is also widely used in many studies [46,47]. According to T0716 in JTG E20-2011 [45], the splitting tensile strength test shown in Figure 3 was performed at 15 °C by the Marshall apparatus, with a load of speed of 50 mm/min until the specimen was broken. The deformation was measured by linear variable differential transformers (LVDT). Then, the splitting tensile strength was calculated by the following equation:

$$R_{T1} = 0.006287 \times P_{T1}/h_1, \quad (3)$$

where $R_{T1}$ is the splitting tensile strength, $P_{T1}$ is the maximum load, and $h_1$ is the specimen height.

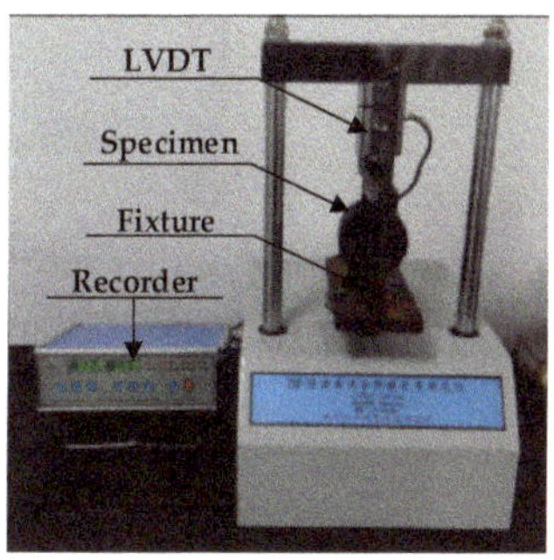


**Figure 3.** Splitting tensile strength test in this paper.

### 2.3.3. Indirect Tensile Stiffness Modulus

The indirect tensile stiffness modulus (ITSM) is an important indicator for evaluating mechanical performance. In this study, the ITSM test at 10 °C was adopted and conducted according to the standard AASHTO TP-31 for evaluating the anti-crack ability of the asphalt mixture [48]. A servo-pneumatic universal testing machine, shown in Figure 4, was used for the ITSM test. Firstly, Marshall specimens were put into the environmental chamber at 10 °C for at least 5 hours. Then, three replicate specimens were measured for ITSM. Then, ITSM ($S_m$) values could be calculated as follows:

$$S_m = F \times (\mu + 0.27)/(h \times Z), \quad (4)$$

where $F$ is the peak load (N), $\mu$ is the Poisson ratio and $\mu = 0.25$ at 10 °C, $h$ is the specimen height (mm), and $Z$ is the horizontal deformation (mm).

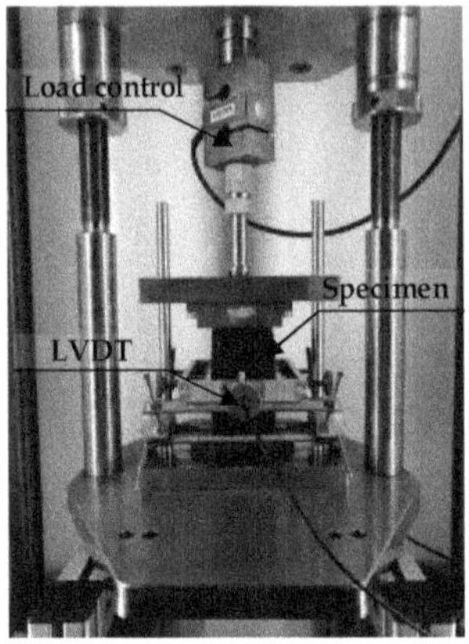


**Figure 4.** Indirect tensile stiffness modulus test in this study.

### 2.4. Theory Background of Damage Model and Prediction Model

#### 2.4.1. Logistic Damage Model

A logistic model is a widely used non-linear statistical model, which is mainly used to discuss and quantify the relationship between the "damage" parameters and the independent variables for the calculation of damage probabilities [24]. The logistic model was developed from Malthus model expressed by Equation (5):

$$dN/dt \times 1/N = r, \tag{5}$$

where $N$ is the dependent variable. $t$ is the independent variable, and $r$ is a constant.

Then the analytical solution of Equation (5) is given by

$$N(t) = N_0 \times e^{rt}, \tag{6}$$

on the basis of Malthus model, assuming that the model is linearly constrained gives

$$dN/dt \times 1/N = -r/Q \times N + r, \tag{7}$$

where $Q$ is the saturation factor and $r$ is the growth rate factor.

Rewriting Equation (7) in the expression of logistic model gives

$$dN/dt = rN \times (1 - N/Q), \tag{8}$$

the analytical solution of logistic model can be solved as

$$N(t) = Q/[1 + (Q - N_0)/(N_0 \times e^{rt})], \tag{9}$$

Equation (9) can be rewritten as

$$N(t) = Q/[1 + e^{(a-rt)}], \tag{10}$$

and the damage degree can be defined as

$$D = (1 - I_i/I_0) \times 100, \tag{11}$$

where $D$ is the damage degree, $I_0$ is the initial index, and $I_i$ is the index of the $i$th freeze-thaw cycle.

Therefore, the damage degree in Equation (12), based on the logistic model, was established in order to analyze the freeze-thaw damage degree of asphalt mixtures.

$$D = (A_1 - A_2)/[1 + (x/x_0)^p] + A_2, \tag{12}$$

where $A_1$ is the minimum value of regression curve, $A_2$ is the maximum value of regression curve, which reflects the degree of damage, and $x_0$ is the x-coordinate value when $D = 0.5A_2$, which reflects the damage grow rate.

### 2.4.2. Multi-Variable Grey Model Represented by GM (1,N)

The grey system is a widely known mathematical theory in various fields such as economics, engineering, etc., which can be used for cases with partially known information or lacking adequate experimental data [49–51]. In general, grey system theory is mainly utilized for two aspects, i.e., grey relational degree analysis (GRA) and grey prediction models (GM). GRA is usually applied for measuring the uncertain relationships between factors. Grey prediction models reveal long-term processes for the development of various factors, and it could be used a priori to predict a property's development with less priori data. In this study, a multi-variable grey model, GM (1,N), was adopted and established to predict the freeze-thaw damage of an asphalt mixture.

Considering that the grey model has $N$ variables, which is denoted by $y_i$ ($i = 1, 2, \ldots, N$) with $m$ initial sequences:

$$y_i^{(0)} = \left\{y_i^{(0)}(1), y_i^{(0)}(2), \ldots, y_i^{(0)}(m),\right\}, (i = 1, 2, \ldots, N), \tag{13}$$

through 1-accumulated generating operation (1-AGO), the 1-AGO sequence of initial sequence can be obtained

$$y_i^{(1)} = \left\{y_i^{(1)}(1), y_i^{(1)}(2), \ldots, y_i^{(1)}(m),\right\}, (i = 1, 2, \ldots, N) \tag{14}$$

$$y_i^{(1)}(t) = \sum_{j=1}^{t} y_i^{(0)}(j), (j = 1, 2, \ldots, m), \tag{15}$$

the whitening differential equation of the grey model could be determined by

$$\frac{dy_1^{(1)}(k)}{dt} + ay_1^{(1)}(k) = \sum_{i=2}^{n} b_i y_i^{(1)}(k), \tag{16}$$

then the grey differential equation is written as follows:

$$y_1^{(0)}(k) + az_1^{(1)}(k) = \sum_{i=2}^{n} b_i y_i^{(1)}(k), \tag{17}$$

$$z_1^{(1)}(k) = \frac{y_1^{(1)}(k) + y_1^{(1)}(k-1)}{2}. \tag{18}$$

In GM (1,N), the grey parameter $P_N$ can be obtained according to the least square method:

$$P_N = (B^T B)^{-1} B^T Y_n, \tag{19}$$

$$B = \begin{bmatrix} -z_1^1(2) & y_2^{(1)}(2) & \cdots & y_n^{(1)}(2) \\ -z_1^1(3) & y_2^{(1)}(3) & \cdots & y_n^{(1)}(3) \\ \vdots & \vdots & & \vdots \\ -z_1^1(m) & y_2^{(1)}(m) & \cdots & y_n^{(1)}(m) \end{bmatrix}, \tag{20}$$

$$y_n = \begin{bmatrix} y_1^{(0)}(2) & y_1^{(0)}(3) & \cdots & y_1^{(0)}(m) \end{bmatrix}^{-1}, \tag{21}$$

substituting $P_N$ in the Equation (19) into the Equation (16) gives

$$\hat{y}_1^{(1)}(k+1) = \left(y_1^{(0)}(1) - \sum_{i=2}^{n-1} \frac{b_i y_i^{(1)}(k+1)}{a}\right)e^{-at} + \sum_{i=2}^{n-1} \frac{b_i y_i^{(1)}(k+1)}{a}, \tag{22}$$

therefore, the $(k+1)$-th predictive value can be obtained by an inverse accumulated generating operation.

$$\hat{y}_1^{(0)}(k+1) = \hat{y}_1^{(1)}(k+1) - \hat{y}_1^{(1)}(k), \tag{23}$$

## 3. Results and Discussion

### 3.1. Logistic Damage Model of Asphalt Mixtures under Freeze-Thaw Cycles

To illustrate the effects of eco-friendly basalt fiber on the performance of an asphalt mixture, a series of experiments was carried out for two kinds of asphalt mixtures, i.e., the control group (AM, asphalt mixtures without basalt fiber) and test group (BFAM, asphalt mixture with basalt fiber). The freeze-thaw damage can be characterized by air voids (*VA*) for the volume parameter, splitting tensile strength (*STS*), and indirect tensile stiffness modulus (*ITSM*) for mechanical parameters. Therefore, the damage degrees of *VA*, *STS*, and *ITSM* were adopted as the evaluating indicators of asphalt mixture under various freeze-thaw cycles.

The *VA*, *STS*, and *ITSM* of AM and BFAM varying with freeze-thaw cycles are plotted in Figure 5a,c,e, and the corresponding damage degree results are shown in Figure 5b,d,f, in which the logistic models are expressed in curves with uncertain data. Based on the logistic damage model theory in the Section 2.4.1, the logistic damage models of *VA*, *STS*, and *ITSM* for AM and BFAM could be established and listed in Table 2. These logistic damage models could be proven effective, due to the higher correlation coefficients $R^2$ above 0.97 and Adj. R-Squared values in Table 2 represent the fitting parameter accuracy. As plotted in Figure 5a,c,e, it can be clearly seen that the *VA* results of asphalt mixtures gradually increased with the increasing of the freeze-thaw cycles, whereas the *STS* and *ITSM* results presented a decreasing trend. Simultaneously, Figure 5b,d,f illustrate that asphalt mixtures were gradually damaged with freeze-thaw cycles. Furthermore, the variation trend of the damage degree of the asphalt mixtures increased rapidly at first and then more slowly. Under the action of freeze-thaw cycles, the internal structure of asphalt mixtures was damaged due to volume expansion and temperature stress. The adhesion capability between asphalt and aggregates became weaker and weaker under the continuous action of freeze-thaw cycles. Before 9 freeze-thaw cycles, the air voids in asphalt mixture first extended and then, adjacent air voids were coalesced, leading to the significant variation trend. However, the expansion and formation of air voids became slow after 9 freeze-thaw cycles.

In addition, by comparative analysis of control and test groups, the addition of basalt fiber can significantly improve the *STS* and *ITSM*, and decline the *VA* of asphalt mixture. Meanwhile, the damage degree of eco-friendly basalt fiber modified asphalt mixture has been as well greatly reduced. This may be owing to the fact that the basalt fiber formed a spatial networking structure, playing the role of reinforcement and toughening. Due to the poor absorption of water, but good adhesiveness with asphalt for basalt fiber, basalt fiber can also limit shrinkage cracking due to temperature stress under freeze-thaw cycles. From Table 2, by comparative analysis of the damage characteristics (i.e., *VA*, *STS*, and *ITSM*) of two kinds of asphalt mixtures, i.e., ordinary asphalt mixture (AM) and basalt fiber modified asphalt mixture (BFAM), it clearly shows that the model parameter "$A_2$" values of the logistic damage model of AM are always more than those of BFAM. Furthermore, the model parameter "$x_0$" values of AM are higher than those of BFAM at all times. The variations of the model parameters between those asphalt mixtures quantificationally demonstrate that the addition of basalt fiber into an asphalt mixture can significantly reduce the damage degree of asphalt mixtures and slow down the damage grow rate.

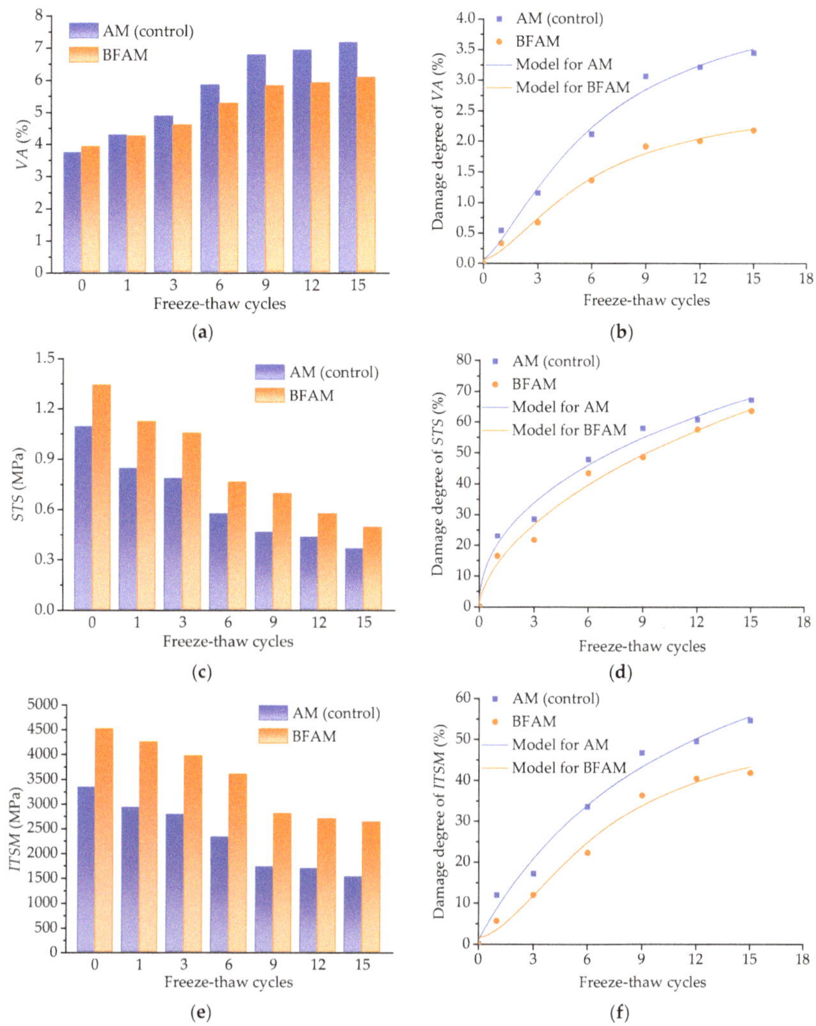

**Figure 5.** The relationships between damage characteristics of asphalt mixture and freeze-thaw cycles: (**a**) air voids (*VA*); (**b**) Damage degree of *VA*; (**c**) splitting tensile strength (*STS*); (**d**) Damage degree of *STS*; (**e**) indirect tensile stiffness modulus (*ITSM*); (**f**) Damage degree of *ITSM*.

**Table 2.** Logistic damage model of asphalt mixtures under freeze-thaw cycles.

| Damage Characteristics | Mixture Type | Logistic Damage Model ($D = (A_1 - A_2)/[1 + (x/x_0)^p] + A_2$) | Adj. R-Squared |
|---|---|---|---|
| *VA* | AM | $D = (0.08 - 4.55)/[1 + (x/6.37)^{1.39}] + 4.55$ | 0.9824 |
| | BFAM | $D = (0.07 - 2.68)/[1 + (x/5.87)^{1.58}] + 2.68$ | 0.9834 |
| *STS* | AM | $D = (0.23 - 395.54)/[1 + (x/360.86)^{0.50}] + 395.54$ | 0.9721 |
| | BFAM | $D = (0.35 - 271.72)/[1 + (x/92.51)^{0.65}] + 271.72$ | 0.9717 |
| *ITSM* | AM | $D = (1.02 - 97.95)/[1 + (x/11.67)^{1.00}] + 97.95$ | 0.9714 |
| | BFAM | $D = (1.63 - 54.53)/[1 + (x/6.96)^{1.70}] + 54.53$ | 0.9746 |

### 3.2. Freeze-Thaw Damage Prediction Model Based on Grey Model by GM (1,N)

In order to illustrate the multi-variable grey model GM (1,N) in this study, the cases for damage characteristics of asphalt mixtures under the actions of 0~15 freeze-thaw cycles have been investigated to validate the feasibility and accuracy of GM (1,N), based on the experimental data of damage characteristics under limited freeze-thaw cycles. Through different combinations of the damage degrees of *VA*, *STS*, and *ITSM*, three GM (1,2) could be established, at the same time, the GM (1,3) could be established according to the damage degrees of *VA*, *STS*, and *ITSM*.

#### 3.2.1. Two-Variable Grey Model GM (1,2)

Based on the damage degrees of *VA*, *STS*, and *ITSM*, three two-variable grey models could be established, i.e., GM (1,2) of *VA* and *STS* (as show in Figure 6), *VA* and *ITSM* (as show in Figure 7), *STS* and *ITSM* (as show in Figure 8). Ren found that the more samples do not necessarily mean the more accurate prediction of a grey model [50,51]. Accordingly, these GM (1,2) were established in terms of the experimental data under 0~9 freeze-thaw cycles, and the experimental data under 12 and 15 freeze-thaw cycles were used to verify the established GM (1,2).

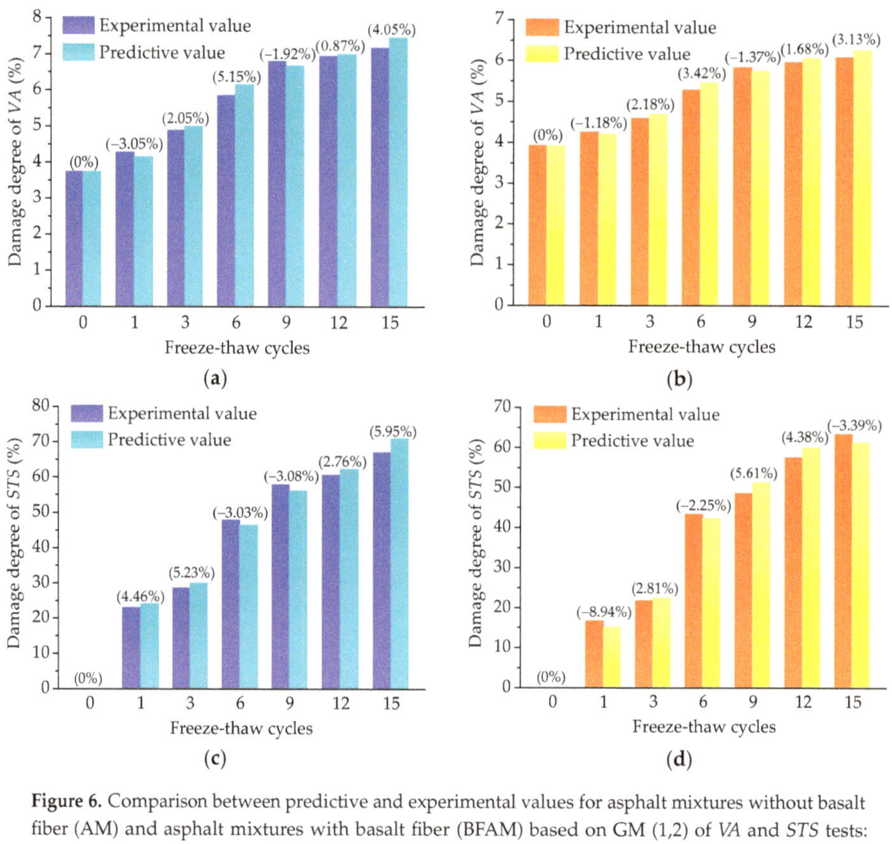


**Figure 6.** Comparison between predictive and experimental values for asphalt mixtures without basalt fiber (AM) and asphalt mixtures with basalt fiber (BFAM) based on GM (1,2) of *VA* and *STS* tests: (**a**) Damage degree of *VA* for AM; (**b**) Damage degree of *VA* for BFAM; (**c**) Damage degree of *STS* for AM; (**d**) Damage degree of *STS* for BFAM.

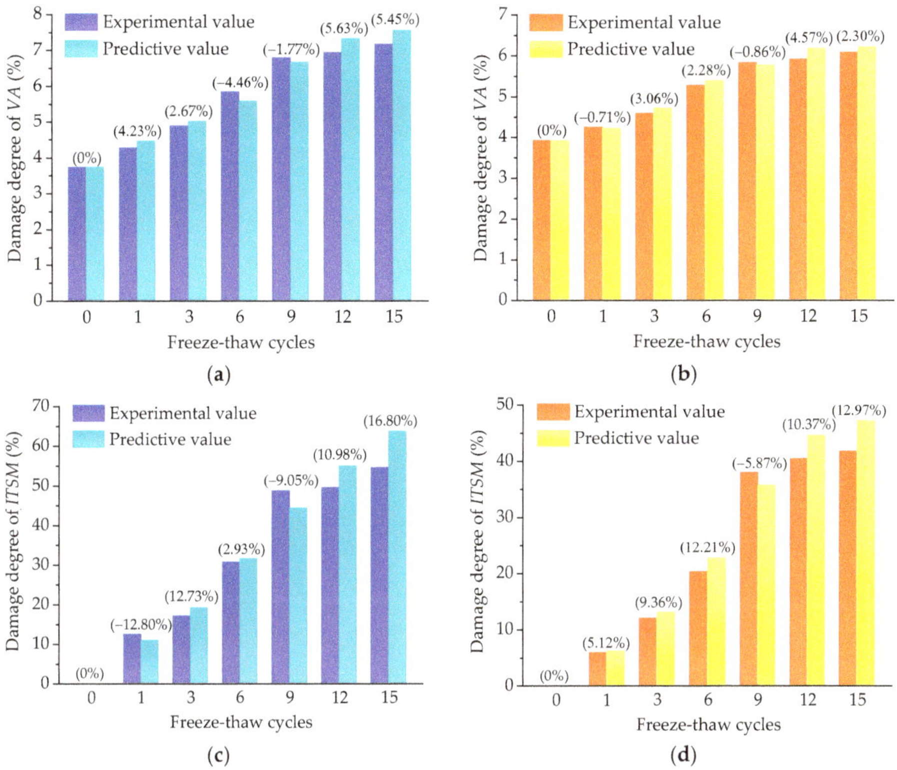


**Figure 7.** Comparison between predictive and experimental values for AM and BFAM based on GM (1,2) of *VA* and *ITSM* tests: (**a**) Damage degree of *VA* for AM; (**b**) Damage degree of *VA* for BFAM; (**c**) Damage degree of *ITSM* for AM; (**d**) Damage degree of *ITSM* for BFAM.

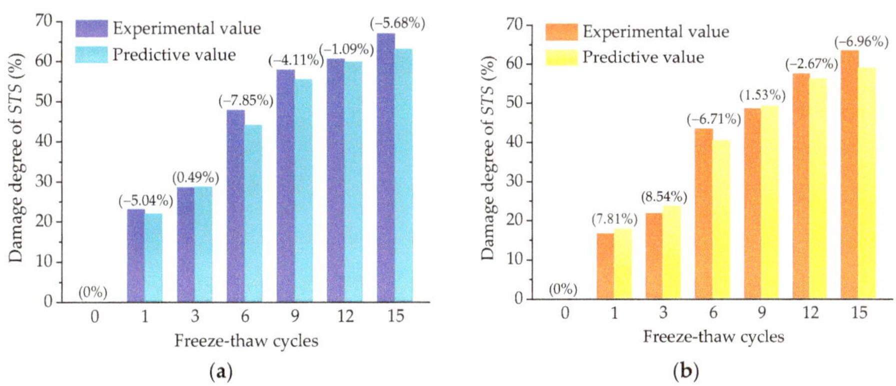


**Figure 8.** *Cont.*

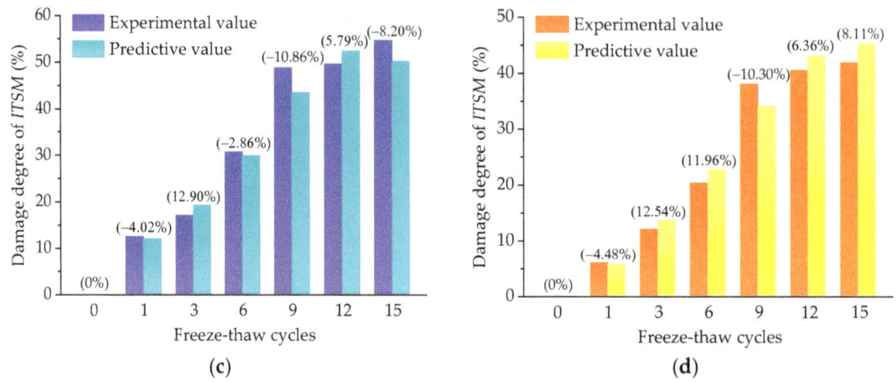


**Figure 8.** Comparison between predictive and experimental values for AM and BFAM based on GM (1,2) of *STS* and *ITSM* tests: (**a**) Damage degree of *STS* for AM; (**b**) Damage degree of *STS* for BFAM; (**c**) Damage degree of *ITSM* for AM; (**d**) Damage degree of *ITSM* for BFAM.

The predictive values of *VA* and *STS* compared with the experimental results are presented in Figure 6, based on the grey model GM (1,2) of *VA* and *STS* tests. It is apparent that the relative errors for *VA* of AM and BFAM are within 6%, and the relative errors of *STS* of AM and BFAM are around 10%.

The predictive values of *VA* and *ITSM* compared with the experimental results are presented in Figure 7, based on the grey model GM (1,2) of *VA* and *ITSM* tests. It is apparent that the relative errors for *VA* of AM and BFAM are around 6%, and the relative errors of *ITSM* of AM and BFAM are within 17%.

The predictive values of *STS* and *ITSM* compared with the experimental results are presented in Figure 8, based on the grey model GM (1,2) of *STS* and *ITSM* tests. It is apparent that the relative errors for *STS* of AM and BFAM are within 9%, and the relative errors of *ITSM* of AM and BFAM are around 13%.

### 3.2.2. Three-Variable Grey Model GM (1,3)

To study the number of independent variables for the prediction accuracy of GM (1,N), the three-variable grey model GM (1,3) was also built according to the experimental data under 0~9 freeze-thaw cycles, to predict the damage characteristics under 12 and 15 freeze-thaw cycles for verification.

The predictive values of *VA*, *STS*, and *ITSM* compared with the experimental results are presented in Figure 9 based on the three-variable grey model GM (1,3) of *VA*, *STS*, and *ITSM* tests. It is apparent that the relative errors for *VA* of AM and BFAM are within 3%, the relative errors for *STS* of AM and BFAM are within 6%, and the relative errors of *ITSM* of AM and BFAM are around 8%. Compared with two-variable GM (1,2) in Section 3.2.1, the results revealed that the relative error of three-variable GM (1,3) is smaller than that of GM (1,2) for the aspect of prediction accuracy. Thus, to some degree, it could be indicated that GM (1,3) has the better prediction accuracy than GM (1,2) under the conditions of the same data.

Through overall consideration, the three-variable grey model GM (1,3) based on *VA*, *STS*, and *ITSM* tests is an effective prediction model for asphalt mixtures under the action of freeze-thaw cycles, which can reflect the evolution of freeze-thaw damage for asphalt mixtures more accurately.

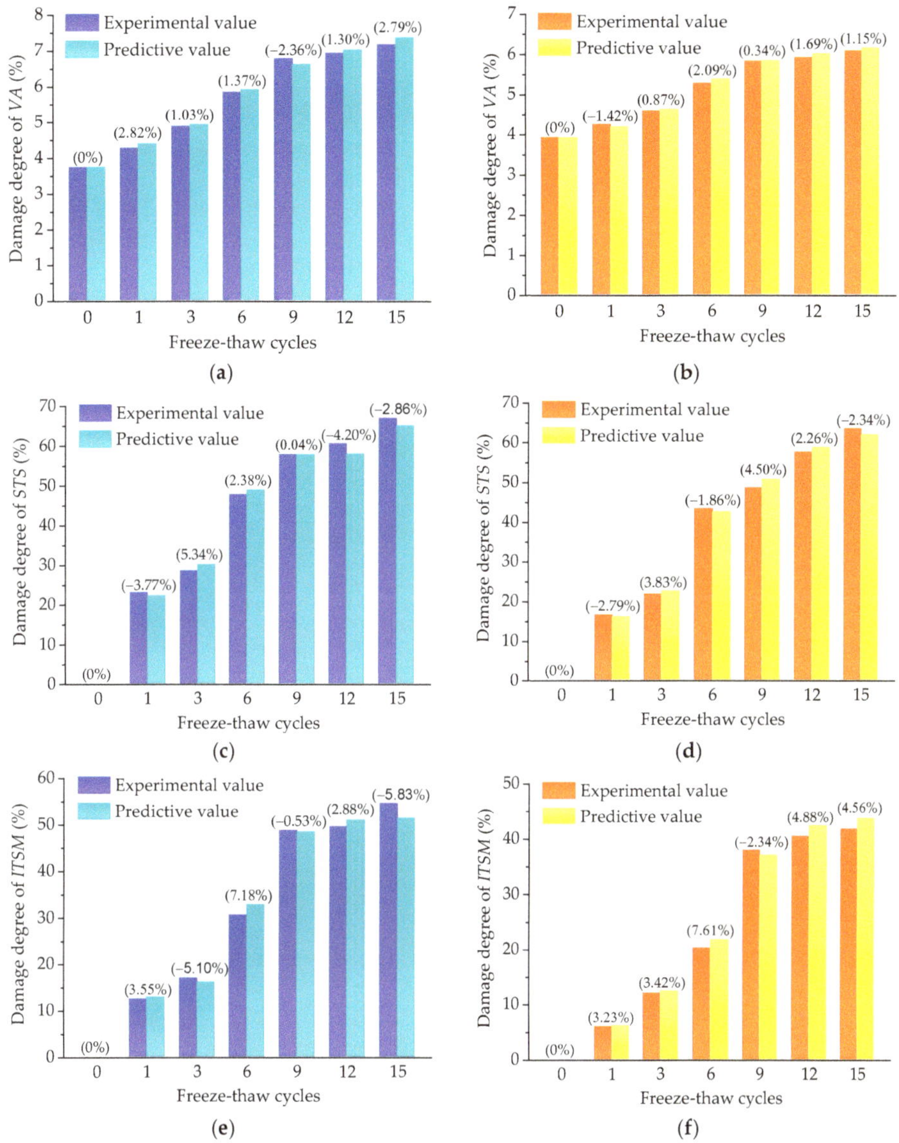


**Figure 9.** Comparison between predictive and experimental values for AM and BFAM based on GM (1,3) of *VA*, *STS*, and *ITSM* tests: (**a**) Damage degree of *VA* for AM; (**b**) Damage degree of *VA* for BFAM; (**c**) Damage degree of *STS* for AM; (**d**) Damage degree of *STS* for BFAM; (**e**) Damage degree of *ITSM* for AM; (**f**) Damage degree of *ITSM* for BFAM.

## 4. Conclusions

This study further explored the freeze-thaw damage characteristics of asphalt mixtures reinforced with eco-friendly basalt fiber through air voids, splitting tensile strength, and indirect tensile stiffness modulus tests. Based on the damage characteristics, a logistic damage model was established to quantificationally analyze the damage degree and damage grow rate of asphalt mixtures. Additionally, multi-variable grey models were established for modelling the freeze-thaw damage characteristics of asphalt mixtures. Thus, the following conclusions can be drawn:

- Freeze-thaw cycles habr a negative effect on the volume and mechanical properties of an asphalt mixture. From air voids, splitting tensile strength, and indirect tensile stiffness modulus, the addition of basalt fiber can significantly improve the freeze-thaw resistance and mechanical performance of an asphalt mixture, leading to a reinforcement mechanism.
- The logistic damage model can quantificationally demonstrate that adding basalt fiber could significantly reduce the damage degree of asphalt mixtures by about 25%, and slow down the damage grow rate by about 45% compared to a control group.
- The ninth freeze-thaw cycle may be the turning point of damage variation of an asphalt mixture, after which the expansion and formation of air voids became slower.
- Results demonstrated that the established multi-variable grey models can accurately predict the variation trend of damage characteristics of asphalt mixtures. GM (1,3) was proven to perform better in prediction accuracy compared to GM (1,2) under the same data. GM (1,3) is an effective prediction model for reflecting the evolution of freeze-thaw damage for asphalt mixtures.
- Appropriate basalt fiber content is recommended for modifying an asphalt mixture. It is necessary to check the dispersion of basalt fibers in an asphalt mixture to ensure the dispersion of basalt fiber. Compared to ordinary asphalt mixtures, the predicted cost is mainly reliant on the basalt fibers.

**Author Contributions:** Conceptualization, W.W. and Y.C.; Methodology, W.W. and G.T.; Validation, Y.C.; Formal Analysis, G.T. and G.M.; Investigation, W.W., G.M., X.S. and S.Y.; Writing–Original Draft Preparation, W.W.; Writing-Review & Editing, Y.C.; Project Administration, G.T.; Funding Acquisition, G.T.

**Funding:** This research was funded by [National Natural Science Foundation of China] grant number [51678271], [Science Technology Development Program of Jilin Province] grant number [20160204008SF], supported by Graduate Innovation Fund of Jilin University grant number [101832018C003], [Transportation Science and Technology Program of Jilin Province] grant number [2018-1-9], [the Education Department's "13th Five-Year" Science and Technology Program of Jilin Province] grant number [JJKH20190015KJ].

**Acknowledgments:** The authors would like to appreciate anonymous reviewers for their constructive suggestions and comments to improve the quality of the paper.

**Conflicts of Interest:** The authors declare no conflict of interest.

## References

1. Chen, M.Z.; Lin, J.T.; Wu, S.P. Potential of recycled fine aggregates powder as filler in asphalt mixture. *Constr. Build. Mater.* **2011**, *25*, 3909–3914. [CrossRef]
2. Zhang, R.; Dai, Q.L.; You, Z.P.; Wang, H.N.; Peng, C. Rheological Performance of Bio-Char Modified Asphalt with Different Particle Sizes. *Appl. Sci.* **2018**, *8*, 1665. [CrossRef]
3. Gao, J.F.; Wang, H.N.; Bu, Y.; You, Z.P.; Hasan, M.R.M.; Irfan, M. Effects of coarse aggregate angularity on the microstructure of asphalt mixture. *Constr. Build. Mater.* **2018**, *183*, 472–484. [CrossRef]
4. Peng, C.; Zhang, H.; You, Z.P.; Xu, F.; Jiang, G.S.; Lv, S.T.; Zhang, R.; Yang, H. Preparation and anti-icing properties of a superhydrophobic silicone coating on asphalt mixture. *Constr. Build. Mater.* **2018**, *189*, 227–235. [CrossRef]
5. Wang, W.S.; Cheng, Y.C.; Tan, G.J.; Tao, J.L. Analysis of Aggregate Morphological Characteristics for Viscoelastic Properties of Asphalt Mixes Using Simplex Lattice Design. *Materials* **2018**, *11*, 1908. [CrossRef] [PubMed]
6. Guo, Q.L.; Bian, Y.S.; Li, L.L.; Jiao, Y.B.; Tao, J.L.; Xiang, C.X. Stereo logical estimation of aggregate gradation using digital image of asphalt mixture. *Constr. Build. Mater.* **2015**, *94*, 458–466. [CrossRef]
7. Yang, X.; You, Z.P.; Hiller, J.; Watkins, D. Correlation Analysis between Temperature Indices and Flexible Pavement Distress Predictions Using Mechanistic-Empirical Design. *J. Cold Reg. Eng.* **2017**, *31*, 04017009. [CrossRef]
8. Lv, S.T.; Liu, C.C.; Zheng, J.L.; You, Z.P.; You, L.Y. Viscoelastic Fatigue Damage Properties of Asphalt Mixture with Different Aging Degrees. *KSCE J. Civ. Eng.* **2018**, *22*, 2073–2081. [CrossRef]

9. Zhang, C.; Wang, H.N.; Yang, X.; You, Z.P. A Combinational Prediction Model for Transverse Crack of Asphalt Pavement. *KSCE J. Civ. Eng.* **2018**, *22*, 2109–2117. [CrossRef]
10. Feng, D.C.; Yi, J.Y.; Wang, D.S.; Chen, L.L. Impact of salt and freeze-thaw cycles on performance of asphalt mixtures in coastal frozen region of China. *Cold. Reg. Sci. Technol.* **2010**, *62*, 34–41. [CrossRef]
11. Pan, P.; Wu, S.P.; Hu, X.D.; Wang, P.; Liu, Q.T. Effect of freezing-thawing and ageing on thermal characteristics and mechanical properties of conductive asphalt concrete. *Constr. Build. Mater.* **2017**, *140*, 239–247. [CrossRef]
12. Badeli, S.; Carter, A.; Dore, G.; Saliani, S. Evaluation of the durability and the performance of an asphalt mix involving Aramid Pulp Fiber (APF): Complex modulus before and after freeze-thaw cycles, fatigue, and TSRST tests. *Constr. Build. Mater.* **2018**, *174*, 60–71. [CrossRef]
13. Zheng, Y.X.; Cai, Y.C.; Zhang, G.H.; Fang, H.Y. Fatigue Property of Basalt Fiber-Modified Asphalt Mixture under Complicated Environment. *J. Wuhan. Univ. Technol.* **2014**, *29*, 996–1004. [CrossRef]
14. Xu, H.N.; Guo, W.; Tan, Y.Q. Internal structure evolution of asphalt mixtures during freeze-thaw cycles. *Mater. Des.* **2015**, *86*, 436–446. [CrossRef]
15. Xu, H.N.; Li, H.Z.; Tan, Y.Q.; Wang, L.B.; Hou, Y. A Micro-Scale Investigation on the Behaviors of Asphalt Mixtures under Freeze-Thaw Cycles Using Entropy Theory and a Computerized Tomography Scanning Technique. *Entropy* **2018**, *20*, 68. [CrossRef]
16. Xu, H.N.; Guo, W.; Tan, Y.Q. Permeability of asphalt mixtures exposed to freeze-thaw cycles. *Cold. Reg. Sci. Technol.* **2016**, *123*, 99–106. [CrossRef]
17. Yan, K.Z.; Ge, D.D.; You, L.Y.; Wang, X.L. Laboratory investigation of the characteristics of SMA mixtures under freeze-thaw cycles. *Cold Reg. Sci. Technol.* **2015**, *119*, 68–74. [CrossRef]
18. Badeli, S.; Carter, A.; Dore, G. Complex Modulus and Fatigue Analysis of Asphalt Mix after Daily Rapid Freeze-Thaw Cycles. *J. Mater. Civ. Eng.* **2018**, *30*, 04018056. [CrossRef]
19. Liu, F.; You, Z.P.; Yang, X.; Wang, H.N. Macro-micro degradation process of fly ash concrete under alternation of freeze-thaw cycles subjected to sulfate and carbonation. *Constr. Build. Mater.* **2018**, *181*, 369–380. [CrossRef]
20. You, L.Y.; You, Z.P.; Dai, Q.L.; Guo, S.C.; Wang, J.Q.; Schultz, M. Characteristics of Water-Foamed Asphalt Mixture under Multiple Freeze-Thaw Cycles: Laboratory Evaluation. *J. Mater. Civ. Eng.* **2018**, *30*, 04018270. [CrossRef]
21. Huang, T.; Qi, S.; Yang, M.; Lv, S.T.; Liu, H.F.; Zheng, J.L. Strength Criterion of Asphalt Mixtures in Three-Dimensional Stress States under Freeze-Thaw Conditions. *Appl. Sci.* **2018**, *8*, 1302. [CrossRef]
22. Tan, Y.Q.; Zhao, L.D.; Lan, B.W.; Meng, L. Research on Freeze-thaw Damage Model and Life Prediction of Asphalt Mixture. *J. Highw. Transport. Res. Dev.* **2011**, *6*, 1–7. (In Chinese)
23. Yi, J.Y.; Shen, S.H.; Muhunthan, B.; Feng, D.C. Viscoelastic-plastic damage model for porous asphalt mixtures: Application to uniaxial compression and freeze-thaw damage. *Mech. Mater.* **2014**, *70*, 67–75. [CrossRef]
24. Zhang, P. Research on Characteristics and Detection Methods of Asphalt Mixture Damage Induced by Water-Temperature-Radiation. Ph.D. Thesis, Jilin University, Changchun, China, June 2015.
25. Ye, Q.S.; Wu, S.P.; Li, N. Investigation of the dynamic and fatigue properties of fiber-modified asphalt mixtures. *Int. J. Fatigu.* **2009**, *31*, 1598–1602. [CrossRef]
26. Wu, S.P.; Ye, Q.S.; Li, N. Investigation of rheological and fatigue properties of asphalt mixtures containing polyester fibers. *Constr. Build. Mater.* **2008**, *22*, 2111–2115. [CrossRef]
27. Tayfur, S.; Ozen, H.; Aksoy, A. Investigation of rutting performance of asphalt mixtures containing polymer modifiers. *Constr. Build. Mater.* **2007**, *21*, 328–337. [CrossRef]
28. Xiong, R.; Fang, J.H.; Xu, A.H.; Guan, B.W.; Liu, Z.Z. Laboratory investigation on the brucite fiber reinforced asphalt binder and asphalt concrete. *Constr. Build. Mater.* **2015**, *83*, 44–52. [CrossRef]
29. Imaninasab, R. Effect of granular polymers on rutting performance of SMA with respect to modification process. *Constr. Build. Mater.* **2017**, *130*, 64–72. [CrossRef]
30. Wang, Y.H.; Chong, D.; Wen, Y. Quality verification of polymer-modified asphalt binder used in hot-mix asphalt pavement construction. *Constr. Build. Mater.* **2017**, *150*, 157–166. [CrossRef]
31. Hajikarimi, P.; Tehrani, F.F.; Nejad, F.M.; Absi, J.; Rahi, M.; Khodaii, A.; Petit, C. Mechanical Behavior of Polymer-Modified Bituminous Mastics. I: Experimental Approach. *J. Mater. Civ. Eng.* **2019**, *31*, 04018337. [CrossRef]
32. Hajikarimi, P.; Tehrani, F.F.; Nejad, F.M.; Absi, J.; Khodaii, A.; Rahi, M.; Petit, C. Mechanical Behavior of Polymer-Modified Bituminous Mastics. II: Numerical Approach. *J. Mater. Civ. Eng.* **2019**, *31*, 04018338. [CrossRef]

33. Cheng, Y.C.; Wang, W.S.; Gong, Y.F.; Wang, S.R.; Yang, S.T.; Sun, X. Comparative Study on the Damage Characteristics of Asphalt Mixtures Reinforced with an Eco-Friendly Basalt Fiber under Freeze-thaw Cycles. *Materials* **2018**, *11*, 2488. [CrossRef] [PubMed]
34. Wang, D.; Wang, L.B.; Gu, X.Y.; Zhou, G.Q. Effect of Basalt Fiber on the Asphalt Binder and Mastic at Low Temperature. *J. Mater. Civ. Eng.* **2013**, *25*, 355–364. [CrossRef]
35. Wang, D.; Wang, L.B.; Christian, D.; Zhou, G.Q. Fatigue Properties of Asphalt Materials at Low In-Service Temperatures. *J. Mater. Civil. Eng.* **2013**, *25*, 1220–1227. [CrossRef]
36. Gu, X.Y.; Xu, T.T.; Ni, F.J. Rheological behavior of basalt fiber reinforced asphalt mastic. *J. Wuhan. Univ. Technol.* **2014**, *29*, 950–955. [CrossRef]
37. Qin, X.; Shen, A.Q.; Guo, Y.C.; Li, Z.N.; Lv, Z.H. Characterization of asphalt mastics reinforced with basalt fibers. *Constr. Build. Mater.* **2018**, *159*, 508–516. [CrossRef]
38. Zhang, X.Y.; Gu, X.Y.; Lv, J.X.; Zhu, Z.K.; Ni, F.J. Mechanism and behavior of fiber-reinforced asphalt mastic at high temperature. *Int. J. Pavement Eng.* **2018**, *19*, 407–415. [CrossRef]
39. Zhang, X.Y.; Gu, X.Y.; Lv, J.X.; Zhu, Z.K.; Zou, X.Y. Numerical analysis of the rheological behaviors of basalt fiber reinforced asphalt mortar using ABAQUS. *Constr. Build. Mater.* **2017**, *157*, 392–401. [CrossRef]
40. Zhang, X.Y.; Gu, X.Y.; Lv, J.X. Effect of basalt fiber distribution on the flexural–tensile rheological performance of asphalt mortar. *Constr. Build. Mater.* **2018**, *179*, 307–314. [CrossRef]
41. Wang, W.S.; Cheng, Y.C.; Tan, G.J. Design Optimization of SBS-Modified Asphalt Mixture Reinforced with Eco-Friendly Basalt Fiber Based on Response Surface Methodology. *Materials* **2018**, *11*, 1311. [CrossRef]
42. Cui, P.D.; Xiao, Y.; Fang, M.J.; Chen, Z.W.; Yi, M.W.; Li, M.L. Residual Fatigue Properties of Asphalt Pavement after Long-Term Field Servic. *Materials* **2018**, *11*, 892. [CrossRef] [PubMed]
43. Wang, W.S.; Cheng, Y.C.; Tan, G.J.; Shi, C.L. Pavement Performance Evaluation of Asphalt Mixtures Containing Oil Shale Waste. *Road. Mater. Pavement* **2018**, 1–22. [CrossRef]
44. Cheng, Y.C.; Wang, W.S.; Tan, G.J.; Shi, C.L. Assessing High- and Low-Temperature Properties of Asphalt Pavements Incorporating Waste Oil Shale as an Alternative Material in Jilin Province, China. *Sustainability* **2018**, *10*, 2179. [CrossRef]
45. JTG E20-2011. *Standard Test Methods of Bitumen and Bituminous Mixtures for Highway Engineering*; Ministry of Transport of the People's Republic of China: Beijing, China, 2011. (In Chinese)
46. Ma, T.; Ding, X.H.; Wang, H.; Zhang, W.G. Experimental Study of High-Performance Deicing Asphalt Mixture for Mechanical Performance and Anti-Icing Effectiveness. *J. Mater. Civ. Eng.* **2018**, *30*, 04018180. [CrossRef]
47. Ma, T.; Wang, H.; He, L.; Zhao, Y.L.; Huang, X.M.; Chen, J. Property Characterization of Asphalt Binders and Mixtures Modified by Different Crumb Rubbers. *J. Mater. Civ. Eng.* **2017**, *29*, 04017036. [CrossRef]
48. Guo, Q.L.; Li, L.L.; Cheng, Y.C.; Jiao, Y.B.; Xu, C. Laboratory evaluation on performance of diatomite and glass fiber compound modified asphalt mixture. *Mater. Des.* **2015**, *66*, 51–59. [CrossRef]
49. Zeng, B.; Luo, C.M.; Liu, S.F.; Bai, Y.; Li, C. Development of an optimization method for the GM(1,N) model. *Eng. Appl. Artif. Intell.* **2016**, *55*, 353–362. [CrossRef]
50. Ren, J.Z. GM(1,N) method for the prediction of anaerobic digestion system and sensitivity analysis of influential factors. *Bioresour. Technol.* **2018**, *247*, 1258–1261. [CrossRef]
51. Ren, J.Z.; Gao, S.Z.; Tan, S.Y.; Dong, L.C. Prediction of the yield of biohydrogen under scanty data conditions based on GM(1,N). *Int. J. Hydrogen Energy* **2013**, *38*, 13198–13203. [CrossRef]

applied sciences

MDPI

Article

# Carbon Nanotubes (CNTs) in Asphalt Binder: Homogeneous Dispersion and Performance Enhancement

Muhammad Faizan ul Haq [1,*], Naveed Ahmad [1], Muhammad Ali Nasir [2], Jamal [3], Murryam Hafeez [1], Javaria Rafi [1], Syed Bilal Ahmed Zaidi [1] and Waqas Haroon [4]

1 Department of Civil Engineering, University of Engineering and Technology Taxila, Taxila 47080, Pakistan; n.ahmad@uettaxila.edu.pk (N.A.); murryamhafeez77@outlook.com (M.H.); javariarafi@outlook.com (J.R.); bilal.zaidi@uettaxila.edu.pk (S.B.A.Z.)

2 Department of Mechanical Engineering, University of Engineering and Technology Taxila, Taxila 47080, Pakistan; ali.nasir@uettaxila.edu.pk

3 Department of Civil Engineering, Royal Melbourne Institute of Technology University, Victoria 3053, Australia; jamal.naasir209@gmail.com

4 Department of Civil Engineering, International Islamic University Islamabad, Islamabad 44000, Pakistan; waqas.haroon@iiu.edu.pk

* Correspondence: Faizan.ul@students.uettaxila.edu.pk; Tel.: +9-231-6188-1117

Received: 9 November 2018; Accepted: 10 December 2018; Published: 17 December 2018

**Abstract:** Conventional binders cannot meet the current performance requirements of asphaltic pavements due to increase in traffic volumes and loads. Nanomaterials, due to their exceptional mechanical properties, are gaining popularity as bitumen modifiers to enhance the performance properties of the asphaltic concrete. Carbon Nanotubes (CNTs) are one of the most widely used nanomaterials because of their strength properties, light weight, small size, and large surface area. CNT addition results in improved substrate characteristics as compared to other modifiers. Due to high length to diameter ratio, dispersion of CNTs in bitumen is a complex phenomenon. In this study, dispersion of CNTs in bitumen was carried out using both dry and wet mixing techniques, the latter was selected on the basis of homogeneity of the resultant asphalt mixture. Scanning Electron Microscopy (SEM) was used to check the dispersion of CNTs in binder while Fourier Transform Infrared Spectroscopy (FTIR) was carried out to ensure the removal of solvent used for wet mixing. Conventional bitumen tests (penetration, softening point, and ductility), dynamic shear rheometer tests, rolling bottle tests, and bitumen bond strength tests were employed to check the improvement in the rheological and adhesion properties of bitumen while wheel tracker test was used to check the improvement in resistance against permanent deformation of asphalt mixtures after addition of CNTs. Results show that CNTs improved the higher temperature performance and permanent deformation resistance in both binder and mixtures. Improvement in bitumen–aggregate adhesion properties and moisture resistance was also observed.

**Keywords:** Carbon Nanotubes (CNTs); wet mixing; bitumen; homogeneous dispersion; rutting; adhesion

---

## 1. Introduction

An efficient transport network plays a key role in the economic development of a country and, therefore, the kilometrage of paved roads existing in a country is often used as an index to assess the extent of its development [1]. Asphalt mixtures have been widely used for construction of road for a long time now and bitumen is its main constituent [2]. Due to a rapid increase in the traffic loading and volume, pavements in Pakistan fail prematurely and the revival of their serviceability normally requires a lot of resources and finance. Pavements constructed utilizing conventional materials (especially virgin bitumen with softening points far less than the temperatures to which the pavements are exposed) fail many times within the design life, putting a lot of burden on the annual maintenance budget and ultimately increasing the life cycle cost of the pavement. If the bitumen were made to last/survive longer, pavement failures could be delayed, which would minimize the life cycle cost of the pavements. Regular wear and tear are expected but total failure before the end of the design life, as often happens in Pakistan, is not desirable. It not only increases the maintenance costs but also adds nuisances for the road users in terms of delays, accidents, and vehicle operating costs. This study aimed to address this issue [3]. Local traditional pavement materials in Pakistan are falling short in meeting the practical demands for present and future highway pavement construction. Thus, higher quality, more sustainable, more reliable and more environment friendly pavement materials are urgently demanded [4].

Civil engineering materials modification through addition of polymers and other additives such as fly ash and lime has often been utilized [5]. The use of Polymer Modified Asphalts (PMAs) allows the construction of safer roads and important reduction in maintenance costs [6–8]. Different modifiers such as crumb rubber and polymers are also used to improve the mechanical properties of conventional asphalts. Although modifiers improve the performance of asphalt, increases in traffic loads and volumes, harsh weather conditions, and the ever-increasing asphalt cost call for more research in this area [9]. Pavement engineers continuously look for innovative additives/modifiers to enhance the performance properties of road materials. Nanotechnology has recently gained popularity in the scientific world. Due to remarkable achievements of nanotechnology in other fields of engineering, researchers have started looking into its utilization in the field of civil engineering. Nanotechnology is often used for the modification and enhancement in properties of cement concrete [10,11]. The effect of nanomaterials on the properties of asphalt concrete is an emerging field that must be explored.

Nanotechnology and its influence on improving the characteristics and performance of asphalt pavements has lately received much attention from researchers [12]. Two specific types of nanomaterials, namely nano-metal oxides and nano-inorganic materials, have been widely investigated. Nano-metal oxides such as nano-$TiO_2$, nano-$SiO_2$, and nano-ZnO can improve the rutting resistance of bitumen, but have little effect on its low temperature cracking resistance [13–15]. Nano-inorganic materials including nano-clay and Carbon Nanotubes (CNTs) hold the potential to redefine the field of traditional materials in terms of both performance and potential applications [16–18]. CNTs are among the most widely used nanomaterials because of their strength properties, light weight, small size, and large surface area. CNTs are basically long hollow cylinders of graphene sheet, which have a diameter starting from approximately 1 nm. They were first discovered by Sumio Iijima in 1991 [19]. CNTs are characterized by superior mechanical properties when compared with other construction materials [20]. Depending on the radius of the tube, the Young's modulus of a CNT can be as high as 1000 GPa [21] and the tensile strength can reach 150 GPa [22]. Because of their small size and large surface area, CNT addition results in improved substrate characteristics as compared to other modifiers. There are two different types of CNTs, i.e., single (called single-walled CNTs) and coaxial tubes (called multiple-walled CNTs) [6]. Multi-walled CNTs (MWCNTs) are less expensive and easier to produce but exhibit lower strength and stiffness than single-wall CNTs [23]. The most common techniques used to produce CNTs are arc discharge, chemical vapor deposition and laser ablation.

Different researchers use different dosages of CNTs as a modifier to enhance the properties of asphalt binder. Gong et al. (2017) investigated the effect of CNTs on performance, chemical and structural properties of asphalt binder [24]. Amin et al. (2016) used MWCNTs as an additive in bitumen and concluded that the modification improved both high and low temperature performance of bitumen [9]. Galooyak et al. (2015) added MWCNTs in asphalt binder and examined the effects on conventional and rheological properties and concluded that CNTs improved the rheological as well as conventional properties of bitumen [25]. Previous studies have also observed that CNTs enhance the resistance of asphalt against ageing, which results in the increase in the pavement life [9,26]. The biggest challenge in developing CNTs reinforced material is the dispersion of CNTs in base binder. To homogeneously disperse the CNTs in bitumen, long chains of CNTs produced by synthesis and agglomerates of CNTs produced by intermolecular van der Walls forces must be broken [27]. Inappropriate dispersion of CNTs may affect the mechanical properties of modified binder. CNTs are dispersed in bitumen using either dry or wet mixing technique. Most previous work has been carried out using dry mixing, while limited work has been done using the wet mixing technique. Faramarzi et al. (2015) used both wet and dry mixing techniques and concluded that wet mixing technique has better ability to homogeneously disperse CNTs as compared to the dry method. They used kerosene as a solvent [28]. The drawback found by Faramazi et al. was that kerosene did not fully evaporate from the bitumen after mixing, which may influence the resulting binder properties. Ziari et al. (2012) compared different mixers and concluded that, with high shear mixer and ultrasonic mixer, homogeneous dispersion of CNTs can be achieved [29]. In the literature, although wet mixing is a better dispersion technique, the selection of appropriate solvent and its complete evaporation after achieving a homogeneous dispersion matters greatly. Therefore, a thorough study from this perspective was carried out in this study.

This study consisted of two parts. The first part of this manuscript focuses on the dispersion of CNTs in asphalt binder, while the second part studies the effect of CNTs on the performance properties of resulting asphalt binder. Dispersion of CNTs in bitumen is a complex phenomenon [24,28,29]. Although different researchers use CNTs in asphalt binder to enhance the properties of asphalt binder, to authors' best knowledge, no comprehensive study is available on the dispersion of CNTs in bitumen. In this research, a detailed methodology was developed for the homogeneous dispersion of CNTs in bitumen. In addition, previous work mostly studies the effect of CNTs on the rheological properties of bitumen, but the effect of CNT addition on bitumen–aggregate adhesion and moisture sensitivity needs to be explored.

## 2. Experimental Work

### 2.1. Materials

A 60/70 penetration grade bitumen supplied by Attock Refineries Limited (ARL), Pakistan was used as a base binder. MWCNTs were imported from US-Nano, USA. Basic properties of base binder and MWCNTs are presented in Tables 1 and 2, respectively.

**Table 1.** Properties of Base Binder.

| Test | Value | Standard |
|---|---|---|
| Penetration Value (0.1 mm, 25 °C) | 64 | ASTM D5 |
| Softening Point (°C) | 49 | ASTM D36 |
| Ductility mm (25 °C) | 102 | ASTM D113 |

**Table 2.** Properties of Multiwall Carbon Nanotubes used for Bitumen Modification.

| Inner Diameter | Outer Diameter | Average Length | Purity | Density | Specific Surface Area | Manufacturing Method |
|---|---|---|---|---|---|---|
| 30–50 nm | 5–12 nm | 0.5–2 μm | >90% | 2.1 g/cm$^3$ | 60 m$^2$/g | Chemical Vapor Deposition |

Scanning Electron Microscopy (SEM) images of CNTs are presented in Figure 1 showing the morphology of the selected CNTs at different magnifications.

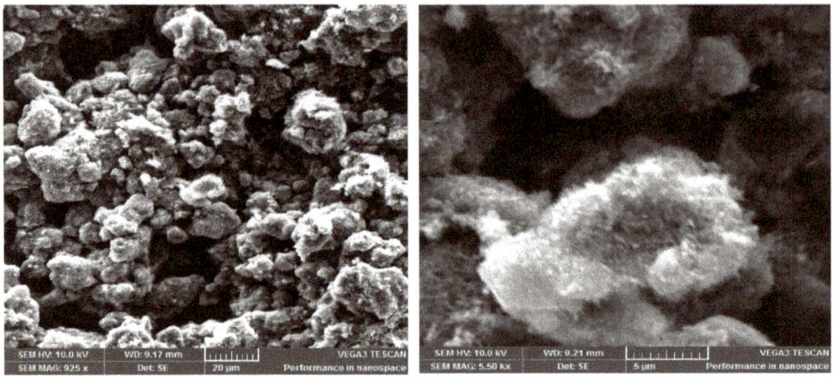


**Figure 1.** SEM images of CNTs.

Aggregates were procured from Margalla quarries, and was basically limestone [30]. The gradation curve for the aggregate material used in this study is presented in Figure 2. The general specifications of the National Highway Authority (NHA) were followed for the selection of material. NHA (a government owned organization for the development of highways) class B gradation for asphalt concrete wearing coarse was selected for the study. Midpoint gradation was used. Mechanical properties of the utilized aggregates are given in Table 3.

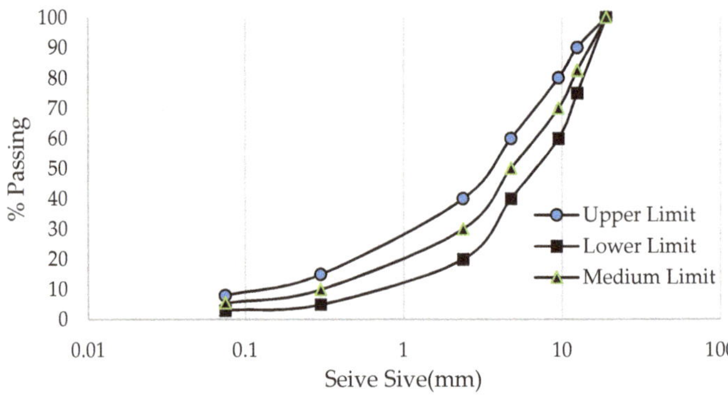


**Figure 2.** Gradation Curve of aggregates used for asphalt mixtures.

**Table 3.** Material Properties of Margalla aggregate used in this research [31].

| Property | Standard | Value | Limit |
|---|---|---|---|
| Los Angeles Abrasion (Coarse) | ASTM C 131 | 15 | ≤15% |
| Soundness (Course) | ASTM C 88 | 7.1 | ≤8% |
| Soundness (Fine) | ASTM C 88 | 4.7 | ≤8% |
| Water Absorption (Coarse) | ASTM C 127 | 1.02 | ≤2% |
| Uncompacted Voids (Coarse) | ASTM C 1252 | 37.5 | ≥45% |
| Flakiness (Coarse) | BS 812.108 | 5 | ≤15% |
| Elongation (Coarse) | BS 812.109 | 11 | ≤15% |
| Fractured Particles (Coarse) | ASTM D 5821 | 100 | ≥90% |
| Sand Equivalent (Coarse) | ASTM D 2419 | 75 | ≥50% |

## 2.2. Sample Preparation

It can be observed from the Scanning Electron Microscopy (SEM) images of CNTs given in the Figure 1 that these tubes are intertwined with each other. This means that higher efforts are required to properly disperse CNTs in bitumen. The most difficult process in bitumen modification with CNTs is its homogeneous dispersion because of their aggregation, which leads to unsatisfactory results. There are two ways to disperse CNTs in bitumen: (i) dry mixing; and (ii) wet mixing. In dry mixing technique, CNTs are directly added into bitumen and dispersed with mixers at high frequencies. In wet mixing technique, CNTs are first dissolved in a solvent. This solution is then dispersed in bitumen with mixers. Different nanomaterials require different mixing conditions for their dispersion in bitumen. For some nanomaterials, dry mixing and mechanical stirrer is sufficient to disperse them in bitumen, while some require high shear mixing and wet mixing for homogeneous dispersion. Thus, a thorough study was required on uniform dispersion of CNTs in asphalt binder.

In this study, sample preparation involved three major steps:

1. Mixer selection
2. Solvent selection
3. Technique (dry/wet) selection

**Mixer Selection**

Mechanical stirrer and high shear mixer (locally fabricated) were used for the dispersion of CNTs in bitumen. Researchers have utilized different mixing techniques for dispersion of nanomaterials in bitumen. The most commonly selected instruments are mechanical stirrer and high shear mixer. Mechanical stirrer uses simple propeller/fan blade configuration for mixing, while high shear mixer generates a vortex type movement within the material to ensure fast and aggressive mixing [29]. This study evaluated both types of mixers. One mixer (mechanical stirrer and high shear mixer) was selected based on dispersion of CNTs in asphalt binder. In many previous studies, mixing rate of around 1500 rpm with mechanical stirrer has been used for the dispersion of CNTs in bitumen [25,28,32] while a mixing speed of around 3000 rpm is generally selected for high shear mixer [29,33]. These mixing rates were selected for this study as well. CNTs content, mixing temperature, and mixing time were kept constant for both mixers. One percent by mass of bitumen was taken as CNT content, while mixing temperature and mixing time were $158 \pm 5$ °C and 45 min, respectively. SEM images were used to compare the dispersion of CNTs in bitumen for both mixers. Microscopy (SEM) specimen was examined point by point directly in a moving electron beam. Electrons reflected by the specimen were used to form a magnified, three-dimensional image on a television screen [34]. Bitumen is a petroleum product and contains volatile compounds, which is why, when subjected to focused electron beam during SEM test, the volatile compounds of bitumen evaporate due to high temperature, which not only contaminates the SEM chamber but also makes the scanning process difficult. To avoid this, sputtering of SEM samples was carried out. During the sputtering process, the bitumen sample was placed in a vacuumed reservoir and coated with thin layers of gold palladium. SEM analysis provided in Section 2.4.1 shows that mechanical stirrer was unable to break the agglomerates of the CNTs, while high shear mixer was able to disperse CNTs more uniformly. Hence, high-shear mixer was selected to disperse CNTs uniformly into the asphalt binder for sample preparation.

**Solvent Selection for Wet Mixing**

The main problem in wet mixing is the selection of solvent that can fully evaporate leaving behind the dispersed CNTs. If solvent remains in bitumen, then it can affect its properties and the real effect of CNTs cannot be judged. Three solvents, i.e., methanol, acetone and toluene, were tested for time of evaporation. Each solvent was separately mixed in bitumen using high shear mixer for 5 min at 3000 rpm. To evaporate the solvent, these mixes were kept in an oven at $158 \pm 5$ °C and were continually (manually) stirred until a constant mass was achieved. The time taken by each solvent to fully evaporate from the modified bitumen was also noted. Methanol was selected as the solvent due to the shorter time taken for its removal from the bitumen.

**Technique (Dry/Wet) Selection**

Samples were prepared using both wet and dry mixing techniques. Hot storage test was used to evaluate the high temperature storage stability of modified bitumen. An aluminum tube (25 mm in diameter and 140 mm in height) was filled with about 50 g of hot CNT-modified bitumen, and then stored vertically in an oven at 163 °C for 48 h. It was then taken out and cooled in a refrigerator at −7 °C for 4 h. Finally, the tube was cut into three equal sections. The difference of softening point value between top and bottom section of tube is known as Separation Index (SI). If SI is less than 2.2 °C, the sample is regarded as storage stable according to ASTM D5892. With dry mixing, maximum SI was about 2.5 °C, while, in wet mixing, it was about 1.3 °C, thus the sample prepared using wet mixing was storage stable. Wet mixing technique was selected for this study. Thus, the samples were prepared using wet mixing technique with methanol as a solvent using high shear mixer.

### *2.3. CNTs Modified Asphalt Binder Preparation*

Four different percentages of CNTs (0.5%, 1%, 1.5% and 3%) by mass of the asphalt binder were used in this study. CNTs were added into the methanol. The Solution was first stirred with the help of magnetic stirrer at 550 rpm for 3 h to break the CNT agglomerates. Then, the solution was kept in a sonication bath for 2 h to disperse the CNTs in solvent. This is important for the proper dispersion of CNTs in the solvent and its stability at room temperature, as shown in Figure 3. During this process, the solvent was covered with aluminum foil so that the methanol did not evaporate during sonication and stirring. A homogeneously mixed black color solution was obtained having high stability at room temperature, as shown in Figure 3.

After homogeneous dispersion of CNTs in methanol, the solution was mixed in bitumen with the help of high shear mixer at 3000 rpm for 45 min to ensure homogeneous mixing as well as to complete evaporation of methanol. Temperature was kept at 158 ± 5 °C during mixing with the help of oil bath.

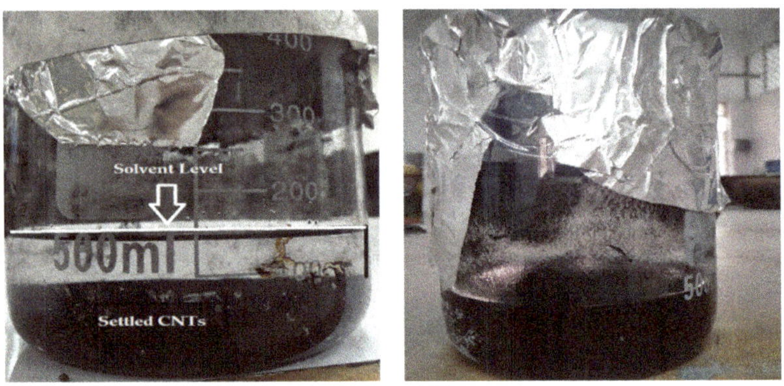


(**a**) (**b**)

**Figure 3.** (**a**) CNTs in solvent before sonication; and (**b**) CNTs in solvent after sonication and stirring.

### *2.4. Investigation into Homogeneous Dispersion*

#### 2.4.1. SEM Analysis

SEM was carried out to check the homogeneous dispersion of CNTs in bitumen. Mixing using high shear technique resulted in better dispersion of CNTs in the bitumen. Figure 4 shows an SEM image of CNT-modified asphalt binder prepared using mechanical stirrer, while Figure 5a,b shows the SEM images of CNT-modified asphalt binder prepared using high shear mixer. Figure 4 clearly shows

CNT agglomerates with mechanical stirrer, while Figure 5a shows evenly distributed CNTs, even at a higher magnification.

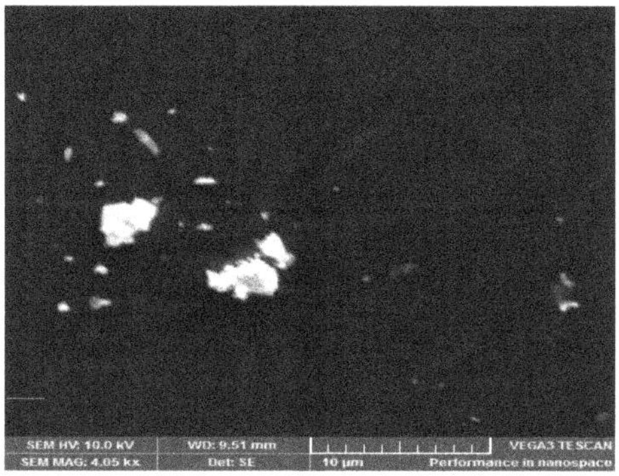


**Figure 4.** CNTs dispersion in bitumen using mechanical stirrer.

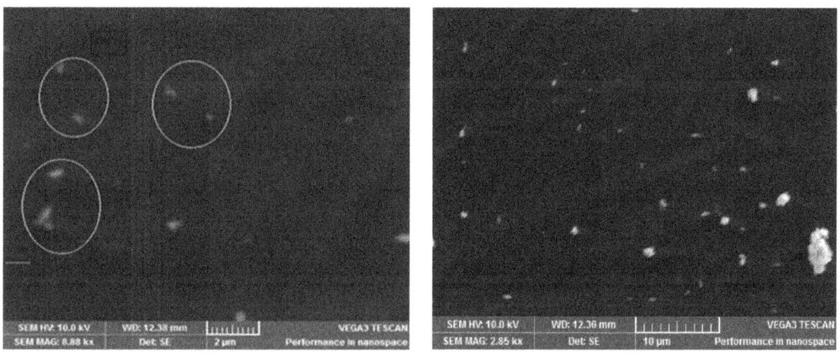


(**a**) (**b**)

**Figure 5.** CNTs dispersion in bitumen using high shear mixer at different magnifications.

### 2.4.2. FTIR Analysis

FTIR was used to check for the complete removal of solvent from the modified binder. The test was performed according to ASTM E1552. For FTIR (6700 Nicolet, Thermo Science, Waltham, MA, USA) spectrometers was used to get the asphalt spectra in the range of 400–4000 $cm^{-1}$ wavenumber. It can be seen from the FTIR spectra shown in Figure 6 that no new chemical functional groups were formed with the addition of MWCNTs, and all different concentrations of MWCNTs showed a similar trend. Thus, the MWCNTs did not chemically react with bitumen. Methanol ($CH_3OH$) was used as a solvent in wet mixing technique to disperse the CNTs in bitumen. In methanol, functional group -OH is present and in FTIR spectrum the peak of -OH alcohol group appears at 3200–3700 $cm^{-1}$ wavenumber spectra. In Figure 6, it is clear that, in the FTIR spectra, no peak appears at that range of -OH functional group, which means the solvent was completely evaporated and no effect of methanol was left. Any change in the properties of asphalt can only be attributed to the addition of CNTs in asphalt binder.

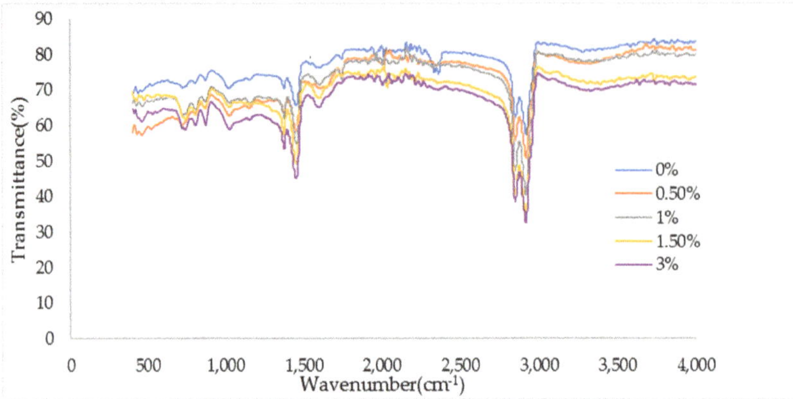


**Figure 6.** FTIR of unmodified and CNT-modified bitumen.

#### 2.4.3. Storage Stability Test

Storage stability test also ensured the homogeneous dispersion of CNTs in the binder. The difference in Softening Point (SP) of the represented sample taken from the top and bottom of the storage test aluminum tube was within the permissible limit (2.2 °C) for wet mixing technique. The results are shown in Table 4.

**Table 4.** Storage stability test data for CNT-modified and unmodified binder.

| CNTs Dosage | 0% | 0.5% | 1% | 1.5% | 3% |
|---|---|---|---|---|---|
| **SP of Top Portion (°C)** | 48.9 | 50.7 | 55.4 | 57 | 57.8 |
| **SP of Bottom Portion (°C)** | 49.1 | 51.7 | 56.7 | 58.5 | 59.7 |
| **Difference (°C)** | 0.2 | 1 | 1.3 | 1.5 | 1.9 |

### *2.5. Engineering Properties Analysis*

#### 2.5.1. Conventional Binder Tests

Penetration, softening point, and ductility tests were performed in accordance with ASTM D5, ASTM D36 and ASTM D113, respectively. The tests were performed to study the change in these properties before and after the CNT addition in asphalt binder.

#### 2.5.2. Dynamic Mechanical Analysis (DMA)

For dynamic mechanical analysis, Dynamic Shear Rheometer (DSR) model MCR101 manufactured by Anton Paar (Gras, Austria) was used and the test was performed according to AASHTO T 315. DSR Plates of 25 mm dimeter with 1 mm gap were used to test the samples at temperatures higher than 46 °C while 8 mm diameter plates with the gap of 2 mm were used for testing the samples at temperatures less than 46 °C. High temperature performance grading (PG) and frequency sweep tests were conducted. In frequency sweep, test strain was kept constant as per superpave criteria (10% for unmodified sample and 2% for modified sample) and frequency ranged 10–0.1 Hz at required temperatures.

#### 2.5.3. Bitumen Bond Strength Test

Bitumen bond strength test was performed to check the adhesion of bitumen with aggregate after both dry and moist conditioning. Pneumatic Adhesion Tensile Tester Instrument (PATTI) manufactured by SEMicro (Derwood, MD, USA) was used to perform this test. Test was performed according to

ASTM D4541. Geometry of the stub used for experiment is shown in Figure 7. BBS test was used to determine the bonding properties between aggregate and bitumen. Apparatus consists of metallic pull-off stub, reaction plate, pressure hose, piston, and a portable pneumatic adhesion tester. During the testing, pull-off force thorough air pressure was used to separate metallic stub from aggregate surface. Failure occurs when applied stresses are more than adhesive strength or cohesive strength.

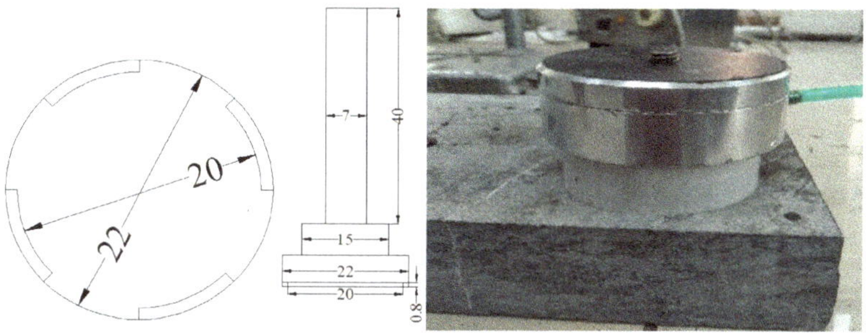


**Figure 7.** Dimensions of stub (mm) used in BBS test and piston setup.

Bitumen samples, limestone aggregate plates and pullout stubs were heated to 150 °C for a minimum of 30 min to remove absorbed water on the aggregate surface and to provide a better bond between the asphalt binder and the aggregate surface. The aggregate plates were brought to an application temperature of 60 °C, whereas the application temperature for the pullout stubs was 60 °C. After a sufficient heat-up time, the molten asphalt samples were carefully poured in 8 mm diameter DSR silicon molds that were then left for 30 min to reach room temperature. Then, these bitumen samples were placed on the heated stubs and the stubs were firmly pressed vertically on aggregate plates until each stub was in contact with the aggregate surface. Excess bitumen was removed with cutter from all the prepared samples. For dry conditioning, prepared samples were kept at room temperature for 24 h before the test. In moist conditioning, samples were kept in deionized water at 25°C for 24 h before testing.

2.5.4. Moisture Susceptibility Analysis

Rolling bottle test was performed to measure the moisture susceptibility of asphalt. The test was performed according to BS EN 12697-11. First, 170 g of aggregate passing through an 8 mm sieve and retained on a 6.3 mm sieve were taken. Then, 8 g modified and unmodified binder samples were mixed with aggregate until all aggregate was fully coated with the bitumen. Then, these coated aggerate particles were placed in bottle and the bottles were filled with deionized water. Bottles were placed in the rolling machine and rotated at 60 rpm. Coating percentage of each sample was checked after 24, 48 and 72 h.

2.5.5. Permanent Deformation Analysis

Wheel tracker test was performed to determine the rutting of the asphalt mixtures under loading cycles. Test was performed according to BS EN 12697-22. Due to shortage of time and less quantity of materials, three slabs were prepared with 0%, 1% and 3% CNT content and the gradation used was NHA Class B. First, Marshall method of mix design was carried out to determine the optimum binder content of all three samples. Marshall method of mix design was performed as per ASTM D1559 using hammer weight to 4.5 lbs., hammer drop height of 18 inches and application of 75 blows on each side (for heavy traffic) of the specimen. OBC for unmodified sample was 4.38% while for 1% and 3% CNT-modified samples it was 4.42% and 4.47%, respectively. Researchers have experienced both increases and decreases in the optimum binder content requirements when utilizing nanomaterials [35,36]. In this study, an increase in optimum binder content was experienced with the increase in CNT dosage. The same trend in OBC by adding nanomaterial was observed by Chelovian and Shafabakhsh [37]. This increase in OBC could be attributed to large surface area of CNTs used in this study. This could also be due to the increase in viscosity of the asphalt binder with the addition of addition of the nanomaterial which made binder stiff. A higher viscosity leads to a thicker binder film of the modified bitumen in the mix, thus increasing the binder volume in the mix [38,39].

Mixing of asphalt was done at 158 ± 5 °C and compaction of asphalt was done at 145 ± 5 °C. A 300 mm × 300 mm × 50 mm slab was prepared and compacted with the help of cooper roller compacter in four different phases with 2.5, 3.5, 4 and 4.5 bar pressure to maintain 5.5 ± 0.5% air voids. The number of passes kept for the specimen was 10,000 and load applied during the test was 700 ± 20 N, while 40 °C temperature was kept constant during test.

All the above-mentioned tests were performed at Taxila Institute of Transportation Engineering, University of Engineering and Technology Taxila, Pakistan.

## 3. Results and Discussions

### 3.1. Conventional Asphalt Binder Properties

Figure 8 shows the effect of MWCNTs on the penetration and softening point value of modified and unmodified bitumen. Penetration value reflects the stiffness and hardening of asphalt binder at moderate temperature. Lower penetration value indicates that binder has become stiff. In Figure 8 it can be seen that, by increasing CNTs in bitumen, the penetration value decreased, which is indicative of a decrease in fluency and increase in the stiffness of bitumen. When 0.5% (by mass of bitumen) CNTs was added, penetration significantly decreased by 11%. By increasing the CNT dosage to 1.5% and 3%, 25% and 28% decreases in penetration value were observed, respectively. Increasing the CNT dosage beyond 1% did not have a significant effect on the penetration value.

The softening point test is commonly used as a standard test for describing an approximate limit between viscous and viscoelastic bitumen behavior, and it reflects the deformation resistance of bitumen at high temperature. A dosage of 1.5% CNTs (by mass of bitumen) increased the softening point by about 7 °C as compared to base binder. When about 3% by mass of CNTs was added, it resulted in an increase in softening point of bitumen of about 8.4 °C. This increase in softening point and decrease in penetration value may be attributed to the large surface energy, high Young's modulus and presence of interaction forces between the CNTs, which make binder stiff [40,41].

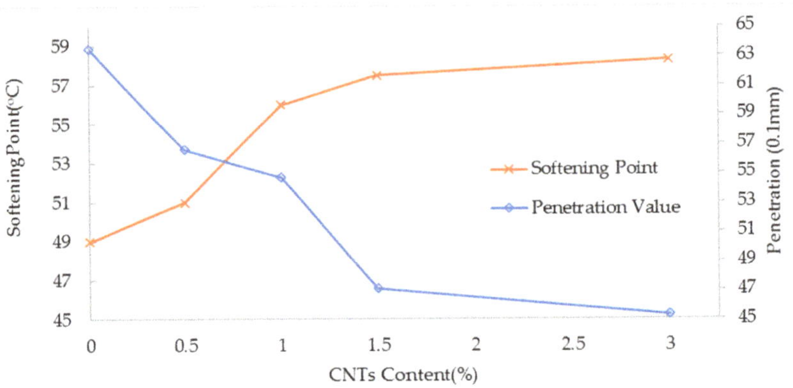


**Figure 8.** Effect on penetration value and softening point value with addition of CNTs.

PI was used to study the effect of the addition of CNTs in the asphalt binder on its temperature susceptibility. The temperature susceptibility evaluation is described below, as illustrated by Ahinola et al. [42], using following equations.

$$PI = \frac{20 - 500A}{1 + 50A} \text{ where } A = \frac{log(800) - log(Pen\ at\ T)}{SP - 25} \quad (1)$$

where $T$ is the temperature at which the penetration test is performed (25 °C), *Pen* at $T$ is the penetration at the required temperature $T$ (0.1 mm), *SP* is the softening point of asphalt (°C), and *Pen* at *SP* is the penetration at the softening point and assumed as 800 (0.1 mm) [43].

Higher value of *PI* indicates lower temperature susceptibility. In addition, it is known that most asphalt binders have *PI* values between −2 and +2. Binders are considered as highly temperature susceptible if their *PI* values go below −2 and they will be more brittle at lower temperatures and experience transverse cracking in cold weather regions [44].

The *PI* results for all the used CNTs percentages, as shown in Table 5 and Figure 9, are within the normal *PI* ranges (from −2 to +2). With the addition of CNTs, the *PI* value of bitumen increased, which means the bitumen temperature susceptibility decreased.

**Table 5.** Penetration Index for modified and unmodified bitumen.

| Sample | 0% CNTs | 0.5% CNTs | 1% CNTs | 1.5% CNTs | 3% CNTs |
|---|---|---|---|---|---|
| PI | −0.7692 | −0.66 | 0.0187 | 0.060 | 0.065 |
| A | 0.04584 | 0.0442 | 0.03758 | 0.0378 | 0.0374 |

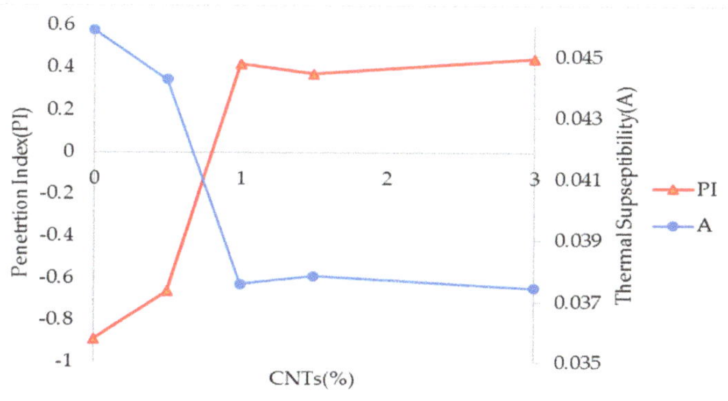


**Figure 9.** Effect on Penetration Index and Thermal Susceptibility with addition of CNTs.

Figure 10 shows that decrease in ductility was observed with addition of CNTs. A 21% decrease in ductility value was observed when CNT content of up to 1.5% was added in bitumen. With 3% addition of CNTs, ductility value reduced by 26% as compared to neat binder. This reduction in ductility value was due to increase in stiffness of bitumen with the addition of CNTs, which accelerated the bitumen fracture process when subjected to tensile stress [11].

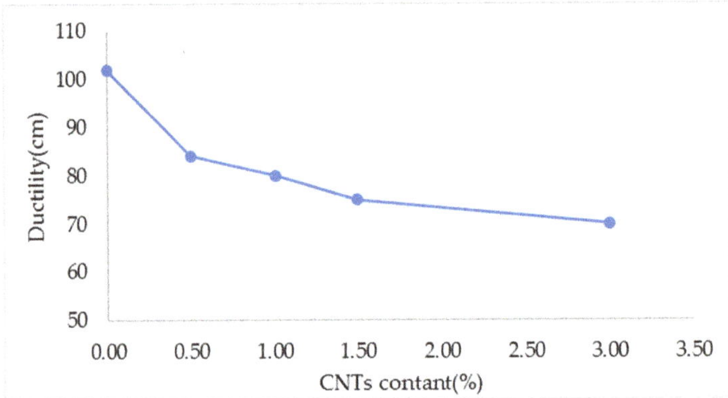


**Figure 10.** Ductility of bitumen with different contents of CNTs.

### *3.2. Dynamic Shear Rheological Properties*

Dynamic shear rheometer test was performed to check the rheological properties of asphalt binder at intermediate to high temperatures for different frequencies. Figure 11 shows that an increase in failure temperature of the bitumen was observed when increasing percentages of MWCNTs were introduced. Failure temperature of the bitumen for unaged samples was determined as per superpave method. Temperature at which G*/Sinδ becomes less than 1 kPa is failure temperature for unaged asphalt binder [45]. The failure temperature of original bitumen was about 61 °C, so the upper performance grade of original asphalt binder was 58°C. By increasing the CNT dosage, the failure temperature of samples was also increased, which is an indication of the reduction in temperature susceptibility. At 0.5% dosage, high PG grade of bitumen was 64, while the PG grades for 1% and 3% CNTs modified bitumens were 70 and 76, respectively. It means that a 3% dosage of CNTs resulted in three grade bumps. The climate in most of Pakistan is hot and summer air temperature rises up to 50 °C and in winter it hardy drops below 0 °C. PG 70 is recommended for most regions of the country [46]. This study aimed at achieving a PG 76 after a great bump in the required PG 70 to accommodate the overloading on the highways of the country.

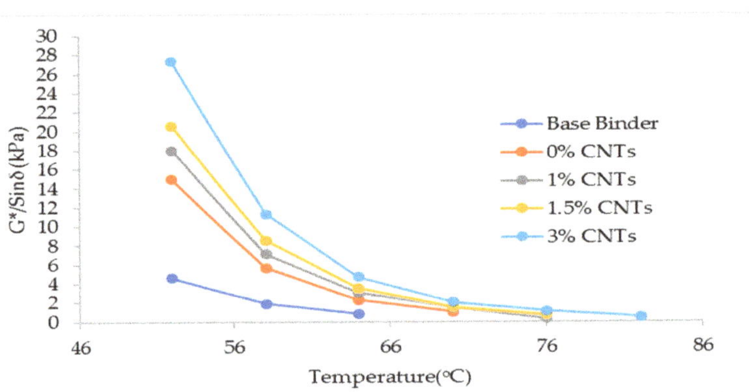


**Figure 11.** Rutting parameter (G*/Sinδ) versus temperature for unaged asphalt binder at 10 rad/s.

Figure 12 shows master curve of Complex Shear Modulus (G*) at 58 °C for unmodified and CNT-modified bitumen. The G* value increased from 3132 Pa to 14,264 Pa at 10 Hz frequency with the addition of 3% of CNTs. By increasing the content of MWCNTs in the original asphalt binder, the

G* value increased, which is an indication of increase in bitumen stiffness and its resistance against permanent deformation at higher temperature. Bitumen is a viscoelastic material that shows viscous behavior at higher temperature and elastic behavior at low temperature. Phase angle is used to describe the bitumen elastic or viscous behavior. In Figure 13, it can be seen that, with the increase of CNTs in bitumen, phase angle of binder decreased, which is an indication of the increase in elastic behavior of bitumen [33]. The smaller is the phase angle, the higher is the elastic recovery when bitumen pavements are subjected to traffic at high temperatures. Thus, for MWCNT-modified bitumen, the chances of permanent deformation at higher temperatures are low.

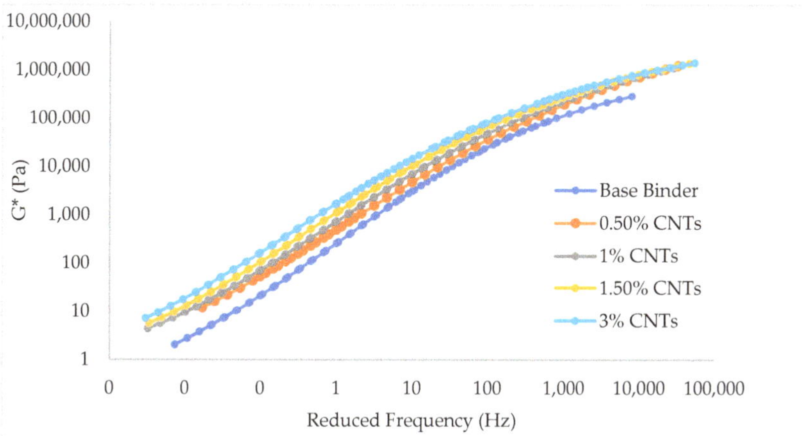


**Figure 12.** Master curve of G* for CNT-modified and unmodified bitumen at 58 °C.

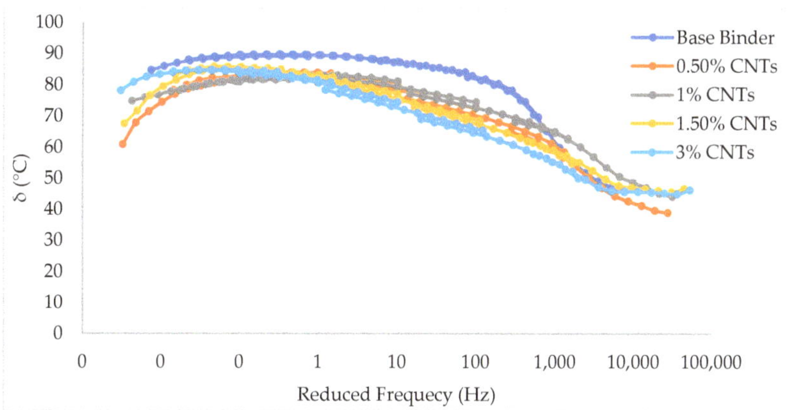


**Figure 13.** Master curve of δ for CNT-modified and unmodified bitumen at 58 °C.

One of the most important problems for asphaltic pavement is the deformation of the pavement in the wheel path due to heavy traffic at high temperature, which is called rutting. The parameter G*/Sinδ is called rut factor and predicts the failure of binder at high temperature. Figure 14 shows that, with the modification of bitumen with MWCNTs, this parameter improved, which means the materials resistance against the rutting is enhanced at higher temperatures. The same effect was observed by Amin et al. (2016) [33].

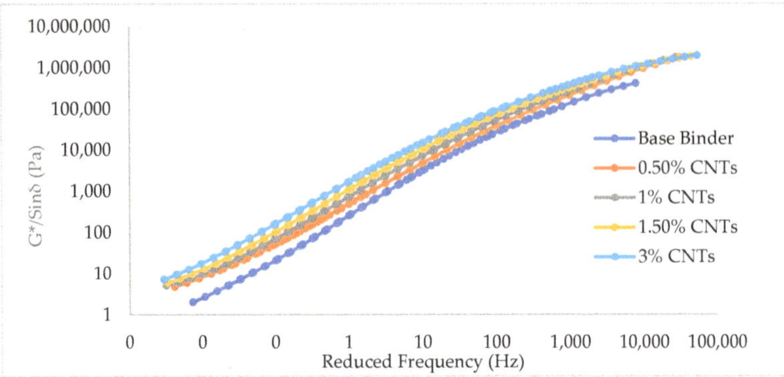


**Figure 14.** Master curve of G*/Sinδ for CNT-modified and unmodified bitumen at 58 °C.

Complex shear modulus (G*) and phase angle (δ) obtained by frequency sweep test are represented in the form of black diagram in Figure 15. Black diagram is free of frequency and temperature. It allows all the data to be presented in a single graph. As shown in Figure 15, the viscoelastic properties of bitumen were very sensitive to CNTs content, as there was a significant change in the curves obtained for unmodified binder in comparison to the binder modified with different dosages of CNTs. It was observed that, with increase in temperature, phase angle of bitumen increased and G* value decreased. After a threshold point in Figure 15, it was observed that, for the CNT-modified asphalt binder, phase angle at high temperature and low frequencies tend to decrease as compared to base binder which means the elastic behavior of bitumen increased at high temperature for these percentages. Thus, resistance against deformation increased with the addition of CNTs in bitumen [16].

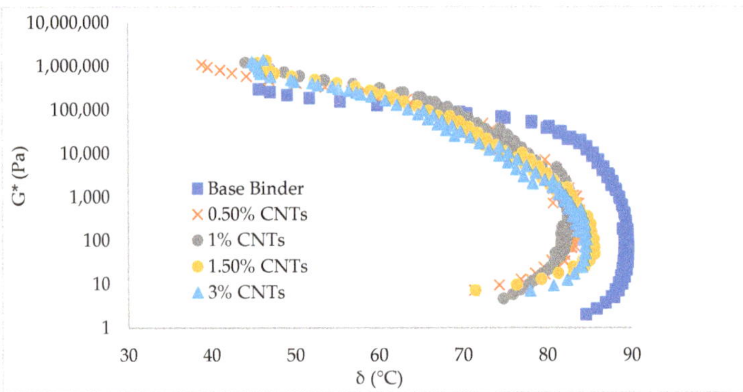


**Figure 15.** Black diagram for unmodified and CNT-modified binder.

### 3.3. Bitumen–Aggregate Bond Strength Analysis

The results of the bitumen bond strength test with aggregate after both dry and wet conditions are shown in Table 6 with failure pattern. In this table, the addition of CNTs improved the adhesion of bitumen with aggregate. Burst pressure obtained by PATTI was converted into pull off tensile strength using following relationship.

$$POTS = \frac{(BP * A_g) - C}{A_{ps}} \tag{2}$$

where $BP$ is burst pressure, $A_g$ is the contact area between gasket and reaction plate, $C$ is piston constant and $A_{ps}$ is area of pull off stub. For F4 type of stub, $A_g$ is 4.06 in$^2$, C is 0.286 and $A_{gs}$ is 0.1963 in$^2$.

**Table 6.** POTS (psi) of unmodified and CNT-modified binders and failure pattern.

| | 24 h Dry Conditions (psi) | | | | | 24 h Wet Conditions (psi) | | | | |
|---|---|---|---|---|---|---|---|---|---|---|
| | 0% | 0.5% | 1% | 1.5% | 3% | 0% | 0.5% | 1% | 1.5% | 3% |
| 1 | 1266.4 (C) | 1493.9 (C) | 1587.0 (C) | 1868.3 (C/A) | 1994.4 (C) | 1119.5 (A) | 1254.0 (A) | 1361.5 (A) | 1514.6 (A) | 1555.9 (C/A) |
| 2 | 1206.4 (C) | 1512.5 (C) | 1570.4 (C) | 1818.6 (C) | 1907.5 (C) | 1059.6 (A) | 1287.1 (A) | 1440.1 (A) | 1496 (C/A) | 1549.7 (C/A) |
| 3 | 1270.5 (C) | 1469.1 (C) | 1560.1 (C) | 1721.4 (C) | 1797.9 (C/A) | 1076.1 (A) | 1295.3 (A) | 1328.4 (A) | 1613.9 (C/A) | 1438.1 (C/A) |
| 4 | 1305.7 (C) | 1485.6 (C) | 1580.8 (C) | 1642.8 (C) | 1924.1 (C/A) | 1080.2 (A) | 1318.1 (A) | 1527 (A) | 1409 (A) | 1630.4 (A) |
| 5 | 1258.1 (C) | 1510.4 (C) | 1607.7 (C) | 1992.3 (C/A) | 1961.3 (C/A) | 1067.8 (A) | 1299.5 (A) | 1384.3 (A) | 1518.7 (A) | 1678.0 (A) |
| Avg | 1261.4 | 1494.3 | 1581.2 | 1808.7 | 1917.1 | 1080.7 | 1290.8 | 1408.3 | 1510.4 | 1570.4 |

C, cohesive failure; A, adhesive failure; C/A, 50% cohesive 50% adhesive failure.

It can be seen from the results that POTS for CNT-modified asphalt binders were higher than for the unmodified asphalt binder in both dry and wet conditions. In dry conditions, most of the failure was cohesive failure, while adhesive failure was observed in most of the samples tested after wet conditioning due to the penetration of water at the interface between bitumen and aggregate, which weakened the bond and resulted in reduction of POTS [47]. CNT-modified bitumen showed less susceptibility to moisture as compared to unmodified bitumen due to the hydrophobic nature of CNTs [48]. In dry condition, addition of 0.5% of CNTs resulted in an increase of POTS by 19% as compared to base binder. As the bitumen content increased, POTS tended to increase. With 3% addition of CNTs, POTS increased by a value of 52%, while, in wet mixing, 3% CNTs increased the POTS by 45%.

### 3.4. Moisture Susceptibility Analysis

Rolling bottle test was used to check the moisture susceptibility of asphalt binder with aggregate. In Figure 16, the percentage loss of aggregate at different durations with unmodified and CNT-modified bitumen is represented. In Figure 17, it can be observed that, with the addition of CNTs in bitumen, the percentage loss of coating was decreased, which means the adhesion between aggregate and bitumen under moist conditions tended to increase in the presence of CNTs. This addition of 1% and 3% CNTs in bitumen improved the coating of binder by 20% and 40%, respectively, at 72 h. According to Liu et al. [49], stiff binder shows better resistance against moisture. This improvement in coating may be attributed to the presence of CNTs in asphalt binder, which made bitumen stiff, as well as to the hydrophobic nature of CNTs [37].

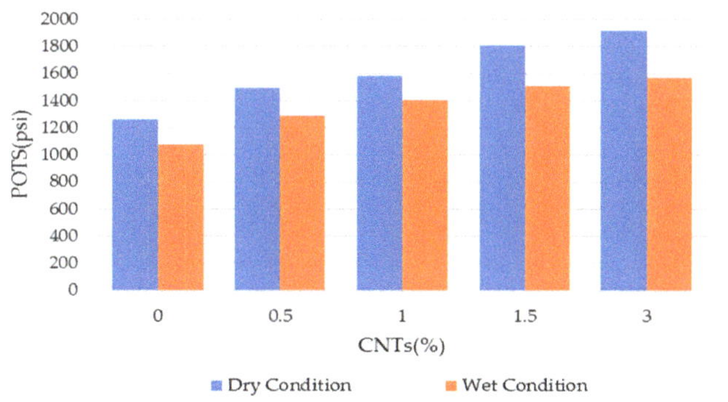


**Figure 16.** Comparison of POTS in 24 h dry and wet conditions.

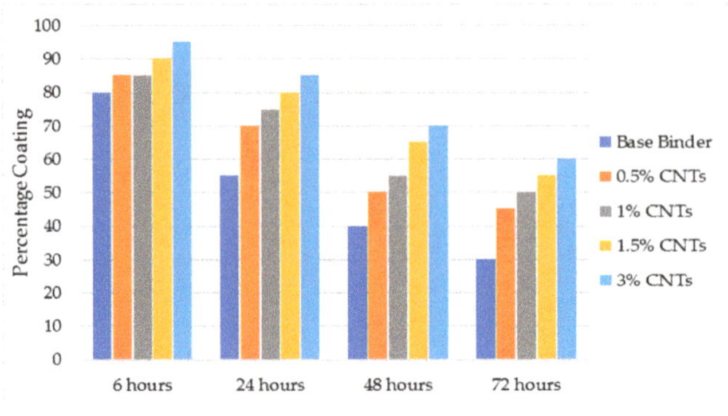


**Figure 17.** Loss of Coating at different durations with various CNTs percentages.

*3.5. Permanent Deformation Analysis*

Cooper wheel tracker test result for the rut depth of different dosages of CNTs is shown in Figure 18. It is clear from the results that, with the addition of CNTs, resistance against rutting increased. Rut depth value decreased with respect to base binder by up to 25% and 37% with the addition of 1% and 3% of CNTs, respectively, at 40 °C. Rut depth decreased because the introduction of CNTs in binder increased its stiffness. Addition of CNTs improved the high temperature performance of asphalt mixtures by improving the elastic response of the bitumen. Moreover, addition of the CNTs to the binder due to the high specific surface can reinforce the bitumen particles and enhance the strength of bitumen particles. The above factor can increase the viscosity and adhesion of asphalt binder and decrease its sensitivity to the permanent deformation [37].

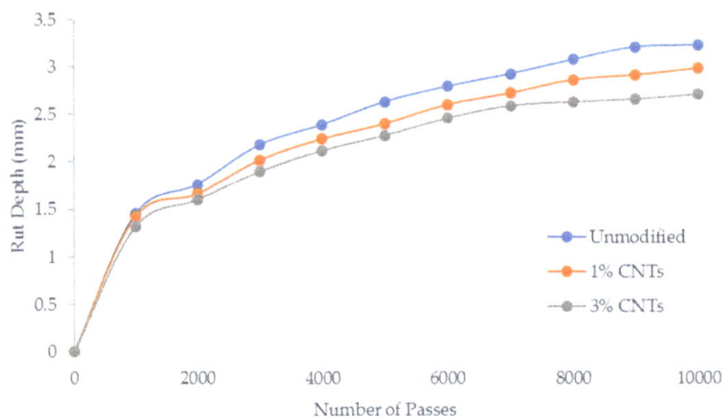


**Figure 18.** Rut depth for unmodified and CNT-modified bitumen mix at 40 °C.

## 4. Conclusions and Recommendations

This study experimentally developed an understanding of the effect of adding carbon nanotubes in asphalt bitumen. Dynamic mechanical analysis and different performance tests were carried out. Different tests were performed on asphalt binder as well as asphalt mixtures. The following conclusions were drawn from the above research

- Wet mixing techniques better helps in achieving homogeneous dispersion of CNTs in bitumen as compared to dry mixing.

- Sonication and magnetic stirring are necessary to improve the stability of CNTs in solvent.
- Introduction of CNTs in asphalt binder resulted in reduction in penetration value and ductility value, while it increased the softening point value of the bitumen, which means stiffness of the bitumen increased with the addition of CNTs.
- With the addition of CNTs, the PI value of bitumen increased, which means the bitumen temperature sensitivity decreased and thermal susceptibility also decreased.
- Complex shear modulus (G*) value increased and phase angle of the bitumen tended to decrease with increase in CNTs, which means the stiffness and elastic behavior of bitumen improved with the addition of CNTs in asphalt binder. Rut factor value also improved, which indicates the increase in resistance against permanent deformation.
- High Performance Grade (PG) of the bitumen increased with an increase in the CNTs dosage. PG 70 is recommended for most of the regions of Pakistan. This study aimed for achieving a PG 76 after a great bump in the required PG 70 to accommodate for the overloading on the highways of the country. The required PG 76 was achieved with 3% addition of CNTs in bitumen. Asphalt samples modified with 3% CNTs showed improved results for all parameters examined in this study. Hence, 3% CNTs dosage was selected as the optimum dosage.
- Bitumen bond strength value improved by adding CNTs in bitumen in both 24 h dry and 24 h moist conditions, while moisture susceptibility of bitumen decreased with addition of CNTs.
- From wheel tracker test results, it was concluded that the addition of 1% and 3% CNTs in bitumen reduced the rut depth by 25% and 37%, respectively, when tested at 40 °C. This is an indication of an increase in the resistance against permanent deformation at higher temperature for CNT-modified binder.
- Although wet mixing is a complicated process, it is preferred over the dry mixing technique because it ensures homogeneous dispersion of CNTs in the binder.

**Author Contributions:** Conceptualization, M.F.u.H. and N.A.; Data curation, M.F.u.H., Jamal, M.H., J.R. and S.B.A.Z.; Formal analysis, M.F.u.H., N.A. and Jamal; Investigation, M.F.u.H., Jamal, M.H. and J.R.; Methodology, M.F.u.H., N.A. and M.A.N.; Resources, M.F.u.H. and W.H.; Supervision, N.A.; Writing—original draft, M.F.u.H.; and Writing—review and editing, N.A., M.A.N., S.B.A.Z. and W.H.

**Acknowledgments:** The authors would like to acknowledge the support of the Civil Engineering department at The University of Lahore, Islamabad Campus in pursuit of this research work.

**Conflicts of Interest:** The authors declare no conflict of interest.

## References

1. Kiel, J.; Smith, R.; Ubbels, B. The Impact of Transport Investments on Competitiveness. *Transp. Res. Procedia* **2014**, *1*, 77–88. [CrossRef]
2. Collins, J.H.; Bouldin, M.G.; Gelles, R.; Berker, A. Improved Performance of Paving Asphalts by Polymer Modification. *J. Assoc. Asph. Paving Technol.* **1991**, *60*, 43–49.
3. Sadeghpour, S.; Dabir, B.; Ehsan, A.; Moeini, A. Rheological properties and storage stability of bitumen/SBS/montmorillonite composites. *Constr. Build. Mater.* **2010**, *24*, 300–307. [CrossRef]
4. Alhamali, D.I.; Wu, J.; Liu, Q.; Hassan, N.A.; Yusoff, N.I.M.; Ali, S.I.A. Physical and Rheological Characteristics of Polymer Modified Bitumen with Nanosilica Particles. *Arab. J. Sci. Eng.* **2016**, *41*, 1521–1530. [CrossRef]
5. Golewski, G.L. Generalized fracture toughness and compressive strength of sustainable concrete including low calcium fly ash. *Materials (Basel)* **2017**, *10*, 1393. [CrossRef] [PubMed]
6. Sengoz, B.; Isikyakar, G. Evaluation of the properties and microstructure of SBS and EVA polymer modified bitumen. *Constr. Build. Mater.* **2008**, *22*, 1897–1905. [CrossRef]
7. Yildirim, Y. Polymer modified asphalt binders. *Constr. Build. Mater.* **2007**, *21*, 66–72. [CrossRef]
8. Behnood, A.; Modiri Gharehveran, M. Morphology, rheology and physical properties of polymer-modified asphalt binders. *Eur. Polym. J.* **2018**. [CrossRef]

9. Amin, I.; El-Badawy, S.M.; Breakah, T.; Ibrahim, M.H.Z. Laboratory evaluation of asphalt binder modified with carbon nanotubes for Egyptian climate. *Constr. Build. Mater.* **2016**, *121*, 361–372. [CrossRef]
10. Raki, L.; Beaudoin, J.; Alizadeh, R.; Makar, J.; Sato, T. Cement and concrete nanoscience and nanotechnology. *Materials (Basel)* **2010**, *3*, 918–942. [CrossRef]
11. Sanchez, F.; Sobolev, K. Nanotechnology in concrete—A review. *Constr. Build. Mater.* **2010**, *24*, 2060–2071. [CrossRef]
12. Li, R.; Xiao, F.; Amirkhanian, S.; You, Z.; Huang, J. Developments of nano materials and technologies on asphalt materials—A review. *Constr. Build. Mater.* **2017**, *143*, 633–648. [CrossRef]
13. Shafabakhsh, G.H.; Ani, O.J. Experimental investigation of effect of Nano TiO2/SiO2 modified bitumen on the rutting and fatigue performance of asphalt mixtures containing steel slag aggregates. *Constr. Build. Mater.* **2015**, *98*, 692–702. [CrossRef]
14. Fang, C.; Yu, R.; Liu, S.; Li, Y. Nanomaterials Applied in Asphalt Modification : A Review. *J. Mater. Sci. Technol.* **2013**, *29*, 589–594. [CrossRef]
15. Li, R.; Pei, J.; Sun, C. Effect of nano-ZnO with modified surface on properties of bitumen. *Constr. Build. Mater.* **2015**, *98*, 656–661. [CrossRef]
16. Shu, B.; Wu, S.; Pang, L.; Javilla, B. The utilization of multiple-walled carbon nanotubes in polymer modified bitumen. *Materials (Basel)* **2017**, *10*, 416. [CrossRef] [PubMed]
17. El-Shafie, M.; Ibrahim, I.M.; Abd El Rahman, A.M.M. The addition effects of macro and nano clay on the performance of asphalt binder. *Egypt. J. Pet.* **2012**, *21*, 149–154. [CrossRef]
18. Amin, G.M.; Esmail, A. Application of nano silica to improve self-healing of asphalt mixes. *J. Cent. South Univ.* **2017**, *24*, 1019–1026. [CrossRef]
19. Amirkhanian, A.N.; Xiao, F.; Amirkhanian, S.N. Characterization of unaged asphalt binder modified with carbon nano particles. *Int. J. Pavement Res. Technol.* **2011**, *4*, 281–286. [CrossRef]
20. Santagata, E.; Baglieri, O.; Tsantilis, L.; Dalmazzo, D. Rheological Characterization of Bituminous Binders Modified with Carbon Nanotubes. *Procedia-Soc. Behav. Sci.* **2012**, *53*, 546–555. [CrossRef]
21. Treacy, M.M.J.; Ebbesen, T.W.; Gibson, J.M. Exceptionally high Young's modulus observed for individual carbon nanotubes. *Nature* **1996**, *381*, 678–680. [CrossRef]
22. de Heer, W.A. Nanotubes and the Pursuit of Applications. *MRS Bull.* **2004**, *29*, 281–285. [CrossRef]
23. Bai, J.B.; Allaoui, A. Effect of the length and the aggregate size of MWNTs on the improvement efficiency of the mechanical and electrical properties of nanocomposites - Experimental investigation. *Compos. Part A Appl. Sci. Manuf.* **2003**, *34*, 689–694. [CrossRef]
24. Gong, M.; Yang, J.; Yao, H.; Wang, M.; Niu, X.; Haddock, J.E. Investigating the performance, chemical, and microstructure properties of carbon nanotube-modified asphalt binder. *Road Mater. Pavement Des.* **2017**, *0629*, 1–24. [CrossRef]
25. Galooyak, S.S.; Palassi, M.; Farahani, H.Z.; Goli, A. Effect of carbon nanotube on the rheological properties of bitumen. *Pet. Coal* **2015**, *57*, 556–564.
26. Omar, H.A.; Yusoff, N.I.M.; Sajuri, Z.; Ceylan, H.; Jakarni, F.M.; Ismail, A. Determining the effects of aging on halloysite nano-tube modified binders through the pull-off test method. *Constr. Build. Mater.* **2016**, *126*, 245–252. [CrossRef]
27. Santagata, E.; Baglieri, O.; Tsantilis, L.; Chiappinelli, G. Fatigue properties of bituminous binders reinforced with carbon nanotubes. *Int. J. Pavement Eng.* **2015**, *16*, 80–90. [CrossRef]
28. Faramarzi, M.; Arabani, M.; Haghi, A.K.; Mottaghitalab, V. Carbon nanotubes-modified asphalt binder: Preparation and characterization. *Int. J. Pavement Res. Technol.* **2015**, *8*, 29–37. [CrossRef]
29. Hasan, Z.; Kamran, R.; Mohammad, F.; Ahmad, G.; Hosein, F. Evaluation of Different Conditions on The Mixing Bitumen and Carbon Nano-Tubes. *Int. J. Civ. Environ. Eng. IJCEE-IJENS* **2012**, *12*, 53–59.
30. Tufail, M.; Ahmad, N.; Mirza, S.M.; Mirza, N.M.; Khan, H.A. Natural radioactivity from the building materials used in Islamabad and Rawalpindi, Pakistan. *Sci. Total Environ.* **1992**, *121*, 283–291. [CrossRef]
31. Hussan, S.; Kamal, M.A.; Hafeez, I.; Farooq, D.; Ahmad, N.; Khanzada, S. Statistical evaluation of factors affecting the laboratory rutting susceptibility of asphalt mixtures. *Int. J. Pavement Eng.* **2017**, *8436*, 1–15. [CrossRef]
32. Ziari, H.; Hosein, F.; Goli, A. Using the Statistical Analysis of Carbon Nano-tubes Dispersion in Bitumen Employing Software MINITAB. *Int. J. Transp. Eng.* **2013**, *1*, 125–136.

33. Amin, I.; El-Badawy, S.M.; Breakah, T.; Ibrahim, M.H.Z. Effect of Functionalization and Mixing Process on the Rheological Properties of Asphalt Modified with Carbon Nanotubes. *Am. J. Civ. Eng. Archit.* **2016**, *4*, 90–97. [CrossRef]
34. Collett, B.M. Scanning Electron Microscopy: A Review And Report Of Research In Wood Science. *Wood Fiber Sci.* **1967**, *2*, 113–132.
35. Ameri, M.; Vamegh, M.; Rooholamini, H.; Haddadi, F. Investigating Effects of Nano/SBR Polymer on Rutting Performance of Binder and Asphalt Mixture. *Adv. Mater. Sci. Eng.* **2018**, *2018*. [CrossRef]
36. Qasim, Z.I.; Qasim, Z.I. Effect of Filler Content on Properties of Asphaltic Mixtures for Marshall and Superpave Gyratory Compactor. *Al-Nahrain J. Eng. Sci.* **2017**, *20*, 183–193.
37. Chelovian, A.; Shafabakhsh, G. Laboratory evaluation of Nano Al2O3effect on dynamic performance of stone mastic asphalt. *Int. J. Pavement Res. Technol.* **2017**, *10*, 131–138. [CrossRef]
38. Hussain, M.H.M.; Ghaly, N.F.; Ibrahim, I.M. Modified Hot Mix Asphalt for Road Maintenance. *World Appl. Sci. J.* **2008**, *5*, 236–245.
39. Rafi, J.; Kamal, M.; Ahmad, N.; Hafeez, M.; Faizan ul Haq, M.; Aamara Asif, S.; Shabbir, F.; Bilal Ahmed Zaidi, S. Performance Evaluation of Carbon Black Nano-Particle Reinforced Asphalt Mixture. *Appl. Sci.* **2018**, *8*, 1114. [CrossRef]
40. Atif, R.; Inam, F. Reasons and remedies for the agglomeration of multilayered graphene and carbon nanotubes in polymers. *Beilstein J. Nanotechnol.* **2016**, *7*, 1174–1196. [CrossRef]
41. Faramarzi, M.; Arabani, M.; Haghi, A.K.; Motaghitalab, V. A Study on the Effects of CNT's on Hot Mix Asphalt Marshal-Parameters. In Proceedings of the 7thSASTech 2013, Bandar-Abbas, Iran, 7–8 March 2013; pp. 1–9.
42. Ehinola, O.A.; Falode, O.A.; Jonathan, G. Softening point and Penetration Index of bitumen from parts of Southwestern Nigeria. *NAFTA* **2012**, *63*, 319–323.
43. Al-Omari, A.A.; Khedaywi, T.S.; Khasawneh, M.A. Laboratory characterization of asphalt binders modified with waste vegetable oil using SuperPave specifications. *Int. J. Pavement Res. Technol.* **2018**, *11*, 68–76. [CrossRef]
44. Enieb, M.; Diab, A. Characteristics of asphalt binder and mixture containing nanosilica. *Int. J. Pavement Res. Technol.* **2017**, *10*, 148–157. [CrossRef]
45. Ali, S.I.A.; Ismail, A.; Karim, M.R.; Yusoff, N.I.M.; Al-Mansob, R.A.; Aburkaba, E. Performance evaluation of Al2O3 nanoparticle-modified asphalt binder. *Road Mater. Pavement Des.* **2017**, *18*, 1251–1268. [CrossRef]
46. Mirza, M.W.; Abbas, Z.; Rizvi, M.A. Temperature Zoning of Pakistan for Asphalt Mix Design. *Pakistan J. Eng. Appl. Sci.* **2011**, *8*, 49–60.
47. Moraes, R.; Velasquez, R.; Bahia, H. Measuring the Effect of Moisture on Asphalt-Aggregate Bond with the Bitumen Bond Strength Test. *Transp. Res. Rec. J. Transp. Res. Board* **2011**, *2209*, 70–81. [CrossRef]
48. Kyakuno, H.; Fukasawa, M.; Ichimura, R.; Matsuda, K.; Nakai, Y.; Miyata, Y.; Saito, T.; Maniwa, Y. Diameter-dependent hydrophobicity in carbon nanotubes. *J. Chem. Phys.* **2016**, *145*. [CrossRef]
49. Liu, Y.; Apeagyei, A.; Ahmad, N.; Grenfell, J.; Airey, G. Examination of moisture sensitivity of aggregate-bitumen bonding strength using loose asphalt mixture and physico-chemical surface energy property tests. *Int. J. Pavement Eng.* **2014**, *15*, 657–670. [CrossRef]

*applied sciences*

MDPI

*Article*

# Characterization of Asphalt Mixes Behaviour from Dynamic Tests and Comparison with Conventional Cyclic Tension–Compression Tests

**Jean-Claude Carret *, Hervé Di Benedetto and Cédric Sauzéat**

LTDS (UMR CNRS 5513), University of Lyon/ENTPE, Rue M. Audin, 69518 Vaulx en Velin, France; herve.dibenedetto@entpe.fr (H.D.B.); cedric.sauzeat@entpe.fr (C.S.)

* Correspondence: jean-claude.carret@entpe.fr

Received: 16 October 2018; Accepted: 29 October 2018; Published: 1 November 2018

**Featured Application: Dynamic tests can be used to characterize asphalt mixes and the results obtained from dynamic tests are in good agreement with the results of conventional cyclic tests.**

**Abstract:** In the presented research, conventional cyclic tension–compression tests and dynamic tests were performed on two types of asphalt mixes (AM). For the tension–compression tests, the complex modulus was obtained from the measurements of the axial stress and axial strain. For the dynamic tests, an automated impact hammer equipped with a load cell and an accelerometer were used to obtain the frequency response functions (FRFs) of the specimens at different temperatures. Two methods were proposed to back-calculate the complex modulus from the FRFs at each temperature: one using the 2S2P1D (two springs, two parabolic elements and one dashpot) model and the other considering a constant complex modulus. Then, a 2S2P1D linear viscoelastic model was calibrated to simulate the global linear viscoelastic behaviour back calculated from each of the proposed methods of analysis for the dynamic tests, and obtained from the tension–compression test results. The two methods of analysis of dynamic tests gave similar results. Calibrations from the tension–compression and dynamic tests also show an overall good agreement. However, the dynamic tests back analysis gave a slightly higher value of the norm of the complex modulus and a lower value of the phase angle compared to the tension–compression test data. This result may be explained by the nonlinearity of AM (strain amplitude is at least 100 times smaller for dynamic tests) and/or by ageing of the materials during the period between the tension–compression and the dynamic tests.

**Keywords:** asphalt mixes; linear viscoelasticity; complex modulus; dynamic measurements; tension–compression tests; frequency response function; back-analysis; finite element method

## 1. Introduction

Asphalt mixes (AM) have a linear viscoelastic (LVE) behaviour in the small strain domain [1] Cyclic tension–compression tests are traditionally used to determine the LVE properties of AM that are strongly dependent on frequency and temperature. However, these tests require expensive experimental devices such as hydraulic presses and are not applicable in situ. An economical alternative is to use non-destructive dynamic tests that are simple to perform and possibly adaptable for measurements on pavement structures. Impulse techniques using impact loadings [2,3] are known to provide accurate characterization of material properties in the case of elastic materials [4,5]. In the case of LVE materials, dynamic tests could be a great alternative to conventional cyclic tension–compression tests. Dynamic tests using wave propagation and measurement of the flying time [6–8] have been applied to LVE materials. Resonance testing considering only the fundamental resonance frequency [9–11] or resonant acoustic spectroscopy (RAS) [12–15] have also been applied to AM but it is not possible to describe accurately the frequency dependency behaviour of AM with these different tests. Recently, measurement of frequency response functions (FRFs) have been performed on LVE materials [16,17] and more specifically on AM [18–21]. Gudmarsson et al. [19,20] and Carret et al. [21] showed that using FRFs measurements to derive the LVE properties of AM is a very promising approach. However, characterizing accurately the LVE behaviour of AM from FRFs is not possible through a simplified analysis [22] and it requires an elaborate approach. In this paper, two different methods using finite element calculations are proposed to obtain the LVE behaviour from FRFs. The first method consists in an optimization of the continuous spectrum 2S2P1D (two springs, two parabolic elements and one dashpot) model constants to back-calculate the complex modulus at each tested temperature while the second method is a more direct back-calculation of the complex modulus at the first resonance frequency. The two methods were applied to two different types of AM representing five specimens that were tested with cyclic tension–compression tests and with dynamic tests. Experimental complex modulus values obtained from tension–compression tests and back-calculated from FRFs with the two proposed methods were used to fit the 2S2P1D model and the Williams-Landel-Ferry (WLF) constants simulating the global LVE behaviour of the material in each case. First, the materials tested in this study are presented. Then, the LVE behaviour characterization with cyclic tests and the modelling with the 2S2P1D model are introduced. Next, dynamic tests are introduced and the two proposed back-analysis methods are explained. Finally, data from tension–compression tests are compared with results from the two methods of back-analysis of the dynamic tests.

## 2. Materials and Methods

Two different types of AM are considered in this paper. The first material is a warm mix that was fabricated in laboratory using bitumen foam and labelled WF for warm foam. It contains 70% of reclaimed asphalt pavement (RAP) after one cycle of recycling. This material was used in a project from the French national research agency called IMPROVMURE [23]. Three specimens of this material were tested with tension–compression and dynamic tests. The second material is a mix with an optimized granular skeleton also fabricated in laboratory and labelled GB5. It contains 30% of RAP and the bitumen used is a polymer modified bitumen (PMB). Two specimens of this material were tested with tension–compression tests and dynamic tests. Table 1 gives some indications on the five studied specimens. The tension–compression tests were performed first and the dimensions listed in Table 1 correspond to the dimensions after the specimens were cut (see Section 3.1) before performing the dynamic tests.

**Table 1.** Specimens used in this study.

| Specimen | Mass (g) | Height (mm) | Diameter (mm) | Density (kg/m$^3$) | Void Ratio (%) | Bitumen Content (%) | RAP Content (%) |
|---|---|---|---|---|---|---|---|
| WF-4 | 1293 | 0.123 | 75 | 2379 | 6.6 | 5.4 | 70 |
| WF-6 | 1320 | 0.123 | 75 | 2431 | 4.2 | 5.4 | 70 |
| WF-8 | 1330 | 0.123 | 75 | 2449 | 3.8 | 5.4 | 70 |
| GB5-3 | 941 | 0.152 | 64 | 2381 | 4.8 | 4.8 | 30 |
| GB5-4 | 951 | 0.152 | 64 | 2378 | 5.1 | 4.8 | 30 |

## 3. Characterization of the Linear Viscoelastic (LVE) Behaviour

### 3.1. Cyclic Tension–Compression Tests

Cyclic tension–compression tests were first performed to determine the complex modulus and complex Poisson's ratio of the five considered cylindrical specimens. A hydraulic press was used in strain-controlled mode to apply cyclic sinusoidal axial loadings with an amplitude of around 50 μm/m. The axial stress $\sigma_Z$ was measured with a load cell, the axial strain $\varepsilon_Z$ was obtained from the average of three extensometers placed at 120° from each other, and the radial strain $\varepsilon_r$ was derived from the measurements of two non-contact sensors. The procedure developed at ENTPE/University of Lyon laboratory is detailed in other publications [24–26]. The complex notation of the axial stress, the axial strain and the radial strain are given in Equation (1):

$$\begin{aligned} \sigma_z^* &= \sigma_{0z} \cdot e^{j\omega t} \\ \varepsilon_z^* &= \varepsilon_{0z} \cdot e^{j(\omega t + \varphi_{\varepsilon z})} \\ \varepsilon_r^* &= \varepsilon_{0r} \cdot e^{j(\omega t + \varphi_{\varepsilon r})} \end{aligned} \tag{1}$$

where $\omega$ is the pulsation ($\omega = 2\pi f$, where f is the frequency), $\sigma_{0z}$ is the norm of the complex axial stress and $\varepsilon_{0z}$ and $\varphi_{\varepsilon z}$ (respectively, $\varepsilon_{0r}$ and $\varphi_{\varepsilon r}$) are the norm and phase angle of the complex axial strain (respectively, complex radial strain). The tension–compression tests were performed at eight loading frequencies (0.003, 0.01, 0.03, 0.1, 0.3, 1, 3 and 10 Hz) and nine temperatures from −25 °C to 55 °C in steps of 10 °C. Details of the experimental set up are shown in Figure 1. The complex modulus (respectively, complex Poisson's ratio) are defined as the ratio between the axial stress and the axial strain (respectively, the opposite of the radial strain and the axial strain) and they are calculated at each temperature and frequency as follow:

$$E^* = \frac{\sigma_z^*}{\varepsilon_z^*} = |E^*| \cdot e^{j\varphi_E} \tag{2}$$

$$\nu^* = -\frac{\varepsilon_r^*}{\varepsilon_z^*} = |\nu^*| \cdot e^{j\varphi_\nu} \tag{3}$$

where $E^*$ is the complex modulus, $\varphi_E$ is the phase angle of the complex modulus, $\nu^*$ is the complex Poisson's ratio and $\varphi_\nu$ is the phase angle of the complex Poisson's ratio.

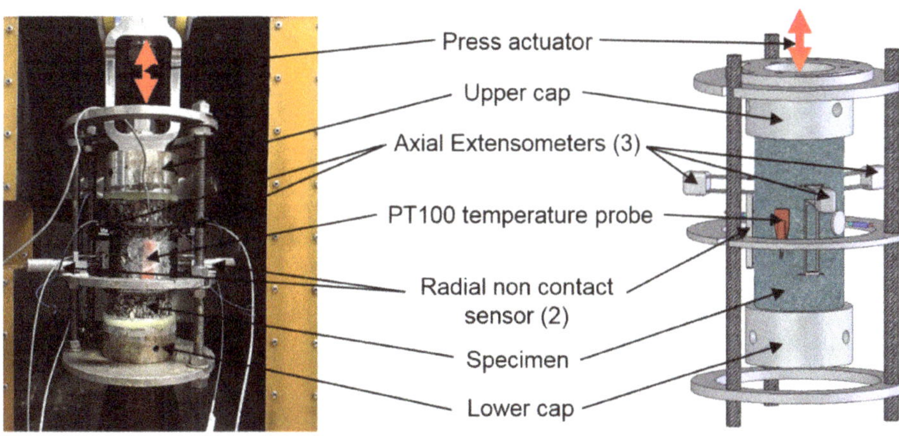


**Figure 1.** Tension–compression test set-up (ENTPE laboratory, University of Lyon).

### 3.2. Modelling of the LVE Behaviour: 2S2P1D Rheological Model

The continuous spectrum 2S2P1D model developed at ENTPE [27–29] was used to model the LVE behaviour of AM. This model is the association in series of two springs, two parabolic creep elements and one dashpot. In the three-dimension case [30], the expressions of the complex modulus and the complex Poisson's ratio, for isotropic behaviour, are given at a given reference temperature (Tref), by Equations (4) and (5), respectively.

$$E^*_{2S2P1D}(\omega) = E_{00} + \frac{E_0 - E_{00}}{1 + \delta(j\omega\tau_E)^{-k} + (j\omega\tau_E)^{-h} + (j\omega\beta\tau_E)^{-1}} \quad (4)$$

$$\nu^*_{2S2P1D}(\omega) = \nu_{00} + \frac{\nu_0 - \nu_{00}}{1 + \delta(j\omega\tau_\nu)^{-k} + (j\omega\tau_\nu)^{-h} + (j\omega\beta\tau_\nu)^{-1}} \quad (5)$$

where $\omega$ is the pulsation ($\omega = 2\pi f$, where f is the frequency), $E_0$ and $\nu_0$ are the high frequency modulus and Poisson's ratio, $E_{00}$ and $\nu_{00}$ are the low frequency modulus and Poisson's ratio, k and h are dimensionless constants such as $0 < k < h < 1$, $\delta$ is a dimensionless constant, and $\beta$ is a dimensionless constant related to Newtonian viscosity $\eta$ by $\eta = (E_0 - E_{00})\,\beta\tau_E$. $\tau_E$ and $\tau_\nu$ are characteristic time constants of the complex modulus and Poisson's ratio linked by a constant ratio. The values of the characteristic times vary only with temperature. The time temperature superposition principle (TTSP) is verified for asphalt mixes in the linear and nonlinear domains [31–33] so it is possible to calculate the characteristic time at any given temperature using Equation (6):

$$\tau(T) = a_T(T)\tau_{ref} \quad (6)$$

where $\tau_{ref}$ is the characteristic time at the reference temperature ($\tau_E$ or $\tau_\nu$) and $a_T$ is the shift factor at the temperature T defined by the Williams–Landel–Ferry (WLF) equation [34]:

$$\log(a_T) = -\frac{C_1(T - T_{ref})}{C_2 + T - T_{ref}} \quad (7)$$

where $C_1$ and $C_2$ are the two constants of the WLF equation and $T_{ref}$ is the reference temperature.

## 4. Dynamic Tests

### 4.1. Measurement of the Frequency Response Functions (FRFs)

First, the specimens used for complex modulus tension–compression test were sawed to separate the glued upper and lower metallic caps before performing the dynamic measurements. An impact hammer equipped with a load cell (PCB model 086E80) was used as an external source of excitation. The order of magnitude of the maximum strain induced in the specimen by the impact is of about 0.1 μm/m [18,21]. The impact hammer was automated with a solenoid piston programmed with a microcontroller (Arduino Uno R3) to improve the repeatability of the test and to allow measurements directly inside a thermal chamber. This automated system was inspired by systems previously developed by Norman et al. in 2012 and Brüggemann et al. in 2015 [35,36]. The response of the materials was recorded with an accelerometer (PCB model 353B15). The impact hammer and the accelerometer were connected to a signal conditioner (PCB model 482C15) and the signal conditioner was connected to a data acquisition device (NI USB-6356) connected to a computer. To achieve free boundary conditions, soft foam was placed under the specimens during the tests. In this study, only the longitudinal compression mode of vibrations was considered. For this mode of vibrations, the impact is applied in the centre of one short side of the cylinder while the acceleration is measured in the centre of the opposite short side. The experimental set up for the dynamic tests corresponding to the longitudinal mode of vibrations is presented in Figure 2.

The measurements were recorded with a sampling frequency of 1 MHz by using a MATLAB application which was specifically developed for this test. Measurements were performed at five temperatures (−20, 0, 15, 35 and 50 °C) and five impacts were applied at each temperature. The applied force and the acceleration were recorded for each impact. The experimental data in time domain were then converted in frequency domain with a 1 Hz resolution using the Fast Fourier Transform (FFT). Figure 3 shows an example of the signals in time and frequency domains for specimen GB5-3 at 14.7 °C (measured temperature with a probe at the surface of the specimen).

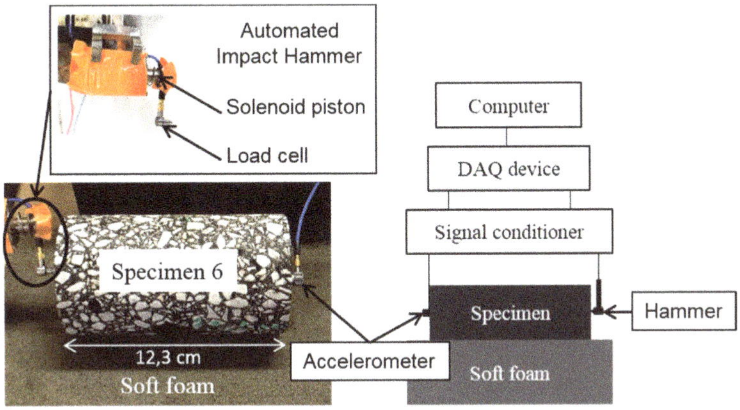


**Figure 2.** Test set up for the dynamic impact tests (example for specimen WF-6, ENTPE laboratory, University of Lyon).

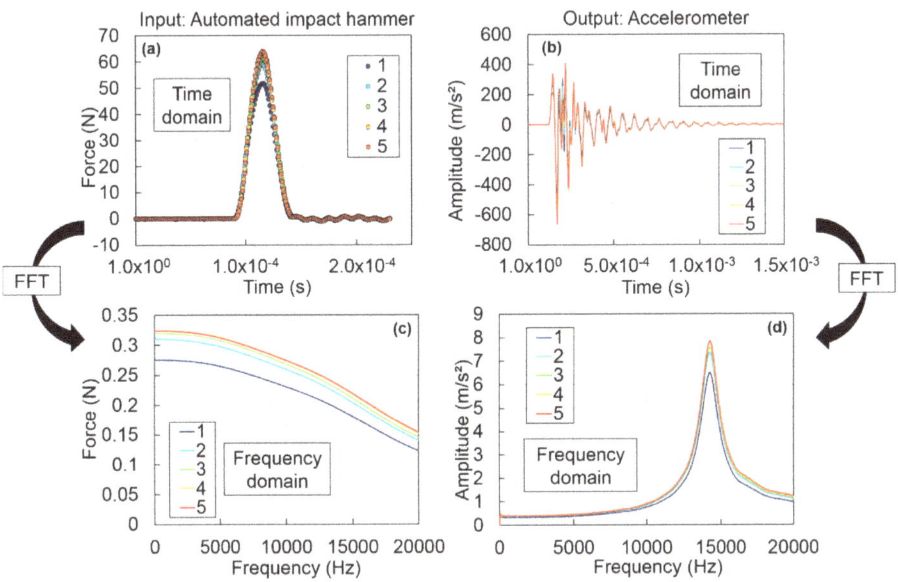


**Figure 3.** Dynamic test experimental data for specimen GB5-3 at 14.7 °C (5 hits): (**a**) force in time domain; (**b**) acceleration in time domain; (**c**) force in frequency domain; and (**d**) acceleration in frequency domain.

As shown in Figure 3, the frequency spectrum of the impact contains energy up to 20 kHz, which is the maximum frequency considered for the calculations of frequency response functions (FRFs) in this study. FRFs were calculated from the frequency domain signals as follow:

$$H(f) = \left(\overline{Y(f) \cdot X*(f)}\right) / \left(\overline{X(f) \cdot X*(f)}\right) \quad (8)$$

where H is the FRF, Y is the FFT of the measured acceleration, X is the FFT of the applied force, X* is the complex conjugate of the applied force and the bar above corresponds to the arithmetic average from the five impacts. The five FRFs corresponding to each of the five impacts and the averaged FRF (Equation (8)) for specimen GB5-3 at 14.7 °C are displayed on Figure 4. Figure 4 shows that the six FRFs overlaps, which confirms the very good repeatability of the test.

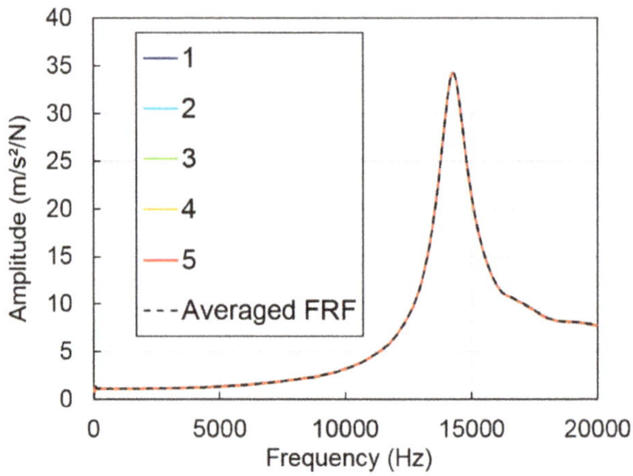


**Figure 4.** FRFs obtained for the 5 hits and averaged FRF (Equation (8)) of specimen GB5-3 at 14.7 °C.

Quality of the measurements was also checked with the coherence function. The coherence is a value between 0 and 1 that indicates how much of the vibratory response recorded with the accelerometer is due to the impact. For a value of 1, the response is fully explained by the impact while decreasing values indicate that something has disrupted the test. Coherence function is calculated according to Equation (9):

$$\mathrm{CF(f)} = \left(\overline{\mathrm{X} * (\mathrm{f})\cdot \mathrm{Y}(f)}\right)^2 / \left(\left(\overline{\mathrm{X(f)}\cdot \mathrm{X} * (\mathrm{f})}\right)\cdot\left(\overline{\mathrm{Y(f)}\cdot \mathrm{Y} * (\mathrm{f})}\right)\right) \quad (9)$$

where CF is the coherence function, Y and Y* are the FFT of the measured acceleration and its complex conjugate, X and X* are the FFT of the applied force and its complex conjugate and the bar above corresponds to the arithmetic average. The coherence functions of specimen GB5-3 for the five tested temperatures are presented in Figure 5. For all temperatures, the coherence function is very good with values close to one for frequencies higher than 1000 Hz. It is therefore recommended to not use the frequencies below 1000 Hz.

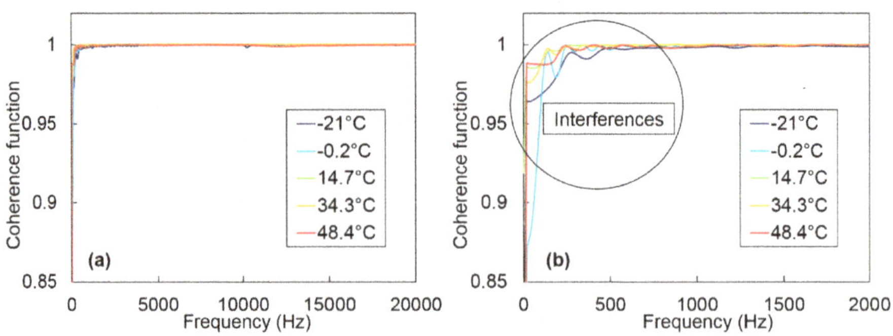


**Figure 5.** Coherence functions obtained for specimen GB5-3 at the five tested temperatures (**a**); and zoom on the lower frequencies (**b**).

### 4.2. Calculation of FRFs with the Finite Element Method (FEM)

Numerical FRFs were calculated with the finite element method (FEM) considering linear viscoelastic behaviour and the dynamic test boundary conditions. Figure 6 shows the FEM mesh and boundary conditions used for the FEM calculation of the FRFS.

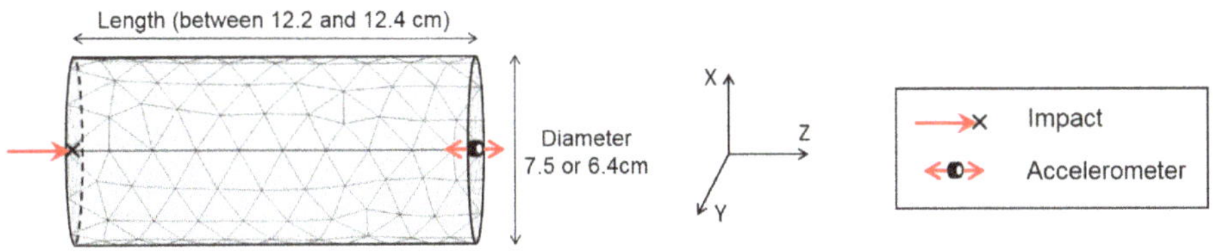


**Figure 6.** Finite element mesh and boundary conditions used for the FEM calculation of the FRFS.

FRFs were calculated at the desired frequencies by resolving the following three-dimensional equation of motion in frequency domain:

$$-\rho\omega^2\mathbf{u} - \nabla\cdot\sigma = 0 \tag{10}$$

where $\rho$ is the bulk density of the material, $\omega$ is the angular frequency, $\mathbf{u}$ is the displacement vector, $\nabla$ is the gradient tensor operator and $\sigma$ is the Cauchy stress tensor. Free boundary conditions are assumed to solve Equation (10) except at the impact point where a cyclic load $e^{i\omega t}$ is applied in the direction of the impact. Since the load amplitude is unity, the calculated FRFs correspond to the calculated acceleration in direction Z (direction of vibration of the accelerometer in physical tests). Back analysis was performed considering two LVE behaviour models successively, as explained below.

### 4.3. Determination of the Material LVE Properties from Dynamic Tests

The LVE properties of the material were determined from the FRFs measured with the dynamic tests in two steps. The first step is a back-calculation of the complex modulus of the material at each tested temperature. Two methods, presented in the next sections, were used for this purpose. They consist in optimizing the constants of the LVE model used to calculate FRFs so that calculated FRFs match the experimental measured FRFs using dynamic tests at the considered temperature. The second step, which is the same for the two proposed methods, consists in using the complex modulus values determined in the first step at each temperature to fit a 2S2P1D model and a WLF law simulating the global LVE behaviour of the material. This operation is similar to what is done with the tension–compression test data.

#### 4.3.1. First Method: Optimization of the 2S2P1D Model Constants to Match Experimental FRFs

Among the 10 constants of the 2S2P1D model, only four constants ($E_0$, k, $\delta$ and $\tau_E$) have a significant influence for the considered range of frequencies involved during dynamic tests and need to be optimized. The complex Poisson's ratio has a very small influence on the calculation of the FRFs below 20 kHz. Poisson's ratio cannot be back-calculated with this procedure; however, it is necessary to assume values for constants $\nu_{00}$, $\nu_0$ and $\tau_\nu$ to back-calculate the complex modulus at each temperature. For each tested temperature, the vector X of the four constants ($E_0$, k, $\delta$ and $\tau_E$) to be identified was optimized iteratively so that the calculated FRFs match the experimental FRFs. The values of the six other constants were fixed to classical values for AM: $E_{00}$ = 100 MPa, $\nu_0$ = 0.19, $\nu_{00}$ = 0.45, h = 0.53, $\beta$ = 250 and $\tau_\nu$ = 31.6$\tau_E$. Only the values of the experimental FRFs at frequencies around the resonance frequencies are used as input in the optimization according to previous studies [19–21] that showed their meaningful importance. Ten frequencies were selected along each resonance peak and the error function to minimize was defined as follow:

$$\text{Error} = \sum_{j=1}^{Npeaks} \sum_{i=1}^{10} \left( \left| \frac{\left| H_{Exp_{ji}} \right| - \left| H_{C_{ji}} \right|}{\left| H_{Exp_{ji}} \right|} \right| \right) \quad (11)$$

where $H_{Exp}$ is the experimental FRF, $H_C$ is the calculated FRF, $N_{peaks}$ is the number of resonance peaks, j is the index of the peak and i is the index of the frequencies. The number of peaks considered for the optimization at each temperature corresponds to the number of peaks that are visible below 20 kHz (maximum considered frequency with our experimental device). This number is given in Table 2 for each temperature.

**Table 2.** Number of peaks considered for the optimization of the 2S2P1D model constants at each temperature.

| Temperature (°C) | −20 | 0 | 15 | 35 | 50 |
|---|---|---|---|---|---|
| Number of peaks | 1 | 1 | 1 | 1 | 2 |

The optimization was performed in MATLAB with the "fminsearch" algorithm and the optimization was stopped when the error and the parameter tolerance of 1% is reached (e.g., when the variation of the error and of all the values of the four constants to be identified is less than 1% between two iterations of the algorithm). The final vector $X_f$ of the four constants ($E_0$, k, $\delta$ and $\tau_E$) was then used with the three fixed constants related to the complex modulus ($E_{00}$, h, and $\beta$) to back-calculate the complex modulus at the resonance frequencies of the peaks used as input for the optimization at the considered temperature. The optimization procedure to identify the four 2S2P1D model constants ($E_0$, k, $\delta$ and $\tau_E$) at each temperature is explained in Figure 7.

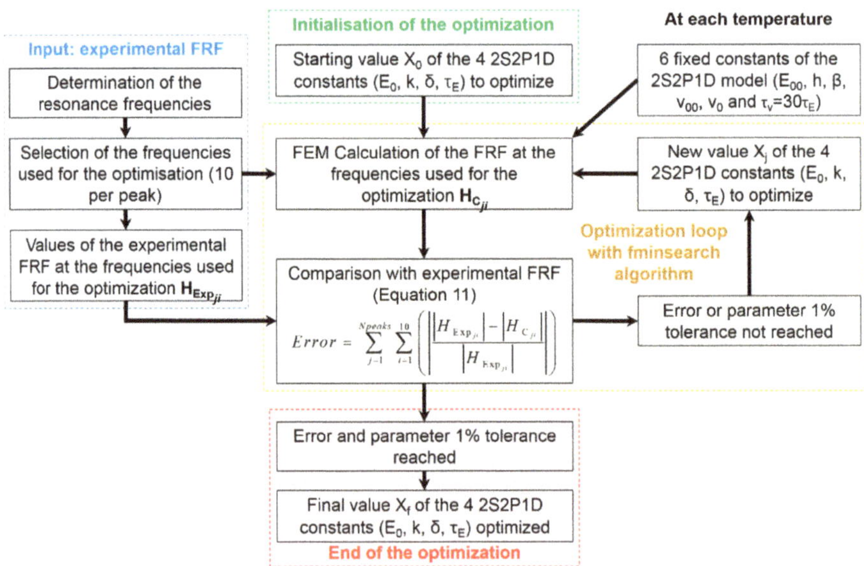


**Figure 7.** Method 1: Principle of the optimization procedure to identify the four 2S2P1D constants ($E_0$, k, $\delta$ and $\tau_E$) at each temperature. $X_f$ is the final vector of the four 2S2P1D constants to identify.

4.3.2. Second Method: Constant Complex Modulus Obtained from the First Resonance Peak Only

The second method is a simplified approach that does not require, in the first step, a rheological LVE model considering the frequency and temperature dependence. At each tested temperature, a constant complex modulus value and a constant real Poisson's ratio of 0.3 were considered. A numerical sensitivity analysis was performed to evaluate the influence of the norm and phase angle of the complex modulus and the real value of the Poisson's ratio on the calculation of FRFs. The influence of each LVE constant was evaluated from FRFs calculated for five values taken in the range of variation of the considered LVE constant while the two others are fixed. Table 2 lists the five values considered for each LVE constant and the corresponding fixed values of the two others. Some results are shown on Figure 8 for a cylinder with similar dimensions than those used in this study and with a density of 2400 kg/m$^3$. It is shown in Figure 8 that the norm of the complex modulus has an important influence on the first resonance frequency but not on the amplitude. It is the reverse for the phase of the complex modulus, while Poisson's ratio has little influence on both the frequency and the amplitude.

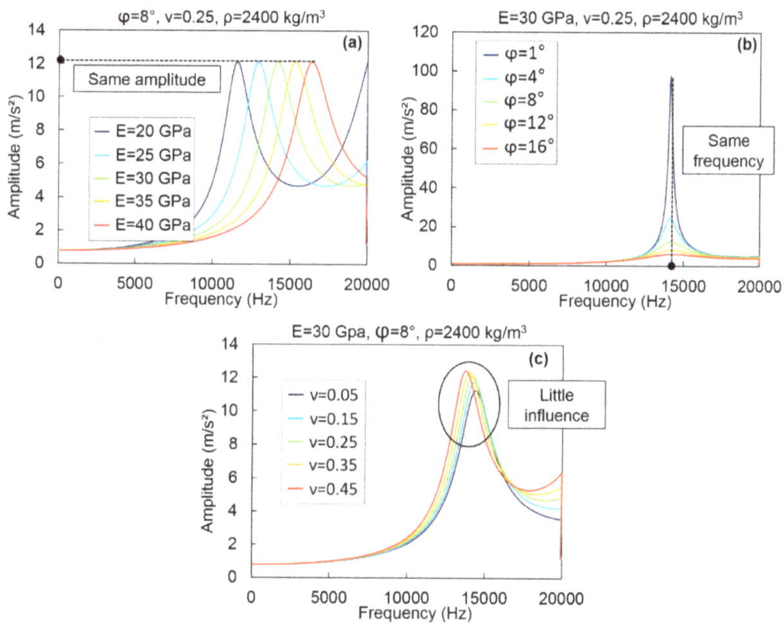


**Figure 8.** Influence of: E (**a**); φ (**b**); and ν (**c**) on the first peak of the FRFs corresponding to the first resonance (example of calculations for a cylinder with a 7.5 cm diameter and a 12.3 cm length).

To confirm the observation raised in Figure 8, the relative standard deviations (RSD) for the first resonance frequency and amplitude were calculated for the three studied constants. Results are given in Table 3. They confirm the previous observations and also indicate that the norm of the complex modulus has really no impact on the peak amplitude while the phase has a very little influence on the frequency. The Influence of the Poisson's ratio can be considered as negligible when compared with the influence of the two other constants. Observations presented in Figure 8 and Table 3 justify the assumption of a constant real value of 0.3 for the Poisson's ratio. In addition, the identification of the norm and phase of the complex modulus can be separated into two steps: the norm can be determined by dichotomy from the first resonance frequency and then the phase can be determined by dichotomy from the corresponding amplitude. This process was repeated iteratively until the error on the amplitude is less than 0.1%. Figure 9 shows the principle of the back-calculation of the complex modulus using the first resonance peak of the FRFs.

**Table 3.** Influence of E, φ and ν on the first resonance frequency and FRF amplitude. When one of the constants varies, the two other constants are fixed at the values listed in the left column.

| | **E (GPa)** | **20** | **25** | **30** | **35** | **40** | **RSD (%)** |
|---|---|---|---|---|---|---|---|
| **φ = 8°**<br>**ν = 0.25** | f (Hz)<br>Amplitude (m/s²) | 11,580<br>12.1 | 12,940<br>12.1 | 14,180<br>12.1 | 15,320<br>12.1 | 16,380<br>12.1 | 13.5<br>$6 \times 10^{-4}$ |
| | **φ (°)** | **1** | **4** | **8** | **12** | **16** | **RSD (%)** |
| **E = 30 GPa**<br>**ν = 0.25** | f (Hz)<br>Amplitude (m/s²) | 14,120<br>97.7 | 14,140<br>24.4 | 14,180<br>12.1 | 14,240<br>8.0 | 14,340<br>5.9 | 0.6<br>130.8 |
| | **ν** | **0.05** | **0.15** | **0.25** | **0.35** | **0.45** | **RSD (%)** |
| **E = 30 GPa**<br>**φ = 8°** | f (Hz)<br>Amplitude (m/s²) | 14,400<br>11.2 | 14,320<br>11.8 | 14,180<br>12.1 | 13,980<br>12.3 | 13,760<br>12.4 | 1.8<br>4.0 |

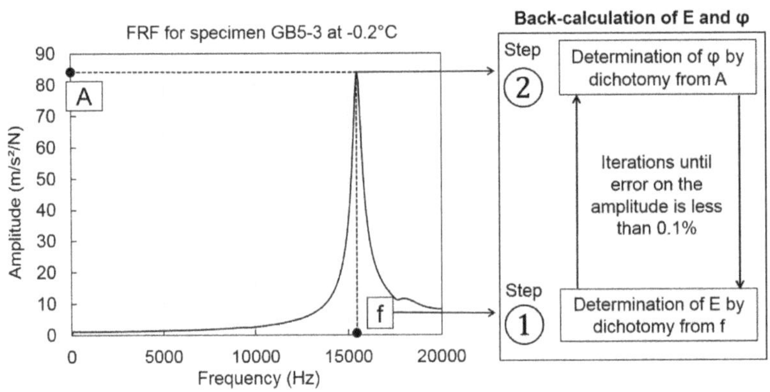


**Figure 9.** Method 2: Principle of the back-calculation of the complex modulus on the first resonance peak at each temperature (example for specimen GB5-3 at −0.2 °C).

#### 4.3.3. Summary and Remarks on the Two Methods

Differences between the two proposed methods concern the first step in which the complex modulus is back-calculated at each temperature, while the second step is identical for the two methods. First, the assumptions on the Poisson's ratio value are different in the two methods. In the first method, the Poisson's ratio is a complex number, which depends on the frequency and on the temperature, and is modelled with the 2S2P1D model assuming the values of constants $\nu_{00}$, $\nu_0$ and $\tau_E/\tau_\nu$. In the second method, a constant real value of Poisson's ratio equal to 0.3 is assumed. Another difference is that all resonance peaks under 20 kHz are considered in the first method while only the first resonance peak is used in the second method. Consequently, the second method gives only one value of the complex modulus at each temperature while the first method gives values for each resonance frequency. This difference is not essential in this study since 50 °C is the only temperature for which two peaks were observed. However, it can be interesting to evaluate more than one value of the complex modulus at each temperature. Finally, in the first method, four constants are evaluated at each temperature using a complex algorithm. In the second method, only two constants are evaluated using a simple dichotomy process. The second method is therefore very easy to apply and time-effective compared to the first method. Figure 10 highlights the main differences between the first step of the two methods and gives the principle of the second step that is identical for the two methods. Note that, even though the same constant h is fixed for the back-calculation at each temperature (in the first step of first method), a different constant h value can be obtained during the calibration process of the second step.

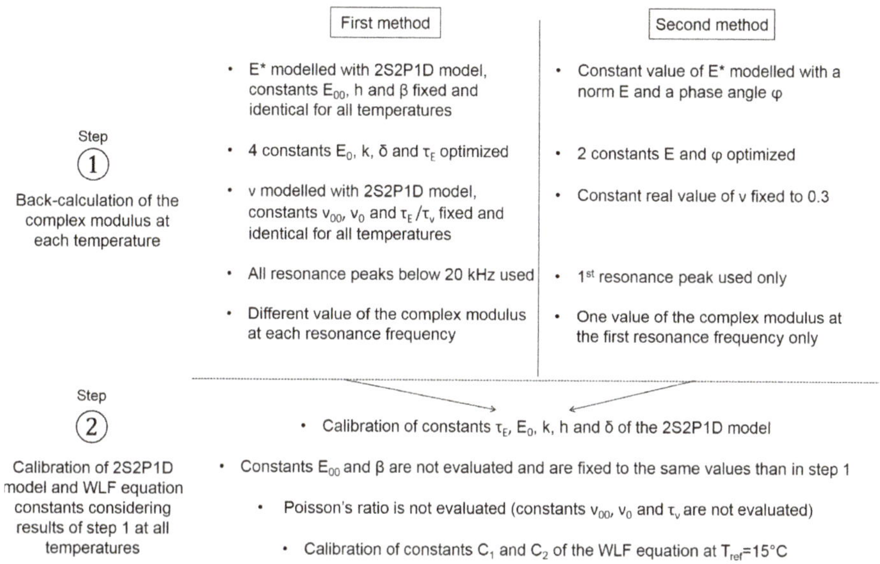


**Figure 10.** Summary and differences between the two proposed methods of back-analysis of the dynamic tests.

## 5. Results

### *5.1. Tension–Compression Tests Results*

Results of the tension–compression tests for specimen WF-8 are plotted in Figure 11. A continuous curve can be seen on the Cole–Cole diagram, which indicates that the material is rheologically simple and that the time–temperature superposition principle (TTSP) is valid. The master curve of the norm of the complex modulus is plotted at a reference temperature ($T_{ref}$) of 15 °C in Figure 11. The 2S2P1D model was fitted to the experimental data and is also plotted in Figure 11. A very good agreement between the experimental data and the 2S2P1D model can be observed.

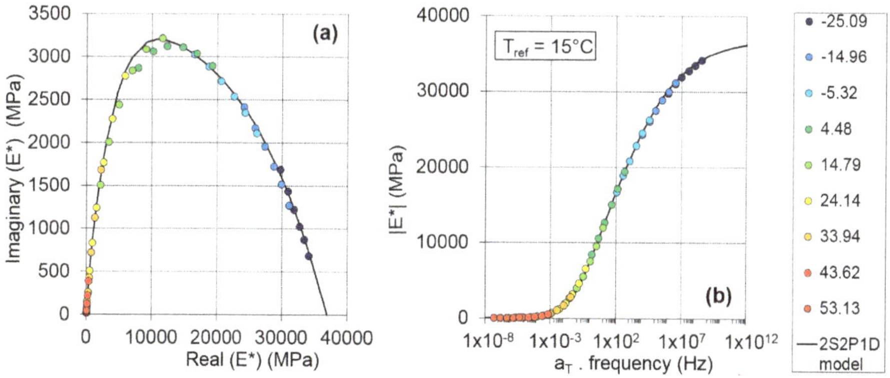


**Figure 11.** Tension–compression test results and fitted 2S2P1D model for specimen WF-8: (**a**) Cole–Cole diagram; and (**b**) master curve for the norm of the complex modulus at 15 °C.

The values of the 2S2P1D model and WLF equation constants obtained from the tension–compression tests are given for all specimens in Table 4. The constants k, h, δ, β, $C_1$ and $C_2$ have the same values for a given material because they are only depending on the bitumen and not on the granular skeleton as shown in previous research [27,37,38].

**Table 4.** Calibrated 2S2P1D constants to match tension–compression measurements.

| Specimen | $E_{00}$ (MPa) | $E_0$ (MPa) | δ | k | h | β | $\tau_{E15°C}$ (s) | $C_1$ | $C_2$ |
|---|---|---|---|---|---|---|---|---|---|
| WF-4 | 28 | 33,400 | 2.28 | 0.177 | 0.57 | 154 | $5.4 \times 10^{-2}$ | 24.9 | 166.6 |
| WF-6 | 40 | 37,500 | 2.28 | 0.177 | 0.57 | 154 | $7.0 \times 10^{-2}$ | 24.9 | 166.6 |
| WF-8 | 56 | 36,900 | 2.28 | 0.177 | 0.57 | 154 | $6.9 \times 10^{-2}$ | 24.9 | 166.6 |
| GB5-3 | 65 | 39,100 | 1.80 | 0.180 | 0.60 | 350 | $7.5 \times 10^{-2}$ | 24.7 | 165.9 |
| GB5-4 | 65 | 39,500 | 1.80 | 0.180 | 0.60 | 350 | $1.5 \times 10^{-1}$ | 24.7 | 165.9 |

The measurements were recorded with a sampling frequency of 1 MHz by using a MATLAB application which was specifically developed for this test. Measurements were performed at five temperatures (−20, 0, 15, 35 and 50 °C) and five impacts were applied at each temperature. The applied force and the acceleration were recorded for each impact. The experimental data in time domain were then converted in frequency domain with a 1 Hz resolution using the Fast Fourier Transform (FFT). Figure 3 shows an example of the signals in time and frequency domains for specimen GB5-3 at 14.7 °C (measured temperature with a probe at the surface of the specimen).

### 5.2. Dynamic Impact Tests Results

The complex modulus values back-calculated at each temperature from the FRFs measurements with the two proposed methods for specimen WF-8 are plotted, as an example, in Figure 12. For the first method of back-calculation, one value of the complex modulus is plotted for temperatures at which only one peak was considered for the optimization (−20, 0, 15 and 35 °C) and two values are plotted at −50 °C because two peaks exist for this temperature. Only one value of the complex modulus is presented for each temperature for Method 2 because the back-calculation is limited to the first resonance frequency, as explained previously. The master curve of the norm of the complex modulus at 15 °C was obtained considering the validity of the TTSP. The 2S2P1D model was fitted to the back-calculated modulus for both methods and the two resulting 2S2P1D model curves are also plotted in Figure 12. Good fitting of the 2S2P1D curves can be observed for both methods, which give only slightly different results.

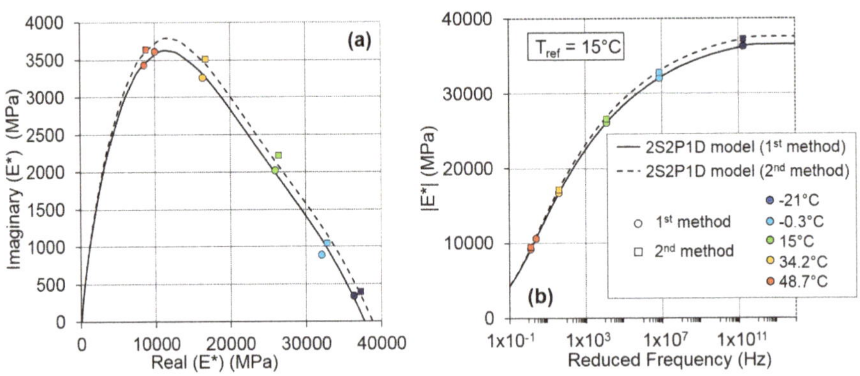


**Figure 12.** Dynamic test results and fitted 2S2P1D model curve for the two analysis methods for specimen WF-8: (**a**) Cole–Cole diagram; and (**b**) master curve for the norm of the complex modulus at 15 °C.

The values of the 2S2P1D model and WLF equation constants fitting the results from the two back-analysis methods are given for all specimens in Table 5 (first method) and Table 6 (second method). Note that similarly to the results of the tension–compression tests, constants k, h, δ, β, $C_1$ and $C_2$ have the same values for a given material. In addition, the same WLF equation constants can be used for the two methods and only constant $E_0$ differs between the first and the second method.

**Table 5.** Calibrated 2S2P1D constants to match complex modulus back-calculated from dynamic tests with the first method using 2S2P1D model at each temperature. Constants $E_{00}$ and β are assumed to be 100 MPa and 250, respectively.

| Specimen | $E_{00}$ (MPa) | $E_0$ (MPa) | δ | k | h | β | $\tau_{E15^\circ C}$ (s) | $C_1$ | $C_2$ |
|---|---|---|---|---|---|---|---|---|---|
| WF-4 | 100 | 34,800 | 1.39 | 0.142 | 0.49 | 250 | $4.0 \times 10^{-2}$ | 18.9 | 133.2 |
| WF-6 | 100 | 38,700 | 1.39 | 0.142 | 0.49 | 250 | $7.0 \times 10^{-2}$ | 18.9 | 133.2 |
| WF-8 | 100 | 37,500 | 1.39 | 0.142 | 0.49 | 250 | $4.0 \times 10^{-2}$ | 18.9 | 133.2 |
| GB5-3 | 100 | 39,100 | 1.17 | 0.130 | 0.442 | 250 | $5.5 \times 10^{-2}$ | 19.2 | 139.5 |
| GB5-4 | 100 | 40,500 | 1.17 | 0.130 | 0.442 | 250 | $7.0 \times 10^{-2}$ | 19.2 | 139.5 |

**Table 6.** Calibrated 2S2P1D constants to match complex modulus back-calculated from dynamic tests measurements with the second method using the first resonance peak at each temperature. Constants $E_{00}$ and β are assumed to be 100 MPa and 250, respectively.

| Specimen | $E_{00}$ (MPa) | $E_0$ (MPa) | δ | k | h | β | $\tau_{E15^\circ C}$ (s) | $C_1$ | $C_2$ |
|---|---|---|---|---|---|---|---|---|---|
| WF-4 | 100 | 36,100 | 1.39 | 0.142 | 0.49 | 250 | $4.0 \times 10^{-2}$ | 18.9 | 133.2 |
| WF-6 | 100 | 39,500 | 1.39 | 0.142 | 0.49 | 250 | $7.0 \times 10^{-2}$ | 18.9 | 133.2 |
| WF-8 | 100 | 38,500 | 1.39 | 0.142 | 0.49 | 250 | $4.0 \times 10^{-2}$ | 18.9 | 133.2 |
| GB5-3 | 100 | 39,800 | 1.17 | 0.130 | 0.442 | 250 | $5.5 \times 10^{-2}$ | 19.2 | 139.5 |
| GB5-4 | 100 | 41,100 | 1.17 | 0.130 | 0.442 | 250 | $7.0 \times 10^{-2}$ | 19.2 | 139.5 |
| GB5-4 | 100 | 40,500 | 1.17 | 0.130 | 0.442 | 250 | $7.0 \times 10^{-2}$ | 19.2 | 139.5 |

It must be highlighted that values of constants $E_{00}$ and β are assumed because they have no influence on the complex modulus values in the frequency range involved during the dynamic tests. The constants governing the value of the Poisson's ratio $\nu_{00}$, $\nu_0$ and $\tau_\nu$ do not appear in Tables 5 and 6 because the Poisson's ratio was not evaluated from the dynamic tests but assumptions on the values of these constants were necessary to back-calculate the complex modulus at each temperature in the first step of the first method. A good proximity between the results obtained with the two methods is shown in Figure 12. The same observation was made for all specimens. To validate this visual impression, the relative difference between the norm (in %) and the phase (in °) of the 2S2P1D simulated complex modulus obtained from the two methods are plotted in Figure 13. In this figure, the relative difference is plotted against the reduced frequency at 15 °C for all specimens. It is seen that the second method considering a constant complex modulus at each temperature and a constant real Poisson's ratio of 0.3 gives a norm slightly higher than the first method with a maximum relative difference of about 3.7%. The phase angle determined with the two methods can be considered equivalent with less than 0.03° of difference, which was expected because only constant $E_0$ of the 2S2P1D model differs between the first and the second method. The slight differences between the two methods may be explained by the different assumptions on the Poisson's ratio value which is modelled with the 2S2P1D model in the first method and taken constant equal to 0.3 in the second method. However, the two proposed methods are in very good agreement. This result is interesting because the second simplified method does not require in the first step the use of an elaborate LVE model. In addition, the back-calculation process is easy to perform as the two calculated constants can be identified using dichotomy process. This new simplified approach considerably reduces the computational time and offers great potential to back-calculate the complex modulus of AM from FRFs measurements using the first resonance peak only.

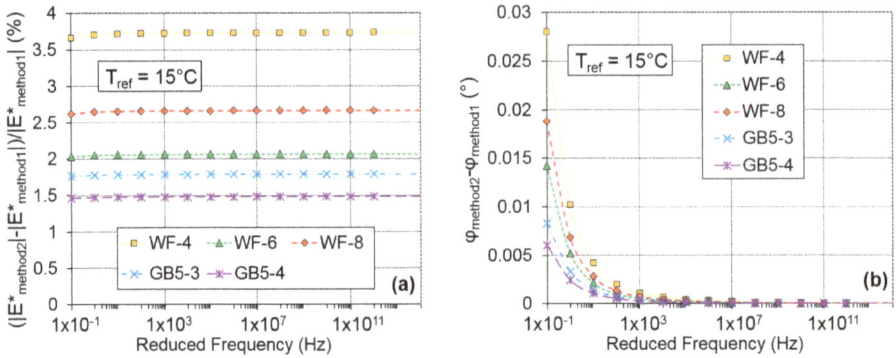


**Figure 13.** Difference between the complex modulus obtained from the two methods of analysis of the dynamic tests plotted at a reference temperature of 15 °C for all specimens: (**a**) relative difference for the norm of the complex modulus (in %); and (**b**) difference for the phase of the complex modulus (in °).

### 5.3. Comparison between Cyclic and Dynamic Tests Results

The reduced frequency range of the tension–compression and dynamic tests is different, as confirmed in Figures 10 and 11. The tension–compression tests cover a reduced frequency range between approximately $10^{-7}$ Hz to $10^9$ Hz at a reference temperature of 15 °C. The dynamic tests cover a reduced frequency range between approximately 1 Hz and $10^{12}$ Hz at the same reference temperature of 15 °C. Therefore, the best fit between the two tests is expected to be for frequencies higher than 1 Hz and lower than $10^9$ Hz for which experimental data from the two tests is available. As the two methods of back analysis of the dynamic tests give similar results, only the complex modulus obtained from the 2S2P1D model calibrated using the second method of analysis was chosen for the comparison with results of the 2S2P1D model calibrated from quasi-static tension–compression tests data. The relative difference between the norm (in %) and the phase (in °) of the complex modulus from the two calibration processes are plotted in Figure 14 where the results are plotted against the reduced frequency at 15 °C for all specimens. It is shown in Figure 14 that the complex modulus simulated from the two tests are in quite good agreement for the high reduced frequencies (>$10^7$ Hz), which was expected when dealing with dynamic measurements. The norm of the dynamic complex modulus is around 3–5% higher than the norm of the complex modulus of the tension–compression tests for this frequency range and there is less than 0.2° of difference between the phase angles from both tests. For lower reduced frequencies (or higher temperatures), the relative difference increases for the norm and reach a value between 12% and 30% at 1 Hz depending on the specimen with an average value of 20.2%. The difference also increases for the phase angle but remains less than 2.5° for reduced frequencies higher than 1 Hz.

The differences observed in Figure 14 show that the dynamic complex modulus has a higher norm and a lower phase angle than the complex modulus obtained from the tension–compression tests. The differences between the two tests increase with temperature. These results are in agreement with results from previous studies using FRFs [18–21]. The differences between the two tests could be due to two phenomena. First, the level of strain applied is different in the two types of tests (about 50 μm/m for the tension–compression tests and about 0.1 μm/m for the dynamic tests). It is known that AM have a nonlinear behaviour showing a strain level dependence even at small strain amplitude [39–41]. The differences observed in this analysis are in the same direction than the observed nonlinearity: increasing of norm and decreasing of phase angle when decreasing strain amplitude. Then, nonlinearity may account for at least a part of the difference between the two tests. Another possibility is ageing of the materials. The tension–compression tests were performed several months before the dynamic tests and this could explain why the dynamic complex modulus is stiffer. However, the overall agreement between the two tests is satisfying.

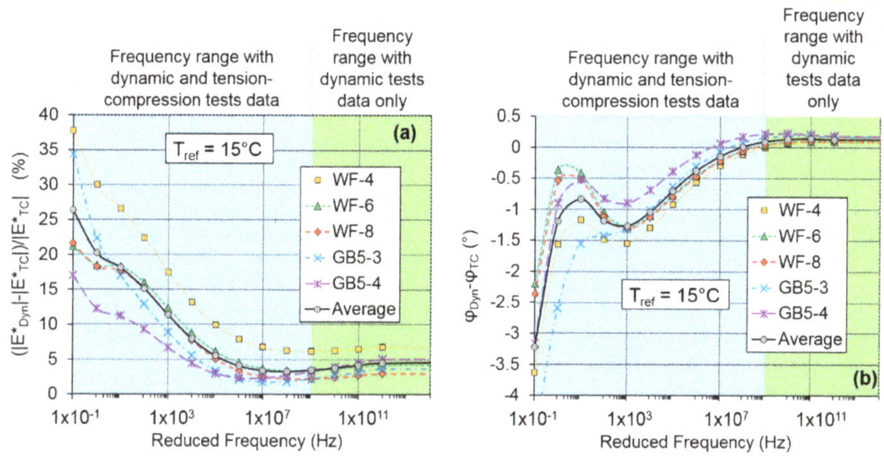


**Figure 14.** Relative difference between the complex modulus simulated with the 2S2P1D model calibrated from back-calculation of the dynamic tests and the complex modulus simulated with the 2S2P1D model calibrated on the data of the tension–compression tests at 15 °C for all specimens: (**a**) relative difference for the norm of the complex modulus (in %); and (**b**) relative difference for the phase of the complex modulus (in °).

The shift factors from the WLF equation obtained from the tension–compression tests and from the dynamic tests are plotted in Figure 15. It is seen that, for the low temperatures, the agreement between the shift factors from both tests is very good. For temperatures higher than 10 °C, the shift factors of the dynamic tests tend to be higher than the shift factors of the cyclic tests and the difference increases with temperature. The difference between the shift factors is more important for material labelled GB5. However, there is no apparent link between the difference observed in Figure 14 for the complex modulus and for the shift factors since the highest difference on the complex modulus evaluation correspond to a specimen of material labelled WF.

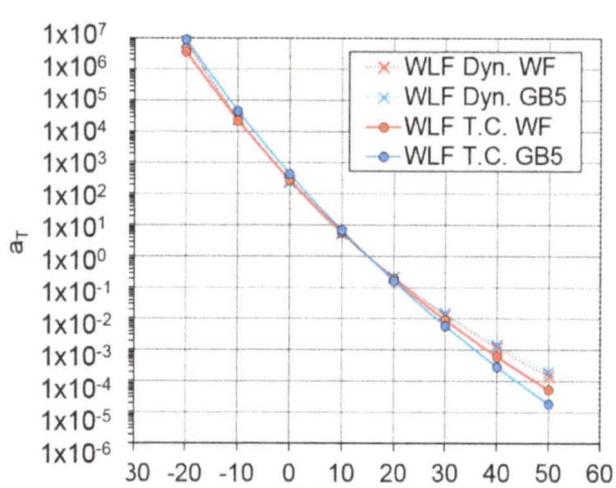


**Figure 15.** Shift factors of the WLF equation (Equation (7)) obtained from the tension–compression tests (WLF T.C.) and from the dynamic tests (WLF Dyn.) for the two tested materials.

## 6. Conclusions

In this paper, conventional cyclic tension–compression tests and dynamic tests were performed to characterize the LVE behaviour of AM on a large range of frequencies and temperatures. Two different AM and five specimens were tested with both tests and results were analysed using the 2S2P1D model.

Two methods were studied to back-calculate the complex modulus of AM from dynamic measurements at each temperature. It is shown that the same shift factors are found with the two methods. Moreover, the two methods give very similar complex modulus values (less than 4% of difference for the norm of the modulus and 0.03° for the phase angle) and the differences observes may be due to the different assumptions on the Poisson's ratio value. Therefore, the second method, which is a new and simpler approach, appears to be a good option to obtain the complex modulus of AM from FRFs.

The results of dynamic tests were also compared to the results of tension–compression tests. The shift factors from both tests are very close for the low temperatures and shift factors from dynamic tests are little higher for temperatures higher than 10 °C. The complex modulus obtained from dynamic tests have a higher norm and a lower phase angle than the ones determined with the conventional approach using cyclic tests. The differences observed between the two tests are very limited for the high frequencies or low temperatures (less than 5% for the norm and 0.2° for the phase angle) and are more important for the low frequencies or high temperatures (around 20% for the norm and 2° for the phase angle at 15 °C and 1 Hz). Since the strain level is approximately 500 times lower in the dynamic tests, the nonlinearity of AM with the level of strain amplitude may explain a part of the differences. Ageing of the materials between the tension–compression and the dynamic tests may also have an impact on the complex modulus evaluation.

The agreement between dynamic tests and the tension–compression tests is still satisfactory on the whole frequency range. The combination of the two tests methods is useful to improve the characterization of the LVE behaviour of AM on a wider frequency range because dynamic tests give access to very high frequencies. The presented research shows that dynamic tests, which have the great advantage of being cheap and rapid, can be back-analysed with a very simple model and can provide accurately the complex modulus of AM on a wide range of frequencies and temperatures.

**Author Contributions:** Methodology, J.-C.C., H.D.B., and C.S.; experimental tests, J.-C.C.; formal analysis, J.-C.C.; writing—original draft preparation, J.-C.C.; and writing—review and editing, H.D.B. and C.S.

**Funding:** This research received no external funding.

**Conflicts of Interest:** The authors declare no conflict of interest.

## References


1. Di Benedetto, H.; Corte, J.F. *Matériaux Routiers Bitumineux 2: Constitution et Propriétés Thermomécaniques des Mélanges*; Lavoisier: Paris, France, 2005; p. 288. (In French)
2. Halvorsen, W.G.; Brown, D.L. Impulse technique for structural response frequency testing. *J. Sound Vib.* **1977**, *11*, 8–21. [CrossRef]
3. ASTM: C215-08. *Standard Test Method for Fundamental Transverse, Longitudinal and Torsional Frequencies of Concrete Specimens*; ASCE: West Conshoshocken, PA, USA, 2008.
4. Migliori, A.; Sarrao, J. *Resonant Ultrasound Spectroscopy-Applications to Physics, Materials Measurements and Nondestructive Evaluation*; Wiley-Interscience Publication: New York, NY, USA, 1997.
5. Renault, A.; Jaouen, L.; Sgard, F. Characterization of elastic parameters of acoustical porous materials from beam bending vibrations. *J. Sound Vib.* **2010**, *330*, 1950–1963. [CrossRef]
6. Di Benedetto, H.; Sauzéat, C.; Sohm, J. Stiffness of bituminous mixtures using ultrasonic wave propagation. *Road Mater. Pavement Des.* **2009**, *10*, 789–814. [CrossRef]
7. Mounier, D.; Di Benedetto, H.; Sauzéat, C. Determination of bituminous mixtures linear properties using ultrasonic wave propagation. *Constr. Build. Mater.* **2012**, *36*, 638–647. [CrossRef]

8. Norambuena-Contreras, J.; Castro-Fresno, D.; Vega-Zamanillo, A.; Celay, M.; Lombillo-Vozmediano, I. Dynamic modulus of asphalt mixture by ultrasonic direct test. *NDT E Int.* **2010**, *43*, 629–634. [CrossRef]
9. Kweon, G.; Kim, Y.R. Determination of the complex modulus of asphalt concrete using the impact resonance test. *J. Transp. Res. Board* **2006**, *1970*, 151–160. [CrossRef]
10. Lacroix, A.; Kim, Y.R.; Far, M.S.S. Constructing the dynamic modulus mastercurve using impact resonance testing. *J. Assoc. Asphalt Paving Technol.* **2009**, *78*, 67–102.
11. Whitmoyer, S.L.; Kim, Y.R. Determining asphalt concrete properties via the impact resonant method. *J. Test. Eval.* **1994**, *22*, 139–148. [CrossRef]
12. Gudmarsson, A.; Ryden, N.; Birgisson, B. Application of resonant acoustic spectroscopy to asphalt concrete beams for determination of the dynamic modulus. *Mater. Struct.* **2012**, *45*, 1903–1913. [CrossRef]
13. Leisure, R.; Willis, F. Resonant ultrasound spectroscopy. *J. Phys. Condens. Matter* **1999**, *9*, 6001–6029. [CrossRef]
14. Ostrovsky, L.; Lebedev, A.; Matveyev, A.; Popatov, A.; Sutin, A.; Soustova, I.; Johnson, P. Application of three-dimensional resonant acoustic spectroscopy method to rock and building materials. *J. Acoust. Soc. Am.* **2001**, *110*, 1770–1777. [CrossRef] [PubMed]
15. Ryden, N. Resonant frequency testing of cylindrical asphalt samples. *Eur. J. Environ. Civ. Eng.* **2011**, *15*, 587–600. [CrossRef]
16. Ren, Z.; Atalla, N.; Ghinet, S. Optimization based identification of the dynamic properties of linearly viscoelastic materials using vibrating beam techniques. *ASME J. Vib. Acoust.* **2011**, *133*, 041012. [CrossRef]
17. Rupitsch, S.J.; Ilg, J.; Sutor, A.; Lerch, R.; Döllinger, M. Simulation based estimation of dynamic mechanical properties for viscoelastic materials used for vocal fold models. *J. Sound Vib.* **2011**, *330*, 4447–4459. [CrossRef]
18. Gudmarsson, A.; Ryden, N.; Birgisson, B. Characterizing the low strain complex modulus of asphalt concrete specimens through optimization of frequency response functions. *J. Acoust. Soc. Am.* **2012**, *132*, 2304–2312. [CrossRef] [PubMed]
19. Gudmarsson, A.; Ryden, N.; Di Benedetto, H.; Sauzéat, C.; Tapsoba, N.; Birgisson, B. Comparing Linear Viscoelastic Properties of Asphalt Concrete Measured by Laboratory Seismic and Tension-Compression Tests. *J. Nondestruct. Eval.* **2014**, *33*, 571–582. [CrossRef]
20. Gudmarsson, A.; Ryden, N.; Di Benedetto, H.; Sauzéat, C. Complex modulus and complex Poisson's ratio from cyclic and dynamic modal testing of asphalt concrete. *Constr. Build. Mater.* **2015**, *88*, 20–31. [CrossRef]
21. Carret, J.-C.; Pedraza, A.; Di Benedetto, H.; Sauzéat, C. Comparison of the 3-dim linear viscoelastic behavior of asphalt mixes determined with tension-compression and dynamic tests. *Constr. Build. Mater.* **2018**, *174*, 529–536. [CrossRef]
22. Carret, J.-C.; Di Benedetto, H.; Sauzéat, C. Multi-modal dynamic linear viscoelastic back analysis for asphalt mixes. *J. Nondestruct. Eval.* **2018**, *37*, 35. [CrossRef]
23. Poirier, J.E.; Pouget, S.; Leroy, C.; Delaporte, B. Projets Mure et Improvmure: Bilan à mi-parcours. *Revue Générale des Routes et de l'Aménagement* **2016**, *937*, 38–41. (In French)
24. Gayte, P.; Di Benedetto, H.; Sauzéat, C.; Nguyen, Q. Influence of transient effects for analysis of complex modulus tests on bituminous mixtures. *Road Mater. Pavement Des.* **2015**, *17*, 271–289. [CrossRef]
25. Graziani, A.; Di Benedetto, H.; Perraton, D.; Sauzéat, C.; Hofko, B.; Poulikakos, L.; Pouget, S. Recommendation of RILEM TC 237-SIB on complex Poisson's ratio characterization of bituminous mixtures. *Mater. Struct.* **2017**, *50*, 142. [CrossRef]
26. Perraton, D.; Di Benedetto, H.; Sauzéat, C.; Hofko, N.; Graziani, A.; Nguyen, Q. 3 Dim experimental investigation of linear viscoelastic properties of bituminous mixtures. *Mater. Struct.* **2016**, *49*, 4813–4829. [CrossRef]
27. Di Benedetto, H.; Olard, F.; Sauzéat, C.; Delaporte, B. Linear viscoelastic behavior of bituminous materials: From binders to mixtures. *Road Mater. Pavement Des.* **2004**, *5*, 163–202. [CrossRef]
28. Olard, F.; Di Benedetto, H. General 2S2P1D model and relation between the linear viscoelastic behaviors of bituminous binders and mixes. *Road Mater. Pavement Des.* **2003**, *4*, 185–224. [CrossRef]
29. Tiouajni, S.; Di Benedetto, H.; Sauzéat, C.; Pouget, S. Approximation of linear viscoelastic model in the 3 dimensional case with mechanical analogues of finite size—Application to bituminous materials. *Road Mater. Pavement Des.* **2011**, *12*, 897–930. [CrossRef]
30. Di Benedetto, H.; Delaporte, B.; Sauzéat, C. Three dimensional linear behavior of bituminous materials: Experiments and modeling. *ASCE Int. J. Geomech.* **2007**, *7*, 149–157. [CrossRef]

31. Nguyen, H.M.; Pouget, S.; Di Benedetto, H.; Sauzéat, C. Time-temperature superposition principle for bituminous mixtures. *Eur. J. Environ. Civ. Eng.* **2009**, *13*, 1095–1107. [CrossRef]
32. Nguyen, M.L.; Sauzéat, C.; Di Benedetto, H.; Tapsoba, N. Validation of the time–temperature superposition principle for crack propagation in bituminous mixtures. *Mater. Struct.* **2013**, *46*, 1075–1087. [CrossRef]
33. Nguyen, Q.T.; Di Benedetto, H.; Sauzéat, C.; Tapsoba, N. Time-temperature superposition principle validation for bituminous mixes in the linear and nonlinear domain. *ASCE J. Mater. Civ. Eng.* **2013**, *25*, 1181–1188. [CrossRef]
34. Ferry, J.D. *Viscoelastic Properties of Polymers*, 3rd ed.; John Wiley & Sons: New York, NY, USA, 1980.
35. Brüggemann, T.; Biermann, D.; Zabel, A. Development of an automatic modal pendulum for the measurement of frequency responses for the calculation of stability charts. *Procedia CIRP* **2015**, *33*, 587–592. [CrossRef]
36. Norman, P.E.; Jung, G.; Ratcliffe, C.; Crane, R.; Davis, C. *Development of an Automated Impact Hammer for Modal Analysis of Structures*; DSTO-TN-1062; DSTO Defence Science and Technology Organisation: Fishermans Bend, Australia, 2012.
37. Mangiafico, S.; Di Benedetto, H.; Sauzéat, C.; Olard, F.; Pouget, S.; Planque, L. New method to obtain viscoelastic properties of bitumen blends from pure and RAP binder constituents. *Road Mater. Pavement Des.* **2014**, *15*, 312–329. [CrossRef]
38. Pouget, S.; Sauzéat, C.; Di Benedetto, H.; Olard, F. Modeling of viscous bituminous wearing course materials on orthotropic steel deck. *Mater. Struct.* **2012**, *45*, 1115–1125. [CrossRef]
39. Airey, G.; Rahimzadeh, B. Combined bituminous binder and mixture linear rheological properties. *Constr. Build. Mater.* **2004**, *18*, 535–548. [CrossRef]
40. Nguyen, Q.T.; Di Benedetto, H.; Sauzéat, C. Linear and nonlinear viscoelastic behavior of bituminous mixtures. *Mater. Struct.* **2015**, *48*, 2339–2351. [CrossRef]
41. Mangiafico, S.; Babadopoulos, L.; Di Benedetto, H.; Sauzéat, C. Nonlinearity of bituminous mixtures. *Mech. Time-Depend. Mater.* **2018**, *22*, 29–49. [CrossRef]

*applied sciences*

MDPI

*Article*

# A Source Pollution Control Measure Based on Spatial-Temporal Distribution Characteristic of the Runoff Pollutants at Urban Pavement Sites

**Changjiang Kou [1,2], Aihong Kang [1,*], Peng Xiao [1], Peter Mikhailenko [2], Hassan Baaj [2], Lu Sun [1,3] and Zhengguang Wu [1]**

1 College of Civil Science and Engineering, Yangzhou University, Yangzhou 225127, China; ckou@uwaterloo.ca (C.K.); xpyzu@163.com (P.X.); sunl@cua.edu (L.S.); zgwu@yzu.edu.cn (Z.W.)
2 Centre for Pavement and Transportation Technology, University of Waterloo, Waterloo, ON N2L 3G1, Canada; p2mikhailenko@uwaterloo.ca (P.M.); hassan.baaj@uwaterloo.ca (H.B.)
3 Department of Civil and Environmental Engineering, The Catholic University of America, Washington, DC 20064, USA
* Correspondence: kahyzu@163.com; Tel.: +86-138-158-13548

Received: 31 August 2018; Accepted: 20 September 2018; Published: 2 October 2018

**Featured Application: It is recommended that the design of source pollution control measures be based on the spatial-temporal distribution characteristics of pavement runoff pollutants.**

**Abstract:** The concentrations of pollutants in urban pavement runoff are normally higher than those in other urban surface runoff, which causes serious problems in protecting the environment of receiving water and soils. The purpose of this study was to propose a source pollution control measure based on the spatial-temporal distribution characteristics of the runoff pollutants at urban pavement sites. Therefore, samples from pavement runoff were collected and tested for analyzing the spatial-temporal distribution characteristics. Then, infiltration tests were conducted on selected purification materials to evaluate their purification ability to the simulated pavement runoff. Results indicated that heavy metals Zn and Pb were at high concentrations near the intersection, the reason being the frequent braking of vehicles at this site. The level of suspended solids was far higher than the limitation in the standard near the site where massive human activities occurred. Besides, the cumulative amounts of all kinds of pollutants tended to be stable with the extension of rainfall duration. The logarithmic function was found to fit the experimental data well. Finally, the pavement runoff was categorized into different situations. The combinations of purification materials were recommended and integrated into a source control measure for the treatments of different pollution situations, which made the most use of each purification material and ensured the high elimination efficiency of different pollutants.

**Keywords:** pavement engineering; runoff pollutants; spatial-temporal distribution; source control measures; purification materials

## 1. Introduction

In the context of stormwater management in the urban area, different practices, including low impact development (LID), sustainable urban drainage systems (SUDS), water sensitive urban design (WSUD) and best management practices (BMPs) have been developed in succession over recent decades [1–3]. The definition of BMPs has since become a more universal term describing best practices related to general pollution prevention [4,5]. Positively, the point source pollutions, such as domestic sewage, industrial emissions and so on have been minimized to a satisfied level with the building of both non-structural and structural control attributes [6]. However, non-point source pollutions caused mainly by the rapid increase of impermeable urban pavement and traffic vehicles are becoming more and more serious [7–9]. As the principal part of non-point source pollutions, pavement runoff pollution has attracted extensive attention from engineers and agencies of pavement engineering. Field investigations carried out in different countries indicated that the pollutants in pavement runoff exceeded the limitation in the local standards to a great extent [10–12]. How to control the pavement runoff pollution more efficiently tended to be a significant problem most engineers pay close attention to. However, most existing investigations focused on the development of pollution control measures (PCMs) and purification materials, ignoring the influence of spatial-temporal distribution characteristics.

Actually, the characteristics of pavement runoff pollutants is of great importance to the design and layout of PCMs. Early in 1995, Barrett reported the main sources of pavement runoff pollutants, including traffic vehicles, atmospheric sedimentation, construction, maintenance and other human activities. The typical characteristics of pavement runoff pollutants are influenced by traffic volume, rainfall parameters, pavement types, existing status of pollutants, ambient environment and climatic features [13,14]. The concept "first flush" is also mentioned in publications, which illuminates that high concentration pollutants exist in the first flush of rainfall events [15,16]. About 60% total suspended solid (TSS) was found in the 30% first flush rainfall [17]. Investigations showed that there are evident correlations between the concentrations of different pollutants. The heavy metal Zn is positively correlated to dissolved organic carbon (DOC) while Pb, Fe and Al are positively correlated to TSS [18]. The nutrients TN and TP are related to TSS, too [19]. All these findings provide a good way to remove pollutants in pavement runoff by filtrating DOC and TSS. As to other factors, conclusions were drawn by Mayer and Winston that the chemical pollutants and biological toxicity of highway runoff with heavy traffic volume are far higher than those with light and medium traffic volume [20], and that the amount of TSS in open graded friction course (OGFC) runoff is less than that in impermeable pavement runoff, meaning the outstanding performance of permeable pavements in pollutant removal [19].

PCMs such as source management, detention ponds, frequent street cleaning, wetlands, sedimentation basins, and percolation treatments are designed for different situations [13]. Gill monitored the removal efficiency of a constructed wetland planted with *Phragmites australis* and *Typha latifolia* to highway pavement runoff, and found that the removal rate of heavy metals Cd, Cu, Pb and Zn is up to 95%, 88%, 86% and 95%, respectively [21]. To make the most of the absorptivity of environmental mineral materials and the decomposition of microorganisms, the combination of zeolite, rock wool and microorganisms is used in the rainwater purification system, which shows good removal ability for COD, SS, $NH_3$-N, TP and TN [22].

Environmental mineral materials were also employed by Hilliges in a three-stage treatment system for highly polluted urban road runoff [23]. In recent years, permeable asphalt pavement, widely paved in the Netherlands, has been accepted broadly as an effective way to reduce the peak of rainfall runoff and detain particle pollutants [24,25]. It has higher air voids ranging from 18% to 25% than traditional hot mix asphalt (HMA), which makes the rainwater infiltrate into pavement structure layers and drain away laterally along the seal coat [26]. The typical structures of permeable asphalt pavement are shown in Figure 1. Other PCMs are also reported in the literature, such as wet bio-filtration and dry detention ponds [27].

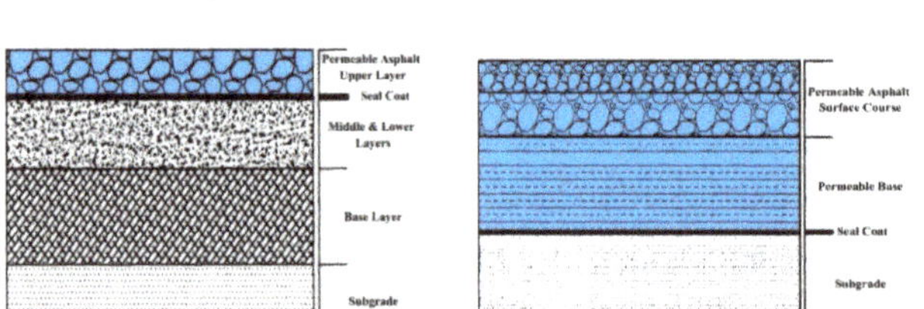


(**a**) Permeable only for upper layer (**b**) Permeable for surface course and base

**Figure 1.** Permeable asphalt pavement structure.

Currently, however, few reports are found about the PCMs based on the spatial-temporal distribution characteristics of pavement runoff pollutants. Generally, studies mainly focus on the composition, concentration and first flush phenomenon of pavement runoff pollutants. The purpose of PCMs is to remove the amount of pollutants as much as possible. Spatial distribution characteristic of pollutants may be different at each section along the road, due to the surrounding environment and human activities. This means that the design and layout of PCMs should be made by the specific distribution characteristics of pollutants. Taking Figure 2 as an example, the concentration levels of A and B are high in Sections 4–6 while that of C is high in Sections 8–10. According to this characteristic, environmental mineral materials with different removal rate for pollutants A, B and C could be placed at corresponding sections.

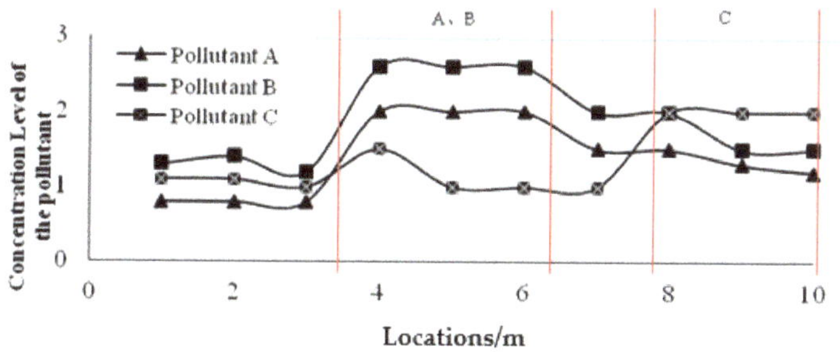


**Figure 2.** Hypothetical distribution characteristics of pollutants along a road.

To have this goal realized, the spatial-temporal distribution characteristic of pollutants are presented in this paper by sampling and analyzing the pavement runoff at different spots along the road. Then, pavement runoff simulations were prepared in the laboratory and used in the infiltration test to obtain the removal rate of six purification materials on different pollutants. Combined with the actual pollution situations of RRS, the combinations of purification materials were recommended and integrated into a source control measure. This paper proposes a specific technology for pavement runoff pollution control based on the actual pollution situations and the removal rate of different purification materials.

In future applications, PCMs containing different purification materials selected for both their removal rate of different pollutants and the actual pollution situations will be placed along the road for on-site treatment, especially where heavily polluted pavement runoff occurs. As an artificial barrier material provides a reliable elimination of pollutants, the infiltration of treated pavement runoff without endangering soil or even groundwater pollution is possible.

## 2. Materials and Methods

### 2.1. Study Site and Sampling

The Runyang Road South (RRS) is the connection line between a highway and a municipal road in Yangzhou, China. It has high traffic volume and is close to the laboratory. Four spots were selected as the target sites for sampling, as shown in Figure 3. The first one is at the intersection. The third one is near the gate of Yangtze campus. The other two are along the road.

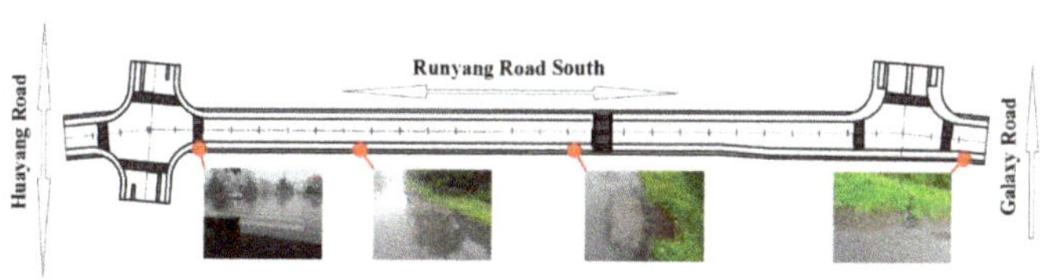


**Figure 3.** Sampling sites of pavement runoff.

Limited by the nature of pavement runoff, manual sampling was adopted in this study. Syringes were used to collect the pavement runoff which was then stored in clean sampling bottles. The entire sampling process started from the formation of runoff and lasted about 1 h. The time points of sampling were set as 1, 3, 5, 8, 10, 15, 20, 30, 45, and 60 min.

From October 2012 to April 2016, samplings were carried out during ten rainfall events in RRS. In this study, a typical pavement runoff collected on 3 April 2016 was chosen as the samples to analyze the spatial-temporal distribution characteristics. The rainfall and sampling conditions are shown in Table 1.

**Table 1.** Rainfall and sampling conditions.

| Date of Sampling | Precipitation (mm) | Duration of Rainfall (min) | Duration of Runoff (min) | Duration of Sampling (min) |
|---|---|---|---|---|
| 3 April 2016 | 14.7 | 153 | 143~148 | 60 |

### 2.2. Sample Analysis and Simulated Pavement Runoff

Pavement runoff samples were promptly sent to the laboratory for water quality analysis after sampling. The analysis indexes included SS, COD, TN, TP, Zn, and Pb. The monitoring methods are shown in Table 2.

**Table 2.** Analysis methods of different pollutants.

| Items | Methods | Referenced Standards |
|---|---|---|
| SS | Gravimetric method | GB/T11901-89 |
| COD | Potassium dichromate method | GB/T11914-89 |
| TN | Alkaline potassium persulfate digestion UV spectrophotometric method | HJ 636-2012 |
| TP | Ammonium molybdate spectrophotometric method | GB 11893-89 |
| Pb | Atomic absorption spectrophotometry method | GB 7475-87 |
| Zn | Atomic absorption spectrophotometry method | GB 7475-87 |

A large volume of pavement runoff was needed in the experiment. However, it was inconvenient and difficult to collect sufficient pavement runoff from the field. Therefore, a pavement runoff simulation was prepared as per the actual concentration of the pollutants by using the chemical reagents listed in Table 3.

**Table 3.** Chemical reagents for simulated pavement runoff.

| Indexes | SS | COD | TN | TP | Zn | Pb |
|---|---|---|---|---|---|---|
| Reagent | Roadside dust | $C_6H_{12}O_6$ | $NH_4Cl$ | $KH_2PO_4$ | $Zn(NO_3)_2$ | $Pb(NO_3)_2$ |

### 2.3. Purification Materials

To make the best use of waste materials, fine sand, zeolite, slag, ceramsite, diatomite and scoria were selected as the purification materials and subjected to simulated pavement runoff.

Zeolite is a kind of microporous aluminosilicate mineral discovered by a Swedish mineralogist, Cronstedt, now widely used as ion-exchange beds in domestic and commercial water purification, softening, and other applications. Slag is the by-product of blast furnace ironmaking. Ground granulated slag is often used in combination with Portland cement as part of a blended cement. Ceramsite is a kind of light aggregate produced by foam technology in a rotary kiln. The honeycomb structure inside endues the ceramsite with satisfactory absorptivity. Diatomite is a typical biomineral with porous structure and large specific surface area [28], which provides an ideal carrier for heavy metal ions. Scoria is a highly vesicular, dark colored volcanic rock. It has low density because of its numerous macroscopic ellipsoidal vesicles. Their macroscopic morphologies are presented in Figure 4.

(a) Zeolite (b) Slag (c) Ceramsite (d) Diatomite (e) Scoria

**Figure 4.** Purification materials.

*2.4. Purification Experiment*

The device shown in Figure 5 was designed and used in the infiltration experiment to test the removal ability of different purification materials to the pollutants listed in Table 2.

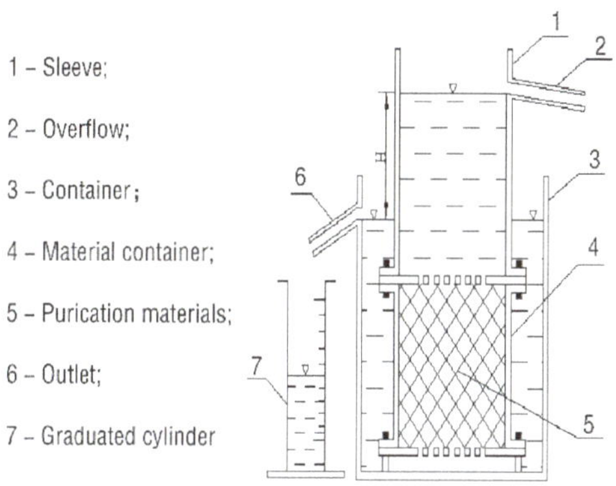


**Figure 5.** Device for the purification experiment.

The simulated pavement runoff was pumped up into the sleeve and then infiltrated through the purification materials into the container. The effluent was collected 2 min after the beginning of the experiment by using a graduated cylinder and immediately sent to the laboratory for water quality analysis. The entire experiment was carried out under constant head.

## 3. Results and Discussion

*3.1. Spatial-Temporal Distribution Characteristics Analysis*

The results of water quality analysis are listed in Table 4. It was evident that the concentration of each pollutant far exceeded the limitation of Grade V required in the Standard of Environmental Quality for Surface Water [29]. Especially, the concentrations of COD and SS were at high levels with the duration of the rainfall at all four sites, which indicated that organic and particle pollution were serious in the entire area. The detailed spatial-temporal distribution characteristics are discussed below.

**Table 4.** Sample analysis results.

| Sites | Rainfall Duration (min) | TN (mg/L) | TP (mg/L) | COD (mg/L) | Zn (mg/L) | Pb (mg/L) | SS (mg/L) |
|---|---|---|---|---|---|---|---|
| Site 1 | 0 | 12.333 | 2.019 | 616 | 7.670 | 1.175 | 496 |
| | 3 | 11.994 | 2.014 | 564 | 6.510 | 1.370 | 711 |
| | 5 | 10.317 | 1.720 | 560 | 5.395 | 1.645 | 674 |
| | 10 | 10.217 | 1.593 | 636 | 4.705 | 1.240 | 700 |
| | 15 | 9.189 | 1.303 | 664 | 4.390 | 0.595 | 633 |
| | 20 | 9.146 | 1.061 | 692 | 3.735 | 2.065 | 490 |
| | 30 | 7.095 | 0.806 | 584 | 3.375 | 1.910 | 316 |
| | 40 | 5.038 | 0.912 | 524 | 2.995 | 0.920 | 139 |
| | 50 | 5.020 | 0.819 | 716 | 2.640 | 0.520 | 72 |
| | 60 | 5.021 | 0.906 | 704 | 2.245 | 0.730 | 104 |
| Site 2 | 0 | 5.132 | 1.403 | 540 | 3.370 | 0.540 | 403 |
| | 3 | 4.183 | 1.017 | 380 | 2.870 | 0.550 | 585 |
| | 5 | 4.095 | 1.010 | 712 | 2.385 | 0.545 | 370 |
| | 10 | 3.048 | 1.002 | 772 | 1.925 | 0.360 | 217 |
| | 15 | 3.013 | 0.934 | 744 | 1.255 | 0.320 | 105 |
| | 20 | 3.037 | 0.709 | 632 | 1.195 | 0.295 | 50 |
| | 30 | 2.018 | 0.708 | 492 | 0.665 | 0.355 | 67 |
| | 40 | 2.031 | 0.710 | 776 | 0.230 | 0.380 | 130 |
| | 50 | 1.048 | 0.805 | 764 | 0.215 | 0.380 | 120 |
| | 60 | 1.007 | 0.716 | 700 | 0.260 | 0.795 | 71 |
| Site 3 | 0 | 10.561 | 4.034 | 552 | 5.105 | 0.775 | 1018 |
| | 3 | 9.313 | 3.827 | 400 | 4.515 | 0.620 | 853 |
| | 5 | 8.327 | 3.523 | 696 | 3.870 | 0.610 | 1437 |
| | 10 | 6.178 | 3.040 | 396 | 3.180 | 0.570 | 858 |
| | 15 | 6.136 | 2.607 | 408 | 2.005 | 0.335 | 661 |
| | 20 | 5.327 | 2.019 | 568 | 1.935 | 0.390 | 663 |
| | 30 | 5.154 | 1.710 | 804 | 1.655 | 0.295 | 309 |
| | 40 | 5.063 | 1.311 | 576 | 1.310 | 0.245 | 184 |
| | 50 | 4.106 | 1.017 | 872 | 0.895 | 0.315 | 215 |
| | 60 | 2.022 | 1.015 | 664 | 0.740 | 0.730 | 11 |
| Site 4 | 0 | 6.304 | 2.129 | 624 | 2.060 | 0.540 | 750 |
| | 3 | 6.135 | 1.718 | 492 | 1.705 | 0.380 | 389 |
| | 5 | 5.110 | 1.813 | 700 | 1.055 | 0.250 | 380 |
| | 10 | 4.110 | 1.508 | 752 | 1.000 | 0.235 | 251 |
| | 15 | 4.028 | 1.004 | 800 | 0.800 | 0.240 | 225 |
| | 20 | 3.125 | 1.013 | 796 | 0.615 | 0.340 | 220 |
| | 30 | 3.072 | 0.863 | 760 | 0.265 | 0.230 | 238 |
| | 40 | 3.153 | 0.906 | 604 | 0.120 | 0.465 | 204 |
| | 50 | 3.033 | 1.008 | 460 | 0.075 | 0.275 | 172 |
| | 60 | 1.008 | 0.729 | 692 | 0.050 | 0.485 | 145 |
| Limitation (Grade V) | | ≤2 | ≤0.2 | ≤40 | ≤2 | ≤0.1 | <30 |

#### 3.1.1. Spatial Distribution Characteristics of Pavement Runoff Pollutants

Box charts of different pollutants are plotted in Figure 6.

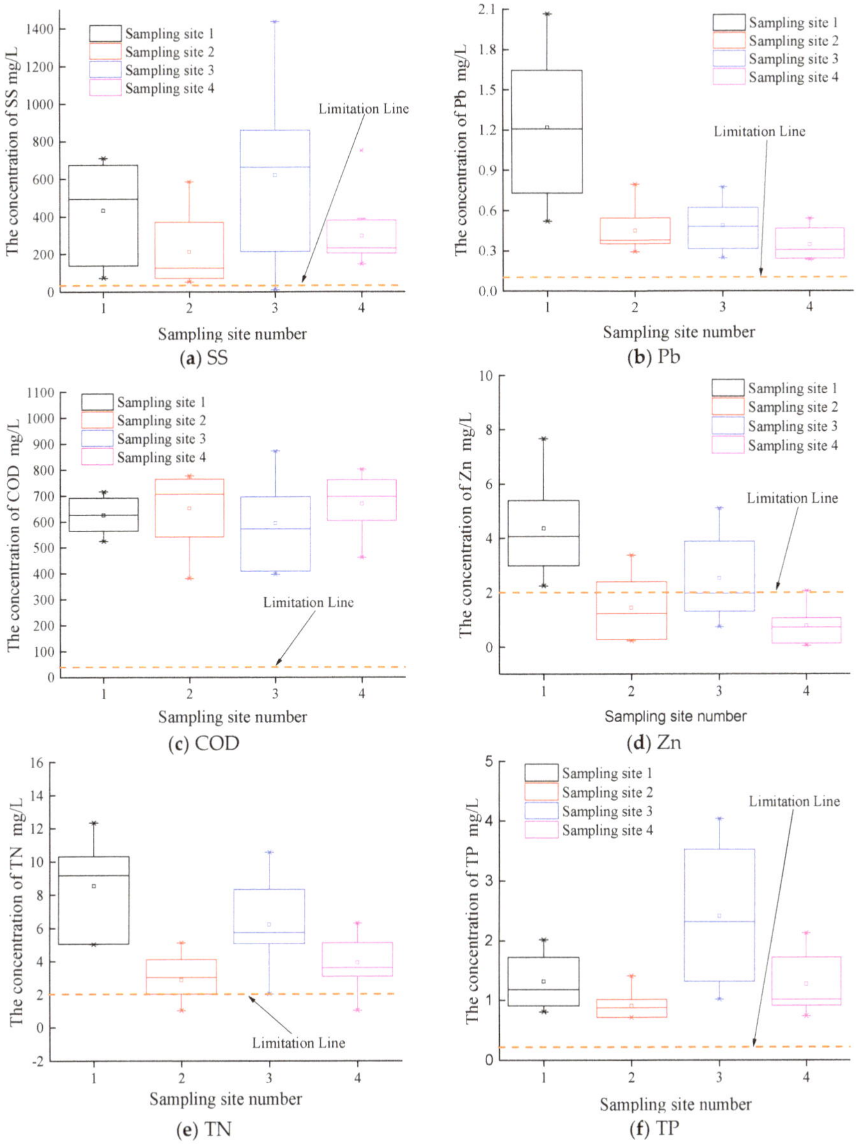


**Figure 6.** Box charts of different pollutants.

The pollutants SS, Zn and Pb showed very clear spatial distribution characteristics. In Figure 6a, the concentrations of SS at Sites 3 and 1 were much higher than those at the other two sites. These two sites were near the intersection and the west gate of Yangtze Campus, respectively. Large amounts of particle dusts or other suspended pollutants brought by vehicles and pedestrians passing by may be the reason for this characteristic. SS is a typical quality index for surface water, the maximum concentration of which was up to 48 times more than the limitation value. Particles of suspended matter are usually visible to human eyes, unstable in water and easy to be removed by porous adsorptive materials. As to the heavy metals Pb and Zn (Figure 6b,d), high concentrations existed in the pavement runoff near the intersection (Site 1). This is due to exhaust emission and frequent braking of vehicles at such sites.

However, COD showed the same concentration level at all four sites, which indicated that there were many organic pollutants in this area. In Figure 6e,f, the spatial distributions of TN and TP were similar to that of SS because dissolved pollutants attach to particles.

### 3.1.2. Temporal Distribution Characteristics of Pavement Runoff Pollutants

Ten samples collected in Site 1 are shown in Figure 7. Visually, the turbidity highly related to the concentration of SS declining with the extension of rainfall duration.

**Figure 7.** Samples of pavement runoff.

To better understand the temporal distribution characteristics of pavement runoff pollutants, different elementary functions were adopted to fit the water quality analysis data.

The natural logarithmic function, shown in Formula (1), was found to fit the cumulative concentration of different pollutants quite well. In the formula, $C_c$ represents the cumulative concentration of different pollutants. $t$ represents the time point of sampling. $a$ and $b$ are the coefficient and constant, respectively. The fitting results for different pollutants at four sites are listed in Table 5.

$$C_c = a \cdot \ln(t) + b \tag{1}$$

**Table 5.** The fitting results for different pollutants at four sites.

| Items | | Fitting Results a | Fitting Results b | $R^2$ | Items | | Fitting Results a | Fitting Results b | $R^2$ |
|---|---|---|---|---|---|---|---|---|---|
| SS | Site 1 | 1057.20 | 435.25 | 0.95 | Zn | Site 1 | 9.4321 | 5.9133 | 0.99 |
| | Site 2 | 404.07 | 588.65 | 0.95 | | Site 2 | 2.8734 | 4.0392 | 0.94 |
| | Site 3 | 1424.6 | 1054.2 | 0.94 | | Site 3 | 5.2462 | 5.3276 | 0.98 |
| | Site 4 | 578.43 | 636.31 | 0.99 | | Site 4 | 1.4852 | 2.5092 | 0.92 |
| Pb | Site 1 | 3.0108 | −0.0563 | 0.96 | TN | Site 1 | 19.384 | 7.7671 | 0.98 |
| | Site 2 | 0.9459 | 0.1947 | 0.97 | | Site 2 | 6.305 | 4.1668 | 0.98 |
| | Site 3 | 0.9987 | 0.5371 | 0.98 | | Site 3 | 13.679 | 7.9176 | 0.99 |
| | Site 4 | 0.7169 | 0.1681 | 0.95 | | Site 4 | 8.6679 | 4.7657 | 0.99 |
| COD | Site 1 | 1464.2 | −162.43 | 0.96 | TP | Site 1 | 2.8676 | 1.5756 | 0.99 |
| | Site 2 | 1572.3 | −381.71 | 0.95 | | Site 2 | 1.9779 | 0.7055 | 0.98 |
| | Site 3 | 1382.71 | −321.74 | 0.93 | | Site 3 | 5.3052 | 3.5971 | 0.98 |
| | Site 4 | 1628.7 | −246.82 | 0.95 | | Site 4 | 2.7355 | 1.5902 | 0.99 |

Taking SS as an example, the correlation curve between time and concentration of SS is plotted in Figure 8.

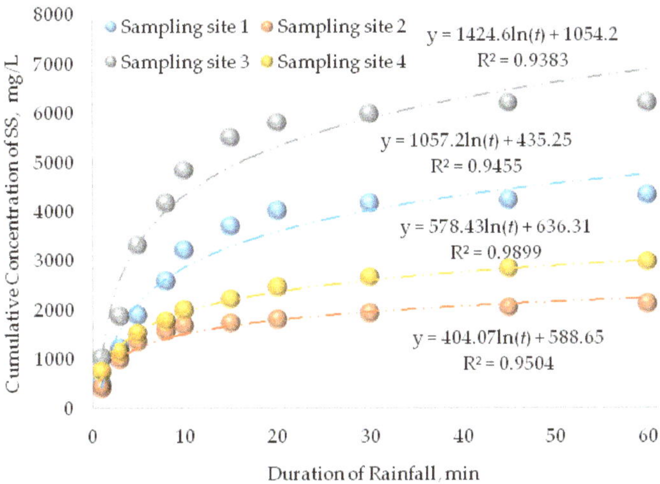


**Figure 8.** The correlation curve between time and concentration of SS.

The increase of the natural logarithmic function proves the continuous input of pollutants during the entire rainfall event. Additionally, the curve increases sharply at the initial stage and then tends to gradually stabilize. This property is consistent with the first flush effect of rainfall events. The cumulative concentration reached a high level in a very short time and then ascended slowly due to the slight input of pollutants. Although the average concentrations of all pollutants are greater than the limitation in the standard, the real-time concentrations become lower and lower, up to a moment when the concentration was lower than the limitation value. This moment, $t_c$, is defined as the critical time point, for pollution control of pavement runoff. Taking TN at Site 2 as an example, the concentration, 1.048 mg/L, meets the requirement of the standard 50 min after the formation of pavement runoff, which means that the critical time point is between 40 and 50 min. It is not necessary to process the TN in the pavement runoff after this point. However, the concentrations of most pollutants analyzed in this study were not lower than the limitation value within 60 min.

### *3.2. A Source Control Measure Based on the Optimal Combinations of Purification Materials*

To develop effective control measures for treating pavement runoff pollution, four pollution situations were determined based on the analysis above. The first situation, denoted as A, contains high concentration SS and small amounts of other pollutants. The second situation, denoted as B, contains high concentration heavy metals Pb and Zn. Pavement runoff with high concentration dissolved pollutants TN and TP belongs to the third situation, C. If the concentrations of all pollutants are at approximate level, it is viewed as the last situation, D. The removal rates, calculated by Formula (2), of six purification materials on different pollutants are listed in Table 6.

$$(|C_{2min} - C_0|/C_0) \times 100\% \tag{2}$$

In Formula (2), $C_0$ is the initial concentration and $C_{2min}$ is the concentration 2 min after the beginning of infiltration experiment.

Because of differences in pore structure, pore size and mineral composition the six materials have different removal rates for different pollutants. Each pollutant corresponds to an optimal purification material. Since it is impossible to use one purification material that has a better effect in treating pavement runoff [30], combinations of purification materials applicable for different pollution situations would be of great significance in processing the pavement runoff more efficiently.

**Table 6.** Results of the infiltration experiment.

| Purification Materials | Removal Rates (%) | | | | | |
|---|---|---|---|---|---|---|
| | SS | COD | TN | TP | Zn | Pb |
| Fine sand | 95.1 | 34.7 | 8.6 | 43.1 * | 36.4 | 47.3 |
| Zeolite | 96.4 | 48.5 * | 45.7 | 40.3 * | 55.9 * | 56.0 |
| Slag | 97.3 * | 45.2 * | 45.3 | 20.8 | 36.4 | 10.8 |
| Ceramsite | 94.6 | 40.1 | 27.4 | 20.8 | 73.4 * | 78.7 * |
| Diatomite | 88.9 | 42.3 | 50.6 * | 31.9 | 21.9 | 26.0 |
| Scoria | 99.5 * | 30.4 | 56.9 * | 15.3 | 34.0 | 84.7 * |

Note: In each column, the top two removal rates are marked with *.

The combination including two purification materials was taken as an example in this paper. In each column of Table 6, the top two removal rates are marked with asterisk. For the pollution situation containing only one kind of high concentration pollutant, the purification materials corresponding to the top two removal rates would be the best for this pollution situation. Therefore, the best combination for A (SS) includes scoria and slag. For the pollution situation containing two or more high concentration pollutants, the combination should be determined by considering mutual complementation of purification materials. Thus, for B (Pb and Zn), zeolite and ceramsite were selected. Finally, the optimal combinations of purification materials applicable for different pollution situations are listed in Table 7.

**Table 7.** Optimal combinations of purification materials applicable for different pollution situations.

| Pollution Situations | A | B | C | D |
|---|---|---|---|---|
| Combination | Slag + Scoria | Zeolite + Ceramsite | Zeolite + Diatomite | Zeolite + Scoria |

Based on the RRS runoff pollution characteristics, the advantages and disadvantages of various control measures, we referred to the three-stage treatment system [23] to integrate different purification materials into a source control measure to process the pavement runoff. The measure was designed as a portable device shown in Figure 9. To ensure the growth of plants, the container cover was made of glass.

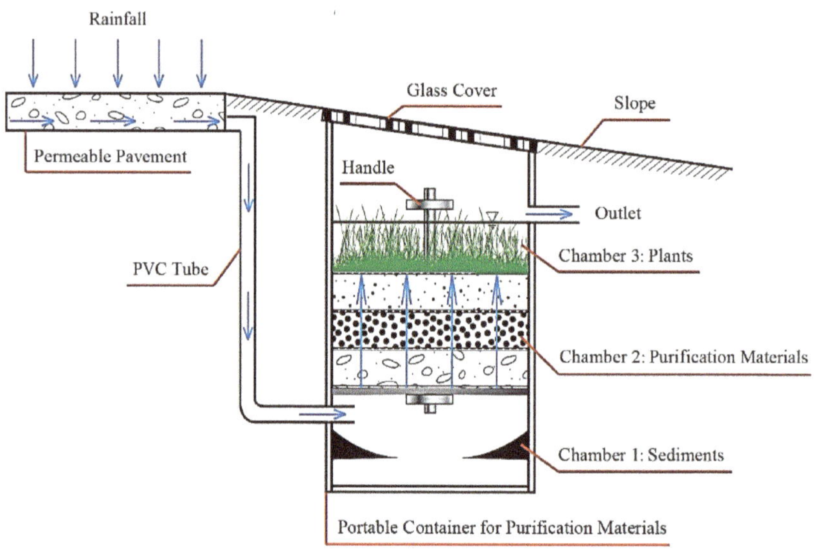


**Figure 9.** A Portable Device of Pollution Control Measures.

The runoff went through the permeable pavement or other media, was collected by PVC tube and flowed into the processing chambers. Bigger particles sank in Chamber 1 while smaller particles and dissolved pollutants continued flowing through Chamber 2 and 3. The flow from the bottom up extended the time that pavement runoff went through the purification materials chamber, which provided full contact between pollutants and materials and helped to improve the removal rate. The purification materials and plants absorbed and detained most of the pollutants. Chambers 2 and 3 were placed on a removable plate with a handle. This made the replacement of old or saturated materials more convenient. Finally, the processed runoff was drained out directly to nearby water bodies or stored into recycling units for reuse in municipal irrigation.

## 4. Conclusions

The spatial-temporal distribution characteristics of the runoff pollutants at urban pavement sites were investigated by analyzing the water quality indexes of pavement runoff samples collected from RRS. The following conclusions can be drawn:

(1) The concentrations of Pb and Zn at the intersection were much higher than those at other sections of the road. The level of suspended solids far exceeded the limitation near the site where frequent human activities occurred. The spatial distributions of dissolved pollutants were similar to that of SS because of their high attachment to particles.
(2) For all pollutants, the cumulative concentration reached a high level in a very short time and then ascended slowly. They were finally stable with the extension of rainfall duration. This feature was consistent with the property of the natural logarithmic function.
(3) Six materials showed different removal rates for different pollutants. Based on the spatial-temporal distribution characteristics of pavement runoff pollutants, combinations of purification materials applicable for different pollution situations were recommended.
(4) Different purification materials were integrated into a source control measure for treating pavement runoff with high pollution potential.
(5) Depending on area locations, surrounding environment and rainfall events, the spatial and temporal distribution characteristics of pavement runoff were different. Only on the premise of understanding the actual pollution situations can engineers integrate different purification materials in the PCMs to make the removal of pollutants more effective and efficient.

**Author Contributions:** Data curation, C.K. and P.M.; Formal analysis, C.K.; Investigation, C.K., P.M., H.B., L.S. and Z.W.; Methodology, A.K. and P.X.; Project administration, A.K. and P.X.; Validation, Z.W.; Visualization, C.K.; Writing—original draft, C.K.; Writing—review and editing, A.K. and P.X.

**Funding:** This research was funded by the Natural Science Foundation Committee of China, grant number 51578481 and 51578480, and the Research and Innovation Plans for Graduates, grant number kyzz15_0363. The APC was funded by the Natural Science Foundation Committee of China".

**Acknowledgments:** The authors would like to acknowledge the financial support of the Natural Science Foundation Committee of China (51578481 and 51578480) and the Research and Innovation Plans for Graduates (kyzz15_0363). We would also like to thank the technical support from the Testing Center of Yangzhou University and the Centre for Pavement and Transportation Technology at the University of Waterloo in Canada.

**Conflicts of Interest:** The authors declare no conflict of interest. The sponsors had no role in the design of the study; in the collection, analyses, or interpretation of data; in the writing of the manuscript, and in the decision to publish the results.

## References

1. Finnemore, E.J.; Lynard, W.G. Management and control technology for urban stormwater pollution. *J. Water Pollut. Control Fed.* **1982**, *54*, 1099–1111. [CrossRef]
2. Van Roon, M. Low impact urban design and development: Catchment-based structure planning to optimize ecological outcomes. *Urban Water J.* **2011**, *8*, 293–308. [CrossRef]

3. Fletcher, T.D.; Shuster, W.; Hunt, W.F.; Ashley, R.; Butler, D.; Arthur, S.; Trowsdale, S.; Barraud, S.; Semadeni-Davies, A.; Bertrand-Krajewski, J.-L.; et al. SUDS, LID, BMPs, WSUD and more—The evolution and application of terminology surrounding urban drainage. *Urban Water J.* **2015**, *12*, 525–542. [CrossRef]
4. Environmental Protection Agency. *National Pollutant Discharge Elimination System (NPDES) Definitions*; 40 C.F.R. § 122.2 (2011a); United States Environmental Protection Agency: Washington, DC, USA, 2011.
5. Environmental Protection Agency. *National Pollutant Discharge Elimination System (NPDES) Establishing Limitations, Standards, and Other Permit Conditions*; 40 C.F.R. §122.4 (2011b); United States Environmental Protection Agency: Washington, DC, USA, 2011.
6. Ghassemi, A. *Handbook of Pollution Control and Waste Minimization*, 1st ed.; Marcel Dekker Inc.: New York, NY, USA, 2002; pp. 389–392.
7. Bannerman, R.T.; Owens, D.W.; Dodds, R.B.; Hornewer, N.J. Sources of pollutants in Wisconsin stormwater. *Water Sci. Technol.* **1993**, *28*, 241–259. [CrossRef]
8. McCarthy, D.T.; Hathaway, J.M.; Hunt, W.F.; Deletic, A. Intra-event variability of Escherichia coli and total suspended solids in urban stormwater runoff. *Water Res.* **2012**, *46*, 6661–6670. [CrossRef] [PubMed]
9. Gorgoglione, A.; Bombardelli, F.A.; Pitton, B.J.L.; Oki, L.R.; Haver, D.L.; Young, T.M. Role of Sediments in Insecticide Runoff from Urban Surfaces: Analysis and Modeling. *Int. J. Environ. Res. Public Health* **2018**, *15*, 1464. [CrossRef] [PubMed]
10. Ju, Y.L.; Kim, H.; Kim, Y.; Han, M.Y. Characteristics of the event mean concentration (EMC) from rainfall runoff on an urban highway. *Environ. Pollut.* **2011**, *159*, 884–888. [CrossRef]
11. Geonha, K.; Joonghyun, Y. Diffuse pollution loading from urban storm water runoff in Daejeon city, Korea. *J. Environ. Manag.* **2007**, *85*, 9–16. [CrossRef]
12. Oshawa, K. *Experimental Study on the Clogging of Voids of Porous Concrete and Porous Asphalt*; Technical Workshop 35th Kanto Branch Society of Civil Engineers: Tokyo, Japan, 2008.
13. Barrett, M.E.; Malina, J.F.; Charbeneau, R.J. *A Review and Evaluation of Literature Pertaining to the Quantity and Control of Pollution from Highway Runoff and Construction*; Bureau of Engineering Research, University of Texas at Austin, J.J. Pickle Research Campus: Austin, TX, USA, 1995.
14. Barrett, M.E.; Malina, J.F.; Charbeneau, R.J.; Ward, G.H. Characterization of highway runoff in Austin, Texas area. *J. Environ. Eng.* **2009**, *124*, 131–137. [CrossRef]
15. Kayhaniana, M.; Stransky, C.; Bayc, S.; Lau, S.L.; Stenstrom, M.K. Toxicity of urban highway runoff with respect to storm duration. *Sci. Total Environ.* **2008**, *389*, 386–406. [CrossRef] [PubMed]
16. Kayhaniana, M.; Fruchtmanb, B.D.; Gulliver, J.S.; Montanaro, C.; Ranieri, E.; Wuertz, S. Review of highway runoff characteristics: Comparative analysis and universal implications. *Water Res.* **2012**, *46*, 6609–6624. [CrossRef] [PubMed]
17. Modugno, M.D.; Gioia, A.; Gorgoglione, A.; Iacobellis, V.; la Forgia, G.; Piccinni, A.F.; Ranieri, E. Build-Up/Wash-Off monitoring and assessment for sustainable management of first flush in an urban area. *Sustainability* **2015**, *7*, 5050–5070. [CrossRef]
18. Herngren, L.; Goonetilleke, A.; Ayoko, G.A. Understanding heavy metal and suspended solids relationships in urban stormwater. *J. Environ. Manag.* **2005**, *76*, 149–158. [CrossRef] [PubMed]
19. Winston, R.J.; Hunt, W.F. Characterizing Runoff from Roads: Particle Size Distributions, Nutrients, and Gross Solids. *J. Environ. Eng.* **2017**, *143*, 1–12. [CrossRef]
20. Mayer, T.; Rochfort, Q.; Marsalek, J.; Parrott, J.; Servos, M.; Baker, M.; McInnis, R.; Jurkovic, A.; Scott, I. Environmental characterization of surface runoff from three highway sites in Southern Ontario, Canada. *Water Qual. Res. J. Can.* **2011**, *46*, 110–136. [CrossRef]
21. Gill, L.W.; Ring, P.; Neil, M.P.; Johnston, P.M. Accumulation of heavy metals in a constructed wetland treating road runoff. *Ecol. Eng.* **2014**, *70*, 133–139. [CrossRef]
22. Zhao, R.H. Accumulation and Purification Materials of Urban Rain Water. Ph.D. Dissertation, Tianjin University, Tianjing, China, 2010.
23. Hilliges, R.; Schriewer, A.; Helmreich, B. A three-stage treatment system for highly polluted urban road runoff. *J. Environ. Manag.* **2013**, *128*, 306–312. [CrossRef] [PubMed]
24. Boogaard, F.C.; Lucke, T.; Giesen, N.; van de Ven, F. Evaluate the infiltration performance of eight Dutch permeable pavements using a new full-scale infiltration testing method. *Water* **2014**, *6*, 2070–2083. [CrossRef]
25. Legret, M.; Colandini, V.; Marc, C.L. Effects of a porous pavement with reservoir structure on the quality of runoff water and soil. *Sci. Total Environ.* **1996**, *189–190*, 335–340. [CrossRef]

26. Fwa, T.F.; Lim, E.; Tan, K.H. Comparison of Permeability and Clogging Characteristics of Porous Asphalt and Pervious Concrete Pavement Materials. *Transp. Res. Rec.* **2015**, *2511*, 72–80. [CrossRef]
27. Hares, R.J.; Ward, N.I. Comparison of the heavy metal content of motorway storm water following discharge into wet bio-filtration and dry detention ponds along the London Orbital (M25) motorway. *Sci. Total Environ.* **1999**, *235*, 169–178. [CrossRef]
28. Selim, A.Q.; El-Midany, A.A.; Ibrahim, S.S. Microscopic evaluation of diatomite for advanced applications: Case study. In *Microscopy: Science, Technology, Applications and Education*; Méndez-Vilas, A., Díaz, J., Eds.; Formatex Research Center: Badajoz, Spain, 2010; pp. 2174–2181.
29. State Environmental Protection Administration of China (SEPA). *Standard of Environmental Quality for Surface Water*; GB 3838-2002; Environmental Science Press: Beijing, China, 2002.
30. Wang, Y.; Gao, C.; Yang, S. Research on runoff pollution characteristics and countermeasures at sensitive highway sites. *Pol. J. Environ. Stud.* **2016**, *25*, 813–821. [CrossRef]

*applied sciences*

MDPI

*Article*

# Micromechanism of the Dispersion Behavior of Polymer-Modified Rejuvenators in Aged Asphalt Material

**Mingyu Zhao [1], Fan Shen [2] and Qingjun Ding [1,*]**

1 State Key Laboratory of Silicate Materials for Architectures, Wuhan University of Technology, Wuhan 430070, China; zmyzhaomingyu@whut.edu.cn

2 School of Material Science and Engineering, Wuhan Institute of Technology, Wuhan 430073, China; shenf@wit.edu.cn

* Correspondence: dingqj@whut.edu.cn; Tel.: +86-186-7230-6262

Received: 1 August 2018; Accepted: 6 September 2018; Published: 8 September 2018

**Featured Application: This study of the dispersion behavior for polymer-modified rejuvenator would lay a foundation for the performance research and further recycling engineering practice of polymer-modified rejuvenated asphalt binders.**

**Abstract:** Polymer-modified rejuvenator has a different composition and dispersion behavior to traditional rejuvenators. The objective of this study was to investigate the micromechanism of polymer-modified rejuvenators on the behavior of aged asphalt binder. Firstly, gel permeation chromatography (GPC) analysis was conducted to determine the dispersion effectiveness. Secondly, the dispersal behavior of polymer-modified rejuvenators was studied by means of atomic force microscopy (AFM) and scanning electron microscopy (SEM). Rheological, toughness-tenacity, and force–ductility analyses of the rejuvenated asphalt binder were additionally performed. The results indicate that the contacted asphaltenic micelles in aged asphalt binder were dispersed by dispersion agent in the polymer-modified rejuvenator, and that the dispersion ability of the polymer-modified rejuvenator was promoted to the commercial rejuvenator level. Additionally, the polymer-modified rejuvenator was found to improve the rejuvenated asphalt binder's resistance to deformation, through the formation of polymeric network structures in the asphalt binder. The results may be used to improve the performance of rejuvenated asphalt binder in recycled-pavement engineering.

**Keywords:** dispersion; aged asphalt binder; modified; rejuvenator; micromechanism; performance

## 1. Introduction

Rejuvenators, due to their rich maltene constituents and high permeation ability, can re-balance the composition of aged asphalt binders and have become the key material in pavement-recycling technology [1–3]. In recent years, many pavement materials have reached the end of their service life and have been abandoned in the city (Figure 1). Therefore, developing more effective and durable recycling technologies to rejuvenate this waste asphalt concrete is important for sustainable urban construction [4–6].

At present, the basic composition of the rejuvenator that is commonly used in asphalt pavement recycling is single component, being comprised of materials such as petroleum by-products, industrial oils, living waste, plant extract oils, etc., [7–12], the common feature of which is a high content of aromatics. Laboratory research and different engineering application results show that rejuvenators can restore the rheological and physical properties of binders [13–16]. Xiaokong Yu et al. studied the effect of two generic rejuvenators effect on asphalt binders through rheological property analyses. The result of this study showed that the two complex modulus master curves of virgin binder and rejuvenated binder almost overlapped at 12% rejuvenator content [17]. Additionally, Martins Zaumanis et al. conducted a comprehensive study to evaluate the use of different recycling agents on the properties of rejuvenated binders. The results of this research revealed that all six different agents selected in the study were able to reduce the kinematic viscosity of the aged binder at an intermediate temperature [18]. Many previous studies have indicated that rejuvenators merely soften aged asphalt binder and recover the rheological and physical properties to the level of the ordinary virgin asphalt, though do not significantly improve its performance [19,20]. Therefore, the function of rejuvenator for the improvement of the viscoelasticity and toughness of asphalt binder would be of significance for the durability of recycled pavement, especially given the present trend of using more reclaimed asphalt pavement in recycled mixtures, and the need for higher performance requirements of recycled pavement due to a likely increase in traffic volume in the future [21,22].

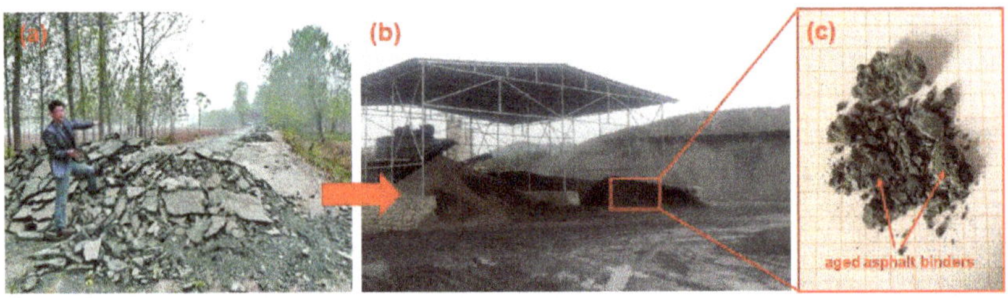


**Figure 1.** Pavement materials which have reached the end of their service life: (**a**) Waste asphalt concrete form old pavement; (**b**) Reclaimed asphalt pavement materials; (**c**) Aged asphalt binders covered with aggregate.

Studies on the modification of aged asphalt binder using a rejuvenating process have recently been receiving research attention. For example, Zhang et al. [23] demonstrated the feasibility of modifying aged asphalt binder by analyzing its features, as well as those of the rejuvenator and modifier. However, this study did not offer specific technical measures. Additionally, a United States patent [24] proposed a method in which a modifier, such as SBS, SBR, or rubber powder, is mixed with a rejuvenator directly during mixing of the asphalt binder in a plant, to achieve the modifying goals. However, in this patent the authors failed to list the technological means by which to ensure the effectiveness of the modifiers on aged asphalt binder. Modifiers require an extended period to disperse and swell in aged asphalt binder at high temperatures, and thus with a short mixing process it was difficult to obtain an anticipative modification effect. To date there are no effective methods to complete the rejuvenating and modifying process simultaneously.

The primary objective of this study is to investigate the effectiveness of polymer-modified rejuvenators on aged asphalt binder, including the two critical aspects of dispersion and performance. The dispersion behavior between the rejuvenator and aged asphalt binder was studied by measuring changes in the molecular weight distribution and micro-colloidal structure along the vertical axis. The effect of rejuvenation and modification on aged asphalt binder was also studied.

## 2. Materials and Methods

### 2.1. Materials

#### 2.1.1. Aged Asphalt Binder

In this study, aged asphalt binder was prepared in the laboratory using a 60/80 penetration grade of virgin asphalt binder (AH-70). Firstly, the virgin asphalt binder was poured into circular plates (diameter 150 mm, depth 30 mm) to produce a film with a thickness of 8–10 mm. The circular plates were then placed in an oven at a temperature of 155 °C for 80 h, being agitated every 30 min to ensure an even aging process, to produce the final aged asphalt binder. To compare the rejuvenating effectiveness of polymer-modified rejuvenators, a commercial modified asphalt binder (Oceanpower Technology Company, Shenzhen, China, Model No 1-D) was used in this study. The basic properties of virgin, aged asphalt binder and modified asphalt binder are presented in Table 1.

**Table 1.** Basic properties of virgin, aged and modified asphalt binder.

| Sample | Penetration (0.1 mm) | Softening Point (°C) | Ductility 25 °C (cm) | Viscosity 60 °C (Pa·s) |
|---|---|---|---|---|
| Virgin asphalt (AH-70) | 72.6 | 47.5 | 162 | 256 |
| Aged asphalt | 35.8 | 65.2 | 12.5 | 1285 |
| Modified asphalt binder (1-D) | 48.2 | 72.5 | 52.3 | 6370 |

### 2.1.2. Rejuvenators

Three different rejuvenators were used in this study, including one commercial rejuvenator (CR) and two polymer-modified rejuvenators (MR-1 and MR-2). The CR, produced by Dongying Mingde Petroleum Technology Co., Ltd. (Dongying, China), is a common rejuvenating agent used for pavement-recycling engineering in China. In this study, CR was used for comparison with the polymer-modified rejuvenators. The polymer-modified rejuvenators were prepared with second-line extract oil (Hengtai Shuangfeng Chemical Factory, Loudi, China), C9 aromatics petroleum resin (Puyang Hengfeng Petrochemical industry, Puyang, China), Linear-type SBS (YH-791, SINOPEC Baling Company, Yueyang, China), stabilizing agent (Dibutyl phthalate) and dispersing agent (Methacrylate quaternary copolymer). MR-2 was the main rejuvenator investigated in this study, while MR-1, which did not contain a dispersing agent, was investigated for comparison with MR-2 in dispersion analysis.

The preparation process for polymer-modified rejuvenator is shown in Figure 2. Firstly, the extracted oil and petroleum resin are blended in a reactor at a temperature of 140 °C for 60 min (Figure 2a). Secondly, the SBS modifier and stabilizing agent are gradually added to a high-speed shearing machine at a temperature of 180 °C for 30 min (Figure 2b). Thirdly, the dispersing agent is added to the same shearing machine, at a lower shearing rate, at a temperature of 170 °C, for 5 min (Figure 2c). Finally, the mixture is kept at a temperature of 150 °C for 30 min (Figure 2d). MR-1 was prepared using steps (a), (b), and (d), as depicted in Figure 2, and MR-2 was prepared using steps (a), (b), (c), and (d).

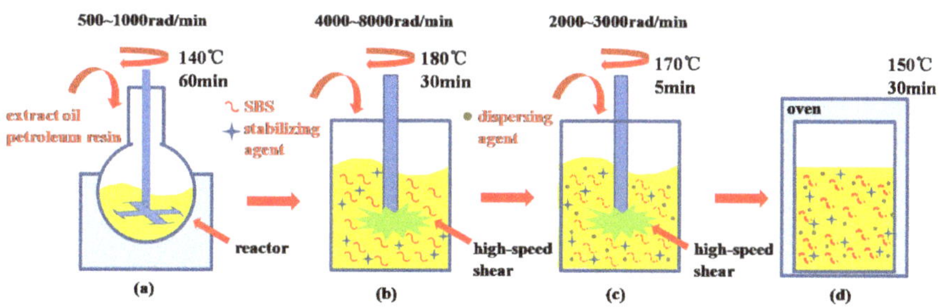


**Figure 2.** Preparation process of the polymer-modified rejuvenators: (**a**) blending the extracted oil and petroleum resin in reactor; (**b**) adding SBS modifier and stabilizing agent in high speed shearing machine; (**c**) adding dispersing agent in high speed shearing machine; (**d**) keeping the samples in oven.

In the aforementioned process, the stabilizing agent was used to prevent the phase segregation of the SBS modifier, and the dispersing agent (in the case of MR-2), which contains carboxyl and amino groups, was used to help evenly disperse the oil, the modifier, and the other rejuvenator components within the aged asphalt binder. The basic properties of the three rejuvenators are presented in Table 2, and the basic composition of the two rejuvenators prepared in the laboratory are presented in Table 3.

**Table 2.** Basic properties of the rejuvenators.

| Sample | Viscosity (60 °C) /Pa·s | Viscosity (135 °C) /Pa·s | Viscosity Ratio by TOFT | Mass Loss by TOFT/% |
|---|---|---|---|---|
| CR | 1.72 | 0.31 | 2.65 | −3.02 |
| MR-1 | 16.31 | 5.36 | 2.03 | −2.18 |
| MR-2 | 8.21 | 2.21 | 1.16 | −2.60 |

**Table 3.** Basic composition of the rejuvenators.

| Sample | Second-Line Extract oil/wt.% | Petroleum Resin/wt.% | Linear-Type SBS/wt.% | Stabilizing Agent/wt.% | Dispersing Agent /wt.% |
|---|---|---|---|---|---|
| MR-1 | 49 | 5 | 40 | 6 | 0 |
| MR-2 | 40 | 5 | 40 | 6 | 9 |

### 2.2. Experiments

#### 2.2.1. Specimen Preparation

To study the dispersal behavior of different rejuvenators in aged asphalt binder, a laboratory method was developed as shown in Figure 3. The preparation process for the rejuvenated specimens at different depths was divided into four steps. Firstly, the aged asphalt binder was poured into a container (Figure 3a) with a depth of 35 mm, and then the rejuvenator (10 g) was deposited evenly on the surface of the asphalt binder. Secondly, the container holding the aged asphalt binder and rejuvenator was placed in an oven at a temperature of 135 °C for 6 h, to initiate the dispersion process (Figure 3b). Thirdly, any rejuvenator that did not penetrate the asphalt binder was removed from the surface using a dropper at a temperature of 105 °C, and the sample was then gradually cooled to a temperature of 5 °C (Figure 3c). Lastly, the samples were removed from the container and cut into three slices—referred to as D1, D2, and D3—each with a thickness of 3 mm, at a temperature of 5 °C (Figure 3d). Following these four steps, the slices were analyzed using gel permeation chromatography (GPC), atomic force microscopy (AFM), and scanning electron microscopy (SEM).

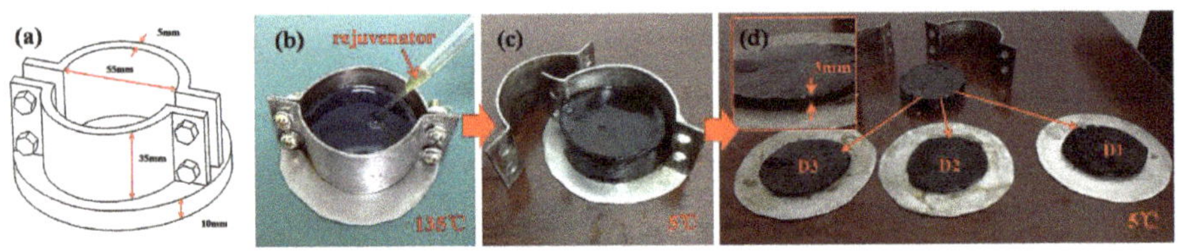


Figure 3. Preparation process of the specimens for dispersal behavior analysis: (**a**) appearance and dimension parameter of the container for aged asphalt binder; (**b**) the asphalt sample at the initial stage of dispersion process; (**c**) the asphalt sample that complete the dispersion process; (**d**) the slices samples which cut at 5 °C.

To assess the performance of the rejuvenated asphalt binder, samples with different rejuvenator contents were also studied. The rejuvenator content of aged asphalt is commonly 4~8 wt.%, and accordingly three different contents—3 wt.%, 6 wt.%, and 9 wt.%—were selected in this study to represent low, medium, and high levels of rejuvenator content, respectively. The aged asphalt binder was mixed with rejuvenator at a temperature of 165 °C for 15 min to ensure a homogeneous system, and rheology, toughness, and tenacity were subsequently assessed.

#### 2.2.2. Gel Permeation Chromatography Analysis

GPC was used to analyze the molecular weight distribution of the rejuvenated asphalt binder at different depths. The measurements were carried out in LT-20A (Shimadzu, Kyoto, Japan) using a RID-10A detector, and two Agilent chromatographic columns (PLgel 5 μm mixed-C, 4.6 × 250 mm) in series. The asphalt binder samples dissolved in tetrahydrofuran (THF), and the polystyrene nanospheres (0.350 mL/min) was used as guide sample in GPC analysis. At least two replicates of each analysis were performed.

#### 2.2.3. Atomic Force Microscopy Analysis

The microstructural characterization of asphalt binders was carried out using AFM, with an Asylum Research MFP-3D-classic AFM instrument (Oxford Instruments, Oxford, UK). A drop of hot, liquid asphalt binder sample was carefully placed onto a 76 × 26 × 1 mm glass slide, and then cooled to ambient temperature before being analyzed by AFM in tapping mode, at a temperature of 30 °C. At least two replicates of each analysis were performed, and 3D height images and 2D phase images (20 × 20 μm) were additionally acquired.

#### 2.2.4. Scanning Electron Microscopy Analysis

SEM was used to characterize the surface morphology of rejuvenated asphalt binder with different rejuvenators. The SEM analysis was performed using a Quanta 450 instrument (FEI, Hillsboro, OR, USA). The asphalt binder sample was placed onto a tin-foil plate at a constant temperature of 135 °C, to ensure that the sample surface was smooth, before being gradually cooled to ambient temperature for SEM analysis. At least two replicates of each analysis were performed.

#### 2.2.5. Rheological Performance Analysis

Dynamic Shear Rheometer (DSR) analyses were conducted to investigate the rheological performance of the rejuvenated asphalt binder. The rheological analyses were performed at a frequency of 10 rad/s (1.59 Hz) at temperatures of between 30 and 80 °C, according to AASHTO PP6 specification. Measurements of the complex shear modulus (G*) and phase angle (δ) were obtained during the analyses, and at least three replicates of each analysis were performed.

#### 2.2.6. Toughness and Tenacity Analysis

Toughness and tenacity performance was analyzed to evaluate the characteristics of internal cohesive force in the asphalt binder during deformation. The samples were imbedded in a container with a hemispherical tension head, and the head was then pulled at a rate of 50 cm/min, at a temperature of 25 °C, to produce a force–deformation curve according to the JTG E20-2011 standard [25]. The analyses were conducted using a Toughness and Tenacity Tester (SYD-0624), and at least three replicates of each analysis were performed. During the test, toughness and tenacity curves for the rejuvenated asphalt binder were obtained.

#### 2.2.7. Force–Ductility Analysis

Force–ductility analysis was conducted to evaluate the deformation resistance capacity of the samples at a temperature of 5 °C, to investigate the durability of recycled pavement material at low in-service temperatures. This analysis was conducted using a Ductility Testing Machine with a dynamometer and data collector (tensile rate of 5 cm/min, sample insulation for 2 h at a temperature of 5 °C before testing) equipped, and at least three replicates of each analysis were performed. During the test, force–ductility curves for the asphalt samples was obtained.

## 3. Results

### 3.1. Gel Permeation Chromatography Analysis

The molecular weight distribution curves for the samples of aged asphalt binder, virgin asphalt binder, and the corresponding rejuvenated samples (by using CR, MR-1 and MR-2), are presented in Figure 4. Compared with the aged samples, the proportion of smaller molecular weight components increased with different degrees in the D1, D2, and D3 slices. Obviously, the molecular weight distribution changes were caused by the dispersion of rejuvenator with small molecular weight into the aged samples. Therefore, the dispersion capacity and rejuvenated effect of aged asphalt binders can be reflected by the changes degree of GPC curves in D1, D2, and D3 slices. Table 4 shows the calculated correlation coefficients of the GPC curves for different sample depths, and aged sample using a statistical method. Larger correlation coefficients correspond to smaller changes in molecular weight. Table 4 shows that the correlation coefficients of MR-2 was close to that of CR, which mean that the dispersion capacity is similar for MR-2 and CR samples. However, the calculated change in molecular component of MR-1 is significantly smaller than that of MR-2 and CR, which mean that the dispersion capacity of MR-1 is lower than MR-2 and CR.

**Table 4.** Correlation analysis results of GPC curves.

| Position | Correlation Coefficient (r) | | |
|---|---|---|---|
| | CR | MR-1 | MR-2 |
| D1 | 0.969758876 | 0.989985163 | 0.970139281 |
| D2 | 0.988691271 | 0.999302387 | 0.987872939 |
| D3 | 0.999597049 | 0.999829042 | 0.999758177 |

The molecular weight of the rejuvenator's major composition is below 800, and accordingly, the proportion of the molecular weight below 800 (expressed as M800) was used to quantify changes in molecular weight more directly in this study. The mean values of M800 and their standard deviations are shown in Table 5.

**Table 5.** Mean values of M800 and their standard deviations of GPC tests.

| Samples | CR | | MR-1 | | MR-2 | |
|---|---|---|---|---|---|---|
| | Mean Value of M800 | Standard Deviation | Mean Value of M800 | Standard Deviation | Mean Value of M800 | Standard Deviation |
| Aged | 28.7 | 1.216 | 28.7 | 1.216 | 28.7 | 1.216 |
| Virgin | 34.1 | 0.925 | 34.1 | 0.925 | 34.1 | 0.925 |
| D1 | 38.8 | 1.053 | 33.7 | 0.931 | 37.7 | 0.832 |
| D2 | 35.0 | 0.921 | 29.9 | 1.823 | 34.8 | 1.056 |
| D3 | 29.8 | 1.336 | 29.4 | 1.521 | 29.2 | 1.228 |

The M800 values for CR in the D1, D2, and D3 slices were 38.8%, 35.0% and 29.8%, respectively. This result indicates that, compared to the aged sample (which has an M800 value of 28.7%), the ratio of substances with a low molecular weight increased, which in turn indicates that the light chemical fractions from the CR rejuvenator diffused into each slice of aged binder to differing degrees. Moreover, the molecular weight distribution of the D2 slice which was rejuvenated by CR is closest to that of the virgin asphalt. The results show that the D2 slice was effectively rejuvenated, and that CR can restore the chemical composition of aged binder to the levels of the virgin binder. Additionally, MR-2 shows a similar effect to CR, with measured M800 values in the D1, D2, and D3 slices of 37.7%, 34.8%, and 29.2%, respectively. This suggests that CR and MR-2 had similar dispersion capacities in the aged asphalt binder. However, the MR-1 rejuvenated sample had significantly lower M800 values in the D2 and D3 slices, which were 29.9% and 29.4%, respectively. The M800 value in the D1 slice is only

33.7%, which indicates that its chemical composition was not restored to the level of the virgin binder. This further indicates that MR-1 did not disperse into the D2 and D3 slices effectively, and that the dispersion ability of MR-1 was lower than that of CR and MR-2.

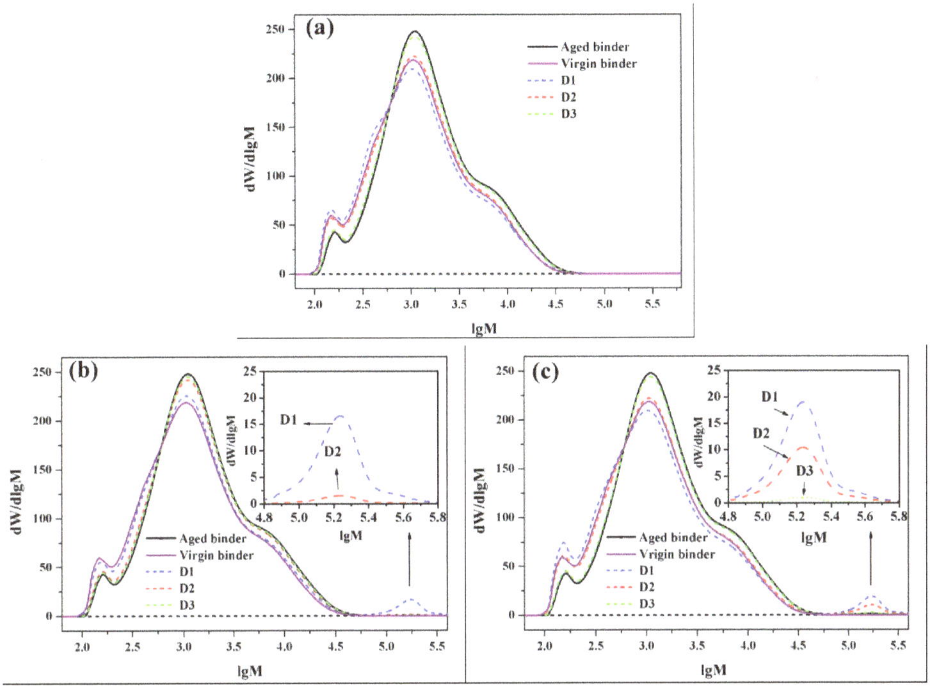


**Figure 4.** The molecular weight distribution curves for asphalt binder samples: (**a**) virgin binder, aged binder, CR rejuvenated binder in D1, D2 and D3 slices; (**b**) virgin binder, aged binder, MR-1 rejuvenated binder in D1, D2 and D3 slices; (**c**) virgin binder, aged binder, MR-2 rejuvenated binder in D1, D2 and D3 slices.

As can be seen in Figure 4, the curves for the rejuvenated samples using MR-1 and MR-2 contain a new peak in the molecular weight (M) range of 80,000 to 250,000 (the corresponding position in Figure 4 is 4.9 to 5.4 in abscissa). These peaks were caused by the presence of the SBS modifier in the rejuvenators [26,27]. Compared to the aged asphalt binder sample, the MR-2 sample had an obvious change in the D1 and D2 slices, and the MR-1 sample only had an obvious change in the D1 slice.

### *3.2. Atomic Force Microscopy Test Results*

As shown in Figure 5a–c, there are numerous continuous peaks in a sequence on the relatively smooth surface. Additionally, as the dispersal depth increases, the peaks increase in intensity. This type of structure is referred to as "bee-like" and is caused by the heterogeneous asphalt colloid structure. This structure can also be distinguished in the phase image (Figure 5d–f), in which the solid asphaltene particles (black and white streaks) appear to be covered by solid particles of resin (light grey areas) surrounded by an oil matrix (yellow areas), in which asphaltene micelles are dispersed [28]. Compared to the D3 slice, the D1 and D2 slices show fewer contact regions, i.e., the contact area of the 'bee-like' structures (Figure 5f) and smaller 'bee-like' structures (<4 μm). Therefore, the addition of the CR rejuvenator increases the development of an oil matrix, and effectively decreases contact phenomena between asphaltene particles. The dispersed asphaltene micelles and continuous oil network lower the cohesive force inside the aged asphalt binder and soften it.

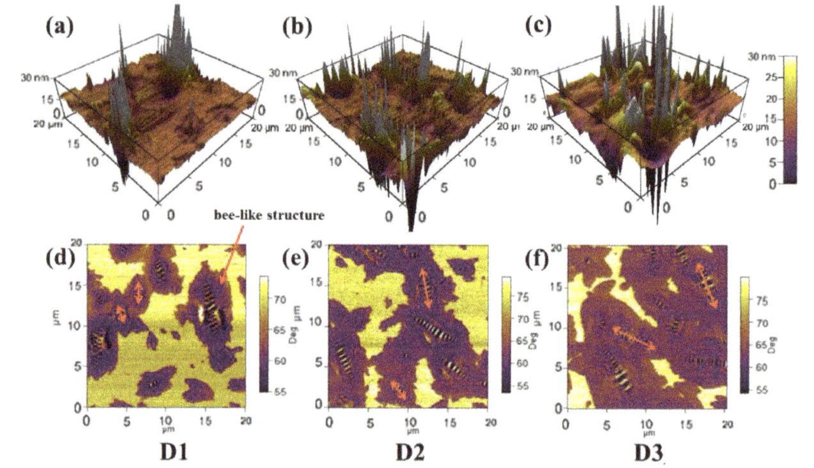


**Figure 5.** AFM images for the CR rejuvenated sample: (**a**) 3D height images of D1 slices; (**b**) 3D height images of D2 slices; (**c**) 3D height images of D3 slices; (**d**) phase images of D1 slices; (**e**) phase images of D2 slices; (**f**) phase images of D3 slices.

Figure 6 shows the obtained AFM micrographs for the D1, D2, and D3 layers of the MR-1 rejuvenated sample. The D2 and D3 layers show the typical structure of aged asphalt binders. However, in contrast to the results shown in Figure 5, the surfaces of the slices from the MR-1 rejuvenated sample (Figure 6a) were relatively rougher than the D2 and D3 slices from the CR rejuvenated sample (Figure 6b,c), where the MR-1 rejuvenator was not permeated in. Previous research [29] has shown that the appearance of a roughened surface structure is caused by the presence of the SBS modifier, due to the formation of a molecular chain network in the asphalt binder.

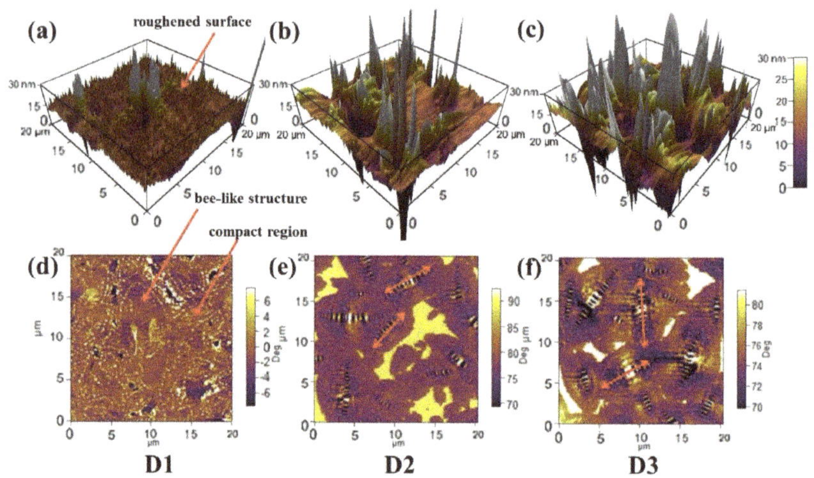


**Figure 6.** AFM images for the MR-1 rejuvenated sample: (**a**) 3D height images of D1 slices; (**b**) 3D height images of D2 slices; (**c**) 3D height images of D3 slices; (**d**) phase images of D1 slices; (**e**) phase images of D2 slices; (**f**) phase images of D3 slices.

Figure 7 shows the AFM micrographs of the D1, D2, and D3 slices from the MR-2 rejuvenated sample. As in the MR-1 rejuvenated sample (Figure 6a), the D1 and D2 slices from the MR-2 rejuvenated sample showed a roughened surface morphology under 3D height mode. Additionally, the 'bee-like' structure in the D1 and D2 slices was smaller (<4 μm) and dispersed, as can be clearly observed in the phase images (Figure 7d,e). Furthermore, the contact phenomena between the asphaltenic micelles are less than for the CR and MR-1 rejuvenators. Therefore, comparing the AFM results between the MR-1 and MR-2 samples, the use of dispersing agents in the polymer-modified rejuvenator can be concluded to have significantly improved the colloid structure of the aged asphalt binder.

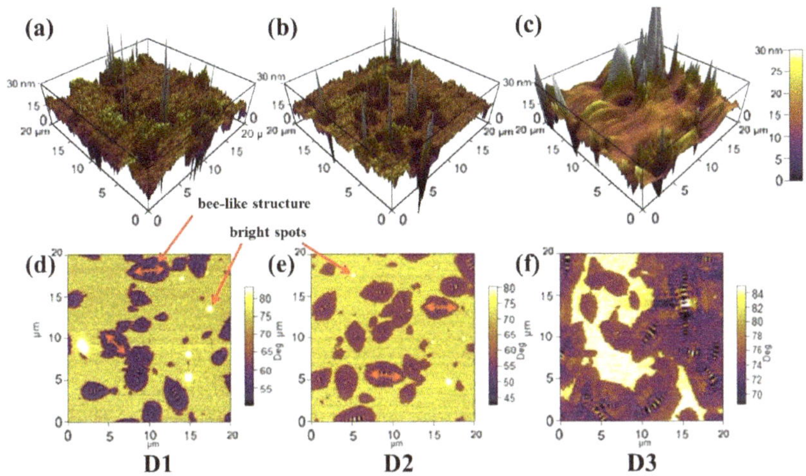


**Figure 7.** AFM images for the MR-2 rejuvenated sample: (**a**) 3D height images of D1 slices; (**b**) 3D height images of D2 slices; (**c**) 3D height images of D3 slices; (**d**) phase images of D1 slices; (**e**) phase images of D2 slices; (**f**) phase images of D3 slices.

In the AFM results, several bright spots can be observed (Figure 7d,e). These are distributed relatively uniformly, are differently sized (0.5–3μm), and have a large phase difference compared to other components. Based on the slice number (D1 and D2) and their large phase difference, these bright spots may be caused by the SBS modifier not completely swelling with the light component of the rejuvenator. SEM was subsequently used to confirm the composition of the bright spots.

### 3.3. Scanning Electron Microscopy Analysis

Figure 8 shows the SEM-derived morphologies of the aged asphalt binder, CR rejuvenated asphalt binder, MR-1 rejuvenated asphalt binder, and MR-2 rejuvenated asphalt binder. It can be observed that the aged and CR rejuvenated samples had flat surfaces. The morphologies of the MR-1 and MR-2 rejuvenated samples showed some light grey patches (0.5~3 μm in size), some of which were raised semicircular particles, and others were as flat as the surface. According to previous SEM research on polymer-modified asphalt binder [30], this appearance is caused by modifier particles, due to the high absorbance of the conglomerate polymer. Additionally, the size of the patches (Figure 8d), and their distribution characteristics, are consistent with the AFM phase images. Therefore, it can be concluded that the bright spots in the AFM phase images were caused by the effect of incomplete swelling with the light component.

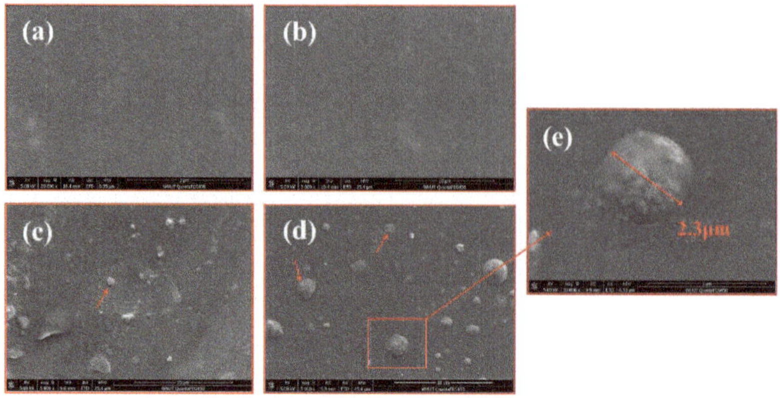


**Figure 8.** SEM-derived morphology for the different asphalt binders: (**a**) aged binder (5000×); (**b**) CR rejuvenated binder (5000×); (**c**) MR-1 rejuvenated binder (5000×); (**d**) MR-2 rejuvenated binder (5000×); (**e**) MR-2 rejuvenated binder enlarged (20,000×).

### *3.4. DSR Analysis*

Figure 9 shows the changes in the complex modulus (G*) and phase angle (δ) of base asphalt binder (AH-70), polymer-modified asphalt binder (1-D), aged asphalt binder, CR rejuvenated asphalt binder (3%, 6%, 9%), and MR-2 rejuvenated asphalt binder (3%, 6%, 9%), with temperature.

In Figure 9a, for both rejuvenators, the rejuvenating resulted in a decrease in the G* for aged asphalt binder. Additionally, as the rejuvenator content increases (3%, 6%, 9%), the decrease range of the G* become greater. This indicates that the addition of the rejuvenators can soften the aged asphalt binder in different degree.

However, it can also be observed that the G* of the MR-2 rejuvenated sample was higher than that of the CR rejuvenated sample at a high temperature (50 to 80 °C) under the same rejuvenator content. Previous studies showed that the asphalt is a temperature sensitive material. Therefore, the test result indicates that the temperature stability of rejuvenated binder with MR-2 is higher than CR samples, and as the increases of MR-2 content, the stability of G* is more obvious.

Accordingly, in Figure 9b, the phase angle shows a similar changes law to the G*, with the different rejuvenator content of CR and MR-2. Nevertheless, there are obvious plateau regions are observable in the phase angle curves, which caused by the stable polymer network structure in MR-2 (9%) and 1-D asphalt samples at high in-service temperature. When the MR-2 rejuvenator content was reduced, the plateau decreased (6% MR-2 content) and ultimately disappeared (3% MR-2 content), which was due to the weakened network structure with the gradual reduction of polymer content in rejuvenator.

Overall, combined with test results of G* and phase angle, the asphalt binder rejuvenated with MR-2 has excellent performance of deformation resistance at a high in-service temperature, as with polymer-modified asphalt. On the contrary, the asphalt binder rejuvenated with CR has a relative general performance of deformation resistance at a high in-service temperature. From these curves, it can also be observed that the changes in complex modulus and phase angle with temperature of the MR-2-9% and CR-9% samples are closest to that of the 1-D and AH-70 samples, respectively.

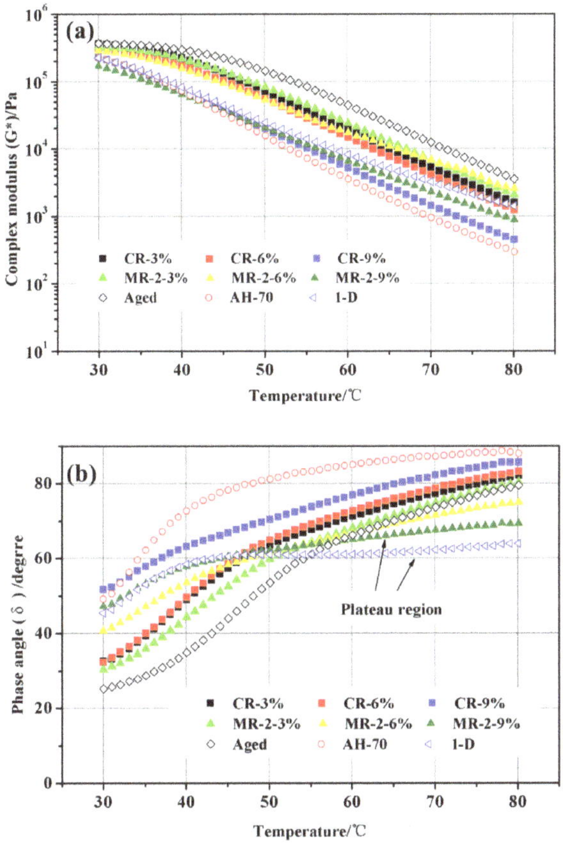


**Figure 9.** DSR analysis results for CR and MR-2 rejuvenated binders at different contents: (**a**) curves of complex modulus as a function of temperature; (**b**) curves of phase angle as a function of temperature.

### *3.5. Toughness and Tenacity Analysis*

The toughness and tenacity curve for the CR rejuvenated asphalt binder is shown in Figure 10a. It has been well established that the shape of such curves reflects internal stress changes with tension. The aged asphalt binder in Figure 10a had the typical brittleness characteristic of the material and exhibited a rapid increase in stress and a direct fracture as the deformation increased. It was observed that the CR rejuvenator effectively softened the aged asphalt binder, which changed with the virgin asphalt binder (AH-70) type. Additionally, as shown in Table 6, an increase of the CR rejuvenator content was associated with a gradual reduction in the toughness of the asphalt binder, and an increase in the tenacity of the asphalt binder.

As shown in Figure 10b, the shape of the toughness and tenacity curve for the MR-2 rejuvenated asphalt binder was similar to that of the modified asphalt binder (1-D). There are two obvious characteristics of the toughness and tenacity curve with increasing MR-2 content: (1) The height of the first peak was reduced; and (2) the yielding phenomenon became more evident. The shape of the first peak on the toughness and tenacity curve varied with the properties of the base asphalt binder, and the shape of the yielding region varied with the properties of the modifier [31,32]. The results indicate that the rejuvenated component of MR-2 can soften the aged asphalt binder, and that the modifying components can improve its ability to resist deformation, especially when the MR-2 content was higher than 6%.

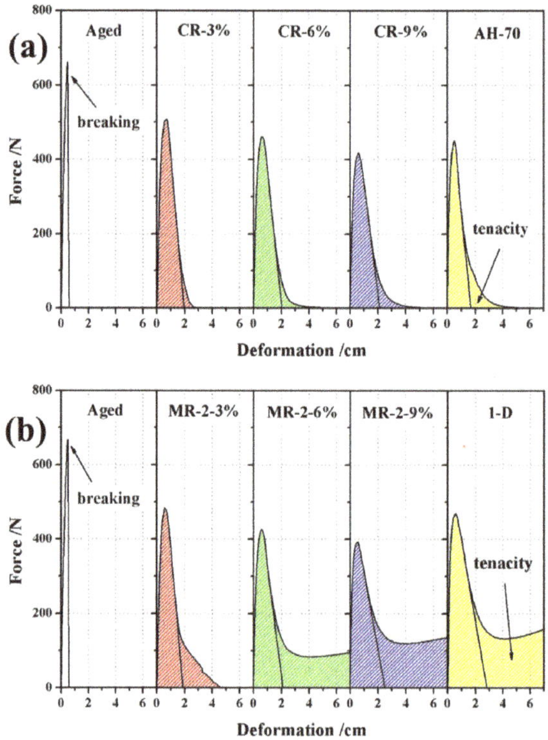


**Figure 10.** Force–deformation curves for asphalt binder samples: (**a**) aged binder, virgin binder (AH-70) and CR rejuvenated binder (3%, 6% and 9%); (**b**) aged binder, polymer-modified binder (1-D) and MR-2 rejuvenated binder (3%, 6% and 9%).

**Table 6.** Toughness and tenacity of asphalt binders.

| Samples | CR | | | MR-2 | | | 1-D | AH-70 |
|---|---|---|---|---|---|---|---|---|
| | 3% | 6% | 9% | 3% | 6% | 9% | | |
| Toughness/N·m | 6.21 | 6.03 | 5.75 | 7.61 | 16.31 | 19.23 | 24.46 | 5.42 |
| Standard Deviation | 0.12 | 0.13 | 0.08 | 0.12 | 0.15 | 0.36 | 0.28 | 0.16 |
| Tenacity/ N·m | 0.32 | 0.72 | 0.91 | 1.63 | 11.23 | 15.06 | 19.72 | 1.21 |
| Standard Deviation | 0.08 | 0.05 | 0.09 | 0.09 | 0.31 | 0.46 | 0.51 | 0.08 |

### *3.6. Force–Ductility Test Result*

The force–ductility curves for the CR rejuvenated binder, MR-2 rejuvenated binder, aged binder, and virgin binders is shown in Figure 11. As can be seen in the enlarged image of the beginning tensile section of the force–ductility curve (Figure 11a), most of the samples (including CR-3%, CR-6%, CR-9%, MR-3%, and aged and virgin samples) experienced brittle fracture immediately after stretching at a temperature of 5 °C, and the ductilities were all less than 1 cm (see Table 7). The results indicate that those asphalt samples partly lost their viscoelastic behavior, and started to show brittle characteristics at a temperature of 5 °C.

The polymer-modified asphalt (1-D) and MR-2 rejuvenated asphalt (with 6% and 9% content) both completed the tensile process at temperatures of 5 °C. The complete force–ductility curve is shown in Figure 11b. In addition to the ductility, the force–ductility curve can be reflected by two significant indices, i.e., the maximum tension force (Point A in Figure 11) and fracture tension force (Point B in

Figure 11). As is shown in Figure 11b and Table 7, the 1-D sample has the highest ductility, tension force, and fracture tension force, which indicates that its deformation resistance ability is improved by a polymer network in the asphalt material structure at a temperature of 5 °C, and that the risk of cracking failure was reduced at low in-service temperatures. The force–ductility curve of the MR-2-9% rejuvenated binder is similar to that of the 1-D sample, and the ductility, maximum tension force, and fracture tension force of the MR-2-9% rejuvenated binder were relatively lower than those of the 1-D sample. Moreover, as the content of the polymer-modified rejuvenator reduced to 6%, the ductility, maximum tension force and fracture tension force were further decreased, and the internal force of resistance to deformation was also gradually decreased before the sample underwent fracture.

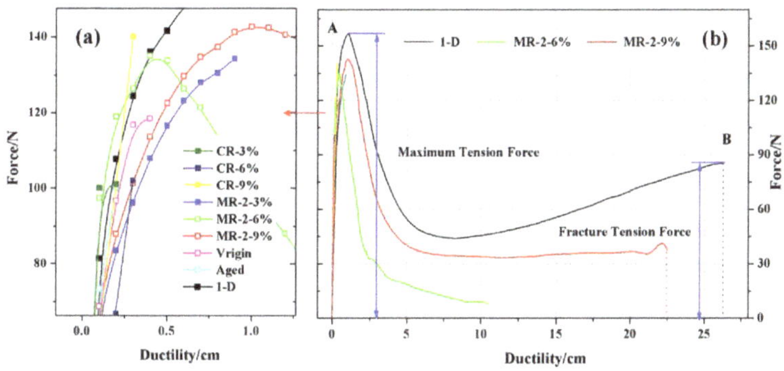


**Figure 11.** Force–ductility curves for the aged asphalt, virgin asphalts, and rejuvenated asphalts at a temperature of 5 °C: (**a**) The enlarged image of the beginning stretch section of the force–ductility curves; (**b**) The complete force–ductility curves image of the 1-D, MR-2-6%, and MR-2-9% samples at a temperature of 5 °C.

**Table 7.** Force–ductility analysis results for the aged, virgin, and rejuvenated asphalt binders.

| Sample | CR | | | MR-2 | | | Aged | Virgin (AH-70) | 1-D |
|---|---|---|---|---|---|---|---|---|---|
| | 3% | 6% | 9% | 3% | 6% | 9% | | | |
| Ductility/cm | <1.0 | <1.0 | <1.0 | <1.0 | 10.5 | 22.5 | <1.0 | <1.0 | 26.3 |
| Maximum tension force/N | 101.2 | 102.5 | 139.8 | 134.1 | 135.1 | 142.6 | 98.8 | 116.5 | 156.6 |
| Fracture tension force/N | 101.2 | 102.5 | 139.8 | 134.1 | 7.6 | 37.3 | 98.8 | 116.5 | 85.2 |

## 4. Discussion

The dispersion behavior of rejuvenators directly affects their rejuvenating effectiveness in aged asphalt binder. There are many factors that can promote or obstruct this dispersion behavior, such as the effective diffusivity, viscosity, interfacial tension, molecular group structure, etc. In this study, we have proposed a multi-component rejuvenator with polymer modifiers to improve the performance of rejuvenated binders. In general, the dispersive ability of CRs is considered to be enough for the process of pavement recycling. However, the dispersive ability of the MR-1 rejuvenator, which directly compounded the SBS modifier with extracted oil, was observed to decrease significantly compared with CR. The GPC analysis showed that, as the dispersing agent was added to MR-1, the dispersive ability of the polymer-modified rejuvenator was improved to the level of CR. Based on the results of the analysis, there are two reasons for this phenomenon: (1) The dispersing agent reduced the viscosity of the rejuvenator from 5.36 Pa·s to 2.21 Pa·s at a temperature of 135 °C, and increased its dispersion rate in the aged asphalt binder; (2) As other studies have shown, the asphaltene micelles which bonded in the asphalt colloid structure provided the main dispersion resistance against the rejuvenator [33]. In asphalt structures, the asphaltene associated by hydroxyl hydrogen bond and agglomerate to form asphaltene micelles. The polar groups in the methacrylate quaternary copolymer (the principal constituent of the dispersing agent) can form carboxyl hydrogen bonds with asphaltene, and the binding force of carboxyl hydrogen bonds is stronger than that of hydroxyl bonds [34]. Thus, the agglomerate structure of asphaltene is partly broken, which is evidenced by the smaller asphaltene micelles observed in the AFM analysis (see Figure 7d,e), and the dispersing resistance from the molecular group structure is reduced.

Based on the results of the DSR, toughness and tenacity, and force–ductility analyses, the asphalt with polymer-modified rejuvenator exhibits the performance features of modified asphalt material, indicating that asphalt binder which is rejuvenated by polymer-modified rejuvenator could improve the performance of recycled pavement material at high and low in-service temperatures. In summary, the polymer-modified rejuvenator not only dispersed the light components into the aged asphalt binder, but also formed a polymer network which reinforced the aged asphalt binder.

## 5. Conclusions

To better understand the effectiveness of polymer-modified rejuvenators on aged asphalt binders, their dispersal behavior, rheological properties, deformation resistance capabilities, and low temperature properties were measured using GPC, AFM, SEM, DSR, toughness and tenacity, and force–ductility (5 °C) analyses. Based on these results, the following conclusions can be drawn:

(1) A new analysis method for assessing the dispersive ability of rejuvenator in aged asphalt binders is proposed, and determined by molecular weight distribution and AFM imaging at different vertical positions within the asphalt binder;
(2) The dispersion ability of polymer-modified rejuvenator on aged asphalt binder is qualified for the pavement-recycling process. The polar groups in the dispersing agent can form carboxyl hydrogen bonds with asphaltene, and partly break the agglomerate structure of asphaltene to reduce the dispersion resistance from molecular group structure;
(3) Asphalt binder treated with polymer-modified rejuvenator could improve the performance of recycled pavement material at high and low in-service temperatures. The complex modulus at a temperature of 60 °C is increased from 5190 to 6640 Pa, the tenacity is increased from 0.91 N·m to 15.06 N·m, and the ductility at a temperature of 5 °C is increased to 22.5 cm.

**Author Contributions:** M.Z. and Q.D.; Data curation: M.Z.; Formal analysis: M.Z. and F.S.; Funding acquisition: F.S.; Methodology: F.S.; Project administration: Q.D.; Writing—original draft: M.Z.; Writing—review and editing: Q.D.

**Funding:** This work was supported by the National Natural Science Foundation of China (NSFC) under Grant number 51302198 and Highway Science and Technology Project of Hubei Province.

**Acknowledgments:** This research was conducted at State Key Laboratory of Silicate Materials for Architectures in China.

**Conflicts of Interest:** The authors declare no conflict of interest.

## References

1. Im, S.; Zhou, F.; Lee, R.; Scullion, T. Impacts of rejuvenators on performance and engineering properties of asphalt mixtures containing recycled materials. *Constr. Build. Mater.* **2014**, *53*, 596–603. [CrossRef]
2. Kou, C.; Xiao, P.; Kang, A.; Mikhailenko, P.; Baaj, H.; Wu, Z. Methods to Evaluate the Aging Grades of Reclaimed Asphalt Binder. *Appl. Sci.* **2017**, *7*, 1209. [CrossRef]
3. Bowers, B.F.; Moore, J.; Huang, B.; Shu, X. Blending efficiency of Reclaimed Asphalt Pavement: An approach utilizing rheological properties and molecular weight distributions. *Fuel* **2014**, *145*, 63–68. [CrossRef]
4. Stimilli, A.; Ferrotti, G.; Conti, C.; Tosi, G.; Canestrari, F. Chemical and rheological analysis of modified bitumens blended with 'artificial reclaimed bitumen. *Constr. Build. Mater.* **2014**, *63*, 1–10. [CrossRef]
5. Han, S.; Chung, H.-W.; Subgranon, T.; Tia, M. Development of Mix Designs for Minimally Refined Reclaimed Asphalt Pavement (RAP) Concrete for Florida Concrete Test Road. *Sustainability* **2018**, *10*, 2598. [CrossRef]
6. An, J.; Nam, B.H.; Youn, H. Investigation on the Effect of Recycled Asphalt Shingle (RAS) in Portland Cement Mortar. *Sustainability* **2016**, *8*, 384. [CrossRef]

7. Zargar, M.; Ahmadinia, E.; Asli, H.; Karim, M.R. Investigation of the possibility of using waste cooking oil as a rejuvenating agent for aged bitumen. *J. Hazard. Mater.* **2012**, *233–234*, 254–258. [CrossRef] [PubMed]
8. Hesp, S.A.; Shurvell, H.F. Quality Assurance Testing of Asphalt Containing Waste Engine oil. In Proceedings of the International Journal of Pavements Conference, Sao Paulo, Brazil, 9–10 December 2013; pp. 3–12.
9. Álvaro, G.; Erik, S.; Martin, V. Properties of capsules containing rejuvenators for their use in asphalt concrete. *Fuel* **2012**, *90*, 583–591.
10. Cao, X.; Cao, X.; Tang, B.; Wang, Y.; Li, X. Investigation on Possibility of Waste Vegetable Oil Rejuvenating Aged Asphalt. *Appl. Sci.* **2018**, *8*, 765.
11. Asli, H.; Ahmadinia, E.; Zargar, M.; Karim, M.R. Investigation on physical properties of waste cooking oil—Rejuvenated bitumen binder. *Constr. Build. Mater.* **2012**, *37*, 398–405. [CrossRef]
12. Silva, H.; Oliveira, J.; Jesus, C. Are totally recycled hot mix asphalts a sustainable alternative for road paving? *Resour. Conserv. Recycl.* **2012**, *60*, 38–48. [CrossRef]
13. Tran, N.H.; Taylor, A.; Willis, R. *Effect of Rejuvenator on Performance Properties of HMA Mixtures with High RAP and RAS Contents*; NCAT Report 12-05; Auburn University: Auburn, AL, USA, 2012.
14. Shen, J.; Amirkhanian, S.; Tang, B. Effects of rejuvenator on performance-based properties of rejuvenated asphalt binder and mixtures. *Constr. Build. Mater.* **2007**, *21*, 958–964. [CrossRef]
15. Cheng, Y.; Wang, W.; Tan, G. Assessing High- and Low-Temperature Properties of Asphalt Pavements Incorporating Waste Oil Shale as an Alternative Material in Jilin Province, China. *Sustainability* **2018**, *10*, 2179. [CrossRef]
16. Walubita, L.F. Comparison of Fatigue Analysis Approaches for Predicting Fatigue Lives of Hot-Mix Asphalt Concrete (HMAC) Mixtures. Ph.D. Thesis, Texas A&M University, College Station, TX, USA, 2006.
17. Yu, X.; Zaumanis, M.; Dos Santos, S.; Poulikakos, L.D. Rheological, microscopic, and chemical characterization of the rejuvenating effect on asphalt binders. *Fuel* **2014**, *135*, 162–171. [CrossRef]
18. Moghaddam, T.B.; Baaj, H. The use of rejuvenating agents in production of recycled hot mix asphalt: A systematic review. *Constr. Build. Mater.* **2016**, *114*, 805–816. [CrossRef]
19. Shen, J.; Amirkhanian, S.; Miller, J.A. Effects of Rejuvenating Agents on Superpave Mixtures Containing Reclaimed Asphalt Pavement. *J. Mater. Civ. Eng.* **2007**, *19*, 376–384. [CrossRef]
20. Romera, R.; Santamaría, A.; Peña, J.J.; Muñoz, M.E.; Barral, M.; García, E.; Jañez, V. Rheological aspects of the rejuvenation of aged bitumen. *Rheol. Acta* **2006**, *45*, 47–48. [CrossRef]
21. Donya, A.; Colina, J.; Bruneaub, D. Reclaimed asphalt concretes with high recycling rates: Changes in reclaimed binder properties according to rejuvenating agent. *Constr. Build. Mater.* **2013**, *41*, 171–181. [CrossRef]
22. Miliutenko, S.; Bjorklund, A.; Carlsson, A. Opportunities for environmentally improved asphalt recycling: The example of Sweden. *J. Clean. Prod.* **2013**, *43*, 56–65. [CrossRef]
23. Zhang, D.; Qu, Y.; Zhao, Y.; Xu, Q. The feasibility analysis of modify the recycled asphalt. *Pet. Asph.* **2010**, *22*, 69–71.
24. Huh, J.D. Temperature-Adjusted and Modified Recycled Ascon Composition for Reusing 100% of Waste Ascon for Road Pavement and Method for Manufacturing Same. U.S. Patent US20120167802A1, 7 March 2012.
25. China Communications Press. *JTG E20: 2011, Standard Test Methods of Bitumen and Bituminous Mixtures for Highway Engineering*; China Communications Press: Beijing, China, 2011.
26. Sun, L.; Wang, Y.Y.; Zhang, Y.M. Aging mechanism and effective recycling ratio of SBS modified asphalt. *Constr. Build. Mater.* **2014**, *70*, 26–35. [CrossRef]
27. Liu, G.; Nielsen, E.; Komacka, J.; Greet, L.; van de Ven, M. Rheological and chemical evaluation on the ageing properties of SBS polymer modified bitumen: From the laboratory to the field. *Constr. Build. Mater.* **2014**, *51*, 244–248.
28. Carrera, V.; Partal, P.; García-Morales, M.; Gallegos, C.; Páez, A. Influence of bitumen colloidal nature on the design of isocyanate-based bituminous products with enhanced rheological properties. *Ind. Eng. Chem. Res.* **2009**, *48*, 8464–8470. [CrossRef]
29. Wu, S.; Pang, L.; Mo, L.; Chen, Y.; Zhu, G. Influence of aging on the evolution of structure, morphology and rheology of base and SBS modified bitumen. *Constr. Build. Mater.* **2009**, *23*, 1005–1010. [CrossRef]
30. Zhu, J.Q.; Birgisson, B.; Kringos, N. Polymer modification of bitumen: Advances and challenges. *Eur. Polym. J.* **2014**, *54*, 18–38. [CrossRef]

31. McNally, T. *Polymer Modified Bitumen-Properties and Characterization*; Woodhead Publishing: Cambridge, UK, 2011.
32. Gao, J.; Wang, H.; You, Z.; Hasan, M.R.M.; Lei, Y.; Irfan, M. Rheological Behavior and Sensitivity of Wood-Derived Bio-Oil Modified Asphalt Binders. *Appl. Sci.* **2018**, *8*, 919. [CrossRef]
33. Kuang, D.L. Preparation of Diffusible Rejuvenator and Its Influence on Performances of Recycled Bitumen and Recycled Asphalt Mixture. Ph.D. Thesis, Wuhan University of Technology, Wuhan, China, 2012.
34. Song, H. Study on Synthesis Technology and Evaluation of Viscosity Reduction Copolymer Based on Long Chain Methacrylate. Master's Thesis, Shandong University, Jinan, China, 2011.

applied sciences

MDPI

Article

# Evaluation of the Durability and the Property of an Asphalt Concrete with Nano Hydrophobic Silane Silica in Spring-Thawing Season

**Wei Guo [1], Xuedong Guo [1], Mingzhi Sun [2] and Wenting Dai [1,*]**

[1] School of Transportation, Jilin University, Changchun 130022, China; guowei17@mails.jlu.edu.cn (W.G.); guoxd@jlu.edu.cn (X.G.)

[2] Research Institute of Highway, Ministry of Transport of China, Beijing 100088, China; mzsun14@mails.jlu.edu.cn

* Correspondence: daiwtl@jlu.edu.cn; Tel.: +86-130-8681-4298

Received: 12 July 2018; Accepted: 24 August 2018; Published: 28 August 2018

**Featured Application: This paper proved that adding Nano Hydrophobic Silane Silica is an effective technique for mitigating freeze-thaw cycle damage of asphalt concrete in spring-thawing season. Moreover, it's found that the freeze factor had a more significant impact on the damage process of asphalt concrete compared with the soak and scour factor, which provides suggestions for pavement construction in seasonal frozen region.**

**Abstract:** In the spring-thawing season, the high frequency of freeze-soak-scour cycles in the short term is the main cause of pavement damage in the frozen region. One of the methods to improve the durability of asphalt concrete in spring-thawing season is to add suitable modifiers and additives which improve adhesion between asphalt binder and aggregate. This study evaluates the effect of nano hydrophobic silane silica (NHSS) on the performance damage of asphalt concrete (AC) in spring-thawing season. The effectiveness of nano hydrophobic silane silica was evaluated by conducting mixture tests after different freeze-soak-scour cycles, and the damage model of NHSS modified asphalt concrete was established based on the logistic damage model. The results showed that adding NHSS is an effective technique for mitigating freeze-soak-scour cycle damage of asphalt concrete in spring-thawing season. Moreover, the influence of scour, soak, and freeze—three separate factors on NHSS-modified AC in spring-thawing season—was discussed based the gray rational degree theory. The results illustrated that the freeze factor had a more significant impact on the damage process of NHSS modified asphalt concrete compared with the soak and scour factor.

**Keywords:** nano hydrophobic silane silica; spring-thawing season; damage evolution; damage model; gray rational degree theory

---

## 1. Introduction

The seasonal frozen region covers a large area in China, accounting for 53.5% of the country's land area. The high frequency of freeze–thaw cycles in the short term is the main cause of pavement damage in the season frozen region. This damage leads to an increase of maintenance cost and decrease in the service life of pavement [1,2]. Despite the developments that have increased the understanding of asphalt mixture behavior and mix design, freeze–thaw cycle damage is still considered a complex problem in asphalt pavements [3–5].

In recent years, the evolution of asphalt mixtures under freeze–thaw cycles was evaluated by various research [6–10] used the logistic model to study the effects of freeze–thaw cycles on the compressibility of mixture [11]. Gong et al. studied the effects of freeze–thaw cycles on the low temperature performance of mixture [12]. Islam et al. studied the effects of freeze–thaw cycles on the strength of mixture [13]. In conclusion, the freeze–thaw cycle damage in asphalt mixtures can be defined as the loss of strength and stiffness of mix because moisture migration and accumulation under the high frequency of freeze–thaw cycles. In the spring-thawing season, free water formed by the melting of snow on the pavement penetrates into the interior of the pavement structure due to the rising daytime temperature. The volume of free water expands after being frozen, which generates internal temperature stress in the mixture, and this effect can enlarge the micropores and original cracks in the asphalt layer. The ice crystalloids are transformed into free water which cause moisture damage with the increase of temperature. Moreover, the free water forms' scouring, squeezing, and pumping effects on the internal structure of pavement under the traffic load cause asphalt pavement distress—such as strength losing, rutting, raveling, and fatigue cracking [14].

Because of the complexity of the freeze–thaw cycle damage phenomenon, using proper additives and modifiers is considered the most cost-effective technique for mitigating freeze–thaw cycle damage. Wei et al. investigated the effect of diatomite and styrene butadiene styrene on antifreezing performance of crumb rubber modified stone mastic asphalt (SMA), the results indicate that crumb rubber and diatomite modified SMA has the good freeze–thaw resistance [15]. Klinsky et al. held that the use of polypropylene and aramid fibers in HMA could enhance the performance of asphalt pavements against common distresses [16]. Zhang et al. found that micro or nano zinc oxide can significantly improve the anti-freeze–thaw performance of asphalt and asphalt mixture [17]. Qian et al. found that phosphorus slag powder modified by TM-P (10% by weight of the powder) improved freeze–thaw resistance of asphalt mixture [18]. Hamedi evaluated the effects of two types of nanomaterials in two different percentages (nano $Al_2O_3$ and $Fe_2O_3$), two types of aggregates (granite and quartzite), and one base asphalt binder. The results showed that asphalt binder modification with nanomaterials decreases the moisture damage susceptibility [19].

Among the existing modifiers, natural and synthetic polymers are widely used to improve the durability and the properties of asphalt concrete in spring-thawing season, but such a modifier presents a serious isolation problem that restricts its application in modifying paving asphalt. Nano hydrophobic silane silica (NHSS) draws much attention at present because of its high compatibility, and its capability to improve base asphalt properties. The nano hydrophobic silane silica (NHSS) is obtained by the surface modification of nanosilica by silane coupling agent, which can greatly improve the dispersibility of modified nanosilica in the organic polymer, so that the whole system has better stability, and the advantages of nanosilica and silane coupling agent can be fully utilized in the mixed system. To the authors' knowledge, the open literature has no experimental studies of nano hydrophobic silane silica modified asphalt concrete subjected to long freeze-soak-scour cycles. In this study, a laboratory freeze-soak-scour cycle test was proposed to simulate the actual condition of asphalt pavement during spring-thawing season. The normal asphalt concrete and NHSS modified asphalt concrete specimens were prepared for the freeze-soak-scour cycle. After different freeze-soak–scour cycles, damaged specimens were collected for mechanical tests and internal structure parameters tests to identify the effectiveness of nano hydrophobic silane silica. Moreover, the damage model of NHSS modified asphalt concrete was established based on the logistic damage model. The influence of freeze, soak, and scour damage factors on the damage process of NHSS modified asphalt concrete in spring-thawing season is analyzed based on the gray rational degree theory.

## 2. Materials

### 2.1. Aggregates

In this study, crushed and sharp-edged aggregates (Jiutai, China, 2018) were used for preparation of AC. Physical properties of the aggregates and the gradation of AC are listed in Table 1.

**Table 1.** Properties of aggregate and gradation of AC-16.

| Sieve (mm) | 16 | 13.2 | 9.5 | 4.75 | 2.36 | 1.18 | 0.6 | 0.3 | 0.15 | 0.075 |
|---|---|---|---|---|---|---|---|---|---|---|
| Apparent density | 2.78 | 2.78 | 2.78 | 2.74 | 2.74 | 2.75 | 2.70 | 2.70 | 2.69 | 2.67 |
| Bulk density | 2.77 | 2.76 | 2.75 | 2.68 | 2.64 | 2.59 | 2.60 | 2.59 | 2.57 | 2.55 |
| Gradation (%) | 95 | 88 | 73 | 46 | 31 | 21.5 | 15.5 | 11.5 | 5.5 | 6 |

### 2.2. Asphalt

The asphalt (Panjin, China, 2018) was used in this study. Laboratory tests were carried out in order to assess the conventional properties of asphalt. The test results of the asphalt used in this study are summarized in Table 2.

**Table 2.** Technical parameters of asphalt.

| Technical Parameters | Penetration | | | 25 °C Ductility | Softening Point | Wax Content | Flash Point | Solubility | Density |
|---|---|---|---|---|---|---|---|---|---|
| | 15 °C | 25 °C | 30 °C | | | | | | |
| Units | | 0.1 mm | | cm | °C | %≤ | °C | %≥ | g·cm$^{-3}$ |
| Test results | 27.5 | 81.3 | 132.6 | >130 | 44.2 | 18 | 340 | 99.9 | 1.003 |

### 2.3. Nano Hydrophobic Silane Silica

Nanosilica is recognized as a modifer of asphalt for its superior stability, thixotropy, reinforcing, and thickening properties [20–22]. Yao et al. found that the antiaging property and rutting and fatigue cracking performance of nanosilica modified asphalt binders are enhanced, and the addition of nanosilica in the control asphalt asphalt mixture significantly improves the dynamic modulus, flow number, and rutting resistance of asphalt mixtures [23]. However, the presence of a large amount of hydroxyl groups on the surface of nanosilica particles causes the nanosilica to exhibit hydrophilicity, and the hydroxyl group of nanosilica has a high surface energy due to its strong polarity. Thus, the nanosilica particles are in a thermodynamic unstable state and are easily attracted to each other to form agglomerates, which makes it difficult to mix well in asphalt and asphalt mixture, leading to poor dispersion and compatibility with asphalt binder.

In order to improve the compatibility of nanosilica in asphalt, the silane coupling agent was grafted onto the surface of nanosilica by chemical coupling to achieve surface modification of nanosilica. The surface modification method of nanosilica is detailed as follows: the X group in the silane coupling agent first undergoes hydrolysis by contacting with water, and then forms a temporary oligomer by dehydration condensation. The hydroxyl groups on the surface of nanosilica can react with the oligomeric structure to generate hydrogen bonds, then the nanosilica and the oligomer continue to undergo condensation and dehydration reaction by heating, drying, etc., and finally the silane coupling agent is successfully grafted onto the surface of the nanosilica by a covalent bond, as is shown in Figure 1.

The nano hydrophobic silane silica (NHSS) is obtained by the surface modification of nanosilica by silane coupling agent, which can greatly improve the dispersibility of modified nanosilica in the organic polymer, so that the whole system has better stability, and the advantages of nanosilica and silane coupling agent can be fully utilized in the mixed system. NHSS was selected from Changtai Micro-Nano Chemical Co.,Ltd (Shouguang, Shandong, China) to be used in this study. The technical parameters of NHSS used in this study are summarized in Table 3.

**Table 3.** Technical parameters of nano hydrophobic silane silica (NHSS).

| Technical Parameters | BET ($m^2/g$) | Average Particle Size (nm) | Loss on Drying (105 °C, 2 h, wt%) | Loss on Ignition (1000 °C, 2 h, wt%) | pH Value | Carbon Content (%) | $SiO_2$ Content (%) |
|---|---|---|---|---|---|---|---|
| **Test results** | 125 ± 20 | 12 | ≤0.5 | 1.5–2.5 | 5.0–8.0 | 2.0–3.5 | ≥99.8 |

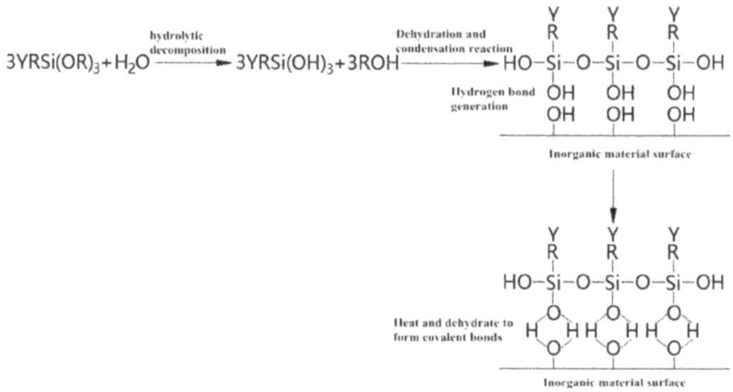


**Figure 1.** Grafting process of silane coupling agent on nanosilica.

The microstructure of NHSS and nanosilica were examined by SU8000 electronic microscopy (Tianmei.co, Tokyo, Japan, 2008). The scanning electron micrographs of nanosilica and NHSS observed at magnifying power of ×10,000 are shown in Figure 2. (a and b). The SEM images of the matrix asphalt with 3% nanosilica (NS-MA) and the matrix asphalt with 3% NHSS (NHSS-MA) are demonstrated in Figure 3a,b and the light particles observed in Figure 3 are NHSS and nanosilica.

(**a**) Nanosilica (**b**) NHSS

**Figure 2.** SEM images of nanosilica and nano hydrophobic silane silica (NHSS) at magnifications of ×10000.

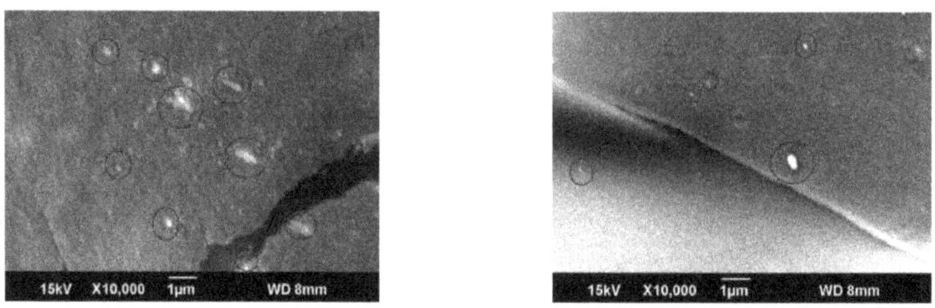


(**a**) NS-MA (**b**) NHSS-MA

**Figure 3.** Scanning electron micrographs of nanosilica modified asphalt and NHSS modified asphalt.

It can be seen from Figure 2 that the surface of aggregates formed by the NHSS particles is rough compared with nanosilica particles, and the NHSS particles combine to form microchains and micro-mesh structures in three-dimensional reticular distribution. The three-dimensional distribution of NHSS particles is due to the formation of an organic film on the NHSS particles by the silane coupling agent, and the organic film reduces the tension on the surface of the particles, exerts a good steric hindrance effect, and effectively inhibits the agglomeration between the particles. Moreover, the mutilayer structure and micro-pores of NHSS particles aggregate contributed greatly to the adhesive property of modified asphalt.

From Figure 3, it can be observed that the size of NHSS particles aggregates is smaller than that of nanosilica particles aggregates in the same random region, which indicates that NHSS has better dispersibility in asphalt. The main reason for the difference of dispersion between nanosilica and NHSS in asphalt is the surface of NHSS contains a large number of atoms, which makes it easier to form a short-range annular diffusion channel in asphalt compared with the nanosilica. Thus, the NHSS has higher diffusivity in the preparation process of the modified asphalt.

### 2.4. NHSS-AC Samples Preparation

Nano hydrophobic silane silica modified asphalt samples with NHSS (1, 3, 5% by mass of base asphalt) were prepared to determine the optimal NHSS content. Due to the high specific surface of NHSS, pre-mixing of the heated NHSS and asphalt binder is necessary for easier mixing. Firstly, base asphalt and prepared NHSS was heated to 165 °C in an oven (2008, Xinxing Test Instruments Co. Ltd, Changchun, China) and preserved for 1h. After pre-mixing, a high-speed mixer (2008, Xinxing Test Instruments Co. Ltd., Changchun, China) was adopted to make the NHSS mixed more uniformly with base asphalt by maintaining a low speed of 1000 r/min for 5 min and proceeding to a higher speed of 3000 r/min for 15 min afterwards. The property tests of NHSS modified asphalt samples were conducted to determine the optimal NHSS content. The results are shown in Table 4.

**Table 4.** Property parameters of modified asphalt with different NHSS.

| Technical Parameters | Penetration (mm) | | | Softening Point (°C) | 10 °C Ductility (cm) | 5 °C Ductility (cm) | Penetration Index | 135 °C Apparent Viscosity (mpa·s) |
|---|---|---|---|---|---|---|---|---|
| | 15 °C | 25 °C | 30 °C | | | | | |
| base asphalt | 27.5 | 81.3 | 132.6 | 44.2 | 45.7 | 6.3 | −0.87 | 485.3 |
| 1% NHSS-asphalt | 26.8 | 71.5 | 113.9 | 47 | 20.9 | 6.5 | −0.32 | 721.3 |
| 3% NHSS-asphalt | 28.4 | 71.2 | 106.9 | 48.6 | 17 | 6.6 | 0.24 | 2219.6 |
| 5% NHSS-asphalt | 24.0 | 68.4 | 105.9 | 50 | 12.8 | 6.1 | −0.52 | 7455.7 |

The results show that the addition of NHSS modifiers improves some properties of the base asphalt. The softening point value, 5 °C ductility value, PI value, and 135 °C apparent viscosity value of 3% NHSS-asphalt are greater than 1% NHSS-asphalt and base asphalt, and the softening point value, 135 °C apparent viscosity of 5% NHSS-asphalt are greater than 3% NHSS-asphalt. Thus, considering that the more the additive is added, the greater the cost is, and the improvement of the overall performance, 3% NHSS was subjectively identified as the optimum content for base asphalt.

The optimum bitumen content of NHSS modified asphalt concrete (NHSS-AC) was determined by Marshall test according to Chinese specifications for design of highway asphalt pavement (JTG D50-2006). Four asphalt mixes were used—4.5%, 5.0%, 5.5%, and 6.0%—and the NHSS-AC were prepared using the standard Marshall mix design procedure with 75 blows on each side of the cylindrical specimens. The Marshall stability (MS), flow (FL), volume of air voids (VV), and voids filled with asphalt (VFA) values of the mixtures with different asphalt content were obtained to determine the optimum asphalt content (OBC) of NHSS-AC. The optimum modified asphalt content was found to be 5.7% for the control mixture. In this study, all NHSS-AC samples were made with 5.7% modified asphalt content in order to maintain consistency throughout the research. Moreover, base asphalt concrete (AC) was prepared for the same tests as a comparative, and the optimum asphalt content was found to be 5.01% for the control mixture. The reason why the OBC of NHSS-AC is higher than AC may be that some of the NHSS has chemically reacted with the asphalt, and the other part of the NHSS and the asphalt are simply physically blended, the physically blended NHSS acts as a mineral powder, so the OBC of NHSS-AC is higher than AC.

## 3. Laboratory Tests

### 3.1. *freeze-soak-scour Cycle Test*

In this paper, a laboratory freeze-soak-scour cycle test is designed to simulate the actual condition of asphalt pavement during spring-thawing season. Pavement Material Dynamic Water Scouring Tester (2014, Suzhou Zhixi Environmental Test Equipment Co., Ltd., Suzhou, China) and High–Low Temperature Experimental Box (2015, Changchun, China) are applied in the laboratory freeze-soak-scour cycle test, as shown in Figures 4 and 5.

**Figure 4.** High–Low Temperature Experimental Box.

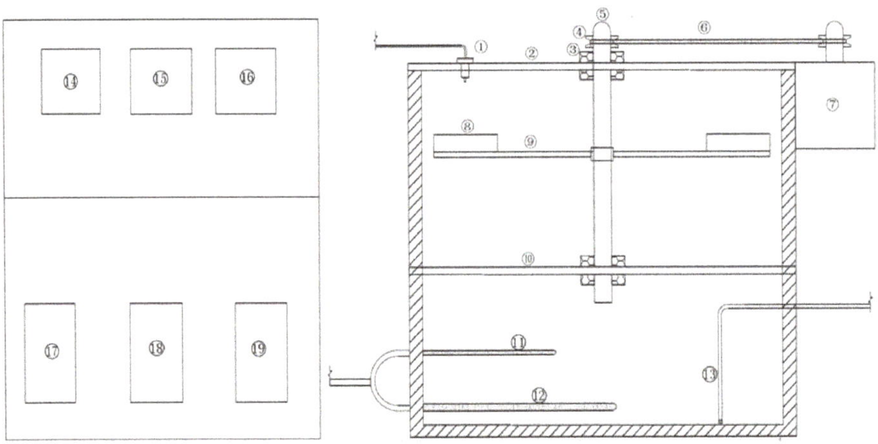

**Figure 5.** Pavement material dynamic water scouring tester: (**1**) water gun; (**2**) support plate; (**3, 10**) bearing; (**4**) pulley; (**5**) disk fixed axis; (**6**) belt; (**7**) motor; (**8**) Marshall samples fixed ring; (**9**) disk; (**11**) temperature sensor; (**12**) U shape heating tube; (**13**) pump intake pipe; (**14**) temperature control dial; (**15**) scouring force control dial; (**16**) disk speed control dial; (**17–19**) pump and its attached equipment.

The Pavement Material Dynamic Water Scouring Tester is self-developed to simulate the dynamic water scouring phenomenon of saturation pavement, as shown in Figure 5. The main parameters of the test system are the pressure of the water gun, the speed of the disk and scouring time. In order to establish the connection between the laboratory scouring test and the actual condition of pavement, a piezoresistive dynamic water pressure sensor is embedded in the Hui-wu expressway (in Songyuan, Jilin, China) to collect the dynamic water scour pressure of pavement, thus adjusting the pressure of water gun and the rotating speed of the disk to realize the laboratory scouring test. The details of a freeze-soak-scour cycle test is described below.

First, the specimens were treated by vacuum saturation in 97.3 kPa for 15 min and submerged in a container containing water, then the container with specimens were placed in the High–Low Temperature Experimental Box at −15 °C and frozen 12 h.

Then, the specimens were soaked in water at 15 °C for 12 h through controlling the High–Low Temperature Experimental Box.

Finally, the specimens were removed from the container and placed in the Pavement Material Dynamic Water Scouring Tester for dynamic water test. The pressure of the water gun is set to 2.56 Mpa, the speed of the disk is set to 420 rad/min, and the scouring time of one cycle is 4.8 min.

As described above, a complete freeze-soak-scour cycle is completed. Then, after 5, 10, 15, and 20 freeze-soak-scour cycles, damaged AC and NHSS-AC specimens were collected for performance testing to explore the influence of NHSS in asphalt concrete under freeze-soak-scour cycles. Three single factor tests were designed to explore the attenuation law of properties of NHSS-AC with the soaking cycle, freezing cycle, and scouring cycle acting seperately.

### 3.2. Durability and Property Test of the Mixture

The value of voids content and weight loss ratio is used to evaluate the durability of asphalt mixture before and after freeze-soak-scour cycles according to Chinese standards JTG T0705-2011. Marshall test, −10 °C splitting test and freeze-soak-scour splitting test were applied to measure the pavement performance of asphalt mixture under freeze-soak-scour cycles. The Marshall test was conducted to measure the high-temperature mechanical properties of the mixture according to Chinese standards JTG T0709-2011. A − 10 °C splitting test was conducted to assess the capacity of asphalt pavement to bear dynamic loads at low temperature according to JTG T0716-2011. A freeze-soak-scour splitting test was conducted to evaluate water stability of asphalt mixture based on modified freeze–thaw splitting test in accordance with JTG T0729-2011. The freeze-soak-scour splitting tensile strength ($R_{FTn}$) and freeze-soak-scour splitting tensile ratio ($TSR_n$) are principal mechanical parameters, which are measured in this test. Specimens subjected to n times of freeze-soak-scour cycles (n = 0, 5, 10, 15, 20) were immersed in water bath at 25 °C for 2 h, and the loading with a constant rate of compression of 50 mm/min was applied. $R_{FTn}$ and $TSR_n$ were calculated as follows.

$$R_{FTn} = 0.006287 P_{FTn}/h_n \tag{1}$$

$$TSR_n = \frac{R_{FT0}}{R_{FTn}} \tag{2}$$

where $P_{FTn}$ is the maximum load-bear of specimens subjected to n times of freeze-soak-scour cycles (N, n = 0, 5, 10, 15, 20) and h is the The height of specimen subjected to n times of freeze-soak-scour cycles (mm, n = 0, 5, 10, 15, 20).

## 4. Results and Discussion

### 4.1. Effect of freeze-soak-scour Cycles on Properties of NHSS-AC

As it is shown in Figures 6 and 7, the voids content of the two mixtures gradually increases as the number of freeze-soak-scour cycles increases. After 20 freeze-soak-scour cycles, the void content of AC is already in line with the Chinese specification (the voids content of AC < 5%), and NHSS-AC is still within the standard range. This is because the adhesion of NHSS modified asphalt and aggregate is greater than that of base asphalt and aggregate. Therefore, the voids content change ratio of NHSS-AC is significantly lower than that of AC after multiple cycles, so NHSS-AC has better durability.

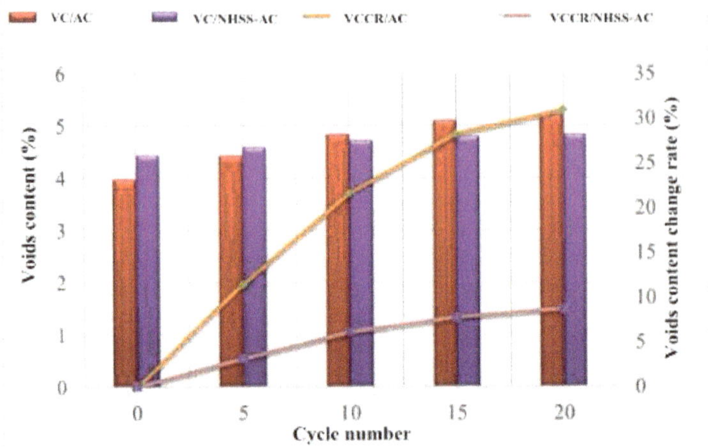


**Figure 6.** Void content changes of mixtures under freeze-soak-scour cycles.

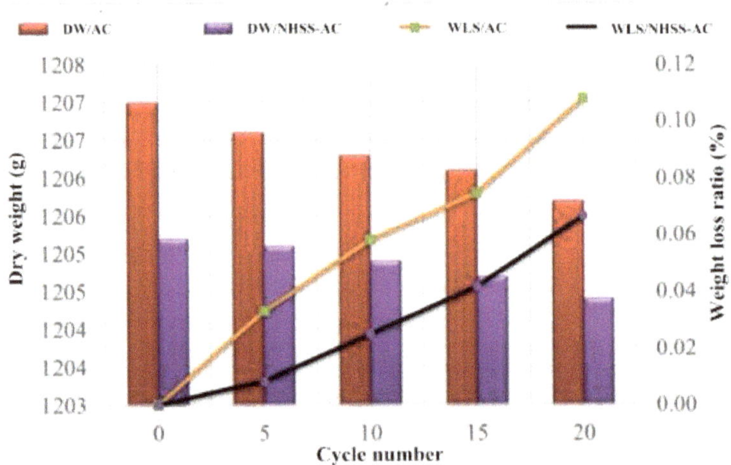


**Figure 7.** Weight loss ratio changes of mixtures under freeze-soak-scour cycles.

It can be seen from Figure 7 that as the number of cycles increases, the weight loss ratio of the two mixtures gradually increases, indicating that the loose fine aggregate of pavement is taken away by the wheel with the increase of freeze-soak-scour cycle in the spring-thawing season, which leads to the reduction of quality of mixtures and the increase of the weight loss ratio of mixtures. Moreover, the NHSS modified asphalt has a better bond strength with the aggregate than the base asphalt so that the weight loss ratio of NHSS-AC is always smaller than the AC after different freeze-soak-scour cycles.

As it is shown in Figure 8, Marshall stability of mixtures decreases with increasing freeze-soak-scour cycles, and the Marshall stability of the AC has been lower than Chinese specification after 15 cycles (Marshall stability of AC $\geq$ 8.0 KN). However, the Marshall stability of the NHSS-AC remained within the scope of the Chinese specification after 20 cycles. The Marshall stability loss ratio of NHSS-AC is far less than that of AC under the same cycles. It can be seen that NHSS has certain advantages in improving the high temperature mechanical properties of asphalt mixtures in spring-thawing season.

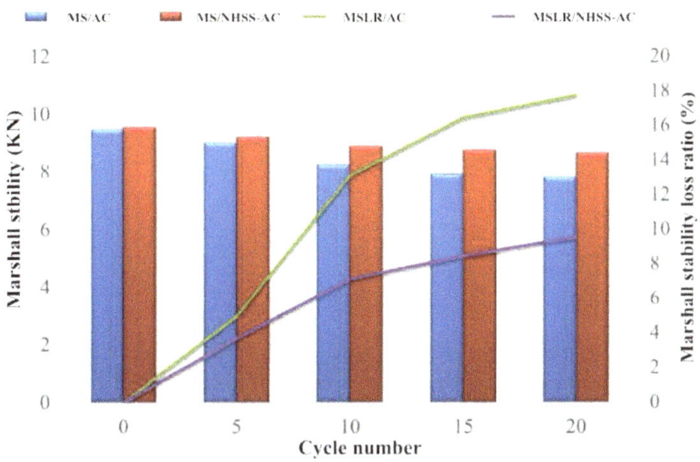


**Figure 8.** Marshall stability changes of mixtures under freeze-soak-scour cycles.

The data of −10 °C splitting test is shown in Table 5. The splitting tensile strength and destruction tensile strain are continuously reduced with the increase of the number of cycles. After 20 cycles, the splitting tensile strength of AC decreased by 9.7% and destruction tensile strain decreased by 10.3%. By contrast, the splitting tensile strength of NHSS-AC decreased by 8.3% and destruction tensile strain decreased by 10.7%. Thus, the low temperature mechanical property of NHSS-AC is better than that of AC in freeze–thawing season. The change of destruction stiffness modulus is disordered, because the attenuation curve of splitting tensile strength and destruction tensile strain is not consistent.

**Table 5.** Results of −10 °C splitting test.

| Mixture Type | Number of Cycles | 0 | 5 | 10 | 15 | 20 |
|---|---|---|---|---|---|---|
| AC | Splitting tensile strength (Mpa) | 4.02 | 3.57 | 3.78 | 3.71 | 3.63 |
| | Destruction tensile strain (με) | 2.06 | 2.00 | 1.97 | 1.88 | 1.85 |
| | Destruction stiffness modulus (Mpa) | 3361 | 3330 | 3305 | 3397 | 3383 |
| NHSS-AC | Splitting tensile strength (Mpa) | 4.33 | 4.22 | 4.15 | 4.07 | 3.97 |
| | Destruction tensile strain (με) | 2.27 | 2.18 | 2.12 | 2.09 | 2.03 |
| | Destruction stiffness modulus (Mpa) | 3282 | 3330 | 3365 | 3353 | 3369 |

The result of freeze-soak-scour splitting test is shown in Figure 9. It illustrates that the $R_{FTn}$ of the mixtures continuously reduced with the increase of the number of cycles, and the $R_{FTn}$ of the AC drastically attenuated during the first five cycles, and the decay rate tends to be slow after that. The $TSR_n$ of NHSS-AC is lower than that of AC at various points in the freeze-soak-scour cycles, and the $TSR_n$ of NHSS-AC is 5% higher than the AC. Thus, the water stability of NHSS-AC is better than that of AC in freeze–thawing season.

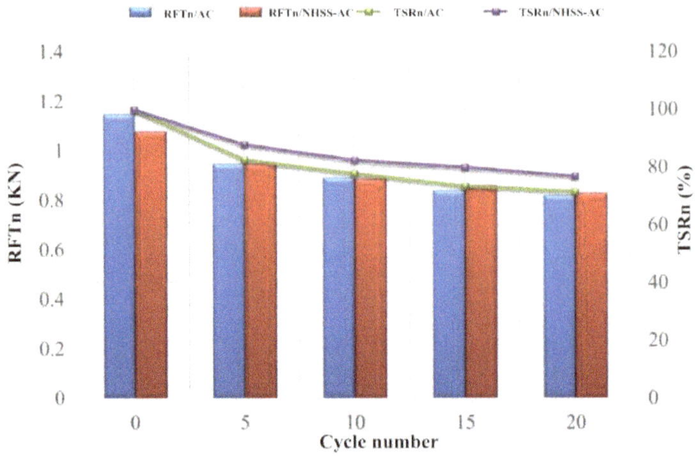


**Figure 9.** TSRn changes of mixtures under freeze-soak-scour cycles.

Overall, after 15 or 20 freeze-soak-scour cycles, the normal asphalt mixture cannot meet the requirements of specifications, and the NHSS modified asphalt mixture still meet the requirements of the specifications, which indicates that the NHSS modified can effectively improve the ability of asphalt concrete to resist the spring environment.

The price of normal asphalt is about 3000 rmb/t, and the price of NHSS modified asphalt is about 3300 rmb/t. The comprehensive unit price analysis of mechanical paving NHSS modified asphalt concrete of 7 cm thick is as follows. (a) Artificial cost: 2.1 rmb/cm$^2$; (b) Materials cost (including main material and auxiliary material): 97.7 rmb/cm$^2$; (c) Mechanical cost: 2.5 rmb/cm$^2$; (d) Other cost (including safe and civilized construction costs, fees, and taxes): 7.4 rmb/cm$^2$. The total cost of NHSS modified asphalt concrete is 109.7 rmb/cm$^2$, and the total cost of normal asphalt concrete is 102 rmb/cm$^2$. The cost of mechanical paving NHSS modified asphalt concrete of 7 cm thick is 7.6% higher than normal asphalt concrete. Considering the ability of NHSS modified asphalt to improve the durability and the property of the asphalt concrete in spring-thawing season, an increase of 7.6% of the cost is still acceptable.

Establishment of the freeze-soak-scour Damage Model of NHSS-AC based on the Logistic Judgment Model

The logistic judgment model is a set of nonlinear regular regression models which are mainly used to describe and infer the relationship between two or more classified dependent variables and a set of variables. Compared with multiple linear regression, logistic regression has many unique advantages. The model does not require the normality and homogeneity of the data variables in the calculation, nor does it limit the specific type of the independent variable, and its statistical coefficient has strong interpretability, so that the logistic regression models can be widely used in metrology research. In this paper, the deterioration process of NHSS-AC under freeze-soak-scour cycles was studied based on logistic judgment model.

The equation of logistic curve is used to study the increasing process of the population by biologist P. F. Verhulst initially. The equation is expressed as

$$y = \frac{K}{1 + ae^{-bt}} \tag{3}$$

The first derivative of the above equation can be found. The resulting equation is the growth velocity function of the described object.

$$\mathrm{v(t)} = \frac{dy}{dt} = \frac{Kabe^{-bt}}{\left(1+ae^{-bt}\right)^2} \tag{4}$$

The growth process of the logistic curve is slow–fast–slow. The first derivative of the growth velocity function can be found, then let the result be 0.

$$\frac{dv(t)}{dt} = \frac{Kabe^{-bt}\left(abe^{-bt}-b\right)}{\left(1+ae^{-bt}\right)^3} = 0 \tag{5}$$

We can get

$$t = \frac{\ln a}{b} \tag{6}$$

So, when $t = \frac{\ln a}{b}$, the object is at the peak of the growth.

Then find the second derivative of the growth velocity function, then let the result be 0.

$$\frac{d^2v(t)}{d^2t} = \frac{Kab^3e^{-bt}\left(1-4abe^{-bt}+a^2e^{-2bt}\right)}{\left(1+ae^{-bt}\right)^4} = 0 \tag{7}$$

We can get that

$$t_1 = \frac{\ln a - 1.317}{b} \tag{8}$$

$$t_2 = \frac{\ln a + 1.317}{b} \tag{9}$$

Re-number the three key points

$$t_1 = \frac{\ln a - 1.317}{b} \tag{10}$$

$$t_2 = \frac{\ln a}{b} \tag{11}$$

$$t_3 = \frac{\ln a + 1.317}{b} \tag{12}$$

$t_1$ is the initial time point of the growth peak of the study, $t_2$ is the peak, $t_3$ is the end. $\varphi_i$ is used as the damage coefficient of the performance of the mixtures after the freeze-soak-scour cycle, then

$$\varphi_i = 1 - \frac{\theta_i}{\theta_0} \tag{13}$$

In the formula, $\theta_i$ is the test index of specimens after i cycles.

According to the tests, the voids content is used to evaluate the durability of mixtures, the Marshall stability is used to evaluate the high-temperature mechanical property of mixtures, −10 °C splitting tensile strength and the destruction tensile strain are used to evaluate the low-temperature mechanical property, the freeze-soak-scour splitting tensile ratio is used to evaluate the water stability. If two indexes are needed to evaluate the damage of a certain mechanical property of mixtures, the damage coefficient takes the mean of the damage coefficient of two indexes

For the convenience of analysis, the damage process represented by the logistic curve equation is transformed into

$$\varphi = \frac{\mathrm{A_{min}} - \mathrm{A_{max}}}{1+\left(\frac{\mathrm{x}}{\mathrm{x_{0.5}}}\right)^{\mathrm{a}}} + \mathrm{A_{max}} \tag{14}$$

In the formula, $A_{min}/A_{max}$ is the min/max value of the regression curve, $x_{0.5}$ is the value of x when $\varphi = 0.5A_{max}$. Then according to the form of the curve, $A_{max}$ can be used to evaluate the damage degree of the properties of asphalt cement, and then $x_{0.5}$ is used to evaluate the damage speed of the properties of mixture.

For the initial condition of the model, when the cycle is not processed, the number of cycles in the damage model is zero, and the growth rate of the damage rate is also zero. Thus, $A_{min} = 0$ and the model is transformed into

$$\varphi = \frac{-A_{max}}{1 + \left(\frac{x}{x_{0.5}}\right)^{a}} + A_{max} \tag{15}$$

For the convenience of the following expression, the above formula is written as

$$y = \frac{-a}{1 + \left(\frac{x}{c}\right)^{b}} + a \tag{16}$$

Then, the first derivative of the damage velocity model is calculated and let it be 0.

$$v'(x) = \frac{\frac{ab}{c^b}x^{b-2}\left[b - 1 - \frac{b+1}{c^b}x^b\right]}{\left(1 + \frac{x^b}{c^b}\right)^3} = 0 \tag{17}$$

Then the second derivative of the damage velocity model is calculated and let it be 0.

$$v''(x) = \frac{\frac{ab(b+1)^2}{c^{2b}}x^{2b} - \frac{ab(b-1)^2 - ab(b+1)(2b-1)c^b - 3ab^2(b-1)}{c^{2b}}x^b + \frac{ab(b-1)^2}{c^b}}{\left(1+\frac{x^b}{c^b}\right)^4 x^{b-2}} = 0 \tag{18}$$

Let

$$A = \frac{ab(b+1)^2}{c^{2b}},\ B = \frac{ab(b-1)^2 - ab(b+1)(2b-1)c^b - 3ab^2(b-1)}{c^{2b}},\ C = \frac{ab(b-1)^2}{c^b} \tag{19}$$

The result can be

$$x_0 = \sqrt[b]{c^b \cdot \frac{b-1}{b+1}},\ x_1 = \sqrt[b]{\frac{-B - \sqrt{B^2 - 4AC}}{2A}},\ x_2 = \sqrt[b]{\frac{-B + \sqrt{B^2 - 4AC}}{2A}} \tag{20}$$

According to the analysis above, it can be known that $x_1$ is the initial time point when some properties of the mixture get into the growth peak period, $x_0$ is the peak time and $x_2$ is the end. According to the experimental data, the damage degree of the pavement properties of mixtures can be got after different cycles. They are shown in Figure 10.

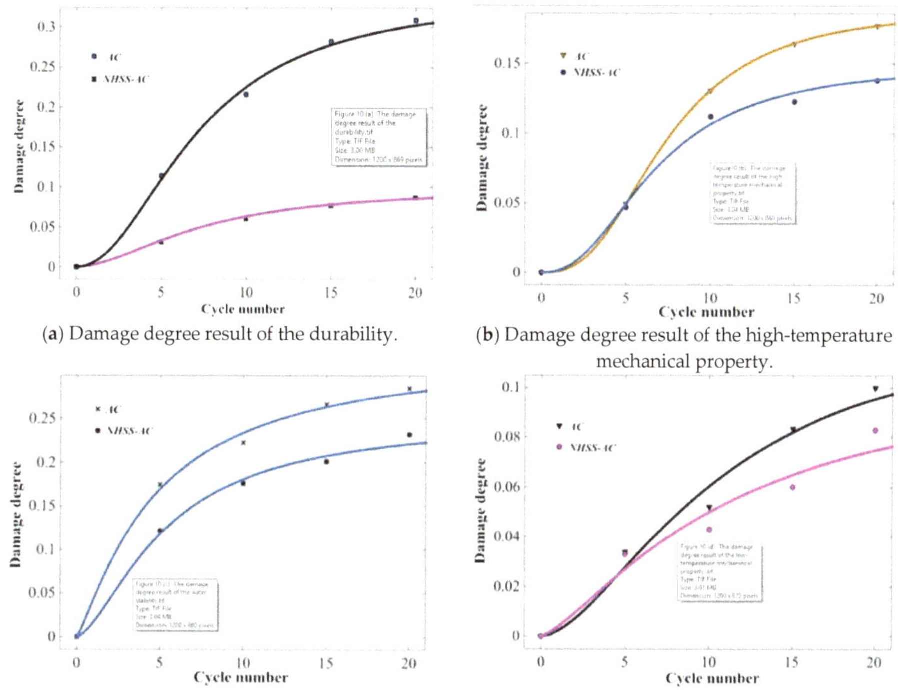


(**a**) Damage degree result of the durability. (**b**) Damage degree result of the high-temperature mechanical property. (**c**) Damage degree result of the water stability. (**d**) Damage degree result of the low-temperature mechanical property.

**Figure 10.** Damage degree result of the various properties of the AC and NHSS-AC.

The regression equation of the damage model and the corresponding value of each model is shown in Table 6. The freeze–thaw–scour damage model is effective, because all $R^2$ of the regression equation are above 0.95.

**Table 6.** Fitting result of the logistic damage model.

| Mixture | Pavement Property | Logistic Damage Model | $SS_e$ | $R^2$ | $R_{MSE}$ | $x_0$ | $x_1$ | $x_2$ |
|---|---|---|---|---|---|---|---|---|
| AC | Durability | $\varphi = -\frac{0.34}{1+\left(\frac{x}{7.2}\right)^{2.044}} + 0.34$ | $1.5 \times 10^{-4}$ | 0.9976 | 0.0063 | 4.3 | 0.35 | 7.3 |
| | High temperature mechanical property | $\varphi = -\frac{0.19}{1+\left(\frac{x}{7.4}\right)^{2.644}} + 0.19$ | $6.7 \times 10^{-8}$ | 0.9999 | 0.0011 | 5.5 | 0.51 | 7.9 |
| | Water stability | $\varphi = -\frac{0.33}{1+\left(\frac{x}{4.8}\right)^{1.203}} + 0.33$ | $1.9 \times 10^{-4}$ | 0.9964 | 0.0069 | 0.7 | 0.03 | 3.4 |
| | Low temperature mechanical property | $\varphi = -\frac{0.13}{1+\left(\frac{x}{10.9}\right)^{1.674}} + 0.13$ | $1.3 \times 10^{-4}$ | 0.9796 | 0.0057 | 4.8 | 0.21 | 10.1 |
| NHSS-AC | Durability | $\varphi = -\frac{0.10}{1+\left(\frac{x}{7.4}\right)^{1.797}} + 0.10$ | $9 \times 10^{-6}$ | 0.9982 | 0.0015 | 3.7 | 0.26 | 7.1 |
| | High temperature mechanical property | $\varphi = -\frac{0.15}{1+\left(\frac{x}{6.8}\right)^{2.31}} + 0.15$ | $8 \times 10^{-5}$ | 0.9941 | 0.0045 | 4.6 | 0.43 | 7.1 |
| | Water stability | $\varphi = -\frac{0.25}{1+\left(\frac{x}{5.4}\right)^{1.546}} + 0.25$ | $2 \times 10^{-4}$ | 0.9940 | 0.0070 | 2.0 | 0.15 | 4.8 |
| | Low temperature mechanical property | $\varphi = -\frac{0.11}{1+\left(\frac{x}{11.5}\right)^{1.354}} + 0.11$ | $1.7 \times 10^{-4}$ | 0.9547 | 0.0066 | 2.8 | 0.08 | 9.1 |

where SSe is residual sum of squares, $R^2$ is determination coefficient, and $R_{MSE}$ is root-mean-square error

In order to compare the model parameters and analyze physical meaning, the contrast figure was given in Figure 11.

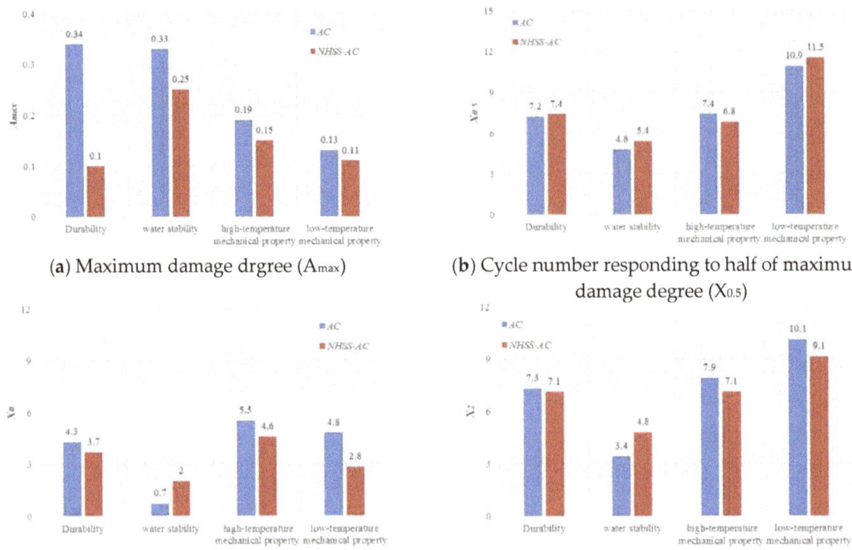


(**a**) Maximum damage drgree ($A_{max}$)

(**b**) Cycle number responding to half of maximum damage degree ($X_{0.5}$)

(**c**) Cycle number responding to the growth peak of damage degree ($X_0$)

(**d**) Cycle number responding to the end of damage degree growing peak ($X_2$)

**Figure 11.** Parameters of the Logistic damage model.

From Figure 11a, it can be seen that after 20 freeze-soak-scour cycles, the durability and water stability are the most damaged of the four properties of the AC, reaching 0.34 and 0.33 respectively. The water stability and the high-temperature mechanical property are the most damaged of the four properties of the NHSS-AC, reaching 0.25 and 0.15 respectively. The difference of the damage degree of the low-temperature mechanical property of AC and NHSS-AC is small. The damage degree of durability and mechanical property of NHSS-AC are less than that of AC under freeze-soak-scour cycles, indicating that the NHSS-AC has significant effects on preventing and controlling the damage of the pavement in spring-thawing season.

Figure 11b shows that water stability, high-temperature mechanical property, durability and low-temperature mechanical property of mixtures were ranked by $X_{0.5}$ from small to big. It is known that under the action of freeze-soak-scour cycles, the damage rate of water stability is the fast, while that of low-temperature mechanical property is the slowest. Thus, the damage degree of water stability of mixtures is the largest and the damage speed is the fastest, so the moisture damage of pavement is most likely to occur in the spring-thawing season. The index $X_{0.5}$ of AC and NHSS-AC are almost identical, indicating that the incorporation of NHSS does not change the pattern of the damage process of the pavement in spring-thawing season.

From Figure 11a,b,d, it can be known that the damage rate of the water stability of the mixture reached the peak first and it was the first to end the damage peak period and enter the stage of stable growth. The index $X_0$ of NHSS-AC is greater than that of AC except water stability.

### 4.2. Analysis of Three Kinds of Damage Factors of NHSS-AC in Spring-Thawing Season Based on the Gray Rational Degree Theory

#### 4.2.1. Soaking Cycles Test

A soaking cycle takes 12 h at a water temperature of 15 °C. After 0, 5, 10, 15, and 20 soaking cycles, damaged NHSS-AC specimens were collected for the durability and performance test. The data after the tests is shown in Table 7.

**Table 7.** Variation of NHSS-AC parameters under soak cycle.

| Number of Cycle | Voids Content (%) | Weight Loss Ratio (%) | Freeze–thaw Splitting Tensile Ratio (%) | Marshall Stability (kn) | −10 °C Splitting Test | | |
|---|---|---|---|---|---|---|---|
| | | | | | Splitting Tensile Strength (Mpa) | Destruction Tensile Strain (με) | Destruction Stiffness Modulus (Mpa) |
| 0 | 4.733 | 0.000 | 100 | 93.5 | 4.32 | 2.27 | 3282 |
| 5 | 4.776 | 0.025 | 89.7 | 8.93 | 4.30 | 2.15 | 3442 |
| 10 | 4.814 | 0.025 | 86.7 | 8.83 | 4.28 | 2.15 | 3422 |
| 15 | 4.861 | 0.034 | 82.6 | 8.78 | 4.24 | 2.12 | 3441 |
| 20 | 4.849 | 0.042 | 80.5 | 8.63 | 4.20 | 2.06 | 3511 |

#### 4.2.2. Freezing Cycles Test

The details of a freezing cycle test are described below. The specimens were treated by vacuum saturation in 97.3 kPa for 15 min and submerged in a container containing water, then the container with specimens were placed in the High–Low Temperature Experimental Box at −15 °C and frozen 12 h. As described above, a complete freezing cycle is completed. Then, after 0, 5, 10, 15, and 20 freezing cycles, damaged NHSS-AC specimens were collected for the performance test after the frozen specimens were removed and placed in a 15 °C environment for 6 h. The results are shown in Table 8.

**Table 8.** Variation of NHSS-AC parameters under freezing cycle.

| Number of Cycle | Voids Content (%) | Weight Loss Ratio (%) | Freeze–thaw Splitting Tensile Ratio (%) | Marshall Stability (kn) | −10 °C Splitting Test | | |
|---|---|---|---|---|---|---|---|
| | | | | | Splitting Tensile Strength (Mpa) | Destruction Tensile Strain (με) | Destruction Stiffness Modulus (Mpa) |
| 0 | 4.009 | 0.000 | 100 | 9.35 | 4.32 | 2.27 | 3282 |
| 5 | 4.207 | 0.025 | 89.4 | 9.02 | 4.28 | 2.18 | 3374 |
| 10 | 4.331 | 0.034 | 85.7 | 8.76 | 4.21 | 2.21 | 3275 |
| 15 | 4.378 | 0.042 | 80.5 | 8.44 | 4.14 | 2.18 | 3267 |
| 20 | 4.425 | 0.050 | 78.8 | 8.21 | 4.04 | 2.15 | 3234 |

#### 4.2.3. Scouring Cycles Test

The scouring cycle test is designed based on Pavement Material Dynamic Water Scouring Tester. The water gun pressure, the speed of the disk, and the scour time are set to 2.56 Mpa, 420 rad/min, and 4.8 min respectively to realize a scouring cycle. Then, after 0, 5, 10, 15, and 20 scouring cycles test, damaged NHSS-AC specimens were collected for the durability and performance test. The results are shown in Table 9.

**Table 9.** Variation of NHSS-AC parameters under scouring cycle.

| Number of Cycle | Voids Content (%) | Weight Loss Ratio (%) | Freeze–thaw Splitting Tensile Ratio (%) | Marshall Stability (kn) | −10 °C Splitting Test | | |
|---|---|---|---|---|---|---|---|
| | | | | | Splitting Tensile Strength (Mpa) | Destruction Tensile Strain (με) | Destruction Stiffness Modulus (Mpa) |
| 0 | 4.370 | 0.000 | 100 | 9.35 | 4.32 | 2.27 | 3282 |
| 5 | 4.474 | 0.008 | 90.6 | 9.32 | 4.31 | 2.24 | 3307 |
| 10 | 4.548 | 0.025 | 85.1 | 9.01 | 4.27 | 2.21 | 3323 |
| 15 | 4.594 | 0.033 | 85.0 | 8.65 | 4.25 | 2.21 | 3304 |
| 20 | 4.678 | 0.042 | 81.5 | 8.58 | 4.22 | 2.18 | 3330 |

#### 4.2.4. Analysis of Results Based on the Gray Rational Degree Theory

The gray rational degree theory focuses on the problems of small samples and poor information which are difficult to solve in probability and statistics and fuzzy mathematics. The theory is based on information coverage, and explores the evolvement rule of things through the action of sequence operators. The main idea of gray rational degree theory is to judge whether the relationship is close according to the similarity of the geometric shape of the sequence curve. The closer the curve develops, the greater the correlation degree between the corresponding sequences will be. The method can provide a quantitative measure for the development and change trend of a system, which is very suitable for the judgment and analysis of dynamic process. The calculation process of grey correlation degree is shown below.

Define the reference sequence and the comparison sequence.

If the reference sequence is set as

$$X_0 = \{x_0(1), x_0(2), \cdots\cdots x_0(n)\} \tag{21}$$

and the comparison sequence is set as

$$X_i = \{x_i(1), x_i(2), \cdots\cdots x_i(n)\} \tag{22}$$

The grey correlation degree value of point $\gamma(x_0(k), x_i(k))$ is calculated as

$$\gamma(x_0(k), x_i(k)) = \frac{\min\limits_i \min\limits_k |x_0(k) - x_i(k)| + \rho \max\limits_i \max\limits_k |x_0(k) - x_i(k)|}{|x_0(k) - x_i(k)| + \rho \max\limits_i \max\limits_k |x_0(k) - x_i(k)|} \tag{23}$$

where $k = 1, 2, \ldots, n$, $\rho$ is the distinguishing coefficient and it value follows the principle of minimum information $\rho \in (0, 1)$.

The grey correlation degree $\gamma(X_0, X_i)$ between $X_i$ and $X_0$ is written as

$$\gamma(X_0, X_i) = \frac{1}{n} \sum_{k=1}^{n} \gamma(x_0(k), x_i(k)) \tag{24}$$

A larger grey correlation degree value indicates the reference sequence has more effect on the comparison sequence, and vice versa.

In this paper, the gray rational degree theory was applied to explore the influence of three separate factors on the pavement properties of NHSS-AC in spring-thawing season. The voids content (durability), freeze-soak-scour splitting tensile ratio (water stability), Marshall stability (high temperature mechanical property), −10 °C splitting tensile strength (low-temperature mechanical property) of NHSS-AC after 0, 5, 10, 15, and 20 freeze-soak-scour cycles are the systematic characteristic behavior sequence. The same of NHSS-AC with 0, 5, 10, 15, and 20 soaking cycles, freezing cycle, and scouring cycle act separately are correlation factor sequence. The correlation degree between the systematic characteristic behavior sequence and correlation factor sequences is calculated according to the gray rational degree theory, the results are as shown in Figure 12.

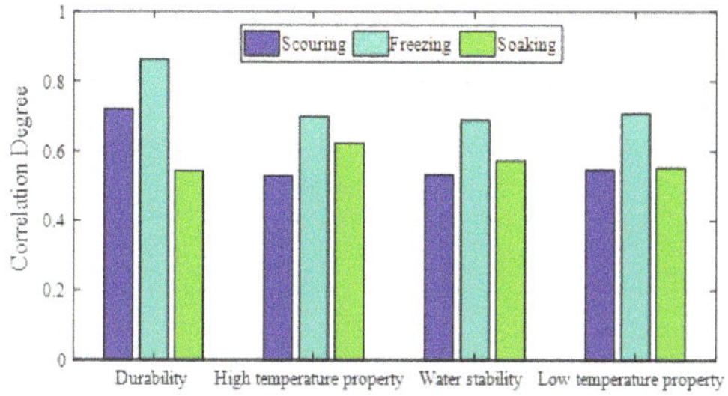


**Figure 12.** Calculation results of NHSS-AC by the gray rational degree theory.

A larger grey correlation degree value indicates the reference sequence has more effect on the comparison sequence, and vice versa. As is shown in Figure 12, the freezing cycles have a more significant impact on the properties of NHSS-AC compared with the soaking and scouring cycles. The impact of the scouring cycles on the durability of NHSS-AC is more significantly than that of the soaking cycle. The effect of scouring and soaking on low-temperature mechanical property of NHSS-AC under freeze-soak-scour cycles is similar. From the perspective of single factor to analyze, the freezing and scouring factor has the highest correlation degree with the durability of NHSS-AC in spring-thawing season, and the soaking factor has the highest correlation degree with the high-temperature mechanical properties of NHSS-AC in spring-thawing season.

It can be seen that the effect of multiple freezing cycles on the internal structure of the NHSS-AC is the most drastic, and the freezing cycles will accelerate the development of microcracks. The soaking cycles is a process of deterioration of material. Especially in the spring-thaw season, various impurities and snow melt agent accelerates the mechanical performance loss of the asphalt binder. The scouring cycle removes the falling mortar particles and accelerates the development of water damage of pavement.

## 5. Conclusions

In this paper, NHSS modified asphalt concrete was prepared to systematically investigate the durability and the performance of NHSS modified asphalt concrete in spring-thawing season according to a self-designed laboratory freeze-soak-scour cycle test, and the freeze-soak-scour damage process of the NHSS modified asphalt concrete is studied by the logistic judgment model. Moreover, the influence of freezing, soaking, and scouring damage factors on the damage process of NHSS modified asphalt concrete in spring-thawing season is analyzed based on the gray rational degree theory. The main conclusions are as follows:

1. The durability and property test results of two mixtures after long-term freeze-soak-scour cycles illustrated that adding NHSS is an effective technique for mitigating freeze–thaw cycle damage of asphalt concrete in spring-thawing season.
2. The freeze–thaw-scour damage model of two mixtures were built and the damage speed and damage degree of the two mixtures were quantified based on the logistic judgment model. Model parameter $A_{max}$ can be used to evaluate the damage degree of the mechanincal properties of asphalt cement, and the parameter $x_{0.5}$ can be used to evaluate the damage speed of the properties of asphalt cement.
3. Through the application of the gray rational degree theory, it found that the freeze factor had a more significant impact on the damage process of NHSS-AC compared with the soak and scour

factor. The impact of the scour factor on the durability of NHSS-AC was more significant than that of the soak factor, and the effect of the scour factor and soak factor on the low temperature mechanical performance was similar.

**Author Contributions:** Conceptualization, X.G.; Data curation, W.G. and M.S.; Formal analysis, W.G., X.G., and W.D.; Funding acquisition, X.G.; Methodology, M.S., and W.D.; Project administration, W.D.; Writing—original draft, W.G.; Writing—review & editing, X.G. and W.D.

**Funding:** This research was funded by the National Nature Science Foundation of China (NSFC) (grant no. 51178204). This financial support is gratefully acknowledged.

**Conflicts of Interest:** The authors declare that there is no conflict of interests regarding the publication of this paper.

## References

1. Badeli, S.; Carte, A.; Dore, G.; Saliani, S. Evaluation of the durability and the performance of an asphalt mix involving Aramid Pulp Fiber (APF): Complex modulus before and after freeze-thaw cycles, fatigue, and TSRST tests. *Constr. Build. Mater.* **2018**, *174*, 60–71. [CrossRef]
2. El-Hakim, M.; Tighe, S.L. Impact of freeze-thaw cycles on mechanical properties of asphalt mixes. *Transp. Res. Rec.* **2014**, *2444*, 20–27. [CrossRef]
3. Shen, A.Q.; Lin, S.L.; Guo, Y.C.; He, T.Q.; Lyu, Z.H. Relationship between flexural strength and pore structure of pavement concrete under fatigue loads and Freeze-thaw interaction in seasonal frozen regions. *Constr. Build. Mater.* **2018**, *174*, 684–692. [CrossRef]
4. Biswas, S.; Hashemian, L.; Bayat, A. Investigation of pothole severity and maintenance methods in canada through questionnaire survey. *J. Cold Reg. Eng.* **2018**, *32*. [CrossRef]
5. Guo, S.C.; Dai, Q.L.; Hiller, J. Investigation on the freeze-thaw damage to the jointed plain concrete pavement under different climate conditions. *Front. Struct. Civ. Eng.* **2018**, *32*, 227–238. [CrossRef]
6. Badeli, S.; Carte, A.; Dore, G.; Saliani, S. Complex modulus and fatigue analysis of asphalt mix after daily rapid freeze-thaw cycles. *J. Mater. Civ. Eng.* **2018**, *30*. [CrossRef]
7. Yang, X.L.; Shen, A.Q.; Guo, Y.C.; Zhou, S.B.; He, T.Q. Deterioration mechanism of interface transition zone of concrete pavement under fatigue load and freeze-thaw coupling in cold climatic areas. *Constr. Build. Mater.* **2018**, *160*, 588–597. [CrossRef]
8. Gong, F.Y.; Ueda, T.; Wang, Y.; Zhang, D.W.; Wang, Z. Mesoscale simulation of fatigue behavior of concrete materials damaged by freeze-thaw cycles. *Constr. Build. Mater.* **2017**, *144*, 702–716. [CrossRef]
9. Ho, C.H.; Linares, C.P.M.; Shan, J.; Almonnieay, A. Material testing apparatus and procedures for evaluating freeze-thaw resistance of asphalt concrete mixtures. *Adv. Civ. Eng. Mater.* **2017**, *6*, 429–443. [CrossRef]
10. Xu, H.N.; Guo, W.; Tan, Y.Q. Internal structure evolution of asphalt mixtures during freeze-thaw cycles. *Mater. Des.* **2015**, *86*, 436–446. [CrossRef]
11. Si, W.; Li, N.; Ma, B. Impact of freeze-thaw cycles on compressive characteristics of asphalt mixture in cold regions. *J. Wuhan Univ. Technol. Mater. Sci. Ed.* **2015**, *30*, 703–709. [CrossRef]
12. Gong, X.; Romero, P.; Dong, Z.; Sudbury, D.S. The effect of freeze-thaw cycle on the low temperature properties of asphalt fine aggregate matrix utilizing bending beam rheometer. *Cold Reg. Sci. Technol.* **2016**, *125*, 101–107. [CrossRef]
13. Islam, M.R.; Asce, S.M.; Tarefder, R.A.; Asce, M. Effects of large freeze-thaw cycles on stiffness and tensile strength of asphalt concrete. *J. Cold Reg. Eng.* **2016**, *30*. [CrossRef]
14. Amini, B.; Tehrani, S.S. Simultaneous effects of salted water and water flow on asphalt concrete pavement deterioration under freeze-thaw cycles. *Int. J. Pavement Eng.* **2014**, *15*, 383–391. [CrossRef]
15. Wei, H.B.; Li, Z.Q.; Jiao, Y.B. Effects of diatomite and SBS on freeze-thaw resistance of crumb rubber modified asphalt mixture. *Adv. Mater. Sci. Eng.* **2017**, *2017*, 1–14. [CrossRef]
16. Klinsky, L.M.G.; Kaloush, K.E.; Faria, V.C.; Bardini, V.S.S. Performance characteristics of fiber modified hot mix asphalt. *Constr. Build. Mater.* **2018**, *176*, 747–752. [CrossRef]
17. Zhang, H.L.; Gao, Y.; Guo, G.H.; Zhao, B.J.; Yu, J.Y. Effects of ZnO particle size on properties of asphalt and asphalt mixture. *Constr. Build. Mater.* **2018**, *159*, 578–586. [CrossRef]

18. Qian, G.P.; Wang, K.; Bai, X.P.; Xiao, T.; Jin, D.Z.; Huang, Q.J. Effects of surface modified phosphate slag powder on performance of asphalt and asphalt mixture. *Constr. Build. Mater.* **2018**, *158*, 1081–1089. [CrossRef]
19. Hamedi, G.H. Evaluating the effect of asphalt binder modification using nanomaterials on the moisture damage of hot mix asphalt. *Road Mater. Pavement Des.* **2017**, *18*, 1375–1394. [CrossRef]
20. Yao, H.; Dai, Q.L.; You, Z.P. Fourier Transform Infrared Spectroscopy characterization of aging-related properties of original and nano-modified asphalt binders. *Constr. Build. Mater.* **2015**, *101*, 1078–1087. [CrossRef]
21. Taherkhani, H.; Afroozi, S. Investigating the creep properties of asphaltic concrete containing nano-silica. *Sādhanā* **2018**, *43*, 24. [CrossRef]
22. Saltan, M.; Terzi, S.; Karahancer, S. Examination of hot mix asphalt and binder performance modified with nano silica. *Constr. Build. Mater.* **2017**, *156*, 976–984. [CrossRef]
23. Yao, H.; You, Z.P.; Li, L.; Lee, C.H.; Wingard, D.; Yap, Y.K.; Shi, X.M.; Goh, S.W. Rheological properties and chemical bonding of asphalt modified with nanosilica. *J. Mater. Civ. Eng.* **2013**, *25*, 1619–1630. [CrossRef]

applied sciences

MDPI

*Article*

# Strength Criterion of Asphalt Mixtures in Three-Dimensional Stress States under Freeze-Thaw Conditions

**Tuo Huang, Shuai Qi, Ming Yang, Songtao Lv *, Hongfu Liu and Jianlong Zheng**

School of Traffic and Transportation Engineering, Changsha University of Science & Technology, Changsha 410114, China; ht@csust.edu.cn (T.H.); 17101030071@stu.csust.edu.cn (S.Q.); myang892018@126.com (M.Y.); lhf0625@csust.edu.cn (H.L.); zjl@csust.edu.cn (J.Z.)

* Correspondence: lst@csust.edu.cn; Tel.: +86-139-7519-7481

Received: 13 July 2018; Accepted: 1 August 2018; Published: 4 August 2018

**Featured Application: This work provides the testing and theoretical reference for material and structure design of asphalt pavement in three-dimensional stress states under freeze-thaw conditions.**

**Abstract:** In order to study the influence of freeze-thaw cycles on the multi-axial strength of AC (Asphalt Concrete)-13 and SMA (Stone Mastic Asphalt)-13 asphalt mixtures which are widely used in China, triaxial tests were carried out in the laboratory. Two nonlinear failure criterions under three-dimensional stress states in octahedral space were established. A linear model for engineering design and its simplified testing method were then presented. The three-dimensional failure criteria of asphalt mixtures after 0, 1, 3, 5, 10, 15, 20 freeze-thaw cycles were also proposed. The results indicated that the multi-axial strength decayed significantly after 20 freeze-thaw cycles. It is noteworthy that the strength degrades rapidly during the first 5 freeze-thaw cycles. Compared with AC-13 asphalt mixture, the SMA-13 asphalt mixture exhibits better performance on the resistance to freeze-thaw damage, and it is recommended as the upper surface layer material of pavement structure.

**Keywords:** asphalt mixtures; three-dimensional stress states; freeze-thaw cycles; triaxial strength; failure criterion

## 1. Introduction

Asphalt pavement is influenced by freeze-thaw cycles in the middle and lower reaches of the Yangtze River and some seasonally frozen areas of China in winter. Many studies have shown that freeze-thaw cycles significantly affect the performance of asphalt mixtures [1–3]. The freeze-thaw process would lead to the loss of adhesion between binder and aggregates, which can also change the properties of aggregate, such as strength, compressibility, porosity, and permeability [2], as well as reduce the strength and stiffness of asphalt mixtures. Therefore, it can cause various forms of premature pavement distress [3]. Some asphalt pavements were discovered to have grave freeze injury during construction right away or after the construction process is completed [4,5]. Hence, effective analysis of the freeze-thaw cycles on the performance of asphalt mixtures is necessary [6–8].

General studies are mainly focused on the influence of freeze-thaw cycles and the macro performance of asphalt mixtures under simple stress states including resistance, stiffness, stability, fatigue life, low-temperature properties, etc. [9–15]. They have made contributions to improve the accuracy of asphalt pavement design parameters. And a number of studies on the damage evolution of asphalt mixtures during freeze-thaw cycles from the basis of micro-level were reported in Ref. [16,17].

Furthermore, a lot of models were established to characterize the mechanical performance of asphalt mixtures after multiple freeze-thaw cycles under simple stress states [18,19].

However, the asphalt mixtures are not only affected by the freeze-thaw cycles, but also by the complex stress states in the site [20,21]. Therefore, the uniaxial tensile test, uniaxial compression test, bending test, and indirect test under one-dimensional or two-dimensional stress states cannot reflect the failure modes of asphalt mixtures in three-dimensional stress conditions in the asphalt pavement structures [22–24]. Generally speaking, the tensile properties of asphalt materials should be emphasized at relatively low temperature. While the conventional triaxial test is available only for the triaxial compressive stress state with confining pressure if the first principal stress $\sigma_1$ equal to the second principal stress $\sigma_2$, the triaxial tensile tests cannot be carried out [25–27]. As it is difficult to study the multi-axial strength properties of asphalt mixtures [28,29], the three-dimensional failure criterion of asphalt mixture after freeze-thaw cycles has not been developed.

The objective in this paper is to study the effects of the complex stress states and freeze-thaw in asphalt pavement in the laboratory, as well as to perform triaxial tests (especially triaxial tensile testing) for AC (Asphalt Concrete)-13 and SMA (Stone Mastic Asphalt)-13 asphalt mixtures after freeze-thaw cycles with the self-developed triaxial test method. In addition, the nonlinear and linear failure criterions under three-dimensional stress states are established to evaluate the impact of stress and freeze-thaw cycles on the performance of these asphalt mixtures. From the strength point of view, the SMA-13 asphalt mixture exhibits better freeze-thaw resistance under complex stress condition compared with the AC-13 asphalt mixture. Therefore, it is recommended as the upper surface layer material of pavement structure.

## 2. Laboratory Experimental Program

### 2.1. Materials

The continuous-graded AC(Asphalt Concrete)-13 and gap-graded SMA(Stone Mastic Asphalt)-13 asphalt mixtures, which are very commonly used for the surface layer on the highway and recommended by Technical Specifications for Construction of Highway Asphalt Pavements (JTG F40-2004) in China, were prepared for the experiment with its gradations listed in Figure 1. The basalt was used as aggregates and the SBS (Styrene Butadiene Styrene) modified bitumen (Zhonghai Asphalt (Taizhou) Co.,Ltd., Taizhou, China) was used as a binder for the preparation of specimens. The basic properties of SBS modified bitumen are as shown in Table 1 and the properties of lignin fiber for the SMA-13 asphalt mixture are shown in Table 2.

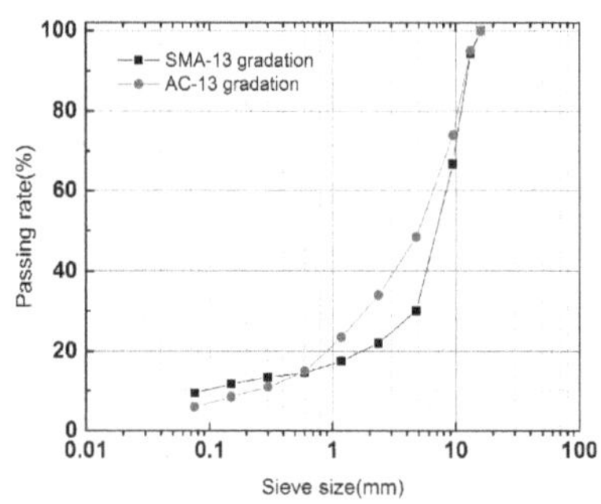


**Figure 1.** Chart of AC(Asphalt Concrete)-13 and SMA (Stone Mastic Asphalt)-13 gradations.

**Table 1.** Properties of SBS (Styrene Butadiene Styrene) modified bitumen.

| Test | SBS Modified Bitumen | Technical Requirements |
|---|---|---|
| penetration (25 °C) (0.1 mm) | 55.9 | 30~60 |
| penetration index | 0.53 | >0 |
| ductility (5 °C /cm, 5 cm/min) | 34.2 | ≥20 |
| softening point (°C) | 79.4 | ≥60 |
| viscosity (135 °C, Pa·s) | 2.30 | ≤3 |

**Table 2.** Properties of fiber.

| Test | Lignin Fiber | Technical Requirements |
|---|---|---|
| fiber length (mm) | 3.5~4.8 | ≤6 |
| ash content (%) | 15.9 | 18 ± 5 |
| pH value | 7.12 | 7.5 ± 1.0 |
| oil absorption (%) | 841 | ≥500 |
| moisture content (%) | 1.29 | ≤5.0 |
| density (g/cm$^3$) | 0.938 | / |

The solid cylindrical specimens were made by the gyratory compactor (TIPTOP China Limited, Shanghai, China) with 100 mm diameter and 102 mm height. The performances of mixtures were extensively tested earlier by Huang [27,28]. The optimum asphalt-aggregates weight ratio of the AC-13 specimens was 5.2% with the air void content of 4.5%. The optimum asphalt-aggregates weight ratio of the SMA-13 specimens was 6.1% with the air void content 3.6%, and the content of the lignin fiber was 0.3%. Moreover, the two ends of the original specimen were polished by the diamond blade (Zhejiang Chenxin Machinery Equipment Co., Ltd, Shaoxing, China) up to the height of 100 mm. The hollow cylinder specimens with a dimension of 10 mm × 50 mm × 100 mm (inner radius × outer radius × height) were prepared for the triaxial tests by coring the solid cylinder specimens.

### 2.2. Testing Conditions and Procedures

#### 2.2.1. Freeze-Thaw Cycles

The freeze-thaw experiments were carried out by repeating the freeze-thaw cycles according to the specification of the Standard Test Methods of Bitumen and Bituminous mixtures for Highway Engineering (JTG E20-2011) in China. Firstly, the specimens were immerged into the tap water in the water tank which was placed in the vacuum drying oven and kept the vacuum for 15 min under the condition of 97.3~98.7 kPa. Then, the valve was opened, the atmospheric pressure restored, and the specimens were placed in the water for 0.5 h. After that, we took out the specimens and put them into a plastic bag, added about 10 mL water and tightened the plastic bag. The specimens were put into the thermostats at the condition of −18 °C for 16 h. Finally, we took out the specimens and immediately put them in the water thermostat tank at 60 °C for 24 h. One freeze-thaw cycle was completed by following the above steps. The triaxial tests were conducted at the end of 0, 1, 3, 5, 10, 15, and 20 cycles.

#### 2.2.2. Triaxial Test

The triaxial test method is used to characterize the mechanical properties of asphalt mixtures under complex stress states, especially in the triaxial tensile stress state, as shown in Figure 2. In this test, a hollow cylinder specimen was placed in the test equipment while the inner and outer surfaces of the specimen were loaded by two independent flexible airbags. Hence, adjustable radial compressive stress and circumferential tensile stress can be generated. Meanwhile, the axial tensile or compressive stresses were produced by a material testing system MTS (Mechanical Testing & Simulation systems company, Minneapolis, USA). Therefore, the three-dimensional unequal stress states can be generated to simulate the complex stress states of asphalt pavement materials in the pavement structures.

According to elastic mechanics, the principal stresses, including axial stress, $\sigma_z$, radial stress, $\sigma_\rho$, and circumferential stress, $\sigma_\varphi$, can be sorted numerically in terms of the principal stresses $\sigma_1$, $\sigma_2$, and $\sigma_3$, respectively [29].

**Figure 2.** Diagram of triaxial test.

In order to simulate the weather of the middle and lower reaches of the Yangtze River and some seasonally frozen areas of China in winter, as shown in Figure 3, 5 °C was selected as the testing temperature. Before the test, the specimens were kept in the temperature control chamber for more than 6 h. The axial loading rate was 2 mm/min, which is the same as the loading rate of the uniaxial compressive test in the current specification.

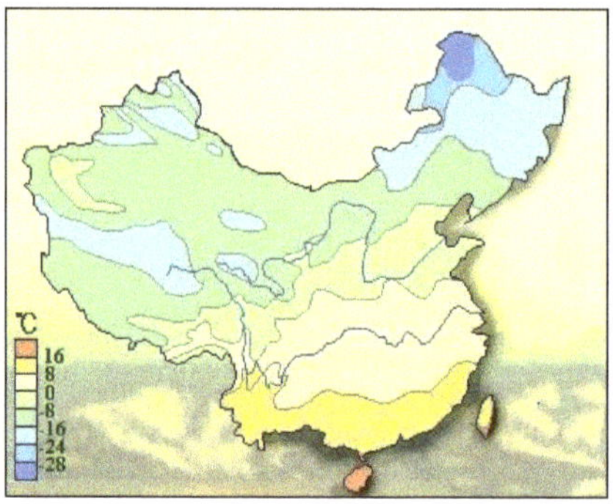


**Figure 3.** Average temperature map of China in January 2018.

Before the triaxial compressive test, some measures have been taken to reduce the friction at the end of the specimen. For example, lubricant oil was smeared on the upper and lower surfaces of the specimen. During the triaxial tensile test, it is necessary to ensure that the specimen is broken without degumming [29]. The triaxial test procedures are as follow:

(1) Tensile meridian and compressive meridian: The tensile meridian/compressive meridian can be obtained by the triaxial tensile/compressive test. Three-direction isobaric stress condition ($\sigma_1 = \sigma_2 = \sigma_3$) of specimens were set at certain stress levels by applying the inner, outer airbags and loading shaft of MTS, and thereafter, the axial tensile/compressive stress $\sigma_1/\sigma_3$ was applied until the failure of the specimen. A series of tensile tests were carried out under different three-direction isobaric stress conditions to obtain the tensile meridian. Likewise, the compressive meridian can be obtained. On the tensile meridian and compressive meridian, the lode angle was equal to 0°and 60°, respectively. The stress path of the tensile meridian and compressive meridians are shown in Figure 4.

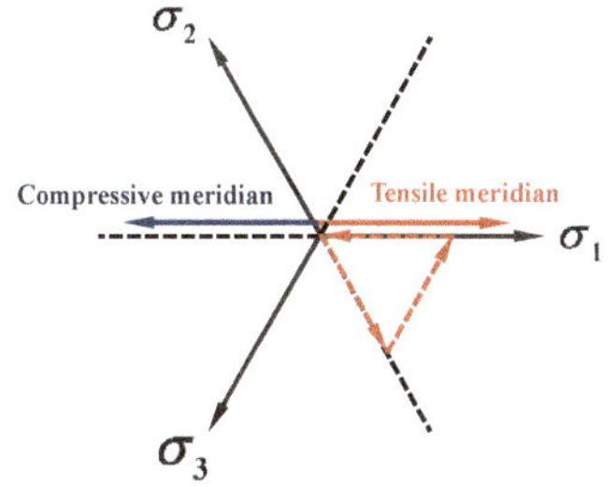


**Figure 4.** Stress path of tensile and compressive meridians.

(2) Strength envelope curve: The strength envelope curve can be obtained by plane tensile and compressive/axial tensile tests. At first, the transverse stresses, $\sigma_2$ and $\sigma_3$, were increased up to the pre-determined values proportionally by applying pressure with the inner airbag. Then, the axial tensile stress, $\sigma_1$, was applied by an MTS material testing machine until specimen failure. On the strength envelope curve, the average stresses were basically the same and the lode angles ranged from 0°~60° gradually. The stress path is shown in Figure 5 [28,29].

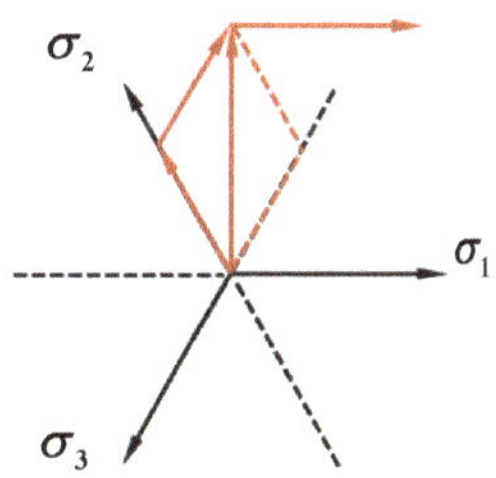


**Figure 5.** Stress path of strength envelope curve.

## 3. Failure Criterions

### 3.1. Nonlinear Failure Criterion

The average value of three effective test results of triaxial compressive/tensile tests, the plane tensile, and compressive/axial tensile tests are presented in Table 3. Based on the test results, the octahedral normal stress, $\sigma_{oct}$, octahedral shear stress, $\tau_{oct}$, and lode angle, $\theta$ can be obtained by the following formulas [29]:

$$\sigma_{oct} = \sigma_m = (\sigma_1 + \sigma_2 + \sigma_3)/3 \tag{1}$$

$$\tau_{oct} = \sqrt{(\sigma_1 - \sigma_2)^2 + (\sigma_2 - \sigma_3)^2 + (\sigma_3 - \sigma_1)^2}/3 \tag{2}$$

$$\theta = \arccos\frac{2\sigma_1 - \sigma_2 - \sigma_3}{3\sqrt{2}\tau_{oct}} \tag{3}$$

**Table 3.** Results of triaxial tests for asphalt mixtures.

| Test | Mixture Type | $\sigma_1$ (MPa) | $\sigma_2$ (MPa) | $\sigma_3$ (MPa) | $\frac{\sigma_{oct}}{f_c}$ | $\frac{\tau_{oct}}{f_c}$ | $\theta°$ |
|---|---|---|---|---|---|---|---|
| triaxial compressive | AC-13 | – | 0.000 | −8.477 | −0.333 | 0.471 | 60 |
| | | −0.200 | −0.200 | −9.372 | −0.384 | 0.510 | 60 |
| | | −0.400 | −0.400 | 9.984 | −0.424 | 0.533 | 60 |
| | | −0.600 | −0.600 | −10.750 | −0.470 | 0.564 | 60 |
| | | −0.800 | −0.800 | −11.806 | −0.527 | 0.612 | 60 |
| | SMA-13 | 0.000 | 0.000 | −9.240 | −0.333 | 0.471 | 60 |
| | | −0.200 | −0.200 | −10.506 | −0.393 | 0.526 | 60 |
| | | −0.400 | −0.400 | −10.983 | −0.425 | 0.540 | 60 |
| | | −0.600 | −0.600 | −11.617 | −0.462 | 0.562 | 60 |
| | | −0.800 | −0.800 | −12.969 | −0.526 | 0.621 | 60 |
| triaxial tensile | AC-13 | 1.110 | 0.000 | 0.000 | 0.044 | 0.062 | 0 |
| | | 1.017 | −0.500 | −0.500 | 0.001 | 0.084 | 0 |
| | | 0.913 | −1.000 | −1.000 | −0.043 | 0.106 | 0 |
| | | 0.700 | −1.500 | −1.500 | −0.090 | 0.122 | 0 |
| | | 0.588 | −2.000 | −2.000 | −0.134 | 0.144 | 0 |
| | SMA-13 | 1.258 | 0.000 | 0.000 | 0.045 | 0.064 | 0 |
| | | 1.176 | −0.500 | −0.500 | 0.006 | 0.086 | 0 |
| | | 1.059 | −1.000 | −1.000 | −0.034 | 0.105 | 0 |
| | | 0.864 | −1.500 | −1.500 | −0.077 | 0.121 | 0 |
| | | 0.748 | −2.000 | −2.000 | −0.117 | 0.140 | 0 |
| plane tensile and compressive/ axial tensile | AC-13 | 1.110 | 0.000 | 0.000 | 0.044 | 0.062 | 0 |
| | | 1.091 | 0.125 | −0.115 | 0.043 | 0.061 | 10.8 |
| | | 1.042 | 0.313 | −0.289 | 0.042 | 0.064 | 26.8 |
| | | 0.991 | 0.442 | −0.408 | 0.040 | 0.068 | 37.8 |
| | | 0.987 | 0.505 | −0.467 | 0.040 | 0.071 | 41.0 |
| | | 0.955 | 0.594 | −0.548 | 0.039 | 0.076 | 46.7 |
| | | 0.896 | 0.780 | −0.720 | 0.038 | 0.087 | 56.3 |
| | | 0.834 | 0.834 | −0.770 | 0.035 | 0.089 | 60 |
| | SMA-13 | 1.258 | 0.000 | 0.000 | 0.045 | 0.064 | 0.0 |
| | | 1.204 | 0.125 | −0.115 | 0.044 | 0.062 | 9.8 |
| | | 1.179 | 0.313 | −0.289 | 0.043 | 0.065 | 24.1 |
| | | 1.128 | 0.442 | −0.408 | 0.042 | 0.068 | 33.5 |
| | | 1.112 | 0.505 | −0.467 | 0.041 | 0.070 | 37.6 |
| | | 1.083 | 0.594 | −0.548 | 0.041 | 0.074 | 43.0 |
| | | 0.945 | 0.783 | −0.720 | 0.036 | 0.081 | 54.9 |
| | | 0.880 | 0.880 | −0.810 | 0.034 | 0.086 | 60.0 |

For triaxial compressive tests, both AC-13 and SMA-13 specimens are mainly represented as shear failure shown in Figure 6. For the triaxial tensile test, the plane tensile, and compressive/axial tensile test, the specimen failures are shown in Figure 7.

**Figure 6.** Shear failure.

**Figure 7.** Tensile failure.

Based on the triaxial test results, the SMA-13 asphalt mixture has a higher strength in three-dimensional stress states (especially in triaxial tensile stress states) compared with AC-13 asphalt mixture. The three-dimensional nonlinear failure criterions of AC-13 and SMA-13 asphalt mixtures were established as follows:

Tensile meridian:

$$\frac{\tau_{oct}^{t}}{f_c} = a + b\frac{\sigma_{oct}}{f_c} + c\left(\frac{\sigma_{oct}}{f_c}\right)^2,\ R_{AC}^2 = 0.98; R_{SMA}^2 = 0.98 \tag{4}$$

Compressive meridian:

$$\frac{\tau_{oct}^{c}}{f_c} = d + e\frac{\sigma_{oct}}{f_c} + f\left(\frac{\sigma_{oct}}{f_c}\right)^2,\ R_{AC}^2 = 0.99; R_{SMA}^2 = 0.98 \tag{5}$$

Strength envelope curve:

$$\tau_{oct}(\theta) = \tau_{oct}^{t} - \left(\tau_{oct}^{t} - \tau_{oct}^{c}\right)\sin^n 1.5\theta,\ R_{AC}^2 = 0.96; R_{SMA}^2 = 0.97 \tag{6}$$

where $f_c$ is the uniaxial compressive strength, $a, b, c, d, e, f$, and $n$ are model parameters as shown in Table 4. The results of the experiments/prediction are presented in Figures 8 and 9, respectively.

**Table 4.** Parameters of the nonlinear failure criterions.

| Mixture Type | $a$ | $b$ | $c$ | $d$ | $e$ | $f$ | $n$ |
|---|---|---|---|---|---|---|---|
| AC-13 | 0.072 | −0.55 | −0.41 | 0.145 | −1.138 | −0.541 | 5 |
| SMA-13 | 0.084 | −0.664 | −0.426 | 0.139 | −1.095 | −0.632 | 7 |

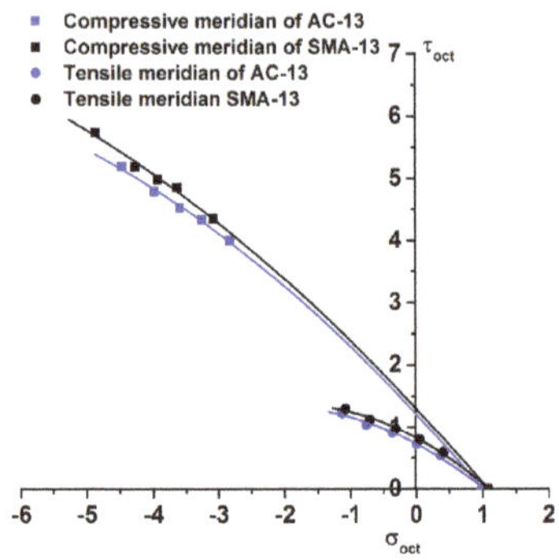


**Figure 8.** Nonlinear tensile and compressive meridians.

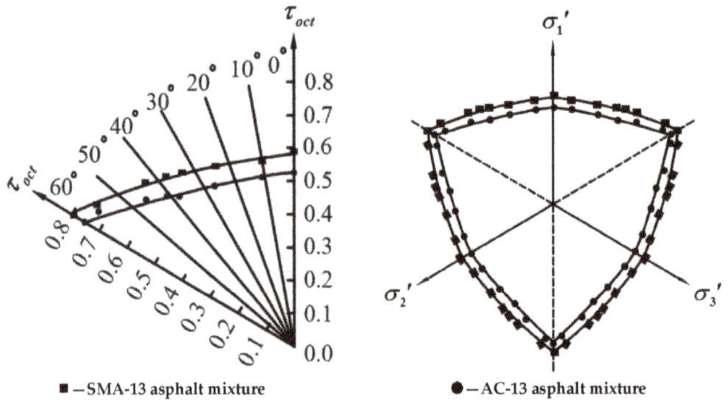


**Figure 9.** Nonlinear strength envelope curves in the $\pi$-plane.

It is shown that these failure criteria are in good agreement with the test results and reflect the difference between the tensile and compressive strength of the asphalt mixtures under complex stress states, as well as the synergistic failure effect of each stress component in the pavement structures.

However, these criteria are complicated and have many fitting parameters. In the process of regression, the tensile meridian can be approximated with a quadratic polynomial. The tensile and compressive meridians must intersect at the triaxial equal tensile point when $\tau_{oct}$ = 0. The tensile meridian and compressive meridian were set at a proportion relationship to simplify the meridians. Furthermore, it is assumed that the strength envelope in the region of 0°~60° is interpolated with the sine function [29,30]. Therefore, the failure criteria can be represented as

Tensile meridian:

$$\frac{\tau_{oct}^{t}}{f_c} = a_1 + b_1\frac{\sigma_{oct}}{f_c} + c_1\left(\frac{\sigma_{oct}}{f_c}\right)^2,\ R_{AC}^2 = 0.94; R_{SMA}^2 = 0.99 \tag{7}$$

Compressive meridian:

$$\frac{\tau_{oct}^{c}}{f_c} = k\left[a_1 + b_1\frac{\sigma_{oct}}{f_c} + c_1\left(\frac{\sigma_{oct}}{f_c}\right)^2\right],\ R_{AC}^2 = 0.94; R_{SMA}^2 = 0.97 \tag{8}$$

Strength envelope curve:

$$\tau_{oct}(\theta) = \tau_{oct}^{t} - \left(\tau_{oct}^{t} - \tau_{oct}^{c}\right)\sin^m 1.5\theta,\ R_{AC}^2 = 0.94; R_{SMA}^2 = 0.97 \tag{9}$$

where $a_1$, $b_1$, $c_1$, $k$, and $m$ are model parameters as shown in Table 5. The comparison with experiments is shown in Figures 8 and 9.

**Table 5.** Parameters of the simplified nonlinear failure criteria.

| Mixture Type | $a_1$ | $b_1$ | $c_1$ | $k$ | $m$ |
|---|---|---|---|---|---|
| AC-13 | 0.085 | −0.67 | −0.32 | 1.69 | 5 |
| SMA-13 | 0.085 | −0.664 | −0.426 | 1.63 | 7 |

### 3.2. Linear Failure Criterion

As it is difficult for the engineering design department to establish the failure criteria, the nonlinear criteria should be further simplified for the convenience of analysis. Therefore, a linear failure criterion can be established by curve fitting as below [29].

Tensile meridian:

$$\frac{\tau_{oct}^t}{f_c} = a_2 + b_2\frac{\sigma_{oct}}{f_c},\ R_{AC}^2 = 0.94; R_{SMA}^2 = 0.96 \tag{10}$$

Compressive meridian:

$$\frac{\tau_{oct}^c}{f_c} = k_1\left(a_2 + b_2\frac{\sigma_{oct}}{f_c}\right),\ R_{AC}^2 = 0.94; R_{SMA}^2 = 0.98 \tag{11}$$

Strength envelope curve:

$$\tau_{oct}(\theta) = \tau_{oct}^t - (\tau_{oct}^t - \tau_{oct}^c)3\theta/\pi,\ R_{AC}^2 = 0.88; R_{SMA}^2 = 0.89 \tag{12}$$

where $a_2$, $b_2$, and $k_1$ are model parameters as shown in Table 6.

**Table 6.** Parameters of the linear failure criteria.

| Mixture Type | $a_2$ | $b_2$ | $k_1$ |
|---|---|---|---|
| AC-13 | 0.088 | −0.733 | 1.4 |
| SMA-13 | 0.08 | −0.579 | 1.6 |

Seeing from Figures 9 and 10, the nonlinear failure envelope curves of asphalt mixtures are transformed from the shape similar to a shield into a hexagon in the $\sigma_{oct} - \tau_{oct}$ space due to linear fitting. With the increase of average stress, the hexagon strength envelope gradually expands along the linear tensile and compressive meridians as shown in Figure 11. Furthermore, the simplified failure criterion under complex stress states can be established by the uniaxial compressive, uniaxial tensile, and the ordinary triaxial tests. Although each of these tests cannot reflect the strength properties of asphalt mixtures under three-dimensional stress states, the synergistic failure effect of each stress component can be considered with the combination of these tests. Therefore, the engineering design department has the ability to complete the related tests.

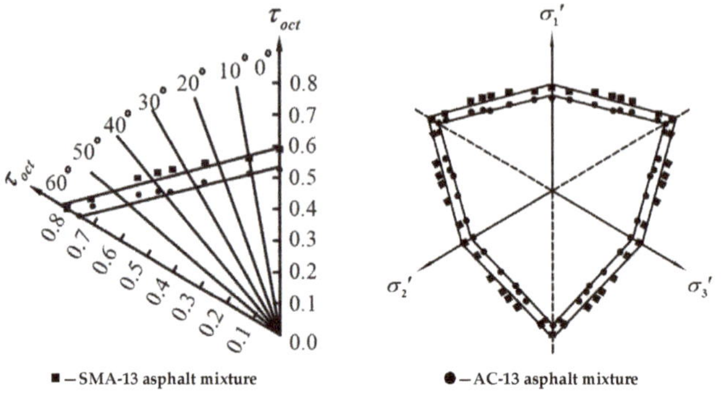


**Figure 10.** Linear strength envelope curves in the $\pi$-plane.

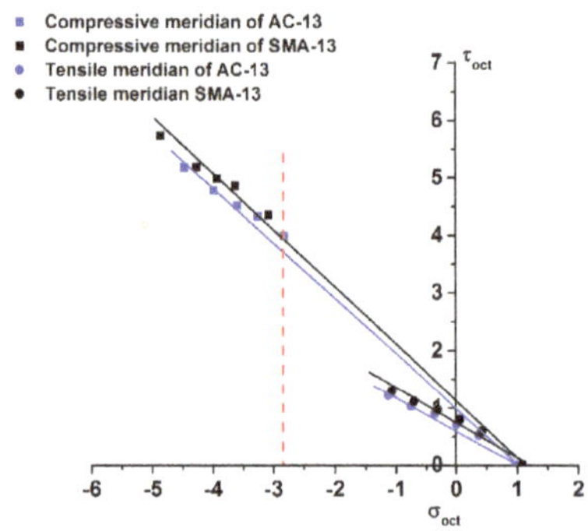


**Figure 11.** Linear tensile and compressive meridians.

## 3.3. Failure Criterion after Freeze-Thaw Cycles

According to the establishment of engineering failure criterion and its testing method, the failure criterion of AC-13 and SMA-13 asphalt mixtures following freeze-thaw cycles can be established by conducting uniaxial compressive, uniaxial tensile, and the conventional triaxial tests. The average value of three effective test results of failure strength after 0, 1, 3, 5, 10, 15, and 20 freeze-thaw cycles are listed in Table 7.

The three-dimensional failure criteria after freeze-thaw cycles can be represented by Tensile meridian:

$$\frac{\tau_{oct}^{t}}{f_c} = a_2 + b_2\frac{\sigma_{oct}}{f_c} - c_2N,\ R_{AC}^2 = 0.85; R_{SMA}^2 = 0.83 \tag{13}$$

Compressive meridian:

$$\frac{\tau_{oct}^{c}}{f_c} = k_1\left(a_2 + b_2\frac{\sigma_{oct}}{f_c} - c_2N\right),\ R_{AC}^2 = 0.98; R_{SMA}^2 = 0.98 \tag{14}$$

Strength envelope curve:

$$\tau_{oct}(\theta) = \tau_{oct}^{t} - (\tau_{oct}^{t} - \tau_{oct}^{c})3\theta/\pi \tag{15}$$

where $N$ is the number of freeze-thaw cycles; $a_2$, $b_2$ $c_2$, and $k_1$ are model parameters as shown in Table 8.

**Table 7.** Failure strength of asphalt mixtures after freeze-thaw cycles.

| Mixture Type | Freeze-Thaw Cycles (N) | Uniaxial Compressive Strength | Triaxial Compressive Strength | | | Uniaxial Tensile Strength |
|---|---|---|---|---|---|---|
| | | $f_c$ (MPa) | $\sigma_1$ (MPa) | $\sigma_2$ (MPa) | $\sigma_3$ (MPa) | $f_t$ (MPa) |
| AC-13 | 0 | −8.477 | −0.2 | −0.2 | −9.372 | 1.110 |
| | 1 | −8.201 | −0.2 | −0.2 | −9.194 | 1.072 |
| | 3 | −7.913 | −0.2 | −0.2 | −8.765 | 0.994 |
| | 5 | −7.460 | −0.2 | −0.2 | −8.451 | 0.936 |
| | 10 | −6.866 | −0.2 | −0.2 | −7.835 | 0.818 |
| | 15 | −6.358 | −0.2 | −0.2 | −7.423 | 0.733 |
| | 20 | −6.104 | −0.2 | −0.2 | −7.301 | 0.692 |
| SMA-13 | 0 | −9.240 | −0.2 | −0.2 | −10.506 | 1.258 |
| | 1 | −9.067 | −0.2 | −0.2 | −10.287 | 1.217 |
| | 3 | −8.634 | −0.2 | −0.2 | −9.862 | 1.109 |
| | 5 | −8.271 | −0.2 | −0.2 | −9.643 | 1.043 |
| | 10 | −7.596 | −0.2 | −0.2 | −8.970 | 0.925 |
| | 15 | −7.128 | −0.2 | −0.2 | −8.553 | 0.844 |
| | 20 | −6.714 | −0.2 | −0.2 | −8.302 | 0.796 |

**Table 8.** Parameters of the linear failure criterions after freeze-thaw cycles.

| Mixture Type | $a_2$ | $b_2$ | $c_2$ | $k_1$ |
|---|---|---|---|---|
| AC-13 | 0.088 | −0.733 | 0.0015 | 1.4 |
| SMA-13 | 0.08 | −0.579 | 0.0012 | 1.6 |

The failure characteristics of AC-13 and SMA-13 asphalt mixtures after freeze-thaw cycles are similar to the previous failure modes. The specimens experienced shear failure and tensile failure and there are a few loose grains on the failure surfaces.

Seeing from Figures 12 and 13, there are multivariate linear relationships between the octahedral shear strength and the octahedral normal strength, as well as the freeze-thaw numbers of AC-13 and SMA-13 asphalt mixtures, the failure envelope curves also shrink during the freeze-thaw process. This criterion provides the testing and theoretical reference for material and structure design of asphalt pavement in three-dimensional stress states under freeze-thaw conditions.

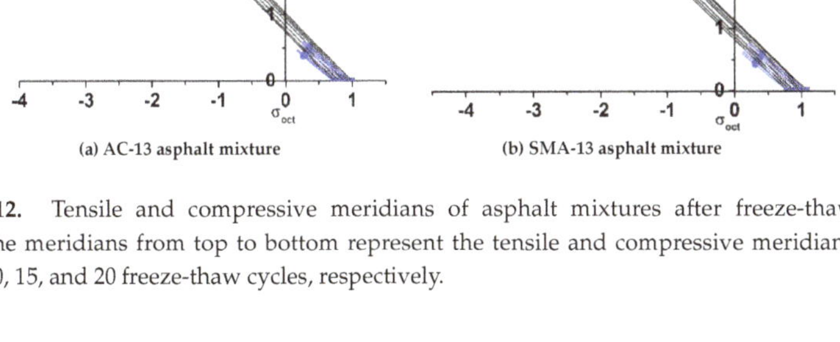


**Figure 12.** Tensile and compressive meridians of asphalt mixtures after freeze-thaw cycles. **Note:** The meridians from top to bottom represent the tensile and compressive meridians after 0, 1, 3, 5, 10, 15, and 20 freeze-thaw cycles, respectively.

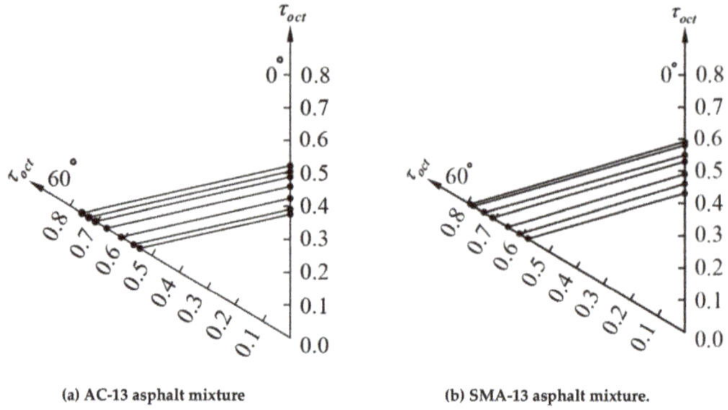


**Figure 13.** Strength envelope curves of asphalt mixtures after freeze-thaw cycles. **Note:** The straight lines from outside to inside represent the failure envelope curves after 0, 1, 3, 5, 10, 15, and 20 freeze-thaw cycles, respectively.

As shown in Figure 14, the freeze-thaw process can degrade the resistance of three-dimensional stress states; the uniaxial compressive strength, triaxial compressive strength, and uniaxial tensile decayed significantly after 20 freeze-thaw cycles for these two asphalt mixtures. Especially, the strength degrades rapidly during the first 5 freeze-thaw cycles. Compared with the continuous-graded AC-13 asphalt mixture, the gap-graded SMA-13 asphalt mixture exhibits better resistance to freeze-thaw damage under complex stress conditions. This is mainly because SMA-13 is a framework-dense structure and AC-13 is a suspend-dense structure, the SMA-13 asphalt mixture consists of a coarse aggregate skeleton which has a high binder content and a low air void content, and there are some fibers in the mixture [31–33]. Since the influence of freeze-thaw cycling and complex stress conditions are inevitable in the middle and lower reaches of the Yangtze River and some seasonally frozen areas of China in winter, the SMA-13 asphalt mixture is recommended to improve the resistance to freeze-thaw erosion.

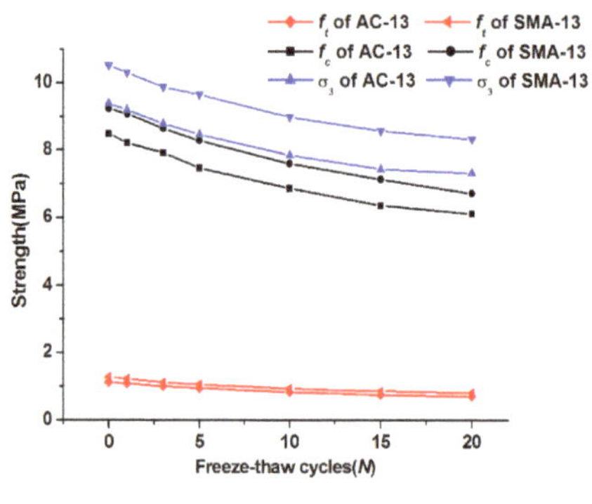


**Figure 14.** Strength of asphalt mixtures after freeze-thaw cycles.

## 4. Summary and Conclusions

Through the triaxial experiments, the nonlinear failure criterion with seven parameters and the simplified nonlinear failure criterion with five parameters have been established to characterize the mechanical behavior of AC-13 and SMA-13 asphalt mixtures under three-dimensional stress states. Furthermore, a linear engineering failure criterion with three parameters is also proposed. This criterion can be obtained by uniaxial compressive, uniaxial tensile, and conventional triaxial compressive tests. With the linear criterion, the synergistic failure effect of each stress components can be considered which is also convenient for engineering analysis.

The three-dimensional failure criteria of AC-13 and SMA-13 asphalt mixtures after 0, 1, 3, 5, 10, 15, and 20 freeze-thaw cycles are presented based on the combination strength tests. There is a multivariate linear relationship between the octahedral shear strength, the octahedral normal strength, as well as the freeze-thaw numbers. Compared with the AC-13 asphalt mixture, the SMA-13 asphalt mixture exhibits better resistance to freeze-thaw damage under complex stress states. Therefore, the SMA-13 SBS asphalt mixture is recommended as the upper surface layer material of pavement structure.

Further analysis is required for asphalt mixtures with varying asphalt content or varying compaction in the lab to improve the three-dimensional resistance following freeze-thaw cycles. Meanwhile, data can be obtained by taking special samples from construction projects to conduct triaxial tests to establish actual failure criteria.

**Author Contributions:** Conceptualization (T.H. and S.L.); Data curation (T.H., S.Q., M.Y., S.L. and H.L.); Formal analysis (T.H. and S.L.); Funding acquisition (T.H., S.L. and J.Z.); Methodology (T.H. and H.L.); Project administration (T.H. and S.L.); Resources (T.H. and S.L.); Supervision (J.Z.); Writing—original draft (T.H., S.Q., M.Y. and S.L.); Writing—review & editing (T.H., S.Q., M.Y. and S.L.).

**Funding:** This research was funded by the National Natural Science Foundation of China [51608055, 51578081, 51608058], Construction Project of Science and Technology of Ministry of Transport of the People's Republic of China [2015318825120], Key Projects of Hunan Province-Technological Innovation Project in Industry [2016GK2096], Hunan Province Natural Science Foundation [2018JJ3550], Open Fund of the Key Laboratory of Highway Engineering of Ministry of Education (Changsha University of Science & Technology) [kfj160201].

**Conflicts of Interest:** The authors declare no conflict of interest.

## References

1. Shang, H.; Song, Y.; Qin, L. Experimental study on strength and deformation of plain concrete under triaxial compression after freeze-thaw cycles. *Build. Environ.* **2008**, *43*, 1197–1204. [CrossRef]
2. Wang, D.; Chen, X.; Stanjek, H.; Oeser, M.; Steinauer, B. Influence of freeze-thaw on the polishing resistance of coarse aggregates on road surface. *Constr. Build. Mater.* **2014**, *64*, 192–200. [CrossRef]
3. Yu, X.; Liu, Y.; Gonzalez, J.; Yu, B. A new TDR sensor for accurate freeze-thaw measurement. *Int. J. Pavement Eng.* **2012**, *13*, 523–534. [CrossRef]
4. Wang, Y.; Ye, J.; Liu, Y.; Qiang, X.; Feng, L. Influence of freeze-thaw cycles on properties of asphalt-modified epoxy repair materials. *Constr. Build. Mater.* **2013**, *41*, 580–585. [CrossRef]
5. Tang, N.; Sun, C.; Huang, S.; Wu, S. Damage and corrosion of conductive asphalt concrete subjected to freeze-thaw cycles and salt. *Mater. Res. Innov.* **2013**, *17*, 240–245. [CrossRef]
6. Yan, K.; You, L.; Wang, X. Test on characteristics of SMA mixtures under freeze-thaw cycle and ultrasonic evaluation method. *China J. Highw. Transp.* **2015**, *11*, 8–14.
7. El-Hakim, M.; Tighe, S. Impact of freeze-thaw cycles on mechanical properties of asphalt mixes. *Transp. Res. Rec. J. Transp. Res. Board* **2014**, *2444*, 20–27. [CrossRef]
8. Kettil, P.; Engström, G.; Wiberg, N.E. Coupled hydro-mechanical wave propagation in road structures. *Comput. Struct.* **2015**, *83*, 1719–1729. [CrossRef]
9. Feng, D.; Yi, J.; Wang, D.; Chen, L. Impact of salt and freeze-thaw cycles on performance of asphalt mixtures in coastal frozen region of China. *Cold Reg. Sci. Technol.* **2010**, *62*, 34–41. [CrossRef]
10. Lv, S.; Liu, C.; Yao, H.; Zheng, J. Comparisons of synchronous measurement methods on various moduli of asphalt mixtures. *Constr. Build. Mater.* **2018**, *158*, 1035–1045. [CrossRef]
11. Kakade, V.B.; Reddy, M.A.; Reddy, K.S. Evaluation of the sensitivity of different indices to the moisture resistance of bituminous mixes modified by hydrated lime and other modifiers. *Road Mater. Pavement Des.* **2017**, *18*, 1395–1410. [CrossRef]
12. You, L.; Yan, K.; Hu, Y.; Ma, W. Impact of interlayer on the anisotropic multi-layered medium overlaying viscoelastic layer under axisymmetric loading. *Appl. Math. Model.* **2018**, *61*, 726–743. [CrossRef]
13. Lv, S.; Zheng, J. Normalization method for asphalt mixture fatigue equation under different loading frequencies. *J. Cent. South Univ.* **2015**, *22*, 2761–2767. [CrossRef]

14. Lv, S.; Wang, S.; Liu, C.; Zheng, J.; Li, Y.; Peng, X. Synchronous Testing Method for Tension and Compression Moduli of Asphalt Mixture under Dynamic and Static Loading States. *J. Mater. Civ. Eng.* **2018**, *30*, 04018268. [CrossRef]
15. Lv, S.; Wang, X.; Liu, C.; Wang, S. Fatigue Damage Characteristics Considering the Difference of Tensile-Compression Modulus for Asphalt Mixture. *J. Test. Eval.* **2018**, *46*, 20170114. [CrossRef]
16. Xu, H.; Guo, W.; Tan, Y. Internal structure evolution of asphalt mixtures during freeze thaw cycles. *Mater. Des.* **2015**, *86*, 436–446. [CrossRef]
17. Shakiba, M.; Darabi, M.K.; Al-Rub, R.K.A.; You, T.; Little, D.N.; Masad, E.A. Three-dimensional microstructural modelling of coupled moisture-mechanical response of asphalt concrete. *Int. J. Pavement Eng.* **2015**, *16*, 445–466. [CrossRef]
18. Tan, Y.; Zhao, L.; Lan, B.; Liang, M. Research on freeze-thaw damage model and life prediction of asphalt mixture. *J. Highw. Transp. Res. Dev.* **2011**, *28*, 1–7.
19. Yi, J.; Shen, S.; Muhunthan, B.; Feng, D. Viscoelastic–plastic damage model for porous asphalt mixtures: Application to uniaxial compression and freeze-thaw damage. *Mech. Mater.* **2014**, *70*, 67–75. [CrossRef]
20. You, L.; Yan, K.; Hu, Y.; Liu, J.; Ge, D. Spectral element method for dynamic response of transversely isotropic asphalt pavement under impact load. *Road Mater. Pavement Des.* **2018**, *19*, 223–238. [CrossRef]
21. Lv, S.; Liu, C.; Chen, D.; Zheng, J.; You, Z.; You, L. Normalization of fatigue characteristics for asphalt mixtures under different stress states. *Constr. Build. Mater.* **2018**, *177*, 33–42. [CrossRef]
22. Su, Y.M.; Hossiney, N.; Tia, M. Indirect tensile strength of concrete containing reclaimed asphalt pavement using the superpave indirect tensile test. *Adv. Mater. Res.* **2013**, *723*, 368–375. [CrossRef]
23. Park, S.W.; Kim, Y.R.; Schapery, R.A. A viscoelastic continuum damage model and its application to uniaxial behavior of asphalt concrete. *Mech. Mater.* **1996**, *24*, 241–255. [CrossRef]
24. Kutay, M.E.; Aydilek, A.H. Dynamic Effects on Moisture Transport in Asphalt Concrete. *J. Transp. Eng.* **2007**, *133*, 406–414. [CrossRef]
25. Tan, S.A.; Low, B.H.; Fwa, T.F. Behavior of asphalt concrete mixtures in triaxial compression. *J. Test. Eval.* **1994**, *22*, 195–203. [CrossRef]
26. Hofko, B.; Blab, R. Enhancing triaxial cyclic compression testing of hot mix asphalt by introducing cyclic confining pressure. *Road Mater. Pavement Des.* **2014**, *15*, 16–34. [CrossRef]
27. Huang, T.B. *Study on Strength Properties of Asphalt Mixture under Complex stress state*; Central South University Press: Changsha, China, 2013.
28. Zheng, J.; Huang, T. Study on triaxial test method and failure criterion of asphalt mixture. *J. Traffic Transp. Eng.* **2015**, *2*, 93–106. [CrossRef]
29. Huang, T.; Zheng, J.; Lv, S.; Zhang, J.; Wen, P.; Bailey, C.G. Failure criterion of an asphalt mixture under three-dimensional stress state. *Constr. Build. Mater.* **2018**, *170*, 708–715. [CrossRef]
30. Song, Y. *Constitutive Relations and Failure Criterions of Various Concrete Materials*; China Water Conservancy and Hydropower Press: Beijing, China, 2002.
31. Cao, X.; Tang, B.; Zhu, H.; Chen, S. Cooling principle analyses and performance evaluation of heat-reflective coating for asphalt pavement. *J. Mater. Civ. Eng.* **2011**, *23*, 1067–1075. [CrossRef]
32. Tarefder, R.; Faisal, H.; Barlas, G. Freeze-thaw effects on fatigue LIFE of hot mix asphalt and creep stiffness of asphalt binder. *Cold Reg. Sci. Technol.* **2018**, *153*, 197–204. [CrossRef]
33. Özgan, E.; Serin, S. Investigation of certain engineering characteristics of asphalt concrete exposed to freeze-thaw cycles. *Cold Reg. Sci. Technol.* **2013**, *85*, 131–136. [CrossRef]

*applied sciences*

*Article*

# Rheology of Asphalt Binder Modified with 5W30 Viscosity Grade Waste Engine Oil

**Touqeer Shoukat [1,2] and Pyeong Jun Yoo [1,2,*]**

1 Department of Smart City and Construction Engineering, Korea Institute of Civil Engineering and Building Technology Campus, University of Science and Technology (UST), 217, Gajeong-ro, Yuseong-gu, Daejeon 34113, Korea; touqeer@ust.ac.kr

2 Department of Infrastructure Safety Research, Korea Institute of Civil Engineering and Building Technology, 283, Goyangdae-ro, Ilsanseo-gu, Goyang-si, Gyeonggi-do 10223, Korea

* Correspondence: pjyoo@kict.re.kr; Tel.: +82-31-9100-175

Received: 4 May 2018; Accepted: 16 July 2018; Published: 20 July 2018

**Featured Application: Authors are encouraged to provide a concise description of the specific application or a potential application of the work. This section is not mandatory.**

**Abstract:** The pavement structure tends to shrink under low temperature conditions and cracks will appear upon crossing threshold binder stiffness. Decreasing the binder viscosity at such low temperatures, by introducing additional oil fraction (aromatics and saturates) in asphalt colloidal systems, may result in improved resistance to thermal cracking. A single multi-grade engine oil (5W30) was used in this study to analyze the rheological properties imparted to binders. Rotational Viscosity (RV) test revealed that after Rolling Thin Film Oven (RTFO) aging, fresh oil and waste oil have a similar effect on decreasing the viscosity of binder and construction temperatures, reducing them by 5~8 °C. Fourier Transform Infrared Spectroscopy (FTIR) test results showed an abrupt increase of carbonyl concertation when fresh engine oil was used for rejuvenation while waste engine oil was less susceptible to oxidative aging. Dynamic analysis of modified binders proved that engine oil has better thermal cracking resistance but relaxation ability of binders and rutting resistance was impaired. Filtered waste engine oil resulted in a 35% decrement in the stiffness of binder compared to virgin asphalt after short term aging but upper Performance Grade (PG) was compromised by 1~3 °C with 2.5% oil inclusion. Unfiltered waste engine oil proved to have the least overall performance compared to fresh and filtered waste engine oil.

**Keywords:** waste engine oil; asphalt; rheological analysis; low temperature stiffness; Discovery Hybrid Rheometer (DHR)

---

## 1. Introduction

Asphalt binder is frequently utilized in hot mix asphalt (HMA) for binding the aggregates mass. Regardless of the global use of asphalt as a binder, its cost is comparatively high. The increased urbanization and surging demand of paved roads led to the need for enhancing asphalt binder intrinsic properties, exclusively the resistance for rutting and thermal cracking. Thermally induced cracking of flexible pavements is critical in cold regions [1], such as South Korea. As the pavement structure tends to shrink under low temperature conditions [2], tensile stresses are developed resulting in cracks or failure. Therefore, asphalt modification is needed as it fundamentally effects the properties of bituminous mixtures. Recently, more attention has been given to improving performance of asphalt binder by modifying it with polymers, resulting in improved engineering properties of asphalt [3–7]. At low temperatures, a decrease in viscosity of the binder is intended to avoid making asphalt too stiff and losing cohesion with aggregates. The viscosity of asphalt can be manipulated by changing the oil fraction in asphalt colloidal system [8].

Engine oil is a by-product of petroleum refining which provides lubrication to rotating bearings and pistons in automobile engines. After being used by vehicles, it gets contaminated with impurities and toxic chemicals like Polycyclic Aromatic Hydrocarbons (PAHs) [9]. Thus, the final waste product, known as Waste Engine Oil (WEO), may cause serious damage to the environment and organic life if it is dumped without proper treatment [10,11]. There is a thriving interest in the pavement industry to utilize waste materials and de-escalate the use of natural resources; this is also cost effective as compared to utilizing new materials. South Korea generated about 370 million liters of waste oil in 2012 [12]. This substantial amount of waste oil is mostly consumed as fuel or re-sold after refining. Chemically, the molecular structure of engine oil is similar to asphalt binder [13,14], therefore waste engine oil can be considered a compatible modifier for asphalt cement [15].

With the increased traffic volume, production of waste engine oil has increased many folds in recent years. Therefore, it is imperative to use the waste engine oil in partial replacement of asphalt binder which will also reduce the cost of Hot Mix Asphalt (HMA) or Reclaimed Asphalt Pavement (RAP). Ayman et al. [16] investigated the potency of different rejuvenators on aging and fatigue cracking resistance. They concluded that paraffinic oils were most suitable to rejuvenate the aged RAP binders without adversely compromising the rut resistance. Research conducted by Zaumanis et al. [17] claimed that waste engine oil with a 12% dose significantly improved the low temperature cracking susceptibility of RAP mixture. A considerable loss of volatiles during RTFO aging was observed during mass loss test which indicates an increased aging susceptibility.

Based on the results by Villanueva et al. [18], modifying asphalt binder with 0–10% of used lubricating oil enhanced the critical cracking temperature to about 2 °C but high temperature PG grade was compromised. Hallizza et al. [19] studied the use of cooking oil as a bitumen rejuvenator. Their research proved that modification of a 40/50 penetration grade aged asphalt binder with 4% waste cooking oil decreased its viscosity to an unaged 80/100 penetration grade condition. Xiaoyang et al. [20] found that a decrement in optimum asphalt content improved fatigue resistance of the mixture and an increased concentration of carbonyl functional groups in asphalt binder after waste engine oil modification. This increase in carbonyl groups left the asphalt binder susceptible to oxidative aging. Oil inclusion also reduced the stiffness at low temperatures but elastic recovery of binder at high temperatures was undermined.

Eriskin et al. [21] focused on reducing the construction cost by decreasing the optimum content of bitumen with partial replacement of waste frying oil. Their study proved that with the inclusion of only 3–5% of frying oil, the bitumen content was reduced by 11%. They further concluded that the fry oil modification enhanced the self-healing properties of HMA at low temperatures. Xavier et al. [22] utilized the used and virgin maize oil for binder modification. Their results showed that with the increase of maize oil content the elasticity was increased while the stiffness was reduced. The mixing and compaction temperatures were decreased by 5~10 °C, proving an efficient modifier for warm mix asphalt (WMA). Zhang et al. [15] investigated the effect of bio-based and refined waste oil on low temperature properties of asphalt binder. They discovered that the glass transition temperature tended to decrease and the fracture energy of asphalt binder could be enhanced up to three folds by using 5% refined waste oil. Simon et al. [23] performed X-Ray fluorescence spectroscopy on asphalt cement modified with waste engine oil. They observed a premature failure in pavement due to loss of strain tolerance and physical hardening of asphalt binder caused by waste oil residues after modification.

Waste engine oil addition can soften and rejuvenate aged asphalt binders and results in an environmentally friendly mixture [24,25]. In this context, the use of waste engine oil in partial replacement of binder in HMA, RAP or WMA can prove to be an acceptable and sustainable solution to managing this waste material. This study was carried out to evaluate the rheological properties of asphalt binder with 2.5% of a single multigrade fresh and waste (filtered and un-filtered) engine oil, with an aim to decrease the viscosity (rejuvenation) and stiffness (thermal cracking) at low temperatures.

*Objective(s)*

Taking into account the previous literature, the objectives of this research work are as follows

1. To assess the properties imparted to asphalt cement with a specific viscosity grade engine oil (5W30) rather than a collective blend of different waste engine oils.
2. To study the effect of metal traces present in waste engine oil on rheological properties of modified asphalt cement.
3. To evaluate the effect of waste oil modification on low and high temperature properties of asphalt cement.

## 2. Materials and Methods

### 2.1. Materials

#### 2.1.1. Asphalt Binder

A PG 64-22 (AP5) asphalt binder frequently used in South Korea was chosen as the base binder for sample preparation and laboratory experimentation. Some standard physical characteristics of virgin asphalt binder are given in Table 1.

**Table 1.** Physical properties of virgin asphalt binder (PG64-22) [26].

| Property | Virgin Asphalt |
|---|---|
| Flash Point, °C | >230 |
| Specific Gravity at 15.6 °C | 1.0386 |
| Absolute Viscosity at 60 °C, poise | 2030 |
| Kinematic Viscosity at 135 °C, centistokes | 362 |

#### 2.1.2. Engine Oil

Waste engine oil (Figure 1) employed in this study was taken from a local automotive repair shop. Engine oils are available in different viscosity grades [27] and can have variable viscosity ranges according to selected viscosity grades [28]. Thus, waste engine oil obtained from different vehicles might have different viscosities. Considering this fact, an SAE 5W30 multigrade engine oil was selected for this study. Some physical characteristics of fresh and waste SAE 5W30 oil have been presented in Table 2.

**Figure 1.** Fresh and waste engine oil.

**Table 2.** Physical properties of fresh and waste engine oil (SAE 5W30) [29,30].

| Property | Fresh Oil | Waste Oil |
|---|---|---|
| Density at 15 °C, (g/cm$^3$) | 0.861 | 0.9116 |
| Viscosity at 40 °C, centistokes | 71 | 107.48 |
| Viscosity at 100 °C, centistokes | 11.75 | 12.93 |

Figure 2 demonstrates and compares the chemical composition of base asphalt and engine oils after FTIR spectroscopy. Each peak in the spectrum signifies a functional group in the medium. Most of the observed peaks in fresh engine oil were similar to asphalt binder which proves its compatibility to bond with asphalt molecules. On the other hand, waste engine oil exhibited additional peaks at 1738 and 1216 wavenumbers indicating the chemical decomposition and aging of fresh oil after vehicle use. Small peaks around 1540 wavenumbers indicated the formation of PAHs waste engine oil, which were negligible in fresh oil. Additional peaks between 2300~2400 were due to $CO_2$ in the atmosphere and should not be considered as a chemical change between asphalt and engine oil.

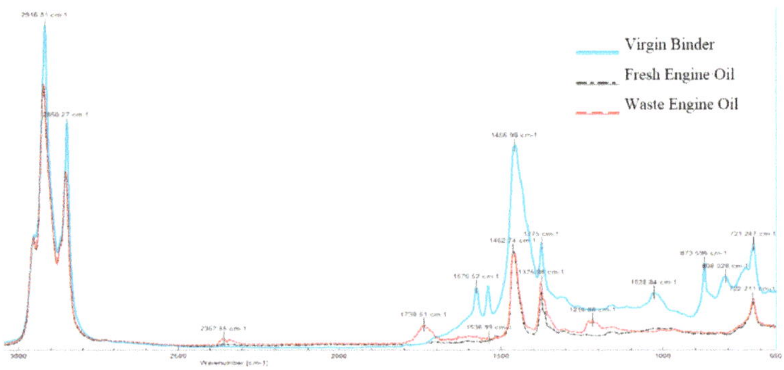


**Figure 2.** Chemical composition of asphalt binder, fresh engine oil and waste engine oil.

### 2.1.3. Filter Paper

Due to wear and tear of the engine during vehicle performance, engine oil was contaminated with metal traces and exposed to repeated heating and oxidation actions [31]. Metallic elements like Potassium, Calcium, zinc and iron were found in weathered motor oil [32]. These metal traces might have changed the rheological properties of asphalt binder after modification. Therefore, to check this rheological difference, waste engine oil was filtered using a Whatman™ grade filter. Some physical properties of filter paper are shown in Table 3.

**Table 3.** Physical properties of Whatman™ filter paper.

| Property | Filter Paper Grade: 1822-047 |
|---|---|
| Material | Glass Fiber |
| Pore Size | 1.2 um |
| Nominal Thickness | 260 um |
| Water Flow Rate | 200 (L/min/cm$^2$ at 100 kPa) |
| Nominal Basis Weight | 53 g/m$^2$ |
| Max. Temperature | 180 °C |

## 2.2. Methods

### 2.2.1. Research Approach

Although there are positive and negative effects of using waste engine oil in asphalt binder, more research is needed to incorporate this waste product efficiently and to avoid direct harmful effects on the environment. Considering these factors and literature available, this research work focused on evaluating the possibility of using a specific viscosity grade waste engine oil as a rejuvenator for asphalt binder and studied its effects on rheological properties considering high and low temperatures. The waste engine oil addition was further assessed on removal of metal traces by filtration process and with short term aging conditions. Rheological tests were performed to check the properties imparted to rejuvenated asphalt. Finally, master curves for a comprehensive characterization of modified asphalt cement were constructed and compared. An outline of the procedure for this study is shown in Figure 3.

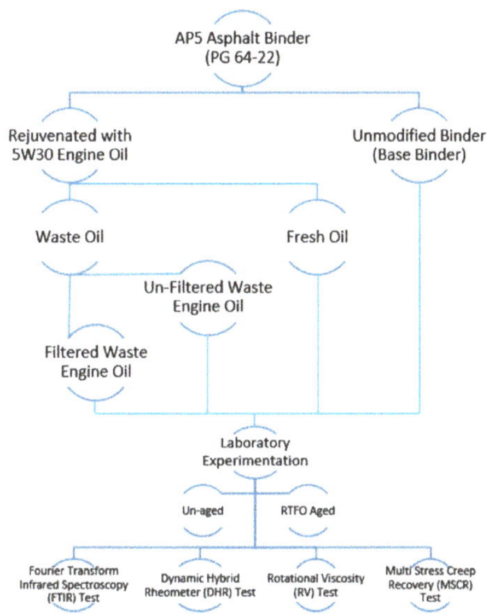


**Figure 3.** Procedural flow chart.

### 2.2.2. Preparation of Modified Asphalt Binder and Sampling

To study the effect of waste engine oil on asphalt performance and to make the waste oil components blend well with asphalt, a high shear mixer along with a heating mantle model MS-DBM604 (MTOPS®, Yangju-si, Kyunggi-do, Korea) was used. Mixing speed was kept at 2000 rpm. The blending time and temperature were controlled at 30 min and 150 ± 5 °C for each mixing turn, respectively.

Samples were prepared by mixing 2.5% (by weight of asphalt) fresh oil, un-filtered waste engine oil and filtered waste engine oil into asphalt binder (Table 4). At least three, and a maximum of five samples for each condition and for each laboratory test were prepared to check the repeatability and results were shown as an average. Virgin binder with the same sheared conditions (2000 rpm/30 min/150 °C) was prepared and termed as Base Binder for better comparison.

**Table 4.** Experimental samples and respective codes.

| Sample | Code |
|---|---|
| Base Binder [1] | BB |
| Asphalt + 2.5% Fresh Engine Oil | FR |
| Asphalt + 2.5% Un-Filtered Waste Engine Oil | UFO |
| Asphalt + 2.5% Filtered Waste Engine Oil [2] | FO |

[1] Base Binder was sheared with same mixing criteria (2000 rpm, 30 min, at 150 °C); [2] Filtration done by Whatman™ filter paper.

### 2.2.3. Rotational Viscosity (RV) Test

This test was employed to check the flow changes on the fresh and waste oil rejuvenated binders and the effect of metal traces present in waste engine oil. Viscosity measurements were taken using Brookfield DV2T Viscometer (Brookfield Engineering Laboratories, Inc., Boston, MA, USA) Dynamic viscosity tests can be conducted at various temperatures, but since manufacturing and construction temperatures are fairly similar regardless of the environment, the test was carried out at a range from 135 °C to 175 °C.

The torque on the apparatus-measuring geometry, rotating in a thermostatically controlled sample holder containing a sample of asphalt, was used to measure the relative resistance to rotation. The torque and speed were used to determine the viscosity of the asphalt. The test was conducted according to AASHTO T 316 and ASTM D 4402 testing standards. Between 11~13 g samples of the base binders and rejuvenated binders were poured into disposable RV testing containers and were tested with SPC4-27 spindle at 20 RPM. At least three samples for each binder configuration were tested and the results were reported as an average. Ideal mixing and compaction temperature for base and modified binder were calculated in unaged and RTFO aged condition by using viscosity data.

### 2.2.4. Fourier Transform Infrared Spectroscopy (FTIR) Test

The FTIR spectroscopy test helps to characterize changes in functional groups and chemical alteration of base asphalt due to engine oil addition. This test was performed using JASCO 4200 spectrometer (TS Science Co., Ltd., Seoul, Korea) with Attenuated Total Reflection (ATR) accessory in the range of 4000 to 650 $cm^{-1}$ wavenumbers. ATR provides reliable and repeatable spectra compared to transmission configuration which is prone to erroneous readings with sample preparation and film thickness variations. Infrared radiations are bombarded on a thin sample of asphalt binder, directly placed on the ATR crystal. This excites some of the molecular bonds within the asphalt molecule according to their natural vibration frequency. Some part of the incident infrared rays is absorbed due to the bond excitations. These molecular vibrations/excitations are then detected by a detector in FTIR. Detector measures the percentage of the transmitted or absorbed radiation to the source radiation and a spectrum is processed and displayed by a computer program [33], imitating the functional groups present in the material. For each spectrum obtained, a total of 36 consecutive scans at a resolution of 4 $cm^{-1}$ were executed and averaged.

Some characteristic functional groups, with their respective wavenumber ranges that can be detected with FTIR, are presented in Table 5. These functional groups can also be present in asphalt molecules which helps us to understand the basic chemistry of base, aged and rejuvenated asphalts.

**Table 5.** Functional groups with their respective range of occurrence in wavenumbers [20,34,35].

| Wavenumber Range ($cm^{-1}$) | Functional Group | Bond Vibration Type (Stretching/Bending/Rocking) |
|---|---|---|
| 3100–3500 | N-H group | Stretching vibration |
| 1550–1640 | N-H group | Bending vibration |
| 2500–3300 | O-H | Stretching vibration, Strong Intensity |
| 2927–2935 | -CH2 | Asymmetric or symmetric stretch |
| 1628–1650 | C=O (1652 $cm^{-1}$)<br>cis C=C (1654 $cm^{-1}$) | Stretching vibrations |
| 1735–1750 | C=O (ester) | Stretching vibrations, strong intensity |
| 1800–1830 | C=O (anhydride) | Stretching vibrations, two bands |
| 1435 | CH2 (1462 $cm^{-1}$) | Bending vibrations |
| 1417 | CH bonds | Rocking vibrations |
| 1404 | CH bonds | Rocking vibrations |
| 1346 | CH3 | Bending vibrations |
| 1000–1300 | C-O group | Stretching vibration, two bands or more |
| 1111 and 1097 | -CH -CH | Bending and Deformation Vibrations |
| 1065–1068 | C-O | Stretching |
| 400–1000 | "Fingerprint region" | – |

### 2.2.5. Discovery Hybrid Rheometer (DHR) Test

A Discovery Hybrid Rheometer (DHR-2) test with Environmental Temperature Control (ETC) system was used for determination of viscoelastic behavior of base and rejuvenated binders. DHR is a stress controlled shear rheometer which uses magnetic bearing to rotate the spindle and provide frictionless application of torque to the asphalt sample. ETC accessory provided fast response and temperature stability and allowed us to achieve any temperature ranging from −160°C to 600°C.

#### Multi Stress Creep Recovery

A Multiple Stress Creep and Recovery (MSCR) test was conducted for predicting the rutting behavior and delayed relaxation response of asphalt binders. The MSCR test was carried out with a repeated loading and unloading of stress on RTFO aged asphalt binders. This test can predict the rut resistance by measuring non-recoverable creep compliance ($J_{nr}$) and binder modifications in the non-linear viscoelastic region. AASHTO T 350 testing guidelines were followed for 1 s creep at a constant stress and 9 s recovery with zero stress level. Ten cycles of creep and recovery were performed at 34, 46, 58 and 64 °C temperatures with 0.1 kPa and 3.2 kPa stress levels.

#### Strain Sweeps

Strain sweep tests were performed on each testing temperature ranging from −15 °C to 95 °C in strain controlled mode. Parallel plates measuring 8mm were used for −15 °C to 35 °C testing temperatures while 25 mm parallel plate geometry was used for a high temperature range of 45 °C to 95 °C. Percent Strain value was swept from 0.1 to 80% at a fixed frequency of 10 rad/s and initial stress value of 3.295 Pa. Complex shear modulus (G*) for all binders were measured and the percent strain value at which G* was reduced to 95% of its initial value and was noted as a threshold for linear viscoelastic (LVE) region. Table 6 lists the target strain values selected for LVE region for frequency sweep testing.

**Table 6.** Target strain values for linear viscoelastic region.

| Temperature (°C) | Target Strain (%) |
|---|---|
| −15 | 1.0 |
| −5 | 1.5 |
| 5 | 3.5 |
| 15 | 5.0 |
| 25 | 7.0 |
| 35 | 9.0 |
| 45–95 * | 12.0 |

* Strain values were greater than 20% for all temperatures above 45 °C, therefore target strain was fixed at 12%.

Frequency Sweeps

Frequency sweep tests were performed on a temperature range of −15 °C to 95 °C in stress controlled mode. Parallel plates measuring 8 mm were used for −15 °C to 35 °C testing temperatures while 25 mm parallel plate geometry was used for a high temperature range of 45 °C to 95 °C. Target strain values were fixed as per Table 6, to establish LVE region of measurement. The testing procedure was followed as per AASHTO T315-10. Frequency was swept from 0.01 Hz to 30 Hz for each testing temperature and complex shear modulus (G*) and phase angle (δ) were obtained for each binder sample.

Master Curve

Master curves for the whole temperature range were developed using frequency sweep isotherms for an overall assessment of rejuvenating the asphalt binder with engine oil. It exhibited the impact of loadings on rheological performance of asphalt cement over a wide range of loading frequency or times. Isotherms obtained from frequency sweep tests were shifted to a reference temperature using the time-temperature superposition principle [36,37]. Shift factor (αT) for superpositioning of isotherms can be obtained by using William-Landel-Ferry (WLF) equation and Arrhenius function [38,39] shown as Equations (1) and (2) respectively.

$$\log(a_T) = -\frac{c_2(T - T_R)}{c_2 + T - T_R} \tag{1}$$

In Equation (1), c1 and c2 are coefficients, T is target temperature, $T_R$ is reference temperature. Values of $C_1$ and $C_2$ are usually taken 19 and 92, respectively [40].

$$\log(a_T) = -\frac{E_a}{2.303R}\left(\frac{1}{T} - \frac{1}{T_{ref}}\right) \tag{2}$$

In Equation (2), Ea is activation energy, T is target temperature and $T_{ref}$ is reference temperature. The shifting of isotherms can also be done using a new dynamic modulus function know as Sigmoidal Model [41]. The equation for sigmoidal model is shown as Equation (3)

$$\log|G^*| = \delta - \frac{\alpha}{1 + e^{\beta+\gamma(\log f_{ref})}} \tag{3}$$

where,

δ = Min. G* of asphalt binder,
α = Max. G*−Min. G*

β and γ are S-shaped function parameters of sigmoidal curve for point of turning and slope of curve respectively. In this study, construction of master curve was done using the sigmoidal model. Shift factors were obtained using Arrhenius equation which wer then used for horizontal shift of isotherms. Data were fitted using the sigmoidal model and master curves were constructed for all asphalt binders.

## 3. Results and Discussions

### 3.1. Flow Behavior

The high temperature flow behavior of asphalt binders is an important property to establish in-plant and in-field temperatures [42,43]. The viscosity of binder depicts its competence for pumping through an asphalt plant, its ability to properly coat the aggregate particles in asphalt mixture and the workability required to place the asphalt mixture in field and appropriate compaction. Therefore, RV test was carried out from 135 °C to 175 °C with an increment of 10 °C.

Figure 4 demonstrates the changes in viscosity of asphalt binders in an unaged condition and after RTFO aging on temperature increments. It can be clearly seen that inclusion of oil into asphalt decreased its viscosity and construction temperatures, which is in line with previous literature.

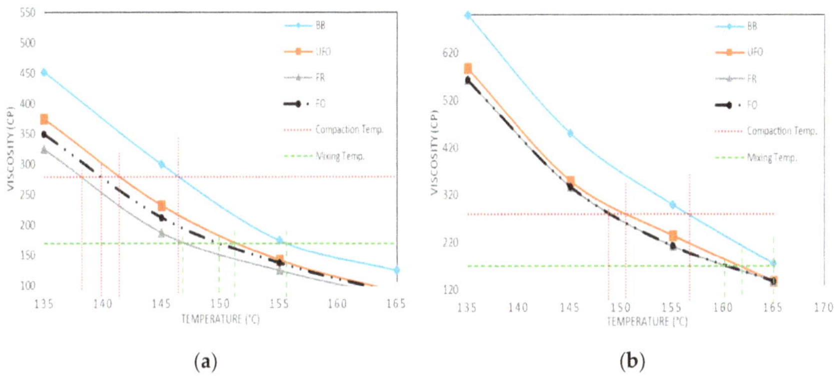

Figure 4. Viscosity variations in different binders: (**a**) Unaged condition and (**b**) RTFO aged.

In-field mixing and compaction temperatures were selected based on the viscosity of binder. As recommended by Asphalt Institute [44], temperature at 170 ± 20 cP is to be taken as mixing temperature while 280 ± 30 cP depicts the compaction temperature of bitumen. It is evident from Figure 4a and Table 7, there was an overall 5~8 °C decrement of compaction and mixing temperature for the unaged binder condition. FR had the highest effect on reducing the mixing and compaction temperatures due the fact that fresh oil had not been aged before and had very low viscosity compared to waste oil.

**Table 7.** Compaction and mixing temperature changes with engine oil addition.

| Binder Property and Condition | | Temperature (°C) | | | |
|---|---|---|---|---|---|
| | | BB | UFO | FR | FO |
| Compaction | Unaged | 146.5 | 141.5 | 138 | 140 |
| | RTFO | 156.5 | 150.5 | 149 | 149 |
| Mixing | Unaged | 155.5 | 151 | 147 | 150 |
| | RTFO | 166 | 162 | 160 | 160 |

After RTFO aging, in Figure 4b, even though the compaction and mixing temperatures increased due to stiffening of binder, oil addition reduced the compaction temperature of BB by 6~7 °C while mixing temperature decreased by up to 5 °C. No significant difference was observed in construction temperatures between FO and FR, while UFO depicted 1 °C increment from other rejuvenated binders.

This 1 °C difference might have been caused by comparatively high viscosity of UFO due to the metal traces present in the asphalt matrix.

### 3.2. Chemical Analysis

In Figure 5, FT-IR spectra of unmodified asphalt binder and after rejuvenating with fresh and waste engine oil has been presented. Noticeable differences can be clearly seen in intensities of functional groups before and after waste oil addition.

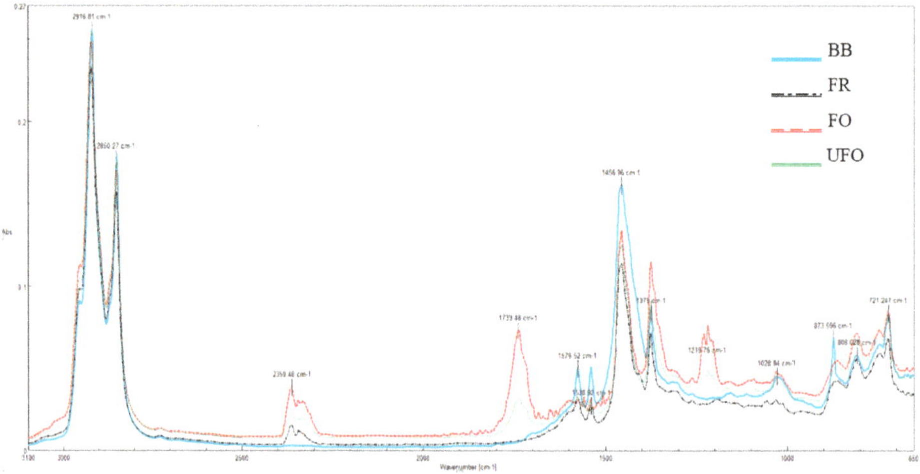


**Figure 5.** FTIR spectra comparison of unaged asphalt binders.

The strong peaks between $\nu \approx 3000$ cm$^{-1}$ to 2800 cm$^{-1}$ represent aliphatic -$CH_3$, -$CH_2$ and CH stretching vibrations. Peaks around $\nu \approx 2359$ cm$^{-1}$ are due to $CO_2$ and its intensity can be controlled by experimental accuracies. Peak in the finger print region from $\nu \approx 1000$ cm$^{-1}$ to $\nu \approx 700$ cm$^{-1}$ indicates benzene rings in asphalt. Signals observed at $\nu \approx 1739$ cm$^{-1}$ and $\nu \approx 1028$ cm$^{-1}$ are representative bands of carbonyl and sulfoxide groups. Intensity changes at these bands are a direct indication of aging binder. In the unaged condition, only FO and UFO exhibited signals at $\nu \approx 1739$ cm$^{-1}$, proving high concentrations of carbonyl functional group C=O due to severe oxidation and heat exposure of waste engine oil during usage in vehicle, which is in line with the previous research [20]. The peak at $\nu \approx 1456$ cm$^{-1}$ is due to $CH_2$ and $CH_3$ bending vibration while $\nu \approx 1375$ cm$^{-1}$ depicts symmetric stretching. The band at $\nu \approx 1216$ cm$^{-1}$ represents C-O bond due to ester molecules in the lubricant oils.

Figure 6 illustrates the chemical profile of the RTFO aged condition for all binders. A significant difference in intensity of carbonyl functional group can be identified at $\nu \approx 1737$ cm$^{-1}$. FR has undergone a severe aging due to the fact that it contains the higher amount of unsaturated hydrocarbons. FO and UFO, containing already aged engine oil, provide less opportunity to produce oxidized products and can be expected to resist aging compared to FR. An increase in peak of FR at $\nu \approx 1216$ cm$^{-1}$ was also observed, which reveals the increased intensity of ester molecules in aged binder.

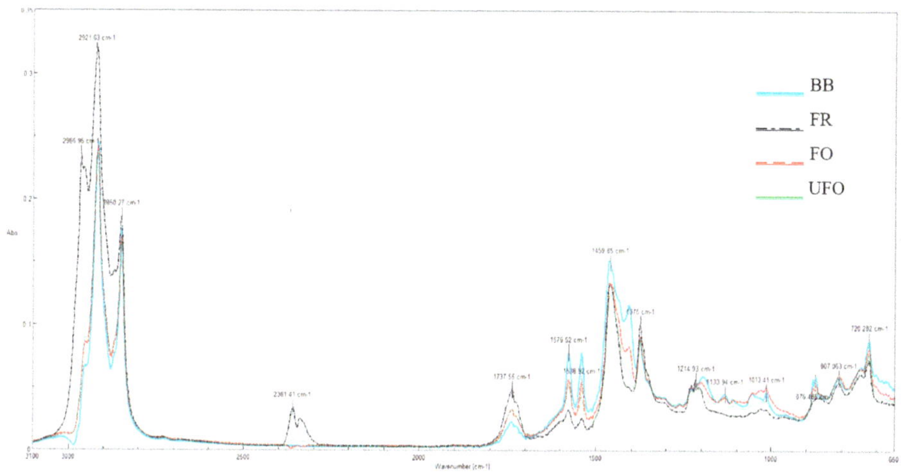


**Figure 6.** FTIR spectra comparison of RTFO-aged asphalt binders.

### 3.3. *Dynamic Mechanical Analysis*

Asphalt binder is a visco-elastic material which behaves as elastic solid at low temperatures while exhibiting fully viscous properties at higher temperatures. In mid-range temperatures, both elastic and viscous properties are shown. Distresses like fatigue and rutting in binder structure depends on the temperature and loading frequency. Dynamic mechanical analysis was done using DHR apparatus and G* and δ were obtained. Complex shear modulus (G*) is the ratio of maximum shear stress to maximum shear strain while the phase angle (δ) is the delay in response of material to applied stress and resulting strain.

#### 3.3.1. Temperature Dependency

Figure 7 explains the behavior of asphalt binders in unaged and RTFO aged conditions at low temperatures at fixed frequency of 10 rad/s. It is obvious from the figure that BB has the highest stiffness at lower temperatures compared to oil rejuvenated binders. In the unaged condition, fresh oil addition worked better than waste oil in reducing the stiffness of base binder.

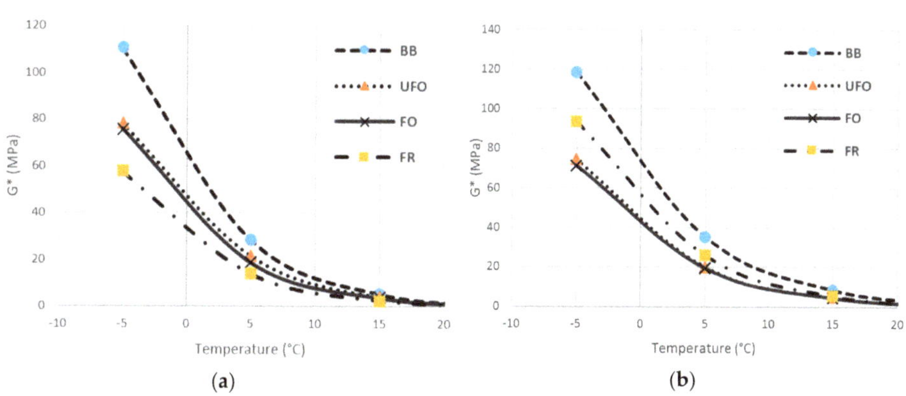


**Figure 7.** Effect on low temperature stiffness of asphalt binders at 10 rad/s for (**a**) unaged and (**b**) RTFO aged.

After RTFO aging, a dramatic rise in stiffness of FR was observed compared to waste oil rejuvenated binders. As FR has more unsaturated bonds to react with oxygen in air compared to FO and UFO, the aging effect is much more severe in fresh oil rejuvenated binder.

Figures 8 and 9 show the effect of waste oil inclusion on high temperature properties of asphalt binder. As per Superpave criteria, unaged binder should have a minimum 1.0 kPa or above value of $G^*/\sin\delta$ to prevent rutting failure. It can be clearly observed that in the unaged condition, there was 1 °C, 2 °C and 3 °C decrease in upper PG grade of UFO, FO and FR binder, respectively. The waste oil in UFO and FO had already been exposed to aging during engine operations which resulted in increased viscosity of rejuvenated binder. FR had the lowest viscosity among all binders due to the unaged fresh oil.

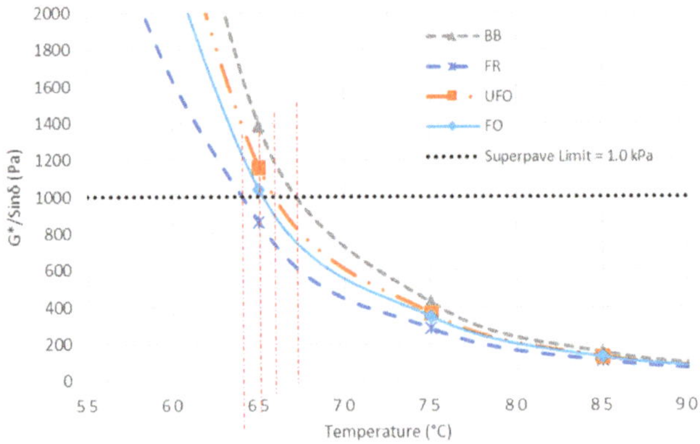


**Figure 8.** Decrement in high temperature performance grade of unaged asphalt binders.

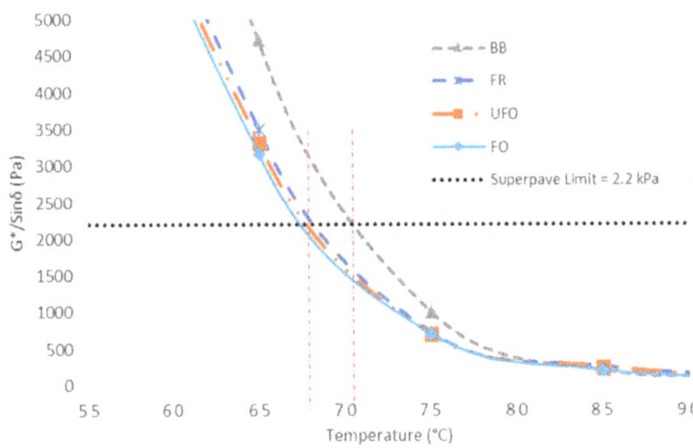


**Figure 9.** Decrement in high temperature performance grade after RTFO aging.

After RTFO aging, stiffness of base binder was increased and there was a 3 °C rise in upper grade of BB. After rejuvenating with oil, the stiffening effect of RTFO aging was relaxed by about 2 °C. There was no considerable difference among the performance FR, FO, UFO binders.

### 3.3.2. Rutting and Delayed Relaxation

Figure 10 compares the first 5 cycles of Creep and Recovery test at 34 °C and two different stress levels. At low stress level (Figure 10a), BB showed less ability to deform due to higher stiffness while all oil rejuvenated binders (FO, FR, UFO) had higher strain accumulations. When the stress was increased up to 3.2 kPa, strain accumulation in FR was highest which depicts a less stiff material compared to FO and UFO. Increasing the temperature to 64 °C (Figure 11) meant binder stiffness decreased and accumulated strains tended to dissipate into the pavement structure causing rutting problems. At higher temperatures and higher stress levels (Figure 11b), no significant difference was observed among FR, FO and UFO binders.

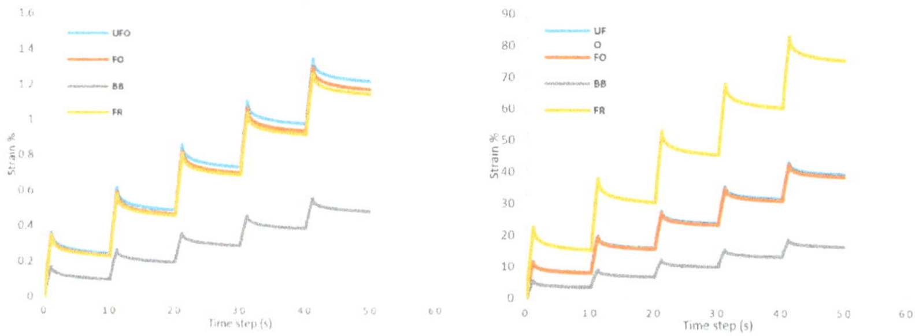


(**a**) (**b**)

**Figure 10.** Five cycles of Creep and Recovery at 34°C (**a**) with 0.1 kPa stress (**b**) with 3.2 kPa stress.

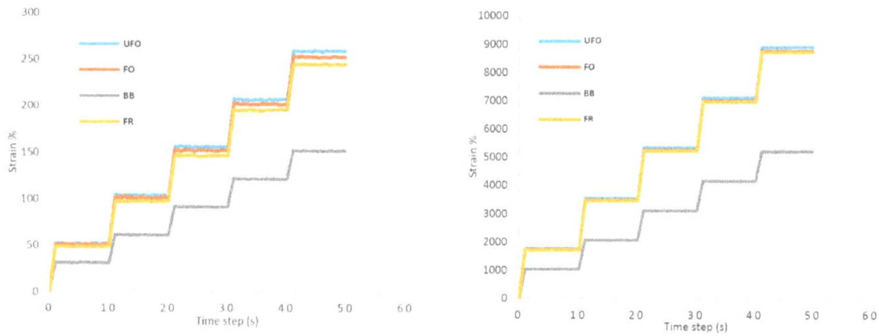


(**a**) (**b**)

**Figure 11.** Five cycles of creep and recovery at 64 °C (**a**) with 0.1 kPa stress and (**b**) with 3.2 kPa stress.

Figure 12 shows the %Recovery, non-recoverable creep compliance and stress sensitivity of all binders. Percent recovery (%R) describes the delayed relaxation ability of asphalt binder to applied stresses while $J_{nr}$ is a direct measure of rut resistance. Recovery increased as we decreased the temperature and stress level while $J_{nr}$ showed an opposite trend on the same criteria. The results from MSCR showed that rut resistance decreased with oil addition. FR had the higher rut resistance compared to FO and UFO while FO performed slightly better than UFO.

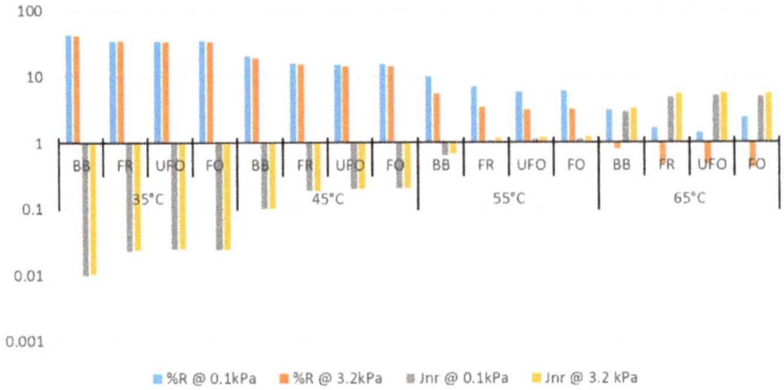


**Figure 12.** %Recovery and $J_{nr}$ at different temperatures.

3.3.3. Frequency Sweep Isotherms

Frequency was swept from 0.01 to 30 Hz for each test temperature and complex shear modulus (G*) and phase angle (δ) were measured. Figure 13 demonstrates the effect of frequency changes on stiffness of base and rejuvenated asphalt binder in the unaged condition. For brevity, G* and Phase angle data for only three testing temperatures (−5, 25, 55 and 95 °C) were shown. It is clear from the figures that with the increase of frequency, G* tended to increase while increasing temperature had an opposite effect on G*. An abrupt behavior in G* at −5 °C was depicted by UFO binder, which happened due to partial slippage or breakage of asphalt material sandwiched between the parallel plate geometry at low temperatures. However, FO samples did not show any breakage of material under the same testing conditions which can be regarded to the presence of metal traces hindering the molecular bonding and causing cracking of binder.

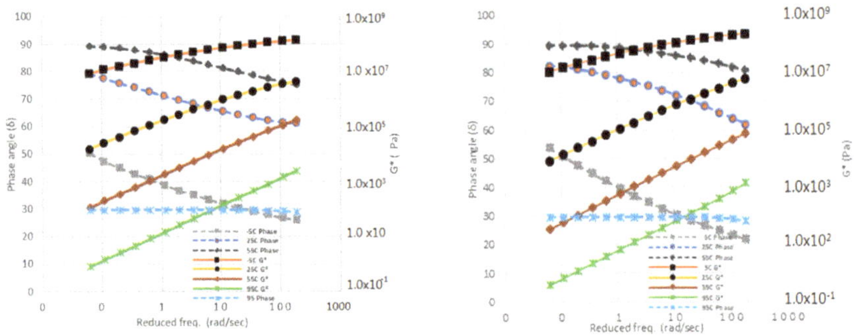


(**a**) (**b**)

**Figure 13.** Isotherms of complex shear modulus and phase angle for (**a**) BB and (**b**) F.O.

Phase angle, ratio of permanent deformation to elastic deformation, increased with increasing temperatures until 70 °C, after which it tended to decrease. At lower temperatures, asphalt binder behaved more elastically which resulted in low phase angle values and higher storage modulus. On the other hand, at higher temperatures, viscous behavior is more prominent and asphalt binder started to behave like a Newtonian fluid and hence resulted in higher phase angle values where almost all the response to applied loading was dissipated within the binder.

### 3.3.4. Master Curve

Isotherms (−15 to 95 °C) obtained after the frequency sweep test were then shifted to a reference temperature (25 °C) using the superposition principle. Figure 14 shows shifting of isotherms for Base binder (BB). Shifting of all other binder configurations was carried out using the same procedure. Frequency was reduced using shift factors which were obtained using Equation (2). Activation Energy ($E_a$) in Equation (2) was calculated while minimizing the sum of square of errors in theoretical and practical measured values of G* (Table 8). After obtaining a reduced frequency axis, G* values measured in the laboratory were plotted against this reduced frequency axis by horizontal translation of the curves obtained at different temperatures to a reference temperature.

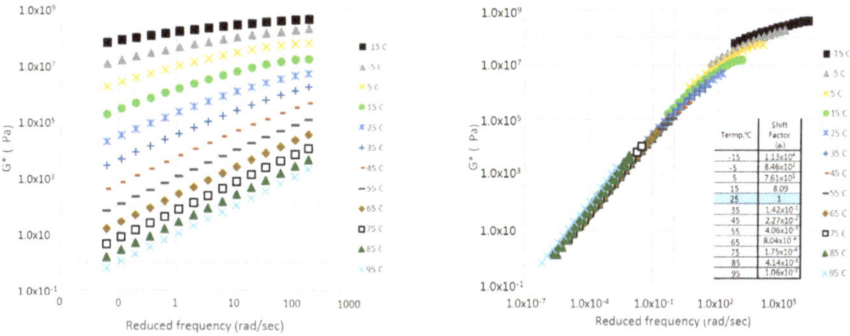


(**a**) (**b**)

**Figure 14.** Superpostioning of unaged base binder (BB) (**a**) isotherms and (**b**) horizontal shifting using shift factors.

**Table 8.** Activation energy $E_a$ of binders.

| Binder Code | Ea (kJ/mol) [1] | |
|---|---|---|
| | Unaged | RTFO Aged |
| BB | 143.293 | 149.361 |
| FR | 137.983 | 146.175 |
| UFO | 141.774 | 137.443 |
| FO | 138.415 | 139.156 |

[1] Activation energies were obtained while minimizing the sum of square of errors in the Sigmoidal Model.

Afterwards, theoretically measured G* values were fitted to the actual measured values of G* using the solver function in excel by minimizing the sum of squares of errors and Mastercurves were obtained for each binder (BB, FR, UFO, FO) as shown in Figures 15 and 16. It is illustrated that waste oil rejuvenated asphalts (FR, UFO, FO) in the unaged condition had lower modulus values as compared to BB on low temperature side and hence were more capable of resisting fatigue or thermal cracking. Reduction of stiffness in case of FO and FR binders were higher compared to UFO. However, on the high temperature end, higher modulus values were desired to resist rutting. Waste-oil rejuvenated binders predicted slightly less rut resistance as compared to base binder.

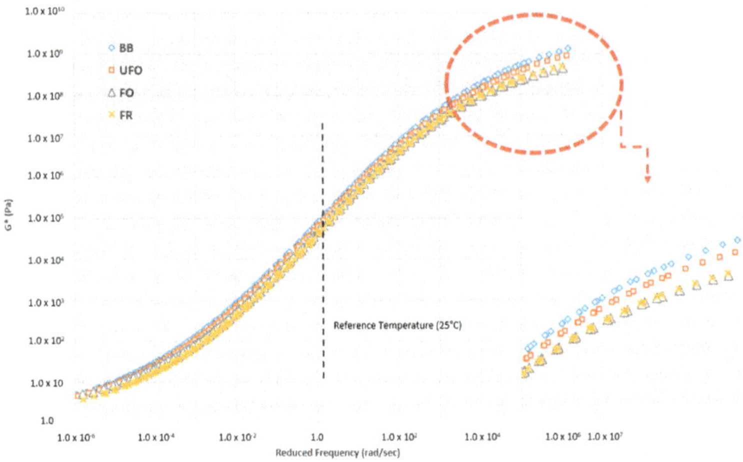


**Figure 15.** Comparison of master curves for unaged samples.

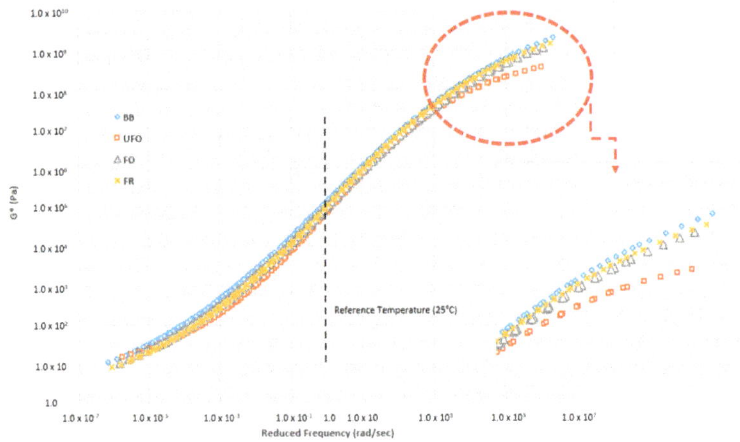


**Figure 16.** Comparison of master curves for RTFO-aged samples.

After RTFO aging, FO showed a slightly lower modulus compared to FR binder. UFO showed a much lower erroneous modulus after RTFO aging which was caused by the material breakage under the same testing criteria and should not be assessed as better fatigue resistance. The material breakage of UFO can be supported by its less activation energy compared to the unaged condition (Table 8). There was no significant difference on the high temperature end, among all binders, even after short term aging.

In summary, this study evaluates the properties imparted to asphalt binder by mixing 2.5% of fresh and waste engine oil and the effect of filtering the waste oil before rejuvenation. Virgin asphalt binder employed in this research is locally used in South Korea for pavement construction. Laboratory experimentation incorporates RV, FTIR and DHR tests for rheological and chemical analysis of binders. RV test results showed that, initially, the decrement in asphalt viscosity after oil addition was more significant in fresh oil. However, after RTFO aging no significant difference was observed between fresh and waste engine oil. An increased amount of carbonyl functional group in rejuvenated asphalt molecules can be visualized through FTIR spectra. FO and UFO binder proved to be less age susceptible considering the deficiency of unsaturated hydrocarbons. DHR test data demonstrated improvement in low temperature properties at the expense of upper PG grade which was compromised by 1~3 °C. DHR data were further analyzed with a Modified Sigmoidal Model to construct Mastercurves for a comprehensive assessment of fatigue cracking and rutting resistance on a wide range of frequency and temperatures.

## 4. Conclusions

Based on the results presented here, the following conclusions can be drawn;

- Infrared spectra demonstrated that engine oil has a similar molecular configuration to asphalt binder. Fresh engine oil is prone to severe plant aging compared to waste engine, indicated by an abrupt increase in carbonyl functional groups.
- Rotational viscosity test showed that inclusion of 2.5% filtered waste engine oil generally resulted in decreased viscosity of asphalt binder and hence imparted better flow properties. No significant difference in rejuvenating the asphalt binder were observed after RTFO aging for fresh oil, filtered waste oil and unfiltered waste oil. Filtered engine oil performed similar to fresh oil in viscosity decrement.
- By adding 2.5% waste engine oil, stiffness of binder was decreased at both temperature ends undermining the upper PG grade by 2~3 °C. Elastic recovery of asphalt binder was compromised and rejuvenated binders were prone to rutting.
- Mastercurves showed that engine oil rejuvenated binders have low modulus values at low temperatures and high frequencies compared to virgin binder, indicating a possibility for better fatigue life. More tests should be performed to verify the improvements in fatigue resistance.

## 5. Recommendations

- Waste engine oil has the properties to rejuvenate the aged binder to its unaged condition. However, more research work is need to compensate for the improvement in high temperature distresses.
- A concentration study to optimize the content of waste engine oil along with other polymer additives is recommended to commercialize the use of waste engine oil as a potential rejuvenator. A study on environmental concerns should also be carried out to quantitatively assess the use of this waste material in asphalt pavements.

**Author Contributions:** This research study was supervised by P.J.Y. All laboratory experimentations, results and analysis and write up of the paper was done by T.S.

**Funding:** This research received no external funding.

**Acknowledgments:** The research study was financially supported by the pothole-free project funded by Korea Institute of Civil Engineering and Building Technology (KICT), South Korea.

**Conflicts of Interest:** The authors declare no conflict of interest.

## References

1. Das, P.K.; Tasdemir, Y.; Birgisson, B. Low temperature cracking performance of WMA with the use of the Superpave indirect tensile test. *Constr. Build. Mater.* **2012**, *30*, 643–649. [CrossRef]
2. Jenny, Z.L.; Sheng, L.; Lin, L.; Peng, S.S. Low temperature cracking analysis of asphalt binders and mixtures. *Cold Reg. Sci. Technol.* **2017**, *141*, 78–85.
3. Kalantar, Z.N.; Karim, R.; Mahrez, A. A review of using waste and virgin polymer in pavement. *Constr. Build. Mater.* **2012**, *33*, 44–62. [CrossRef]
4. Naskar, M.; Chaki, T.K.; Reddy, K.S. Effect of waste plastic as modifier on thermal stability and degradation kinetics of bitumen/waste plastics blend. *Thermochim. Acta* **2010**, *509*, 128–134. [CrossRef]

5. Wen, G.; Zhang, Y.; Zhang, Y.; Sun, K.; Fan, Y. Rheological characterization of storage-stable SBS-modified asphalts. *Polym. Test.* **2002**, *21*, 295–302. [CrossRef]
6. Yan, Y.; Chun, S.; Roque, R.; Kim, S. Effects of alternative polymer modifications on cracking performance of asphalt binders and resultant mixtures. *Constr. Build. Mater.* **2016**, *121*, 569–575. [CrossRef]
7. Yan, Y.; Roque, R.; Cocconcelli, C.; Bekoe, M.; Lopp, G. Evaluation of cracking performance for polymer-modified asphalt mixtures with high RAP content. *Road Mater. Pavement Des.* **2017**, *18*, 450–470. [CrossRef]
8. Didier, L. The colloidal structure of bitumen: Consequences on the rheology and on the mechanisms of bitumen modification. *Adv. Colloid Interface Sci.* **2009**, *145*, 42–82.
9. Shakoane, L.N. Characterization and Quantification of Polycyclic Aromatic Hydrocarbons (PAHs) in Used and Fresh Engine Oil by Chromatography and Spectroscopy Techniques. Master's Thesis, University of the Witwatersrand, Johannesburg, South Africa, 2015.
10. Baker, J.M. The effects of oils on plants. *Environ. Pollut.* **1970**, *1*, 27–44. [CrossRef]
11. Pitchel, J. *Waste Management Practices: Municipal Hazardous and Industrial*; CRC Press: Boca Raton, FL, USA, 2014.
12. Ministry of Environment. Available online: http://eng.me.go.kr/eng/web/main.do (accessed on 2 May 2018).
13. DeDene, C.D. Investigation of Using Waste Engine Oil Blended with Reclaimed Asphalt Materials to Improve Pavement Recyclability. Master's Thesis, Michigan Technological University, Houghton, MI, USA, 2011.
14. Litvishkova, V.A.; Bukhter, A.I.; Nepogod'ev, A.V.; Bezhannidze, A.M. Chemical composition of used motor oils. *Chem. Technol. Fuels Oils* **1974**, *10*, 962–965. [CrossRef]
15. Zhang, L.; Bahia, H.; Yi-qiu, T. Effect of bio-based and refined waste oil modifiers on low temperature performance of asphalt binders. *Constr. Build. Mater.* **2015**, *86*, 95–100.
16. Ayman, W.A.; Yusuf, A.M.; Aaron, N.; Caitlin, P.; Thomas, B. Investigation of the impacts of aging and RAP percentages on effectiveness of asphalt binder rejuvenators. *Constr. Build. Mater.* **2016**, *110*, 211–217. [CrossRef]
17. Zaumanis, M.; Mallick, R.B.; Poulikakos, L.; Frank, R. Influence of six rejuvenators on the performance properties of Reclaimed Asphalt Pavement (RAP) binder and 100% recycled asphalt mixtures. *Constr. Build. Mater.* **2014**, *71*, 538–550. [CrossRef]
18. Villanueva, A.; Ho, S.; Zanzotto, L. Asphalt modification with used lubricating oil. *Can. J. Civ. Eng.* **2008**, *35*, 148–157. [CrossRef]
19. Halliza, H.; Ahmadinia, E.; Zargar, M.; Karim, M.R. Investigation on physical properties of waste cooking oil – Rejuvenated bitumen binder. *Constr. Build. Mater.* **2012**, *37*, 398–405.
20. Xiaoyang, J.; Baoshan, H.; Bowers, B.F.; Sheng, Z. Infrared spectra and rheological properties of asphalt cement containing waste engine oil residues. *Constr. Build. Mater.* **2013**, *50*, 683–691.
21. Eriskin, E.; Karahancer, S.; Terzi, S.; Saltan, M. Waste frying oil modified bitumen usage for sustainable hot mix asphalt pavement. *Arch. Civ. Mech. Eng.* **2017**, *17*, 863–870. [CrossRef]
22. Xavier, A.C.; Christiane, L.; Elísio, A.; Costa, D.B.; Araujo, K. Rheological properties of asphalt binders prepared with maize oil. *Constr. Build. Mater.* **2017**, *152*, 1015–1026. [CrossRef]
23. Simon, A.M.; Shurvell, H.F. X-ray fluorescence detection of waste engine oil residue in asphalt and its effect on cracking in service. *Int. J. Pavement Eng.* **2010**, *11*, 541–553.
24. DeDene, C.D.; You, Z. The performance of aged asphalt materials rejuvenated with waste engine oil. *Int. J. Pavement Res. Technol.* **2014**, *7*, 145–152.
25. Kuczenski, B.; Geyer, R.; Zink, T.; Henderson, A. Material flow analysis of lubricating oil use in California. *Resour. Conserv. Recycl.* **2014**, *93*, 59–66. [CrossRef]
26. BP Performance Grade–PG Asphalt Binders Unmodified. Available online: https://www.bp.com/content/dam/bp-country/en_us/PDF/Asphalt/pg-asphalt-binder-bp-unmodified.pdf (accessed on 3 May 2018).
27. Sander, J. Putting the Simple Back into Viscosity. Lubrication Engineers, Inc., 2011. Available online: https://www.lelubricants.com/lit/news/White%20Papers/simple_viscosity.pdf (accessed on 3 May 2018).
28. Sabiha, T.; Umesh, C.S.; Ram, P. Rheology of multigrade engine oils. *Indian J. Chem. Technol.* **2006**, *13*, 180–184.
29. Vicente, M.; Bernardo, T.; Santiago, R.; Guillermo, M. Low viscosity engine oils: Study of wear effects and oil key parameters in a heavy duty engine fleet test. *Tribol. Int.* **2016**, *94*, 240–248.

30. Osman, D.I.; Attia, S.K.; Taman, A.R. Recycling of used engine oil by different solvent. *Egypt. J. Pet.* **2017**. [CrossRef]
31. Dominguez, R.E.; Pichtel, J. Chemical characterization of fresh, used and weathered motor oil via GC/MS, NMR and FIRT techniques. *Proc. Indiana Acad. Sci.* **2003**, *112*, 109–116.
32. Grzegorz, Z.; Joanna, S.B.; Tomasz, S.; Andrzej, K.; Agnieszka, K. Designation of chosen heavy metals in used engine oils using the XRF method. *Pol. J. Environ. Stud.* **2015**, *24*, 2277–2283.
33. Brain, C.S. *Fundamentals of Fourier Transform Infrared Spectroscopy*, 2nd ed.; CRC Press: Boca Raton, FL, USA, 2011.
34. Silverstein, R.M.; Bassler, G.C.; Morrill, T.C. *Spectrometric Identification of Organic Compounds*; Wiley: Hoboken, NJ, USA, 1981.
35. CHrysanthi, C.; Agapios, A.; Rebecca, K. Use of FTIR spectroscopy and chemometrics for the classification of carobs origin. *J. Adv. Res.* **2018**, *10*, 1–8.
36. Abbas, B.; Walla, S.M.; Ramon, B. How to construct an asphalt binder master curve and assess the degree of blending between RAP and virgin binders. *J. Mater. Civ. Eng.* **2013**, *25*, 1813–1821.
37. John, D.; Don, P. Time-temperature superpostion—A users guide. *Rheol. Bull.* **2009**, *78*, 16–31.
38. Anderson, D.; Christensen, D.; Bahia, H. Physical properties of asphalt cement and the development of performance related specifications. *J. Assoc. Asph. Pav. Technol.* **1991**, *60*, 437–475.
39. Van det Put, T.A.C.M. Theoretical derivation of the WLF- and annealing equations. *J. Non-Cryst. Solids* **2010**, *356*, 394–399. [CrossRef]
40. Binder Characterization and Evaluation. Available online: http://onlinepubs.trb.org/onlinepubs/shrp/SHRP-A-369.pdf (accessed on 3 May 2018).
41. Abdulhaq, H.A. Construction of a Complex Shear Modulus Master Curve for Iraqi Asphalt Binder using a Modified Sigmoidal Fitting. *Int. J. Sci. Eng. Technol. Res.* **2015**, *4*, 682–690.
42. Asphalt Institute. *Performance Graded Asphalt Binder Specification and Testing, SP-1*; The Asphalt Institute: Lexington, KY, USA, 2003.
43. Al-Khateeb, G.G.; Al-Akhras, N.M. Properties of Portland cement-modified asphalt binder using superpave tests. *Constr. Build. Mater.* **2010**, *25*, 926–932. [CrossRef]
44. Asphalt Institute. *Superpave Level 1 Mix Design. Asphalt Institute Superpave Series No. 2 (SP-2)*; The Asphalt Institute: Lexington, KY, USA, 1995.

applied sciences

MDPI

Article

# Performance Evaluation of Carbon Black Nano-Particle Reinforced Asphalt Mixture

**Javaria Rafi *, Mumtaz Ahmed Kamal, Naveed Ahmad, Murryam Hafeez, Muhammad Faizan ul Haq, Syeda Aamara Asif, Faisal Shabbir and Syed Bilal Ahmed Zaidi**

Department of Civil Engineering, University of Engineering and Technology Taxila, Taxila 47080, Pakistan; dr.kamal@uettaxila.edu.pk (M.A.K.); n.ahmad@uettaxila.edu.pk (N.A.); murryamhafeez77@outlook.com (M.H.); faizan.ul@students.uettaxila.edu.pk (M.F.u.H.); amara.asifshah@yahoo.com (S.A.A.); faisal.shabbir@uettaxila.edu.pk (F.S.); bilal.zaidi@uettaxila.edu.pk (S.B.A.Z.)

* Correspondence: javariarafi@outlook.com; Tel.: +92-3476878275

Received: 24 May 2018; Accepted: 6 July 2018; Published: 10 July 2018

**Abstract:** Applications of nanotechnology in the pavement industry have increased rapidly during the last decade in order to enhance a pavement's sustainability and durability. Conventional asphalt binder generally does not provide sufficient resistance against rutting at high temperatures. Carbon black nano-particles (CBNPs, produced by perennial mountain trees' carbonization) were mixed into the performance grade (PG) 58 asphalt binder in this study. Conventional asphalt binder tests (penetration, ductility and softening point), frequency sweep, performance grading, and bitumen bond strength tests were conducted to study the enhancement in the properties of asphalt binder. Dynamic modulus and wheel tracking tests were also performed to investigate the effect of CBNPs on asphalt mixture properties. Experimental results demonstrated that preferred dosage of CBNPs in asphalt is 10% by weight of the bitumen. Results of scanning electron microscopy (SEM) and storage stability tests validated homogenous and stable dispersion of CBNPs in the asphalt binder. Asphalt mixtures became stiffer and resistant to rutting at high temperatures by addition of CBNPs in asphalt binder. Significant improvement in bitumen aggregate bond strength was also observed by incorporating CBNPs. It is concluded that CBNPs can be used to effectively enhance the high-temperature performance and consequently the sustainability of flexible pavements.

**Keywords:** carbon black nano-particles (CBNPs); asphalt binder; performance grading; scanning electron microscopy (SEM); dynamic modulus; rutting

## 1. Introduction

Poor flexible pavement performance at high temperatures is the major problem that is faced by the pavement industry in Pakistan. Asphalt pavements in the country fail prematurely because of mix rutting. Mix rutting is observed only in the high-temperature areas of the country. The lower softening point values of the locally produced unmodified binders make them susceptible to rutting during the summers. The other most commonly faced distress in the country is the moisture damage of the asphaltic pavements especially during the rainfall/monsoon season.

Pavement engineers are always in search of modifiers that could not only improve the rheological properties of the binders but could also concurrently enhance their adhesion capabilities. Asphalt mixtures have been modified in the past by carbonaceous materials, as they are believed to be intrinsically compatible with the asphalt mixture [1]. Among the carbonaceous materials, different researchers have utilized carbon black because of its easy availability. It has been learnt that high dosages of carbon black in asphalt can increase the rutting resistance and lessen the temperature susceptibility [2–11]. It is also said that the performance of carbon black depends upon its structure, particle size and surface area [12,13]. Nanotechnology is an emerging field and is extensively used by pavement engineers worldwide. Different types of carbon black nano-particles (CBNPs) have been manufactured and incorporated to modify asphalt properties [14–16] but, to the authors' best knowledge, the one made with natural sources has never been used by pavement engineers. The selected CBNPs come from a natural source and their effect on the rheology of the modified binder, bitumen-aggregate adhesion, and moisture susceptibility needs to be investigated.

The pavement industry in Pakistan is still using penetration grading for bitumen selection while the rest of the world is fast adopting the latest performance grading (PG). Performance grading is based on the concept that the properties of the asphalt binder should be related to the conditions under which it is used. According to temperature zoning done by Mirza et al. [17], PG 70-10 would be sufficient for most parts of the country. Base asphalt binder generally used in the country has PG 58-22 which is softer than the required PG 70-10. As mentioned earlier, the problematic areas are only the high temperature areas of the country where, even in winters, temperatures hardly fall below 0 °C. Therefore, high PG values are of more concern and the main aim is to use a dosage of nano-material that would help achieve the 70 °C threshold.

The paper aims to investigate the effect of CBNPs (150 nm) produced by perennial mountain trees' carbonization on the rheological properties of asphalt binder, aggregate and asphalt binder bond strength, and rutting resistance of asphalt.

## 2. Experimental Work

### 2.1. Materials

**Aggregates**

Aggregates (limestone) used in this study were procured from Margalla in Punjab, Pakistan. Different material properties of Margalla aggregates are displayed in the Table 1.

**Table 1.** Margalla aggregates material properties [18].

| Material Properties | Percent (%) | Recommended Range by NHA * |
|---|---|---|
| Fractured particles | 100 | 90% (min) |
| Flakiness | 5 | 10% (max) |
| Elongation | 11 | 10% (max) |
| Sand equivalent | 75 | 50% (min) |
| Los Angeles abrasion | 15 | 15% (max) |
| Water absorption | 1.02 | 2% (max) |
| Soundness (Coarse) | 7.1 | 8% (max) |
| Soundness (Fine) | 4.7 | 8% (max) |
| Uncompacted voids | 37.5 | 45% (min) |

* National Highway Authority of Pakistan.

**Bitumen**

In this study, 60/70 penetration bitumen was used. It was obtained from Attock Oil Refinery Limited (ARL) which is the most commonly used source for bitumen procurement in Pakistan.

**Carbon Black Nano-Particles (CBNPs)**

Carbon black nano-particles (stock # US1067) procured from US Research Nanomaterials, Inc. (Houston, TX 77084, USA) were used in this study. These are produced by carbonization and extreme fine grinding of perennial trees at high temperature of up to 1300 °C. Features of procured CBNPs are shown in Table 2.

**Table 2.** Carbon black nano-particles (CBNPs) feature.

| Morphology | Spherical |
|---|---|
| **Colour** | Black |
| **Purity** | Greater than 95% |
| **Average particle size (APS)** | 150 nm |
| **$H_2O$** | Less than 5% |
| **Ash** | Less than 3.2% |
| **PH** | 9.8 |
| **Density** | 0.38 g/mL |
| **Electrical Resistivity** | 0.30 Ω·cm |

## 2.2. Preparation of CBNPs Modified Bitumen

Two percentages of CBNPs were used in this study i.e., 5% and 10% by weight of asphalt binder. The maximum CBNPs content of 10% was selected, as performance grade (PG) 70 was achieved at this dosage. PG 70 is the desired performance grade of binder in Pakistan as per the temperature zoning [17]. PG 70-10 asphalt binder covers almost 70% areas of Pakistan. Asphalt binder (60/70) procured from Attock Oil Refinery has PG 58-22, therefore CBNPs were mixed with the asphalt binder to achieve the desired performance grade. All modified binders were prepared using a high shear mixer. Asphalt binder was heated to 158 ± 5 °C until it became fluid. Then CBNPs were added steadily into the asphalt binder and the mixture was stirred at 2800 rpm for 45 min to ensure homogenous dispersion of nanoparticles. Scanning electron microscopy (SEM) was used to check the homogenous dispersion of CBNPs in the asphalt binder. Figure 1a shows the spherical shape of CBNPs used in this study whereas homogenous dispersion of CBNPs in the asphalt binder is displayed in Figure 1b. Figure 1b depicts that the mixing operation of applying 2800 rpm for 45 min to disperse CBNPs in the asphalt binder is acceptable.

(a) (b)

**Figure 1.** (**a**) Scanning electron microscope (SEM) image of carbon black nano-particles (CBNPs); (**b**) SEM image of carbon black nano-particles in asphalt binder.

Homogenous dispersion was further ensured by storage stability analysis and repeatable results of PG during trial blending techniques. The storage stability analysis (BS EN 13399-2017) of CBNPs modified samples showed less than 2.5 °C difference in softening point among the top and bottom portions of the sample, which is proof of stable and homogenous CBNPs-modified asphalt binder.

A number of CBNPs-modified binder samples were prepared by using different mixing durations and blending techniques (hand mixing, and shear mixing utilizing different blender RPMs). All these samples were tested under dynamic shear rheometer (DSR) and their high PG temperatures were determined. Mixing duration of 45 min and 2800 blender RPMs were finally selected as these two conditions provided with a homogeneous sample with repeatable results. DSR test results for the samples that were taken from three different portions of the sample container (top, bottom, and middle) during PG test also provides proof of homogenous dispersion of CBNPs in the asphalt binder. Results are presented in Table 3.

**Table 3.** Performance grades showing homogeneous dispersion.

| Sample Container | 5% CBNPs | 10% CBNPs |
|---|---|---|
| **Top Portion** | PG 64 (64.5 °C) | PG 70 (71.0 °C) |
| **Middle Portion** | PG 64 (64.2 °C) | PG 70 (70.9 °C) |
| **Bottom Portion** | PG 64 (64.7 °C) | PG 70 (70.6 °C) |

The traceability of particles through SEM is largely related to the size of the particles. As the utilized CBNPs had an average diameter of 150 nm, they were relatively easily traceable. However, some trials were still required to be carried out, and it was necessary to zoom-in to the required resolution to study the dispersion of CBNPs in bitumen. SEM is generally sufficient to study the dispersion of particles [19], however, for more in depth analysis into the size and internal structure of the particles, transmission electron microscopy (TEM) may be preferred.

### *2.3. Preparation of Asphalt Mixtures*

Midpoint aggregate gradation of National Highway Authority (NHA) class B (19.5 mm nominal maximum aggregate size) was used in this study. It is specified as a finer gradation and is generally used by the local pavement industry as an asphalt wearing course. The midpoint gradation curve for NHA class B is shown in Figure 2. Marshall mix design was performed for two CBNPs percentages (5% and 10% by weight of asphalt binder) to get the optimum binder content for asphalt mix design. In this test, a standard hammer of 4.5 lbs was used and it was dropped from 18 in height; 75 blows of the hammer were applied on each side of the specimen (specified for heavy traffic in the specifications of the Asphalt Institute). Midpoint of the in-place air voids range (3 to 8 percent) as specified in MS-2 series, i.e., 5.5% of air voids were maintained in the design. Optimum bitumen contents for the two CBNPs percentages used in asphalt performance testing are shown in the Table 4.

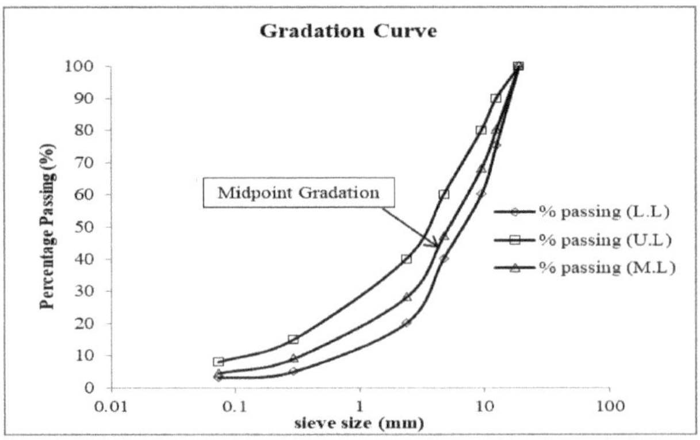


**Figure 2.** Aggregate gradation curve.

**Table 4.** Optimum binder content (OBC).

| Asphalt Sample with | OBC (%) |
|---|---|
| **Base binder** | 4.33 |
| **5% CBNPs modified binder** | 4.48 |
| **10% CBNPs modified binder** | 4.63 |

Generally an increase in the filler content in an asphalt mixture results in the reduction in the optimum binder content requirement for that mix because filler material acts as an 'extender' and requires less bitumen in order to fulfil the asphalt volumetric requirements [20]. However, researchers have experienced both an increase and decrease in the optimum binder content requirements while utilizing nano materials. Previous studies have shown that different fillers differently affect the properties of asphalt mixtures. Such changes in test properties can be associated with changes in viscosity of the filler-binder containing different fillers. However, it is difficult to get a direct correlation between filler-binder viscosity and compacted mixture properties. The reason for this could be attributed to the variable effects of different fillers on the compaction and volumetric properties of different asphalt mixtures [21].

An increase in optimum binder content was experienced with the increase in CBNPs dosage. Chelovian and Shafabakhsh [22] observed the same trend. An increase in optimum binder content requirement could be attributed to large surface area of CBNPs used in this study. This could also be attributed to the increase in viscosity of the binder after addition of the nano material. CBNPs, because of their small size, enhance the viscosity of bitumen making it stiffer. A higher viscosity leads to a thicker binder film of the modified bitumen in the mix thus increasing the binder volume in the mix [23].

## 2.4. Tests Performed

### 2.4.1. Conventional Binder Tests

Conventional binder tests; penetration, ductility and softening point were performed in conformity with ASTM D5-13, ASTM D113-99 and ASTM D36-76 respectively.

### 2.4.2. Storage Stability

The storage stability test was performed as per BS EN 13399 (2017). This technique was used to study CBNPs-modified asphalt binder's storage stability at high temperatures. CBNPs-modified asphalt binder was poured into a glass tube of 25 mm diameter and 140 mm height. The tube was placed vertically in an oven for 48 h at 163 °C and then it was cooled in refrigerator at −7 °C for 4 h. The tube was finally cut in three equal sections and, after melting the samples, a softening point test was performed for top and bottom portions. A difference of softening point less than 2.5 °C is a widely used criterion for considering the blend to be stable at high temperatures [5].

### 2.4.3. Performance Grading and Frequency Sweep Tests

Performance grading and frequency sweep tests were performed in conformity with AASHTO T 315 using Anton Paar Dynamic Shear Rheometer (DSR). Pavement performance at high and intermediate temperatures can be assessed by using DSR [24]. Performance grading was carried out at 10 Hz frequency using 25 mm geometry and failure temperatures were recorded when $G^*/\sin\delta$ fell below 1 KPa.

The frequency sweep test was performed under strain controlled conditions at six different temperatures (20 °C, 30 °C, 40 °C, 50 °C, 60 °C and 70 °C); 8 mm and 25 mm plate geometries were used and with the gap of 2 mm and 1 mm, respectively, between the parallel plates. Frequency was varied from 0.1 Hz to 10 Hz during the test; 10% strain limit was applied for base asphalt binder and 0.45% for CBNPs-modified asphalt binder. Rheological parameters; complex shear modulus ($G^*$), rutting factor ($G^*/\sin\delta$) and phase angle ($\delta$) were determined.

### 2.4.4. Bitumen Bond Strength Test

Bitumen aggregate bond strength was studied by a Pneumatic Adhesion Tensile Testing Instrument (PATTI) in accordance with ASTM D 4541. Margalla limestone aggregate plate (15″ × 6″ × 1.5″) was heated at 150 °C in the oven for one hour before the start of the test to remove all absorbed moisture. F-4 pullout stubs (0.5 inch diameter) were also heated for 30 min at 65 °C. Samples were tested after 24 h of dry conditioning and moisture conditioning. Pneumatic pressure was applied on the samples and burst pressure was recorded at which the stub was detached. The burst pressure was converted to pull-off tensile strength (*POTS*) by the equation given below. Five samples were tested for each category.

$$POTS = \frac{(BP \times Ag) - C}{Aps} \tag{1}$$

where;

*BP* = Burst pressure, *Ag* = Contact area of the gasket with the reaction plate, *C* = Piston constant and *Aps* = Pullout stub area.

For F-4 stub type; $Ag = 4.06\ \text{in}^2$, $C = 0.286\ \text{in}^2$ and $Aps = 0.193\ \text{in}^2$.

### 2.4.5. Cooper Wheel Tracking Test

**Preparation of Specimen**

Slab specimens with dimensions 300 mm × 300 mm × 50 mm were used for the study. The material's weight that is required for slab manufacturing is determined on the basis of the volume of the mould and the required specific gravity ($G_{mb}$). Temperature values maintained during the mixing and compaction operations were 158 ± 5 °C and 145 ± 5 °C, respectively. Slab specimens were compacted using a Cooper Roller Compactor in four phases with 2.5, 3.5, 4.0 and 4.5 bar pressure. Trial specimens were used to calibrate for the required number of compactor passes. All the specimens achieved the required specific gravity at almost 10 passes per phase. Compacting head speed was controlled automatically on the basis of the required compaction effort (varies up to 10 cycles per min). Air voids of 5.5 ± 0.5% were maintained during the compaction of the slabs.

**Sample Testing**

The test was performed in conformity with BS EN 12697-25. Slabs were placed in the Cooper Wheel Tracking equipment two hours before the start of test for temperature equilibrium. A load of 700 N was applied through the wheel at a frequency of 26.5 rpm. Tests were conducted at 40 °C and 55 °C (under 10,000 loading cycles) in order to cover the temperatures encountered by pavements in Pakistan. Three samples were tested for each test condition.

### 2.4.6. Dynamic Modulus Test

**Sample Preparation**

Aggregates weighing 6500 g were used to prepare the cylindrical specimen (170 mm height and 150 mm diameter) using a Superpave Gyratory Compactor. Mixing of heated aggregates and asphalt binder was carried out at a temperature of 158 ± 5 °C. Loose mixture was then placed in a flat pan for short-term conditioning as per AASHTO R-30. The compaction mould was preheated in the oven for 30 min and the mixture was brought to the compaction temperature of 145 ± 5 °C before placing in the Superpave Gyratory Compactor. A pressure of 600 kPa was applied to the specimen while the internal angle of gyration was 1.16 ± 0.02°. A target height of 170 mm was given as input to the machine and gyrations were applied by the machine until the desired height and 5.5 ± 0.5% air voids content was achieved. A specimen of 101.6 mm diameter was cored from the compacted sample. The specimen was further trimmed to get 150 mm height for the dynamic modulus testing.

#### Specimen Testing

A Cooper NU-14 was used to determine the dynamic modulus of asphalt in accordance with AASHTO TP 62-03. Cylindrical specimens having 101.6 mm diameter and 150 mm height were placed in an environmental chamber to achieve the required temperature and then placed into the equipment frame for dynamic modulus analysis. Sinusoidal loading of 195 kPa and 53 kPa was applied to samples at 40 °C and 55 °C respectively at six different loading frequencies i.e., 25, 10, 5, 1, 0.5 and 0.1 Hz. 195 kPa and 53 kPa is the midpoint of the dynamic stress levels provided in the standard for 40 °C and 55 °C respectively Three samples were tested for each test condition.

## 3. Results and Discussion

### 3.1. Conventional Asphalt Binder Properties

The results of penetration, softening point and ductility tests of base asphalt binder and CBNPs-modified asphalt binder are presented in Table 5. Addition of CBNPs in base asphalt binder has decreased the penetration and ductility values whereas elevation in softening point has been observed. The resultant effect is more pronounced with the 10% CBNPs content i.e., 24.5% reduction in penetration, 40% reduction in ductility and 12.5% increase in the softening point. Reduction in penetration along with ductility and increase in the softening point shows the increase in stiffness and decrease in high-temperature susceptibility by the introduction of CBNPs in asphalt binder. The large surface area of CBNPs and high degree of dispersion of CBNPs in the asphalt binder also influence the conventional properties. This indicates that the high-temperature performance of asphalt binder has been improved by incorporating CBNPs.

**Table 5.** Conventional testing results.

| Tests | Base Asphalt Binder | CBNPs Content (%) | |
|---|---|---|---|
| | | 5 | 10 |
| Penetration (1/10 of mm) | 61 | 53 | 46 |
| Softening Point (°C) | 48 | 51.5 | 54 |
| Ductility (cm) | 100 | 75 | 60 |

### 3.2. Effect of CBNPs on Storage Stability of Asphalt Binder

The difference in the softening point of the top and bottom sections of the test tube indicates the storage stability of CBNPs-modified bitumen. The results of the storage stability test are displayed in Table 6. Results indicate that difference of only 1 °C has been observed in case of 5% CBNPs content and 1.3 °C by incorporating 10% CBNPs content into asphalt binder. This indicates that 5% and 10% CBNPs content in asphalt binder meet the storage stability criteria i.e., the softening point difference should be less than 2.5 °C for good high-temperature storage stability. Thus, CBNPs-modified asphalt binder is a stable material and can be utilized in road construction.

**Table 6.** Difference in softening point of CBNPs-modified bitumen.

| Softening Point (°C) | Base Asphalt Binder | CBNPs Content (%) | |
|---|---|---|---|
| | | 5 | 10 |
| Top Portion | 49.5 | 52 | 54.4 |
| Bottom Portion | 49.9 | 53 | 55.7 |
| Difference | 0.4 | 1 | 1.3 |

### 3.3. Dynamic Shear Rheological Properties

The effect of CBNPs on rheological parameters was studied by a frequency sweep test. The relationship between $G^*$, $\delta$, $G^*/\sin\delta$ and reduced frequency are shown in Figure 3a–c at a reference temperature of 50 °C. Figure 3a shows that CBNPs-modified asphalt binder has higher $G^*$ values as compared to the base asphalt binder. Addition of 5% CBNPs has increased stiffness at high temperatures/low frequencies and this effect became pronounced with the increase in CBNPs content to 10%. This illustrates that asphalt binder turns stiffer and more resistant to permanent deformation at high temperatures by adding CBNPs. The large surface area of CBNPs and its homogenous dispersion in asphalt binder contribute towards the increase in $G^*$ value at high temperatures. Figure 3b shows a gradual increase in phase angle values with the increase in temperature or decrease in loading frequency. The addition of CBNPs in asphalt binder has reduced the phase angle values at all loading frequencies and this reduction is more prominent at 10% CBNPs content. For instance at 10 Hz frequency, the values of the phase angle shifted from 83.44 degrees to 62.45 degrees by the addition of 10% CBNPs content. Thus, the results indicate that CBNPs have enhanced the elastic behaviour of the asphalt binder and have influenced the asphalt binder's rheology. It has also been reported in literature that carbon black-modified asphalt binder exhibits more elastic behaviour than conventional asphalt binders [5].

The rutting factor ($G^*/\sin\delta$) can be related to resistance against permanent deformation and has been reported in the analysis of pavement performance at high temperatures [13]. Figure 3c shows that $G^*/\sin\delta$ behaviour is similar to that of complex shear modulus ($G^*$). Modified asphalt binders have shown higher rut resistance as compared to base asphalt binder. The high $G^*/\sin\delta$ value of modified binders is due to the increase in stiffness by addition of CBNPs. Thus, it can be stated that modifying asphalt binder with CBNPs effectively improves the resistance against rutting as well as the elastic behaviour of asphalt binder, making it suitable for high-temperature areas.

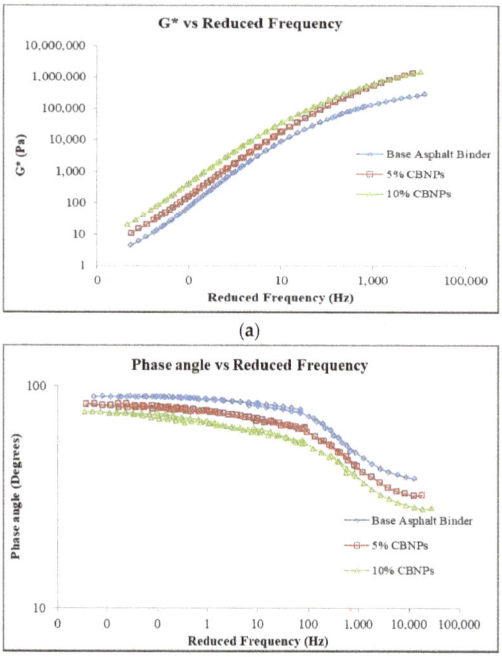


(a)

(b)

**Figure 3.** *Cont.*

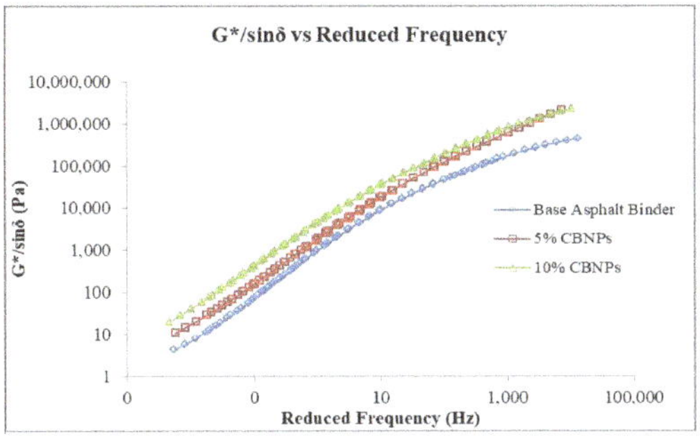


(c)

**Figure 3.** (**a**) Graphical representation of $G^*$ and reduced frequency; (**b**) graph of reduced frequency and phase angle; (**c**) graph of $G^*/\sin\delta$ versus reduced frequency.

### 3.4. Performance Grading (PG) of CBNPs-Modified Asphalt Binder

The effect of CBNPs on the high-temperature PG of asphalt binder is displayed in Table 7. According to the results, addition of 5% CBNPs has changed the PG 58 to PG 64 by improving failure temperature from 62.5 °C to 64.5 °C. Failure temperature has increased to 8.4 °C by increasing the amount of CBNPs to 10% in the asphalt binder. The improvement in PG illustrates the enhanced pavement resistance against permanent deformation (rutting) by addition of CBNPs. Based on this, 10% CBNPs modified asphalt binder is suitable for pavement construction to meet the high-temperature requirements of Pakistan.

**Table 7.** Effect of CBNPs on the performance grade (PG).

| Sample | Failure Temperatures (°C) |
|---|---|
| **Base Asphalt Binder** | 62.5 (PG 58) |
| **5% CBNPs** | 64.5(PG 64) |
| **10% CBNPs** | 70.9(PG 70) |

As mentioned earlier in the problem statement, flexible pavements at high temperature are more susceptible to mix rutting which becomes the cause of their premature failure. Due to extreme hot weather conditions, the premature failure of flexible pavements is the major problem faced by the pavement industry in Pakistan. According to temperature zoning done by Mirza et al. [17], PG 70-10 is considered suitable for most parts of the country. High-temperature areas in Pakistan are problematic, whereas in winters temperature hardly goes below 0 °C. It has been observed that base binder (PG-58-22) is more sensitive to temperature than PG 70-10. Therefore, it has been evaluated that high PG values obtained at 10% CBNPs dosage are more suitable.

### 3.5. Effect of CBNPs on Bitumen Aggregate Bond Strength

The effect of CBNPs on bitumen aggregate bond strength was studied using a pneumatic adhesion tensile testing instrument (PATTI) and the results are displayed in Table 8a,b. Figure 4 shows the comparison of POTS after 24 h of dry and moisture conditioning in the form of a bar chart. Samples modified with CBNPs have higher POTS values than the base asphalt binder after 24 h of dry and moisture conditioning. The addition of CBNPs has increased the POTS value from 7.5 MPa to 17 MPa after 24 h of dry conditioning. The mode of failure is cohesive (breakage of bond at bitumen–bitumen interface) post 24 h of dry conditioning and changed to adhesive failure (breakage of bond at bitumen–aggregate interface) after 24 h of moisture conditioning. The change of failure mode could be attributed to water penetration through the aggregate (porous material) which weakens the bond at the bitumen–aggregate interface and results in low pull-off tensile strength values [25]. After 24 h of moisture conditioning, modified binders performed better than the base binder and 10% CBNPs-modified binder has shown the highest POTS values. Addition of nano material has increased the wetting potential of the modified binder promoting better wetting of the bitumen onto the aggregate surface, thus enhancing the bitumen–aggregate bond strength. This could also be attributed to the increase in stiffness of the asphalt binder because of the addition of nano particles [26]. Moreover, significant improvement in bond strength of a nano-hybrid material could be attributed to hydrogen bonds and van der Waals interaction [27].

According to these results, it can be concluded that CBNPs have the potential to improve the adhesive/cohesive bond strength of asphalt.

**Table 8.** (**a**) Pull-off tensile strength (POTS) values (MPa) and failure modes after 24 h of dry conditioning. (**b**) POTS values (MPa) and failure modes after 24 h of moisture conditioning.

| Replicates | 24 h Dry Conditioning | | |
|---|---|---|---|
| | Base Asphalt Binder | 5% CBNPs | 10% CBNPs |
| 1 | 7.67 (C) | 11.80 (C) | 16.54 (C/A) |
| 2 | 8.00 (C) | 11.80 (C) | 16.50 (C) |
| 3 | 7.67 (C) | 12.00 (C) | 16.73 (C) |
| 4 | 7.11 (C) | 12.00 (C) | 17.34 (C) |
| 5 | 6.82 (C) | 11.80 (C) | 17.80 (C) |
| Average | 7.50 | 11.90 | 17.00 |

C = Cohesive failure and *C*/*A* = 50% Cohesive and 50% Adhesive failure.

(**a**)

| Replicates | 24 h Moisture Conditioning | | |
|---|---|---|---|
| | Base Asphalt Binder | 5% CBNPs | 10% CBNPs |
| 1 | 6.55 (*A*) | 10.53 (*C*/*A*) | 15.43 (*A*) |
| 2 | 6.80 (*A*) | 10.53 (*A*) | 15.37 (*A*) |
| 3 | 6.55 (*A*) | 10.50 (*A*) | 15.43 (*A*) |
| 4 | 6.42 (*A*) | 10.69 (*A*) | 16.20 (*A*) |
| 5 | 6.42 (*A*) | 10.69 (*A*) | 14.48 (*A*) |
| Average | 6.55 | 10.60 | 15.40 |

*C*/*A* = 50% Cohesive and 50% Adhesive failure and *A* = Adhesive failure.

(**b**)

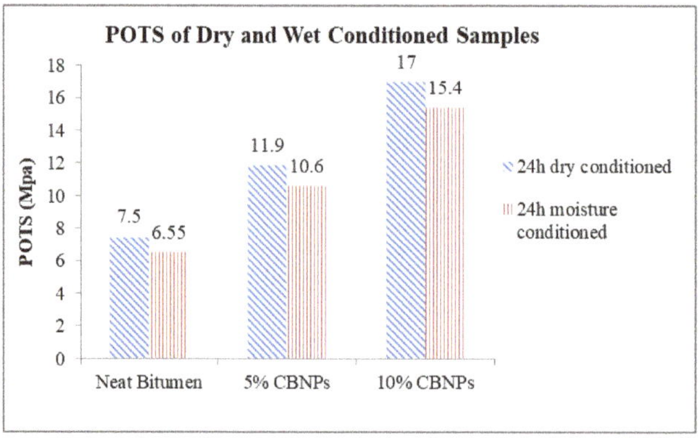


**Figure 4.** POTS of dry and moisture-conditioned samples.

### 3.6. Study of CBNPs Effect on Rut Depth

A wheel tracking test was performed to study the effect of CBNPs on the rut depth. Graphs between number of wheel passes and rut depth at 40 °C and 55 °C are shown in Figure 5a,b. Base asphalt showed the highest rut depth values against the standard 10,000 wheel passes. Addition of 5% CBNPs reduced the rut depth to 3 mm and 5.6 mm at 40 °C and 55 °C respectively (after 10,000 wheels passes). Furthermore, 10% CBNPs-modified asphalt has shown the maximum reduction in rut depth at both temperatures and after 10,000 passes i.e., 20% reduction at 40 °C and 37.8% reduction at 55 °C. Nano-particles increase the surface area of the asphalt binder and may enhance the interaction within the asphalt which results in more resistance against rutting [28]. Zhao et al. experienced an increase in rutting resistance when nano-carbon black modified asphalt was subjected to 8000 loading cycles during an asphalt pavement analyzer rutting test. This illustrates that CBNPs successfully enhances the high-temperature performance of asphalt by increasing its resistance against rutting.

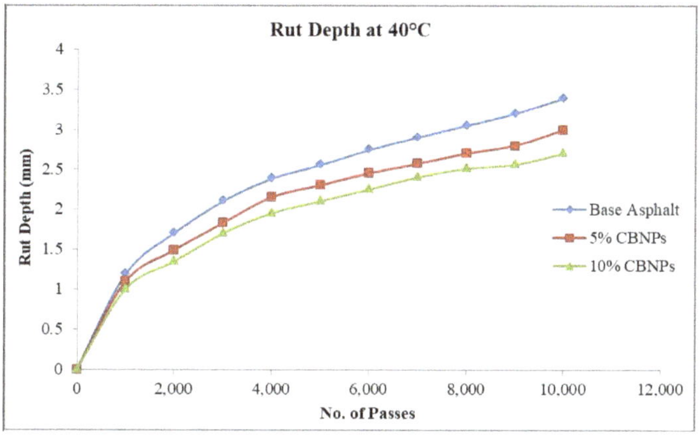


(a)

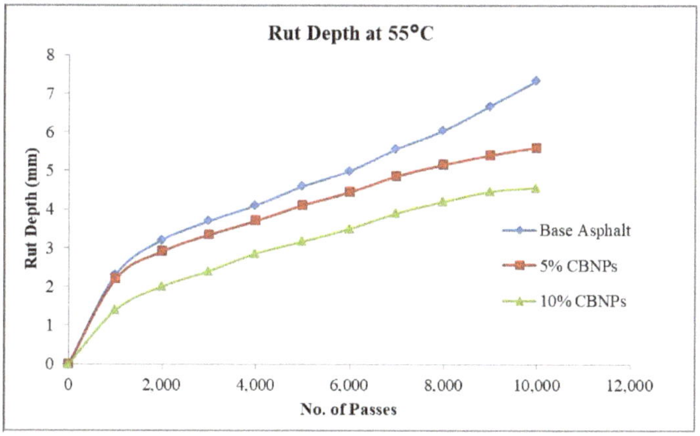


(b)

**Figure 5.** (**a**) Graph of rut depth and number of passes at 40 °C; (**b**) graph of rut depth and number of passes at 55 °C.

### 3.7. Effect of CBNPs on Dynamic Modulus of Asphalt

The dynamic modulus indicates the stiffness of asphalt and can be used as a measure to study the resistance of asphalt against permanent deformation. The higher the dynamic modulus value, the higher the resistance against permanent deformation. Figure 6a,b show the relationship of dynamic modulus with loading frequency at 40 °C and 55 °C. As shown, the dynamic modulus of asphalt decreases with the decrease in loading frequency and increase in temperature. This is due the fact that asphalt is temperature-dependent and becomes soft with an increase in temperature which leads to a decrease in dynamic modulus [29]. Modified samples have greater values of dynamic modulus than the base asphalt samples. Moreover, 10% CBNPs-modified asphalt has shown the highest dynamic modulus values at 40 °C and 55 °C and at all loading frequencies. Nano-modified asphalt binder creates more contact forces with aggregate particles. Scattered nano-particles fill the gap between the aggregates which induces an improvement in dynamic modulus of asphalt and rutting resistance as well [28]. Hence, CBNPs have proven to be a promising modifier for enhancing the high-temperature performance of asphalt.

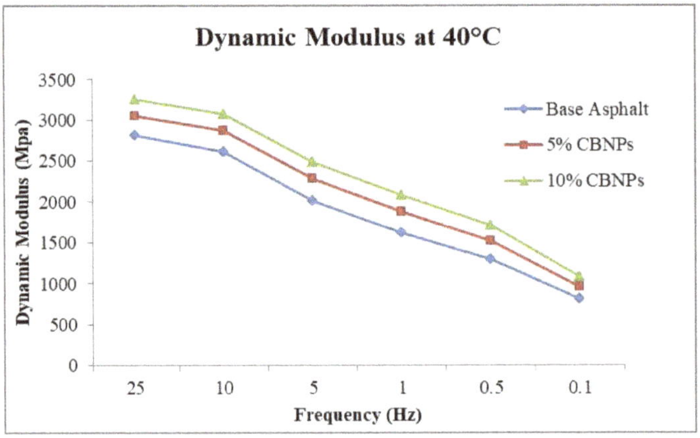


(a)

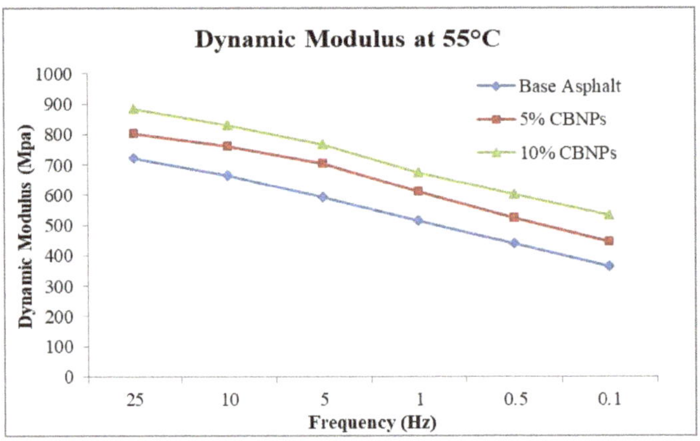


(b)

**Figure 6.** (**a**) Graph of dynamic modulus and frequency at 40 °C; (**b**) graph of dynamic modulus and frequency at 55 °C.

According to the Federal Highway Administration (FHWA), the structural and functional capacity of a pavement and its useful service life are key performance indicators and greatly impact the overall sustainability of a pavement [30]. This is because improved pavement performance leads to lower maintenance costs and longer service life, consequently conserving resources. It can be concluded from the results of the CBNPs-modified asphalt's performance testing that the addition of CBNPs in asphalt could help in building sustainable pavements.

## 4. Conclusions

The following conclusions have been drawn through careful analysis of the test results:

1. Addition of carbon black nano-particles has enhanced the conventional asphalt binder properties. For instance, 10% CBNPs content in conventional binder has reduced the penetration and ductility value by 24.5% and 40%, respectively. Furthermore, an elevation in softening point has been observed by the addition of CBNPs which leads to the reduction in high-temperature susceptibility of the asphalt binder. The large surface area of CBNPs and high degree of dispersion has contributed towards the change in the conventional asphalt binder properties, especially the stiffness of the binder.
2. A storage stability test has been used to check the permanence of the dispersion of CBNPs in the asphalt binder through identifying the difference in softening points of the samples obtained from the top and bottom sections of the test tube. The difference in softening point that has been obtained during the test is less than 2.5 °C i.e., 1 °C and 1.3 °C for 5% and 10% CBNPs-modified binder, respectively. This confirms that CBNPs-modified asphalt binder is a stable material and can be utilized by the pavement construction industry.
3. The complex shear modulus and performance grade of asphalt binder is significantly enhanced by incorporating CBNPs. Addition of 10% CBNPs has increased the complex shear modulus values from 8847 Pa to 34,679 Pa at 10 Hz frequency and 50 °C temperature. Also, 10% CBNPs addition has changed the PG 58 to PG 70 by improving the failure temperature from 62.5 °C to 70.9 °C. The large surface area of CBNPs and its homogenous dispersion influences the asphalt binder's rheology and helps achieve the desired stiffness improvement. Thus, it can be concluded that CBNPs is a promising modifier when it comes to improving resistance against permanent deformation.
4. Bitumen bond strength test results show significant improvement in the cohesive and adhesive bond strength of the binder through the incorporation of CBNPs. During 24 h of dry conditioning, and moisture conditioning, addition of 10% CBNPs has increased the POTS values from 7.5 MPa to 17 MPa and 6.55 MPa to 15.40 MPa respectively. The increase in adhesive and cohesive bond strength could be attributed to the electrostatic contribution in nano-modified material. The addition of nano material has increased the wetting potential of the modified binder promoting better wetting of the bitumen on to the aggregate surface thus enhancing the bitumen–aggregate bond strength. This could also be attributed to the increase in stiffness of the asphalt binder because of addition of nano particles [26].
5. Bond strength is an important parameter in measuring the ability of asphalt binder to withstand moisture damage. Tests that are carried out directly on different bitumen–aggregate combinations after moisture conditioning can efficiently measure their moisture sensitivity [31]. Moisture sensitivity directly affects the durability of an asphalt mixture [32]. Significant improvement in POTS after 24 h of moisture conditioning shows that CBNPs have the potential to improve the durability of asphalt.
6. CBNPs have clearly shown potential for improving the permanent deformation resistance of the modified asphalt mixture. During a wheel tracking test, almost 37.8% reduction in rut depth at 55 °C, and 20% reduction in rut depth at 40 °C has been observed through the addition of 10% CBNPs in asphalt. Dynamic modulus values at 40 °C and 55 °C have been observed to be

enhanced by introducing CBNPs. For instance, at 10 Hz and 55 °C, CBNPs have increased the dynamic modulus values from 664.4 MPa to 830.67 MPa. The improvement in dynamic modulus and improved rutting resistance is due to the fact that CBNPs-modified asphalt binder has higher stiffness and a better ability to bond with the aggregates.

CBNPs produced by perennial mountain trees (a natural source) have been used in this study to modify base asphalt binder properties. Similar natural resources are easily available in the country and it would be easy for it to utilize CBNPs in asphalt pavements. It has been found in the literature that CBNPs can effectively improve asphalt pavement performance at 15% or 20% dosage [12,14,33], while the CBNPs used in this study have 10% desirable dosage in asphalt. Thus, the CBNPs produced through natural sources have more potential to influence the asphalt properties and can help in conserving resources. As the raw material (perennial mountain trees) is a natural source and it is required in lesser quantity, the production and utilization of CBNPs in the pavement industry can also prove beneficial from an economic point of view as well. The high-temperature performance of asphalt binder is found to be considerably enhanced by incorporating CBNPs. Moreover, it has also improved the cohesive and adhesive bond strength of the asphalt binder, making it a strong candidate to be used as a modifier.

Any modifier gifted with the dual capability of promoting adhesion and the rheological properties of the asphalt, would be considered ideal for most pavement industry applications. CBNPs seem to have this potential, which could help address the problems of the local pavement construction industry.

**Author Contributions:** The idea was conceived by N.A. The methodology was proposed by J.R. and M.A.K. Experimental work was performed by J.R., M.H. and M.F.u.H. Results were analyzed by J.R., M.A.K., N.A., M.H., S.A.A. and M.F.u.H. The paper was written by J.R. The paper was reviewed and edited by M.A.K., N.A., S.B.A.Z., S.A.A. and F.S. All authors approved and studied the final paper.

**Funding:** This research received no external funding.

**Acknowledgments:** The authors thank Muhammad Ali Nasir for his support and guidance.

**Conflicts of Interest:** The authors declare no conflict of interest.

## References

1. Abtahi, S.M.; Sheikhzadeh, M.; Hejazi, S.M. Fiber-reinforced asphalt-concrete—A review. *Constr. Build. Mater.* **2010**, *24*, 871–877. [CrossRef]
2. Rostler, F.S.; White, R.M.; Dannenberg, E.M. Carbon black as a reinforcing agent for asphalt. In Proceedings of the Association of Asphalt Paving Technologists Proc, San Antonio, TX, USA, 21–23 February 1977; Volume 46, pp. 376–410.
3. Khosla, N.P. Effect of the use of modifiers on performance of asphaltic pavements. *Transpor. Res. Rec.* **1991**, 1317.
4. Yao, Z.; Monismith, C.L. Behaviour of Asphalt Mixtures With Carbon Black Reinforcement (With Discussion). In Proceedings of the Association of Asphalt Paving Technologists Proc, St. Paul, MN, USA, 1986; Volume 55, pp. 564–585.
5. Feng, Z.; Rao, W.; Chen, C.; Tian, B.; Li, X.; Li, P.; Guo, Q. Performance evaluation of bitumen modified with pyrolysis carbon black made from waste tyres. *Constr. Build. Mater.* **2016**, *111*, 495–501. [CrossRef]
6. Ahmedzade, P.; Geckil, T. Influence of carbon black on the mechanical and electrical properties of asphalt mixtures. *Indian J. Eng. Mater. Sci.* **2007**, *14*, 358–364.
7. Rodgers, W.J.; Gunay, B.; Woodside, A. Rheological and electrical properties of modified bitumen. *Proc. Inst. Civ. Eng. Transp.* **2010**, *163*, 175–182. [CrossRef]
8. Martin, K.G. Preliminary microviscometer studies of carbon black/rubber/bitumen dispersions. In Proceedings of the Australian Road Research Board (ARRB) Conference, Canberra, Australia, 1962; Volume 1, pp. 895–911.
9. Park, T.; Lovell, C.W. Using pyrolized carbon black (PCB) from waste tires in asphalt pavement (part 1, limestone aggregate). *Jt. Transp. Res. Progr.* **1996**, 222. [CrossRef]
10. Chebil, S.; Chaala, A.; Roy, C. Use of softwood bark charcoal as a modifier for road bitumen. *Fuel* **2000**, *79*, 671–683. [CrossRef]

11. Zhang, J.; Wang, J.; Wu, Y.; Wang, Y.; Wang, Y. Evaluation of the improved properties of SBR/weathered coal modified bitumen containing carbon black. *Constr. Build. Mater.* **2009**, *23*, 2678–2687. [CrossRef]
12. Li, C.; Fan, Z.; Wu, S.; Li, Y.; Gan, Y. Effect of carbon black nanoparticles from the pyrolysis of discarded tires on the performance of asphalt and its mixtures. *Appl. Sci.* **2018**, *8*, 624. [CrossRef]
13. Cong, P.; Xu, P.; Chen, S. Effects of carbon black on the anti aging, rheological and conductive properties of SBS/asphalt/carbon black composites. *Constr. Build. Mater.* **2014**, *52*, 306–313. [CrossRef]
14. Bai, B.C.; Park, D.-W.; Vo, H.V.; Dessouky, S.; Im, J.S. Thermal Properties of Asphalt Mixtures Modified with Conductive Fillers. *J. Nanomater.* **2015**, *16*, 255. [CrossRef]
15. Zhao, S.; Huang, B.; Shu, X.; Ye, P. Laboratory Investigation of Biochar-Modified Asphalt Mixture. *Transp. Res. Rec.* **2014**, *2445*, 56–63. [CrossRef]
16. Mirbaha, B.; Abdi, A.; Zarei, M.; Zarei, A.; Akbari nia, F. Experimental determination of the optimum percentage of asphalt mixtures reinforced with Nano-carbon black and polyester fiber industries. *Eng. Solid Mech.* **2017**, *5*, 285–292. [CrossRef]
17. Mirza, M.W.; Abbas, Z.; Rizvi, M.A. Temperature Zoning of Pakistan for Asphalt Mix Design. *Pak. J. Eng. Appl. Sci.* **2011**, *8*, 49–60.
18. Hussan, S.; Kamal, M.A.; Hafeez, I.; Farooq, D.; Ahmad, N.; Khanzada, S. Statistical evaluation of factors affecting the laboratory rutting susceptibility of asphalt mixtures. *Int. J. Pavement Eng.* **2017**, 1–15. [CrossRef]
19. Faramarzi, M.; Golestani, B.; Lee, K.W. Improving moisture sensitivity and mechanical properties of sulfur extended asphalt mixture by nano-antistripping agent. *Constr. Build. Mater.* **2017**, *133*, 534–542. [CrossRef]
20. Huang, B.; Shu, X.; Chen, X. Effects of mineral fillers on hot-mix asphalt laboratory-measured properties. *Int. J. Pavement Eng.* **2007**, *8*, 1–9. [CrossRef]
21. Kallas, B.F.; Puzinauskas, V.P.; Krieger, H.C. Mineral Fillers in Asphalt Paving Mixtures. *Highw. Res. Board Bull.* **1962**, *329*.
22. Chelovian, A.; Shafabakhsh, G. Laboratory evaluation of Nano $Al_2O_3$ effect on dynamic performance of stone mastic asphalt. *Int. J. Pavement Res. Technol.* **2017**, *10*, 131–138. [CrossRef]
23. Hussain, M.H.M.; Ghaly, N.F.; Ibrahim, I.M. Modified Hot Mix Asphalt for Road Maintenance. *World Appl. Sci. J.* **2008**, *5*, 236–245.
24. Zaniewski, J.P.; Pumphrey, M.E. Evaluation of Performance Graded Asphalt Binder Equipment and Testing Protocol. *Asph. Technol. Progr.* **2004**, *107*.
25. Moraes, R.; Velasquez, R.; Bahia, H. Measuring the Effect of Moisture on Asphalt-Aggregate Bond with the Bitumen Bond Strength Test. *Transp. Res. Rec.* **2011**, *2209*, 70–81. [CrossRef]
26. Bahia, H.; Moraes, R.; Velasquez, R. The Effect of Bitumen Stiffness on the Adhesive Strength Measured by the Bitumen Bond Strength Test. In Proceedings of the 5th Eurasphalt and Eurobitume Congress, Istanbul, Turkey, 2012.
27. Valadares, L.F.; Linares, E.M.; Bragança, F.C.; Galembeck, F. Electrostatic adhesion of nanosized particles: The cohesive role of water. *J. Phys. Chem. C* **2008**, *112*, 8534–8544. [CrossRef]
28. Yao, H.; You, Z. Effectiveness of Micro- and Nanomaterials in Asphalt Mixtures through Dynamic Modulus and Rutting Tests. *J. Nanomater.* **2016**, *2016*, 10. [CrossRef]
29. Li, P.; Zheng, M.; Wang, F.; Che, F.; Li, H.; Ma, Q.; Wang, Y. Laboratory Performance Evaluation of High Modulus Asphalt Concrete Modified with Different Additives. *Adv. Mater. Sci. Eng.* **2017**, *2017*. [CrossRef]
30. Van Dam, T.J.; Harvey, J.T.; Muench, S.T.; Smith, K.D.; Snyder, M.B.; Al-Qadi, I.L.; Ozer, H.; Meijer, J.; Ram, P.V.; Roesler, J.R.; et al. Chapter 5. Construction Considerations to Improve Pavement Sustainability. In *Towards Sustainable Pavement Systems: A Reference Document*; US Department of Transportation, Federal Highway Administration: Washington, DC, USA, 2015.
31. Canestrari, F.; Cardone, F.; Graziani, A.; Santagata, F.A.; Bahia, H.U. Adhesive and cohesive properties of asphalt-aggregate systems subjected to moisture damage. *Road Mater. Pavement Des.* **2010**, *11*, 11–32. [CrossRef]

32. Behiry, A.E.A.E.M. Laboratory evaluation of resistance to moisture damage in asphalt mixtures. *Ain Shams Eng. J.* **2013**, *4*, 351–363. [CrossRef]
33. Zarei, M.; Zahedi, M. Effect of nano carbon black on mechanical properties of asphalt mixtures. *J. Fundam. Appl. Sci.* **2016**, *8*, 2996–3008.

*applied sciences*

MDPI

*Article*

# Effects of Titanate Coupling Agent on Engineering Properties of Asphalt Binders and Mixtures Incorporating LLDPE-$CaCO_3$ Pellet

**Mohd Rosli Mohd Hasan [1,*], Zhanping You [2,*], Mohd Khairul Idham Mohd Satar [3], Muhammad Naqiuddin Mohd Warid [3], Nurul Hidayah Mohd Kamaruddin [4], Dongdong Ge [2] and Ran Zhang [5]**

1 School of Civil Engineering, Universiti Sains Malaysia, Engineering Campus, 14300 Nibong Tebal, Pulau Pinang, Malaysia

2 Department of Civil and Environmental Engineering, Michigan Technological University, 1400 Townsend Drive, Houghton, MI 49931, USA; dge1@mtu.edu

3 Department of Geotechnics and Transportation, Faculty of Civil Engineering, Universiti Teknologi Malaysia, 81310 Johor Bahru, Johor, Malaysia; khairulidham@utm.my (M.K.I.M.S.); naqiuddin@utm.my (M.N.M.W.)

4 Faculty of Civil and Environmental Engineering, Universiti Tun Hussein Onn Malaysia, 86400 Parit Raja, Johor, Malaysia; hidayahk@uthm.edu.my

5 School of Highway, Chang'an University, South Erhuan Middle Section, Xi'an 710064, Shaanxi, China; ranzhang@mtu.edu

* Correspondence: cerosli@usm.my (M.R.M.H.); zyou@mtu.edu (Z.Y.); Tel.: +60-604-599-6288 (M.R.M.H.); +1-906-487-1059 (Z.Y.)

Received: 31 May 2018; Accepted: 21 June 2018; Published: 24 June 2018

**Abstract:** This study was initiated to evaluate the performance of asphalt binders and mixtures incorporating linear low-density polyethylene- calcium carbonate (LLDPE-$CaCO_3$) pellet, either with or without titanate coupling agent. The detailed manufacturing process of modifier pellets was displayed. The coupling agent was used to enhance the cross-linking between materials by means of winding up covalent bonds or molecule chains, thus improving the performance of composites. In the preparation of modified bitumen, the preheated asphalt binder was mixed with the modifiers using a high shear mixer at 5000 rpm rotational speed for 45 min. Experimental works were conducted to evaluate the performance of asphalt binders in terms of volatile loss, viscosity, rutting potential, and low temperature cracking. Meanwhile, the asphalt mixtures were tested using the flow number test and tensile strength ratio (TSR) test. The addition of LLDPE-$CaCO_3$ modifiers and coupling agent does not significantly affect the volatile loss of modified asphalt binders. The addition of modifiers and coupling agent has significantly improved the resistance to permanent deformation of asphalt binders. Even though, the addition of LLDPE-$CaCO_3$ modifier and coupling agent remarkably increased the mixture stiffness that contributed to lower rutting potential, the resistance to low temperature cracking of asphalt binder was not adversely affected. The combination of 1% coupling agent with 3% PECC is optimum dosage for asphalt binder to have satisfactory performance in resistance to moisture damage and rutting.

**Keywords:** asphalt modification; coupling agent; rheological behavior; plastic; calcium carbonate powder

## 1. Introduction

Over decades, a wide range of modification at macro, meso-, micro- and nano-scales have been conducted to improve the performance of asphalt pavement [1]. This is essential to design durable, safe and efficient asphalt pavement that can function efficiently across a wide range of temperatures and distresses. Polymer modified binders have got increasing use on improving the resistance to thermal cracking and permanent deformation, as well as reducing the fatigue cracking potential of asphalt pavement [2,3]. Brown et al. [4] mentioned that an ideal asphalt condition should exhibit both: (a) Relatively high stiffness at high temperature to prevent rutting and shoving and; (b) great adhesion between asphalt binder and aggregates matrices in the presence of water to reduce moisture induced damage, for example stripping. Additionally, Chen et al. [5] mentioned that the asphalt is continuously exposed to a wide range of climates, loading rate and time, but it does not have essential engineering properties to sustain those situations, where the binder is soft under hot summer days and brittle in the freezing condition.

Various types of polymer have been used to chemically or mechanically improve the properties of asphalt binder, which can be classified into three different categories: elastomer, plastomer and reactive polymers [4,6,7]. The application of polymers in the asphalt pavement has been growing rapidly since the early 1970s. Based on previous studies, modifications of asphalt binder using polymer materials can significantly improve the properties of asphalt material, such as reducing the temperature susceptibility, providing better rheological characteristics, as well as enhancing the material durability [3,8,9]. Polyethylene (PE) and polyethylene-based copolymers (new or recycled) has been used to modify the performance of asphalt binder and mixture over many years [2,10]. The PE and polypropylene (PP) are categorized as plastomers which can bring a high rigidity to the material and improve the resistance to permanent deformation under traffic load [11,12]. A study was performed by Habib et al. [13] to evaluate the rheological properties and interaction of asphalt binder with different thermoplastics, such as high density polyethylene (HDPE), linear low density polyethylene (LLDPE) and PP. Based on their study, the viscoelastic behavior of asphalt binder was significantly affected by the modifier concentration, bitumen grade and the temperature. In addition, based on the overall studies, the best results occurred when the polymer concentration was limited to 3%, which resulted from the thermodynamically stable structure condition. This had significantly improved the resistance to rutting, fatigue, and temperature susceptibility.

Limestone ($CaCO_3$) is an inert material which has been used as an additive in asphalt mixture for more than 100 years. However, the hydrated lime came into regular use only just in the 1980s. Several states including Georgia, Nevada, Texas, Virginia, and Utah used lime to solve the water susceptibility issue on asphalt pavement [14]. $CaCO_3$ is the most widely used filler in thermoplastics because of its low cost and superior mechanical properties. Much effort has been focused to increase mechanical properties such as tensile and flexural strength, impact resistance of $CaCO_3$-filled PP, HDPE, LDPE, and LLDPE composites [15]. The term "lime" is used referring to either quicklime or hydrated lime which comes originally from limestone. Calcinations process of limestone dissociated the calcium from carbon dioxide, leaving calcium oxide. This is known as quicklime, it is then combined with water to form hydrated lime. In the asphalt industry, the lime generally refers to hydrated lime or calcium hydroxide [16]. The fineness and high surface area of hydrated lime contributes to a high speed of chemical reaction. Hydrated lime is the only form of lime which has been shown to be useful in controlling stripping. Many aggregates are quite acidic and the asphalt binder also contains acids, the ions present at the interface of both materials repelled each other electrically. The presence of lime could neutralize the acidic aggregates and asphalt binder, which provide opposite-charge ions to enhance adhesion [16]. Thus, lime in asphalt mixture is not only used as an anti-stripping agent, but it may improve mixture stiffness, reduce plasticity index when clays are present and reduce the oxidation rate [14,16].

Lu and Isacsson [17] revealed that, even though the thermoplastic modifiers had improved the viscosity and stiffness of asphalt binder, it did not significantly help in terms of elastic behavior. The embrittlement makes asphalt susceptible to fracture, especially when subjected to high levels of stress [18]. The polymer should be homogeneously dispersed into asphalt to ensure proper adhesion of the asphalt binder. However, incompatibility between the binder and the polymer is sometimes inevitable. As a result, it is difficult to establish the bond, either in terms of a physical or chemical manner [18,19]. Prior to gaining a better performance, an application of coupling agent in organic materials or treating inorganic fillers has been used. This enhances the materials cross-linking by means of winding up covalent bonds or molecule chains, thus improving the performance of composites [20,21]. It was found that a small amount of coupling agent could increase shear resistance in mechanical properties [20]. There are various types of coupling agents that have been used to improve the bonding of composite materials such as: Silane, zirconate, dicarboxylic anhydry-dem, titanate and phosphate ester. However, the most important commercial coupling agents are formed by silane and titanate [22]. A titanate coupling agent has been used in enhancing the bonding in $CaCO_3$–thermoplastic composites [23]. A study conducted by Atikler et al. [15] showed that a silane coupling agent had significantly ameliorated the mechanical properties of the HDPE–fly ash composites. Sae-oui et al. [24] reported that excessive use of a silane coupling agent could cause a negative effect on certain properties such as modulus and hardness due to plasticizing effect. Additionally, Chen et al. [25] also concluded that too much of a titanate coupling agent resulted in polymer bridging, and reduced the phase boundary condition. It would also influence the economic aspect, whereby the excess of a coupling agent could increase the cost of production. Former studies [26–28] had recommended using the titanate-coupling agent in a range of 0.1 wt % to 0.5 wt % to attain the best performance based on the adhesion test.

Even though various types of polymers have been used to enhance the performance of asphalt composite, only a few of them are considered as satisfactory based on the performance and economic standpoints [29,30]. In this study, newly manufactured asphalt modifiers that comprised of LLDPE and $CaCO_3$ with or without titanate coupling agent was used to enhance the engineering properties of asphalt binder and asphalt mixture durability towards various distresses.

## 2. Materials and Methods

### 2.1. Materials

#### 2.1.1. Asphalt Binder and Aggregate

The basic materials that used in this study were obtained from a local source in Hancock, Michigan. The PG 58-28 was used as a control binder. The aggregate gradation used was based on Michigan Department of Transportation (MDOT) specifications for Upper Peninsula region. The nominal maximum aggregate size of the gradation is 9.5 mm and the designed traffic level is less than three million equivalent single axles loads (ESALs) based on the Superpave asphalt mixture design procedure.

#### 2.1.2. Coupling Agent

For this project the coupling agent Ken-React® CAPS® L® 12/L was used [31]. Ken-React® CAPS® L® 12/L is a neoalkoxy titanate, with a specific gravity of 0.95, and is in the form of an off-white/beige solid pellet. Ken-React® CAPS® L® 12/L was used as-received. Figure 1 shows the chemical structure of the Ken-React® LICA® 12 (20% Active Portion of CAPS® L® 12/L). The chemical name for Ken-React® CAPS® L® 12/L is titanium IV 2,2 (bis-2-propenolatomethyl) butanolato, tris(dioctyl) phosphate-O [32]. Figure 2 shows the appearance of coupling agent.

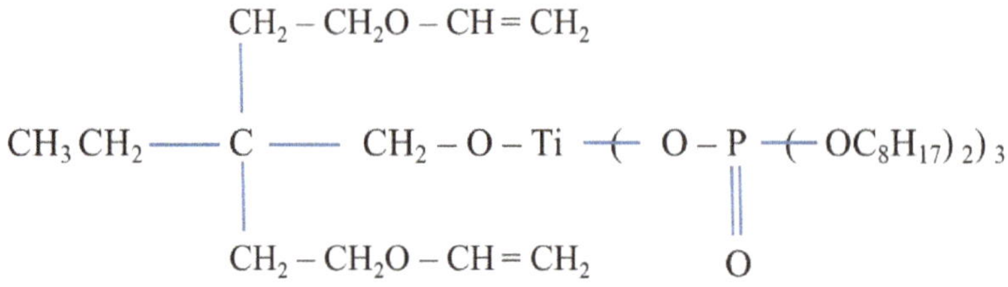

**Figure 1.** Chemical structure of Ken-React® LICA® 12 (20% active portion of CAPS® L® 12/L).

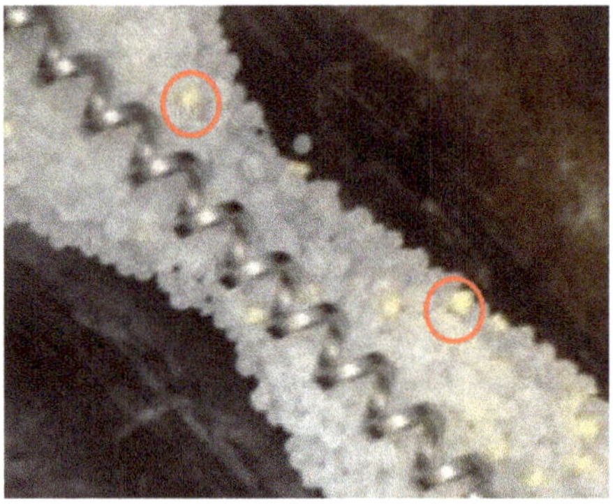

**Figure 2.** Titanate coupling agent (shown in red circle) mixed with LLDPE pellets.

2.1.3. Manufacturing Process of Modifier Pellets

A V-cone mixer was used operating at 24 rpm for four minutes to mix the Ken-React® CAPS® L® 12/L and LLDPE pellets. The modifier pellets were prepared using extrusion equipment in the Chemical Engineering Department at Michigan Technological University. An American Leistritz Extruder Corporation, model ZSE 27 has a 27 mm co-rotating intermeshing twin-screw extruder with ten heating zones, a length/diameter ratio of 40 and two stuffers was used to produce the modifier pellets. Two Schenck AccuRate gravimetric feeders were used to accurately control the amount of LLDPE, $CaCO_3$ and the coupling agent supplied into the extruder. Figure 3 illustrates the screw design that used during manufacturing of pellet modifiers [33]. The LLDPE and LLDPE/Ken-React® CAPS® L® 12/L coupling agent was added to zone 1. Meanwhile, $CaCO_3$ was added in Zone 5.

After passing through the extruder, the polymer strands (3 mm in diameter) enter a 3 m long water bath (Sterling Blower model WT-1008-10) and then a pelletizer (Accu-grind Conair Model 304) produced nominally 3 mm diameter by 3 mm long pellets. The final product of the manufacturing process is shown in Figure 4. After compounding, the pelletized composite resin was dried in an indirect heated dehumidifying oven (Bry Air RD-20) at 60 °C for 7 h. It was then stored in sealed moisture barrier bags prior to mixing with asphalt binder [34].

The extrusion conditions for different modifiers are shown in Table 1. Three different formulations were extruded. The first was called PECC, which contained 70 wt % LLDPE and 30 wt % $CaCO_3$. The second was called PECC-1CA, which contained 69.3 wt % LLDPE, 29.7 wt % $CaCO_3$, and 1 wt % Ken-React® CAPS® L® 12/L coupling agent. The third was called PECC-2CA, which contained 29.4 wt % $CaCO_3$, 68.6 wt % LLDPE, and 2 wt % Ken-React® CAPS® L® 12/L coupling agent.

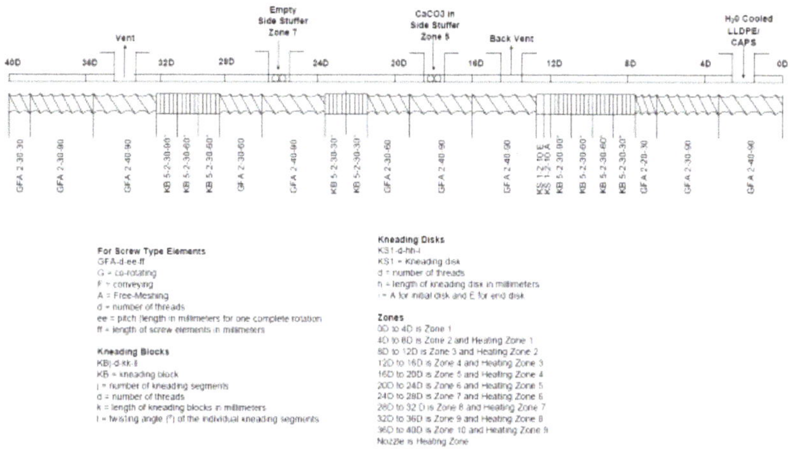


**Figure 3.** Extruder screw design.

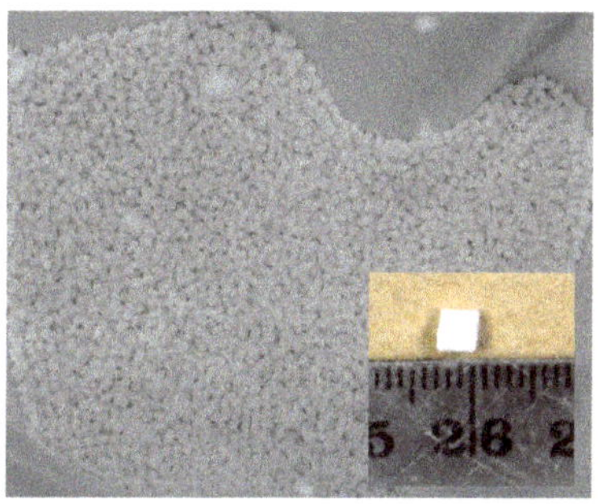

**Figure 4.** Final product of manufactured pellets.

**Table 1.** Extrusion conditions.

| Material Number | PECC | PECC-1CA | PECC-2CA |
|---|---|---|---|
| Extrusion Date | March 12 2013 | March 12 2013 | March 12 2013 |
| Extruder, rpm | 250 | 250 | 250 |
| Motor Amperage, % | 27 | 27 | 27 |
| Melt Temperature, °C | 244 | 244 | 244 |
| Melt Pressure, psig | 80 | 80 | 80 |
| #2 Feeder Setting, lb/h | 6.0 | 6.0 | 6.0 |
| Material in Feeder #2 | $CaCO_3$ | $CaCO_3$ | $CaCO_3$ |
| #3 Feeder Setting, lb/h | 14 | 14 | 14 |
| Material in Feeder #3 | LLDPE | LLDPE + CAPS | LLDPE + CAPS |
| Vacuum Port | 1 atm | 1 atm | 1 atm |
| Zone 5 Side Stuffer Setting, rpm | 300 | 300 | 300 |
| Feeder at Zone 5 | $CaCO_3$ | $CaCO_3$ | $CaCO_3$ |
| Zone 7 Side Stuffer setting, rpm | 300 | 300 | 300 |
| Feeder at Zone 7 | none | none | none |
| Feed Section Temperature | $H_2O$ Cooled | $H_2O$ Cooled | $H_2O$ Cooled |
| Zone 1 Temperature, °C | 175 | 175 | 175 |
| Zone 2 Temperature, °C | 195 | 195 | 195 |
| Zone 3 Temperature, °C | 210 | 210 | 210 |

**Table 1.** *Cont.*

| Material Number | PECC | PECC-1CA | PECC-2CA |
|---|---|---|---|
| Zone 4 Temperature, °C | 220 | 215 | 215 |
| Zone 5 Temperature, °C | 220 | 220 | 215 |
| Zone 6 Temperature, °C | 220 | 220 | 215 |
| Zone 7 Temperature, °C | 220 | 220 | 215 |
| Zone 8 Temperature, °C | 220 | 220 | 215 |
| Zone 9 Temperature, °C | 220 | 220 | 215 |
| Zone 10 Temperature, °C | 220 | 220 | 220 |
| Die Type and Gap | 3 × 3 mm | 3 × 3 mm | 3 × 3 mm |
| Pelletizer Setting at 8/Bath | $H_2O$ bath | $H_2O$ bath | $H_2O$ bath |
| Output Rate, lbs/h | 20 | 20 | 20 |

Notes: (1) LLDPE and Ken-React® CAPS® L® 12/L mixed in V cone blender for 4 min at 24 rpm in 2 lb batches and then placed in Feeder 3. (2) Approximately 20 lbs of each extruded material was produced. (3) Feeder 3 helix: 0.5″ open helix with end stub and 0.75″ nozzle side discharge. (4) Feeder 3 Toshiba laptop feeder3_VectraA950RX0.5inopen0.75insidedischargejak.par. (5) Feeder 2 helix: 0.75″ open helix with 0.75″ ID polyliner. (6) Feeder 2 NEC laptop Thermocarb file. (7) Extruder screw 5-14-05 design used from American Leistritz. Extruder screw cleaned in sand bath prior to use. (8) Purged with LLDPE at end of extrusion run to clean out extruder.

### 2.2. Methods

#### 2.2.1. Preparation of Modified Asphalt Binder

The modified asphalt binders were prepared using a high shear mixer. The binder and modifiers were properly mixed to ensure the materials were evenly dispersed in asphalt binder. The temperature that used for the production of modified asphalt binder is at 170 °C. In the asphalt binder preparation process, about 500 g of asphalt binder PG 58-28 was poured into a one liter metal container. Then, an adequate amount of modifier was added to the same container and heated up in an oven for about two hours prior to the mixing process. After two hours of inducing the melting process, the binder and modifier were stirred using a high shear mixer at 5000 rpm rotational speed for 45 min. Based on literature review, LLDPE-related modifiers were used as asphalt modifiers in a range of 2% to 6% based on the weight of asphalt binder [11,13,35–38]. Based on the range, the amounts of modifier incorporated in the asphalt binder were decided at 3% and 5% based on the asphalt binder weight. At least three replicate specimens were used for all the asphalt binder and asphalt mixture tests.

#### 2.2.2. Preparations of Asphalt Mixture Specimen

A bucket mixer was used to blend the aggregates and asphalt binder. The sample was compacted using a gyratory compactor at 86 gyrations. Prior to compaction, the mixture was heated in an oven for two hours to simulate the short-term aging condition that occurs during preparation of asphalt mixture in the field. The Superpave specifications [39–41] were referred during the preparation of asphalt mixture.

#### 2.2.3. Asphalt Binder Test Method

The rolling thin film oven (RTFO) was used to quantify the volatiles lost (mass loss) during the short-term aging process of asphalt binder. Based on the Superpave Specification, the mass loss of asphalt binder should be less than 1 wt % to ensure the asphalt binder not to lose a significant number of volatiles over its life.

The rotational viscometer was used to determine the viscosity of asphalt binders at high temperature. During the sample preparation, about 10.5 g asphalt binder is required for each sample, and spindle #27 was used in this test. This test measures the required torque value to maintain a constant rotational speed (20 rpm) of a cylindrical spindle under a constant temperature. The results were recorded in centipoises (cP) at one-minute intervals for a total of three readings.

In this study, the multiple stress creep recovery (MSCR) was conducted by introducing the RTFO aged asphalt binder specimen to the repeated creep and recovery process at high temperature. The test has been conducted in accordance with AASHTO TP 70-13 at 58 °C, which is the high temperature grade of selected asphalt binder, PG 58-28. The 25 mm diameter with 1 mm thickness circular disk-shaped asphalt binder sample was used. Two stress levels were introduced to the sample, which were 0.1 kPa and 3.2 kPa at one second loading time and nine seconds recovery time while performing the test [42,43]. The test started with 0.1 kPa stress for ten cycles without time lags, and proceeded with 3.2 kPa stress under the same number of cycles. The new MSCR test, which is conducted based on AASHTO T 350-14, conducted 20 cycles 0.1 kPa stress loading and unloading. The MSCR test result may be different if conducted in accordance with AASHTON T 350 [44,45].

The bending beam rheometer (BBR) test was performed in accordance with AASHTO T 313; a simply supported beam of asphalt binder was subjected to a constant load of 980 mN for four minutes. The test was conducted at −12 °C, −18 °C and −24 °C to define the critical cracking temperature of control and modified binders.

#### 2.2.4. Asphalt Mixture Test Method

The flow number test was referred to a dynamic creep or repeated load testing. Basically, a 0.1 s loading followed by a 0.9 s dwell (rest time) was applied to the specimen. Additionally, an effective temperature of 45 °C, often referred to as rutting temperature was used in this test [46,47]. Prior to the testing, the specimens were conditioned at 45 °C.

The tensile strength ratio (TSR) was used to evaluate the moisture susceptibility of asphalt mixture. The moisture susceptibility was evaluated by comparing the indirect tensile strength (ITS) of asphalt mixtures in dry and wet conditions. The ITS test was performed according to AASHTO T283. The specimens were tested at the room temperature and constant loading speed, 0.085 mm/s. The specimen was subjected to compression loads which act parallel to the vertical diameter plane.

## 3. Characterization of Asphalt Binder

### 3.1. *Volatile Loss*

At elevated temperature, the smaller molecules from asphalt binder are driven off, resulting in an increase of the asphalt's viscosity. The effects of heat and flowing air on a thin film of semi-solid asphaltic material are considered in this procedure. Figure 5 presents the mean mass loss values of each specimen tested using short-term aging protocol. Based on the test results, incorporation of modifiers and coupling agent do not significantly affect the volatile loss of modified asphalt binders compared to control binder PG 58-28.

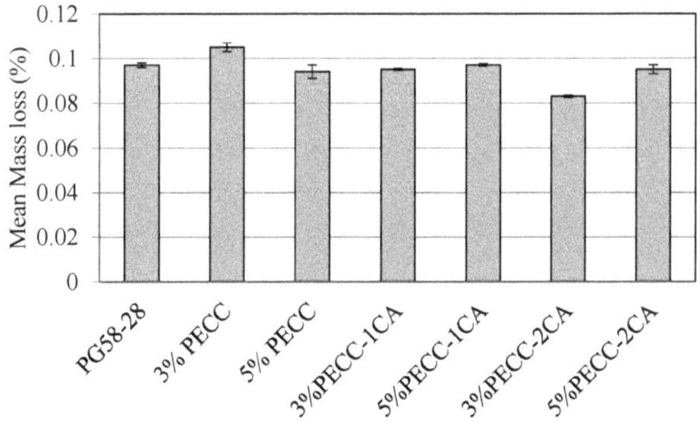


**Figure 5.** Mass loss test results of each asphalt binder.

### 3.2. Rotational Viscosity

Figures 6 and 7 show the results of asphalt binders modified using PECC and PECC-CA modifiers. Modified asphalt binders have higher viscosity value compared to the control asphalt binder. The addition of a titanate coupling agent has slightly increased the viscosity and consistency of asphalt binders, except for the specimen prepared using 3% PECC-1CA. The increaments are ranging from 4% to 40% depending on the percentage of coupling agent and test temperature.

Hypothetically, the viscosity of asphalt binder could also be used as an early indicator of resistance to permanent deformation. Whereas, higher viscosity could sustain higher temperature before the binder flow or change its physical behavior. In the field, the melting temperature can be related to the atmospheric ambient temperature. In this study, the addition of modifiers and coupling agent has significantly improved the resistance to permanent deformation of asphalt binders.

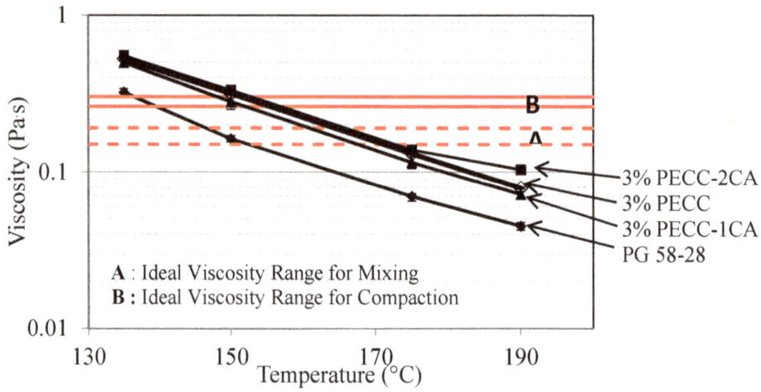


**Figure 6.** Modified Asphalt Binder Viscosity at Low Modifier Percentage.

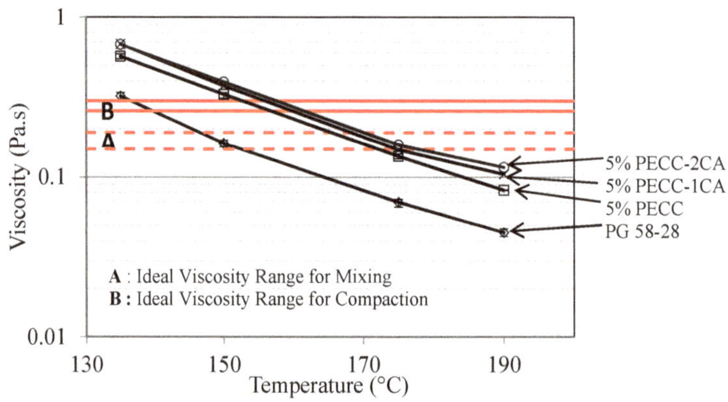


**Figure 7.** Modified asphalt binder viscosity at high modifier percentage.

### 3.3. Multiple Shear Creep Recovery

The DSR with $G^*/\sin\delta$ (AASHTO M320) is the typical parameter for rutting prediction of asphalt pavement. However, this method has been revised to provide a better prediction on the rutting performance of modified asphalt binder by MSCR. This method measures the permanent strain accumulated in the binder after designated cycles of shear loading and unloading. In which, lower permanent shear strain indicates higher rutting resistance of the pavement.

Subsequently, the rutting resistance of asphalt binder is characterized using non-recoverable compliance ($J_{nr}$) which is considered as the best approach to replace the current Superpave testing method, $G^*/\sin \delta$ ($\omega$ = 10 rad/s) [43,48]. Additionally, the percent recovery (R) is also determined in order to understand the high temperature viscoelastic deformation properties [43], where higher R value indicates a better resistance to rutting. Meanwhile, a lower $J_{nr}$ value shows a better resistance to permanent deformation (rutting).

The mean recovery percentage ($n$ = 3) of the tested asphalt binders are presented in Figures 8 and 9. Based on the non-recoverable compliance criteria, the results show that the modified binders have lower rutting potential as compared to the control binder, PG 58-28. Additions of newly manufactured pellets have significantly increased the resistance to permanent deformation by at least 27% greater than PG 58-28 binder. Asphalt binder modified using 5% PECC-1CA has shown the best performance, in terms of non-recoverable compliance criteria and percent recovery after continuously multiple loading action on the sample. Referring to the specimens prepared using 3% PECC-1CA and 3% PECC-2CA, application of 1% coupling agent in the PECC material represents better elastic response compared to 2% coupling agent, which is consistent to the $J_{nr}$ analysis. Samples tested using lower stress levels have resulted in smaller non-recoverable compliance and superior in terms of percent recovery. This is clearly simulated the condition in the field, where heavy vehicles (e.g., lorries and trucks) cause more severe permanent deformation compared to other vehicles, as we can see the rut depths in the slow lane are typically more severe than the fast lane of a highway.

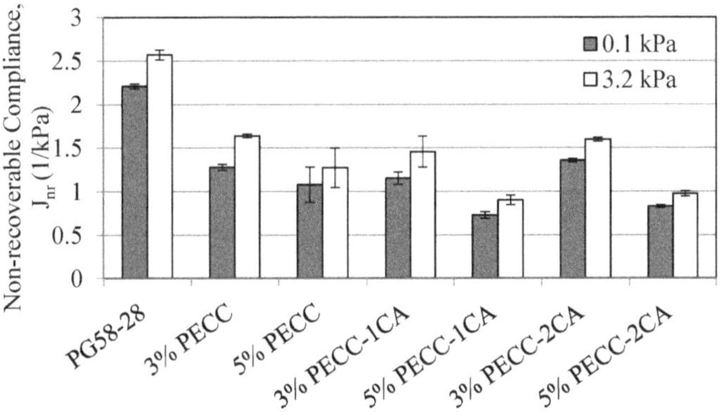


**Figure 8.** Non-recoverable compliance for evaluation of rutting potential.

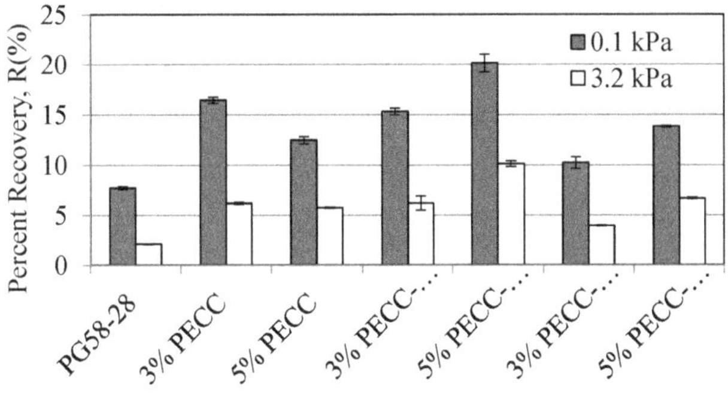


**Figure 9.** Recovery percentage of the tested asphalt binders.

Overall, without the presence of coupling agent, asphalt binder modified using 3% PECC has shown a better recovery percentage compared to 5% PECC sample. With the addition of coupling agent, a higher percentage could be adopted in the modification process of asphalt binder. However, the amount of coupling agent should be limited to 1% to avoid adverse effects on its resistance to rutting.

### 3.4. Low Temperature Cracking Using BBR Test

The BBR test was conducted to evaluate the low temperature stiffness and relaxation properties of asphalt binders, based on the function of load and duration. These parameters give an indication of an asphalt binder's ability to resist low temperature cracking.

The data was then analyzed to calculate the critical cracking temperatures ($T_{cr}$) of the asphalt binders based on the measured creep stiffness and m-values. Figure 10 shows the limiting low temperature or $T_{cr}$ for each binder. Overall, all the modified asphalt binders have shown comparable performance in terms of resistance to low temperature cracking. It was found that incorporating 3% PECC-1CA modifier had contributed to the low temperature performance of the asphalt binder, where it could resist the thermal cracking at −33.2 °C, compared to the control asphalt binder that may only resist the thermal cracking at temperature as low as −30.5 °C.

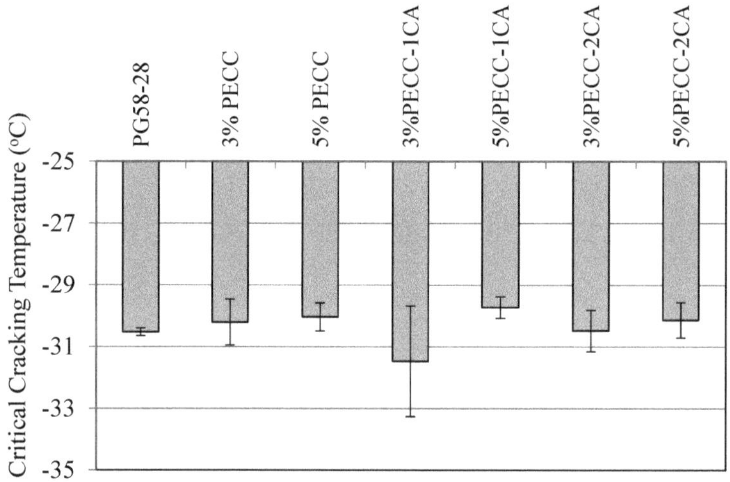


**Figure 10.** Critical cracking temperature of PECC-based asphalt binders compared to control specimen.

## 4. Performance of Asphalt Mixtures

### 4.1. Resistance to Permanent Deformation

Flow number test was conducted to evaluate the rutting resistance of asphalt pavement. The test was typically used to assess the resistance to permanent deformation for the past several years [49,50]. Faheem et al. [51] mentioned that flow number test has a strong correlation to the Traffic Force Index (TFI), which represents the densification loading by the traffic during its service life. It was also found that this test has a good correlation with the field rutting performance [52,53]. The test is performed by introducing repeated traffic loading (loading and unloading) on the cylindrical asphalt specimen and the permanent deformation is recorded as a function of load cycles at the minimum permanent strain rate.

Figure 11 shows the flow number test result. The specimens prepared using modified asphalt binders have significantly higher resistance to rutting compared to the control sample. Greater amount of modifier has resulted in a higher flow number, which indicated a better resistance to rutting. The addition of coupling agent also has remarkably increased the mixture stiffness that contributes to lower rutting potential.

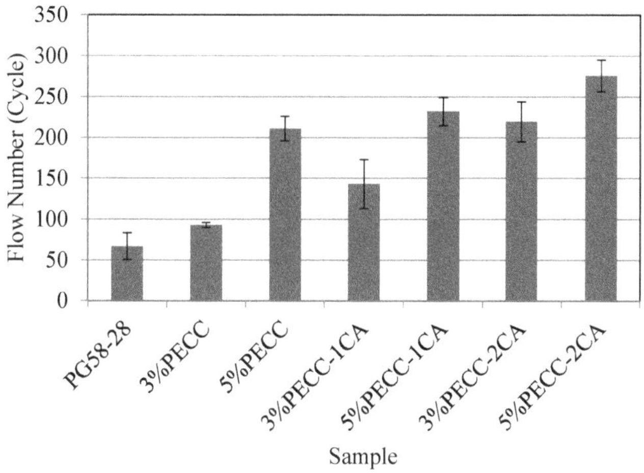


**Figure 11.** Flow number test results.

*4.2. Moisture Susceptibility*

Figure 12 shows the ITS values of tested samples. Overall, the addition of modifiers does not significantly alter the ITS of the sample in the dry condition, except 5% PECC-2CA. There are no significant effects of using different compositions of modifiers and the coupling agent in terms of indirect tensile strength results as indicated by the error bars presented. However, the modifiers help in enhancing the ITS values of the wet samples. The samples incorporated lower amount of modifiers (3%) have a better ITS value compared to sample prepared using 5% modifiers. Incorporating 1% and 2% coupling agents also do not have significant differences between them in term of wet samples' ITS values.

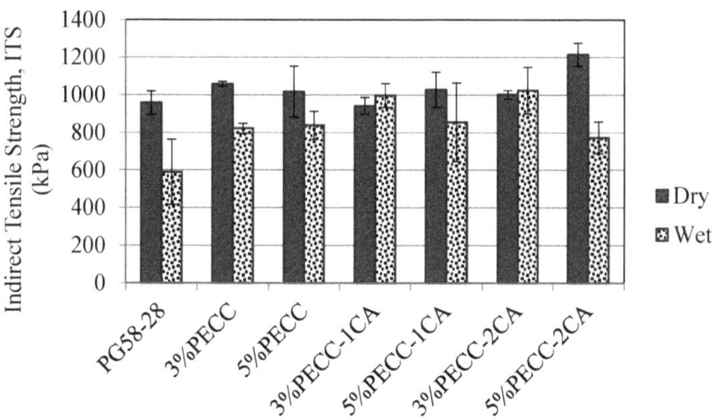


**Figure 12.** ITS test results.

Based on Figure 13, additions of modifiers have remarkably improved the TSR of modified asphalt mixtures compared to the control mixture, except specimen incorporating 5% PECC-2CA. Combination of 3% PECC and coupling agent at 1% and 2% has greatly enhanced the resistance to moisture damage of asphalt mixture. However, the combination of the coupling agent with 3% PECC is optimum for this study in order to avoid adverse due to moisture damage. The combination of 5% PECC and 2% coupling agent yield mixtures with worse moisture susceptibility characteristic as compared to other modified mixtures.

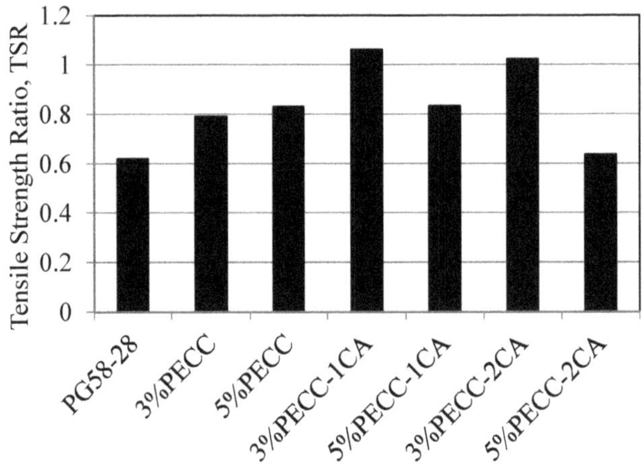


**Figure 13.** Moisture susceptibility of asphalt mixtures.

## 5. Conclusions

Based on the outcome of this study, several conclusions can be made as follows:

1. Addition of LLDPE-$CaCO_3$ modifiers and coupling agent do not significantly affect the volatile loss of modified asphalt binders.
2. Modified asphalt binders have higher viscosity value compared to the control asphalt binder. The addition of a titanate-coupling agent has slightly increased the viscosity and consistency of asphalt binders depending on the percentage of coupling agent and test temperature. The viscosity of asphalt binder could also be used as an early indicator of resistance to permanent deformation. The addition of modifiers and coupling agent has significantly improved the resistance to permanent deformation of asphalt binders.
3. With the addition of coupling agent, a higher percentage could be adopted in the modification process of asphalt binder. However, the amount of coupling agent should be limit to 1% to avoid adverse effects on its resistance to rutting.
4. Overall, all the modified asphalt binders have shown comparable performance in terms of resistance to low temperature cracking.
5. A greater amount of modifier has resulted in a higher flow number, which indicated a better resistance to rutting. The addition of coupling agent also has remarkably increased the mixture stiffness that contributes to lower rutting potential.
6. The modifiers help in enhancing the ITS values of the wet samples. The combination of the coupling agent with 3% PECC is optimum for this study to avoid adverse due to moisture damage.

**Author Contributions:** Mohd Rosli Mohd Hasan (Data curation, Writing- original draft, Methodology) Zhanping You (Supervision, Project administration, Writing- review & editing), Mohd Khairul Idham Mohd Satar (Formal analysis, Writing- review & editing), Muhammad Naqiuddin Mohd Warid (Formal analysis, Writing- review &

editing), Nurul Hidayah Mohd Kamaruddin (Formal analysis, Writing- review & editing), Dongdong Ge (Formal analysis, Writing- review & editing), Ran Zhang (Formal analysis, Writing- review & editing).

**Acknowledgments:** The authors grateful to express their appreciation to Kenrich Petrochemicals, Inc. (Bayonne, NJ, USA), Specialty Minerals Inc. (Bethlehem, PA, USA), Payne & Dolan Inc. (Waukesha, WI, USA), and Dow Chemical Company (Midland, MI, USA) for donating test materials. The authors would like to acknowledge the research assistantships to Mohd Rosli Mohd Hasan, Mohd Khairul Idham Mohd Satar, Muhammad Naqiuddin Mohd Warid, Nurul Hidayah Mohd Kamaruddin, Ran Zhang, and Dongdong Ge. The authors also want to acknowledge Julia A. King from Department of Chemical Engineering of Michigan Technological University for her significant contributions in materials preparation, test design, and paper revision. It is impossible for the authors to complete the work without her effort. Any opinions, findings and conclusions expressed in this paper are those of the authors' and do not necessarily reflect the views of the official views and policies of any institution or company.

**Conflicts of Interest:** The authors declare no conflict of interest.

## References

1. You, Z.; Mills-Beale, J.; Foley, J.M.; Roy, S.; Odegard, G.M.; Dai, Q.; Goh, S.W. Nanoclay-modified asphalt materials: Preparation and characterization. *Constr. Build. Mater.* **2011**, *25*, 1072–1078. [CrossRef]
2. Ge, D.; Yan, K.; You, Z.; Xu, H. Modification mechanism of asphalt binder with waste tire rubber and recycled polyethylene. *Constr. Build. Mater.* **2016**, *126*, 66–76. [CrossRef]
3. Diab, A.; You, Z. Small and large strain rheological characterizations of polymer-and crumb rubber-modified asphalt binders. *Constr. Build. Mater.* **2017**, *144*, 168–177. [CrossRef]
4. Brown, E.R.; Kandhal, P.S.; Roberts, F.L.; Kim, Y.R.; Lee, D.Y.; Kennedy, T.W. *Hot Mix Asphalt Materials, Mixture Design, and Construction*; NAPA Research and Education Foundation: Auburn, AL, USA, 2009.
5. Chen, J.-S.; Liao, M.-C.; Shiah, M.-S. Asphalt modified by styrene-butadiene-styrene triblock copolymer: Morphology and model. *J. Mater. Civ. Eng.* **2002**, *14*, 224–229. [CrossRef]
6. Polacco, G.; Stastna, J.; Biondi, D.; Zanzotto, L. Relation between polymer architecture and nonlinear viscoelastic behavior of modified asphalts. *Curr. Opin. Colloid Interface Sci.* **2006**, *11*, 230–245. [CrossRef]
7. Kalantar, Z.N.; Karim, M.R.; Mahrez, A. A review of using waste and virgin polymer in pavement. *Constr. Build. Mater.* **2012**, *33*, 55–62. [CrossRef]
8. Munera, J.; Ossa, E. Polymer modified bitumen: Optimization and selection. *Mater. Des.* **2014**, *62*, 91–97. [CrossRef]
9. Jan, H.; Aman, M.Y.; Tawab, M.; Ali, K.; Ali, B. *Performance Evaluation of Hot Mix Asphalt Concrete by Using Polymeric Waste Polyethylene, Modeling, Simulation, and Optimization*; Springer: Newyork, NY, USA, 2018.
10. Nakhaei, M.; Naderi, K.; Nasrekani, A.A.; Timm, D.H. *Chemical, Rheological, and Moisture Resistance Properties of Warm Mix Asphalt Modified with Polyethylene-Way and Ethylene-Bis-Stearamide Additives*; Sage: Washington, DC, USA, 2018.
11. Polacco, G.; Berlincioni, S.; Biondi, D.; Stastna, J.; Zanzotto, L. Asphalt modification with different polyethylene-based polymers. *Eur. Polym. J.* **2005**, *41*, 2831–2844. [CrossRef]
12. Stastna, J.; Zanzotto, L.; Vacin, O. Viscosity function in polymer-modified asphalts. *J. Colloid Interface Sci.* **2003**, *259*, 200–207. [CrossRef]
13. Habib, N.Z.; Kamaruddin, I.; Napiah, M.; Tan, I.M. Rheological properties of polyethylene and polypropylene modified bitumen. *Int. J. Civ. Environ. Eng.* **2011**, *3*, 96–100.
14. Little, D.N.; Epps, J.A. *The Benefits of Hydrated Lime in Hot Mix Asphalt*; National Lime Association: Arlington, VA, USA, 2001.
15. Atikler, U.; Basalp, D.; Tihminlioğlu, F. Mechanical and morphological properties of recycled high-density polyethylene, filled with calcium carbonate and fly ash. *J. Appl. Polym. Sci.* **2006**, *102*, 4460–4467. [CrossRef]
16. Silva, R.V.; de Brito, J.; Dhir, R.K. Tensile strength behaviour of recycled aggregate concrete. *Constr. Build. Mater.* **2015**, *83*, 108–118. [CrossRef]
17. Lu, X.; Isacsson, U. Compatibility and storage stability of styrene-butadiene-styrene copolymer modified bitumens. *Mater. Struct.* **1997**, *30*, 618–626. [CrossRef]
18. Estevez, M. Use of coupling agents to stabilize asphalt–rubber–gravel composite to improve its mechanical properties. *J. Clean. Product.* **2009**, *17*, 1359–1362. [CrossRef]
19. Marzocchi, A.; Roberts, M.G.; Bolen, H.E.; Harrington, E.R. Chemically modified Asphalt Compositions. Patent US4301051A, 24 February 1978.

20. Hristov, V.; Vlachopoulos, J. Influence of coupling agents on melt flow behavior of natural fiber composites. *Macromol. Mater. Eng.* **2007**, *292*, 608–619. [CrossRef]
21. Yao, H.; You, Z.; Li, L.; Lee, C.H.; Wingard, D.; Yap, Y.K.; Shi, X.; Goh, S.W. Properties and chemical bonding of asphalt and asphalt mixtures modified with nanosilica. *J. Mater. Civ. Eng.* **2012**, *25*, 1619–1630. [CrossRef]
22. Martí-Ferrer, F.; Vilaplana, F.; Ribes-Greus, A.; Benedito-Borrás, A.; Sanz-Box, C. Flour rice husk as filler in block copolymer polypropylene: Effect of different coupling agents. *J. Appl. Polym. Sci.* **2006**, *99*, 1823–1831. [CrossRef]
23. Albano, C.; Gonzalez, J.; Ichazo, M.; Rosales, C.; de Navarro, C.U.; Parra, C. Mechanical and morphological behavior of polyolefin blends in the presence of $CaCO_3$. *Compos. Struct.* **2000**, *48*, 49–58. [CrossRef]
24. Sae-oui, P.; Sirisinha, C.; Thepsuwan, U.; Hatthapanit, K. Roles of silane coupling agents on properties of silica-filled polychloroprene. *Eur. Polym. J.* **2006**, *42*, 479–486. [CrossRef]
25. Chen, W.-S.; Chang, Y.-L.; Hsiang, H.-I.; Hsu, F.-C.; Yen, F.-S. Effects of titanate coupling agent on the dielectric properties of NiZn ferrite powders–epoxy resin coatings. *Ceram. Int.* **2011**, *37*, 2347–2352. [CrossRef]
26. Zhang, A.Q.; Wang, Y.M.; Zhuang, C.Y. Discussion the effect on improving granite bituminous mixture performance by adding coupling agent. *Appl. Mech. Mater.* **2012**, *204*, 4115–4118. [CrossRef]
27. Zhang, A.Q.; Li, J. Surface modification research for asphalt mixture filler. *Appl. Mech. Mater.* **2012**, *174*, 821–825.
28. Zhang, A.; Wang, Y.; Sun, S. Improving water stability of granite-asphalt interface by adding titanate ester coupling agent. *Int. Conf. Transp. Eng.* **2009**. [CrossRef]
29. Pinnavaia, T.J.; Beall, G.W. *Polymer-Clay Nanocomposites*; Wiley Online Library: Indianapolis, IN, USA, 2000.
30. Jahromi, S.G.; Khodaii, A. Effects of nanoclay on rheological properties of bitumen binder. *Constr. Build. Mater.* **2009**, *23*, 2894–2904. [CrossRef]
31. Sugerman, G.; Monte, S.J. Neoalkoxy Organo-Titanates and Organo-Zirconates Useful As Coupling and Polymer Processing Agents. U.S. Patent 4623738, 1986.
32. Kenrich Petrochemicals, Inc. *Product Data Sheet: Ken-React® CAPS® L® 12/L-Titanate Coupling Agent*; Kenrich Petrochemicals, Inc.: Bayonne, NJ, USA, 2011.
33. Krueger, Q.J.; King, J.A. Synergistic effects of carbon fillers on shielding effectiveness in conductive nylon 6, 6-and polycarbonate-based resins. *Adv. Polym. Technol.* **2003**, *22*, 96–111. [CrossRef]
34. Clingerman, M.L. *Development and Modelling of Electrically Conductive Composite Materials*; Michigan Technological University: Houghton, MI, USA, 2001.
35. Al-Abdul Wahhab, H.I. Effect of modifiers and additives on fatigue behavior of asphalt concrete mixes in the gulf. *Int. J. Pavement Res. Technol.* **2012**, *5*, 326–332.
36. Wahhab, H.A.-A.; Abaker, I. Performance modification of asphalt binders using thermoplastic polymers. *J. Eng. Res.* **2004**, *1*, 19–28. [CrossRef]
37. Fang, C.; Li, T.; Zhou, S.; Wang, X.; Heiliger, L. Study on the modification mechanism of asphalt with waste packaging polymer. *Transp. Dev. Innov. Best Pract.* **2008**. [CrossRef]
38. Yousefi, A.A. Rubber-polyethylene modified bitumens. *Iran. Polym. J.* **2004**, *13*, 101–112.
39. Asphalt Institute. *Superpave Performance Graded Asphalt Binder Specification and Testing*, 3rd ed.; Asphalt Institute: Lexington, KY, USA, 2003.
40. Asphalt Institute. *Superpave Performance Graded Asphalt Binder Specification and Testing*; Asphalt Institute: Lexington, KY, USA, 1995.
41. Asphalt Institute. *Superpave*; Asphalt Institute: Lexington, KY, USA, 1996.
42. Greene, J.; Choubane, B.; Chun, S. Evaluation of a heavy polymer modified binder through accelerated pavement testing. In Proceedings of the Transportation Research Board 92nd Annual Meeting, Washington, DC, USA, 13–17 January 2013.
43. Wang, C.; Zhang, J.X.; Song, P.P. Short-term ageing influence on the creep and recovery characteristics of asphalt binders. *Appl. Mech. Mater.* **2013**, *275*, 2045–2050. [CrossRef]
44. Behnood, A.; Olek, J. Rheological properties of asphalt binders modified with styrene-butadiene-styrene (SBS), ground tire rubber (GTR), or polyphosphoric acid (PPA). *Constr. Build. Mater.* **2017**, *151*, 464–478. [CrossRef]
45. Behnood, A.; Olek, J. Stress-dependent behavior and rutting resistance of modified asphalt binders: An MSCR approach. *Constr. Build. Mater.* **2017**, *157*, 635–646. [CrossRef]

46. Witczak, M.W.; Kaloush, K.; Pellinen, T.; El-Basyouny, M.; Quintus, H.V. Simple performance test for superpave mix design. In *National Cooperative Highway Research Program*; NCHRP Report 465; National Academy Press: Washington, DC, USA, 2002.
47. Carpenter, S.H.; Vavrik, W.R. Repeated triaxial testing during mix design for performance characterization. *Transp. Res. Rec.* **2001**, *1767*, 76–84. [CrossRef]
48. D'Angelo, J.A. The relationship of the MSCR test to rutting. *Road Mater. Pavement Des.* **2009**, *10*, 61–80. [CrossRef]
49. Zhou, F.; Scullion, T.; Sun, L. Verification and modeling of three-stage permanent deformation behavior of asphalt mixes. *Am. Soc. Civ. Eng.* **2004**. [CrossRef]
50. Monismith, C.L.; Ogawa, N.; Freeme, C.R. Permanent deformation characteristics of subgrade soils due to repeated loading. *Transp. Res. Rec.* **1975**, *537*, 1–17.
51. Faheem, A.F.; Bahia, H.U.; Ajideh, H. Estimating results of a proposed simple performance test for hot-mix asphalt from superpave gyratory compactor results. *Transp. Res. Rec.* **2005**, *1929*, 104–113. [CrossRef]
52. Witczak, M.W. Specification criteria for simple performance tests for rutting. In *National Cooperative Highway Research Program*; NCHRP Report 580; Transporation research board: Washington, DC, USA, 2007.
53. Dongre, R.; D'Angelo, J.; Copeland, A. Refinement of flow number as determined by asphalt mixture performance. *J. Transp. Res. Rec.* **2009**, *2*, 127–136. [CrossRef]

applied sciences

Article

# Rheological Behavior and Sensitivity of Wood-Derived Bio-Oil Modified Asphalt Binders

Junfeng Gao [1,2], Hainian Wang [1,*], Zhanping You [2,*], Mohd Rosli Mohd Hasan [3], Yong Lei [1] and Muhammad Irfan [4]

1 South Erhuan Middle Section, Chang'an University, School of Highway, Xi'an 710064, China; gaojunfeng@chd.edu.cn (J.G.); leiyong@chd.edu.cn (Y.L.)
2 Department of Civil and Environmental Engineering, Michigan Technological University, 1400 Townsend Drive, Houghton, MI 49931, USA
3 School of Civil Engineering, Universiti Sains Malaysia, Engineering Campus, Nibong Tebal, Pulau Pinang 14300, Malaysia; cerosli@usm.my
4 College of Civil Engineering, National University of Sciences and Technology (NUST) NUST Campus, Risalpur 24080, Pakistan; mirfan@mce.nust.edu.pk
* Correspondence: wanghainian@aliyun.com (H.W.); zyou@mtu.edu (Z.Y.)

Received: 21 April 2018; Accepted: 30 May 2018; Published: 3 June 2018

**Featured Application: This study of rheological behavior and sensitivity of wood-derived bio-oil modified asphalt binders would lay a foundation for the performance research and further engineering practice of bio-asphalt binders.**

**Abstract:** The demand for bituminous materials is continuously growing; crude oil-based asphalt binders are non-renewable, and are facing rapid depletion. With the increase of petroleum-based asphalt prices, seeking an alternative, renewable material such as bio-asphalt has become a hot research topic. However, shortcomings in this research area have been identified, notably concerning the high-temperature performance of bio-asphalt at present. This research aims to comprehensively apply conventional tests to, and study the rheological behavior of, the high-temperature performances of bio-asphalt binders, i.e., by temperature and frequency sweeps, using a dynamic shear rheometer (DSR). It will also assess the chemical functional groups of specimens prepared by different aging conditions. Fifty penetration grade base asphalt binder (50#), bio-oil modified asphalt binders with 0%, 5%, 10%, and 30% bio-oil contents by mass, and bio-oil modified asphalt binder with combinations of 5% bio-oil-1% SBS, and 10% bio-oil-1% SBS were used in this study. The conventional performance of bio-asphalt binders was tested using penetration, ductility, and softening point, before and after short-term aging conditioning. The temperature sweep and frequency sweep of bio-asphalt under different bio-oil contents were carried out via DSR. Two-logarithmic equations of rutting factor and temperature were established, and the temperature sensitivity of bio-asphalt was analyzed. The master curves of virgin asphalt and bio-asphalt were constructed at 64 °C. The results indicate that the incorporation of bio-oil reduced the anti-rutting performance of asphalt, and the bio-oil content had a significant effect on the mass loss of the bio-asphalt binder. The performance of bio-oil modified asphalt binders using 5% bio-oil, 5% bio-oil-1% SBS, and 10% bio-oil-1% SBS, could meet the requirements of 50# grade asphalt. The temperature sensitivity of bio-asphalt did not show obvious change before and after short-term aging, whereas the temperature sensitivity of bio-asphalt with 5% bio-oil was relatively small. With an increase in temperature, the phase angle increased gradually. In contrast, the storage modulus, loss modulus, and complex modulus decreased progressively. The complex modulus and rutting factor of bio-asphalt with 5% bio-oil steadily increased with the increase in testing frequency. Otherwise, chemical reactions were detected in the 50# base asphalt modified with the bio-oil.

**Keywords:** road engineering; bio-oil; asphalt binder; high-temperature performance; master curve

## 1. Introduction

Nowadays, the binding agent for road materials is mainly petroleum asphalt extracted from the fossil fuel. With the gradual reduction of oil reserves, oil prices are rising, which results in a higher cost of bitumen. With the growing demand for asphalt, the construction and maintenance costs of highways have also escalated. Therefore, the performance of traditional asphalt must be improved, and the need for alternatives is great [1–3]. Biomass energy has a wide range of sources, large yields, and low prices. In recent years, a significant number of researchers have turned their attention to bio-oil extracted from biomasses by high-temperature pyrolysis [4–6].

Many researchers mainly use bio-oil as an additive to modify the base binder. However, most were only focused on the performance of bio-asphalt containing a mixture of base binder and bio-oil. Fini et al. [7–11] modified petroleum asphalt with 2%, 5%, and 10% bio-oil by mass of the asphalt binder. Bio-oil was extracted from pig manure through a process of pyrolysis. Bio-oil modified binders were tested for high-temperature viscosity, and the low and high temperature rheological behavior was examined using a rotational viscometer (RV), a dynamic shear rheometer (DSR), and a bending beam rheometer (BBR). Mills-Beale et al. [12] tested 5% bio-oil extracted from pig manure to modify petroleum asphalt PG 64-22; performance was tested by RV, DSR, and BBR, before and after short-term aging. Raouf et al. [13,14] obtained bio-oil from oak wood, and pretreated it for 2 h, then mixed 2% and 4% dosages of polyethylene to modify it. They then determined the viscosity within a temperature range of 40~160 °C, and analyzed the effect of temperatures on viscosity. Tang et al. [15] applied three types of bio-oil extracted from corn stalks, oak trees, and shredded grass, to modify base asphalt and SBS modified asphalt. The dosages of each bio-oil were 3%, 6%, and 9%, based on binder mass. The high temperature performances of bio-asphalt before and after short-term aging were then tested, and the anti-aging properties were analyzed. Yang et al. [16] tested high-temperature viscosity, density, and the rheological properties of bio-asphalt with different contents of bio-oil extracted from cedar sawdust. The 25% and 50% bio-oils were utilized to replace part of the AC-20 base binder, while the 2% and 8% bio-oil were used for the modification of a similar binder. Wang, Gao et al. [17–19] summarized and compared the bio-oil production procedure and fabrication process of bio-asphalt. They then analyzed the performance and characterization of different bio-binders. They also studied variations in viscosity at different temperatures and with different contents of bio-oil. The results showed that the viscosity of the bio-binder decreased with an increase in the content of bio-oil when the test temperature was under 135 °C. He et al. [20] investigated the conventional performances of bio-asphalt by penetration, softening point, and ductility. The test results indicated that, using a mixture of bio-oil and modified asphalt, the penetration declined with an increase in the proportion of modified asphalt, and ductility improved greatly, which corrected with the disadvantages of hard and brittle bio-asphalts at low temperatures.

At present, the research of bio-asphalt mainly focuses on the binder materials, and tests mainly focus on the conventional indexes of the binder materials; such studies lack systematic analysis. The study of bio-asphalts at different frequencies is also rarely reported. In this study, bio-oil was extracted from wood chips, and a bio-asphalt binder was prepared with petroleum asphalt with different contents of bio-oil. The conventional performance indexes of the bio-asphalt binder were tested. The temperature and frequency sweep tests of bio-asphalt with different bio-oil contents were carried out by DSR. The two-logarithmic equations, rutting factor and temperature, were established, and the temperature sensitivity of bio-asphalt was analyzed. The master curves of the asphalt and bio-asphalt matrix were recorded at 64 °C; this was selected as the reference temperature. Changes of complex modulus and rutting factors of bio-asphalt under different loading frequencies were studied. This study may provide important fundamental information for the further study of the high-temperature performance of bio-asphalt binders and the selection of evaluation indices, and would lay the foundation for the further engineering applications of bio-asphalts.

## 2. Objective

The specific objectives of this study are as follows:

(1) To develop a new, environmentally friendly, and sustainable asphalt material for road engineering.
(2) To systematically study the high-temperature performance of bio-asphalt binders.
(3) To determine the changes of complex modulus and rutting factors of bio-asphalts at a wide range of loading frequencies.

The experimental plan followed in this study is shown in Figure 1.

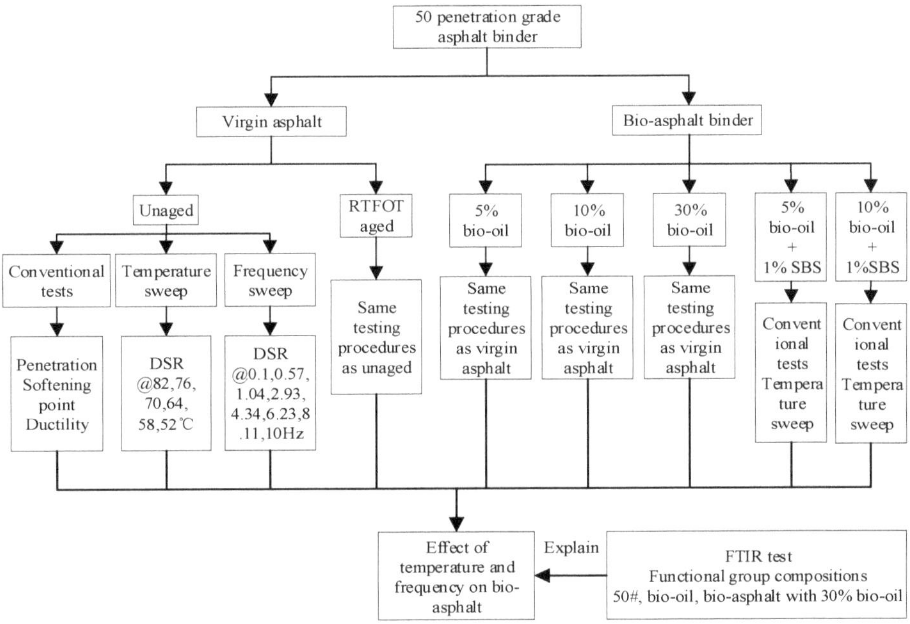


**Figure 1.** Experimental design map.

## 3. Materials and Test Program

### 3.1. Materials and Preparation of Bio-Asphalt Binder

#### 3.1.1. Materials

Fifty-penetration grade asphalt (hereafter referred to as 50#) produced by Sinopec Group Maoming Petrochemical Company, Mao Ming, China, was selected as the base asphalt binder. Bio-oil was provided by Toroyal New Energy Company, Dongying, China; it was extracted from wood chip, and is dark brown in color and shows plasticity at room temperature, with a certain mobility at a higher temperatures. The density of the bio-oil is 1.1 g/cm$^3$, and the pH is 2.6. Its elemental composition is: 54–56% C, 35–45% O, 5.5–7.2% H, and 0–0.2% N. The bio-oil is shown in Figure 2. The SBS additive is produced by Yuehua Company, Yueyang, China; it has a 1301-1 linear structure.

**Figure 2.** Bio-oil derived from the wood chip.

#### 3.1.2. Preparation of Bio-Asphalt Binders

A preparation program was determined according to a previous study [19] and former tests. Asphalt binders with bio-oil/ SBS additives were prepared by different processes.

Asphalt binder with bio-oil: The 50# grade matrix asphalt was heated to 140 to 145 °C. Different contents of bio-oil were added to the base asphalt. The contents of the bio-oil were 5%, 10%, and 30% of the total binder by weight. The base binder and bio-oil were mixed with a high shear mixer for 20 min, with a rotation speed of 5000 r/min.

Asphalt binder with SBS additive and bio-oil: The 50# grade matrix asphalt and SBS were mixed at 180 °C through a high-speed shear machine for 15 min. Then, the bio-oil was added to the mixed binders, and mixed for 20 min; the temperature was kept at 140~145 °C. The speed of the shear machine was 3000 r/min [21]. A 1% content by of SBS total binder by weight was selected. Bio-oil contents were 5% and 10% of the total binder by weight, and the prepared bio-asphalt with SBS additive were named 5%-S and 10%-S.

The asphalt binder with bio-oil, and the asphalt binder with SBS additive and bio-oil were both named bio-asphalt binder.

### *3.2. Test methods and Master Curve Generation Method*

#### 3.2.1. Conventional Test Method

Conventional tests of the prepared bio-asphalt binders were carried out, including penetration, softening point, and ductility, before and after Rolling Thin Film Oven (RTFO) aging, and mass loss. The tests were conducted based on the Standard Test Methods of Bitumen and Bituminous Mixtures for Highway Engineering (JTG E20-2011).

#### 3.2.2. Temperature Sweep Test Method

The temperature sweep of the bio-asphalt binders with 5%, 10%, and 30% bio-oil, and 5%-S, 10%-S, before and after Rolling Thin Film Oven (RTFO) aging, was carried out using a dynamic shear rheometer DHR-1, manufactured by the TA Company. Continuous sinusoidal alternating load and strain control mode were selected; the strains were 12% and 10% before and after RTFO. The temperature sweep range was 52~82 °C; the interval was 6 °C. Test frequency was 1.59 Hz. The diameter of the asphalt sample fixture was 25 mm, while the test spacing of asphalt sample was 1 mm. The complex shear modulus $G^*$, phase angle $\delta$, and the rutting factors $G^*/\sin\delta$ of the bio-asphalt binder, were measured and analyzed.

#### 3.2.3. Frequency Sweep Test Method

After conventional tests and temperature sweep, the matrix asphalt and the selected bio-asphalt were subjected to a frequency sweep test. The test conditions, including strain control mode, the diameter of asphalt sample fixture, the test spacing of asphalt sample, were the same as those of the temperature sweep test. Temperature range for the frequency sweep was 40~76 °C; the interval was 6 °C. Test frequencies were 0.1 Hz, 0.57 Hz, 1.04 Hz, 2.93 Hz, 4.34 Hz, 6.23 Hz, 8.11 Hz, and 10 Hz, respectively.

#### 3.2.4. Master Curve Generation Method

The viscoelasticity of viscoelastic materials has a certain dependence on temperature and loading frequency. At different temperatures and frequencies of action, the viscoelastic materials might exhibit the same mechanical behaviors, that is, the effects of time and temperature on the viscoelastic material are equivalent. The viscoelastic curve obtained at different temperatures could be converted into the viscoelastic master curve at the reference temperature through the time-temperature equivalent principle; in this way, the scanning results could be extended to the broadband range.

In this paper, a sigmoidal function was used to construct the master curve of complex modulus and rutting factor, by means of Excel solver; the formula is shown in Equation (1).

$$\log(A) = \delta + \frac{\alpha}{1 + e^{\beta - \gamma \log(\xi)}} \tag{1}$$

where $A$ is complex modulus or rutting factors, which is the minimum value of complex modulus or rutting factor, $\delta$ is the reduced frequency at the reference temperature, $\alpha$ is the difference between the maximum and minimum values of complex modulus or rutting factor, and $\beta$, $\gamma$ are the shape parameters.

#### 3.2.5. Fourier-Transform Infrared Spectroscopy

The Fourier-transform infrared spectroscopy (FTIR) is a machine used to determine the spectroscopy; it is widely used to analyze the chemical functional groups of asphalt materials. The TENSOR27, produced by the Bruker Optics Company, was used to research the asphalts and bio-asphalts in this study. The resolution was 4 $cm^{-1}$, the scanning speed was 32 sheets per second, and the selected scanning range was 4000~600 $cm^{-1}$. The functional groups were identified and compared for further analysis.

## 4. Results and Discussion

### *4.1. Conventional Tests*

The results of conventional performance indicators of bio-asphalt binders with 5%, 10%, and 30% bio-oil, and 5%-S, 10%-S, were analyzed.

The penetration ratio and residual penetration ratio of unaged and RTFO-aged bio-asphalt are shown in Figure 3. For the unaged bio-asphalt, with an increase of bio-oil content, the penetration of bio-asphalt increases. This showed that, with the incorporation of bio-oil, asphalt became soft; its high temperature performance was reduced to some extent. Meanwhile, the penetration of 5%-S and 10%-S were decreased by 5% and 10% respectively, due to the incorporation of SBS. Compared with the matrix asphalt 50#, the penetration increments of unaged bio-asphalt with 5%, 10%, and 30% bio-oil were 5%, 12.3%, and 79.2%, respectively. This indicated that a higher bio-oil content produces a greater change in penetration. After short-term aging, the bio-asphalt became harder, and resistance to rutting increased for the RTFO-aged bio-asphalt. The change in the residual penetration ratio also illustrated this point. Compared with RTFO aged bio-asphalt with 5% and 10% content bio-oil, the penetration of RTFO-aged bio-asphalt with 30% bio-oil was much smaller, and the residual penetration ratio was 36.73%. This was due to the effects of the volatility of the light components of bio-oil during RTFO aging. An increase in bio-oil content would lead to an increase in the aging and variability of bio-asphalt. The residual penetration ratio of bio-asphalt with 5% content bio-oil and 5%-S, 10%-S could meet the requirements of 50# asphalt binder, as outlined in Technical Specifications for Construction of Highway Asphalt Pavements (JTG F40-2004). However, the residual penetration ratio of bio-asphalts with 10% and 30% content bio-oil would not meet the requirements.

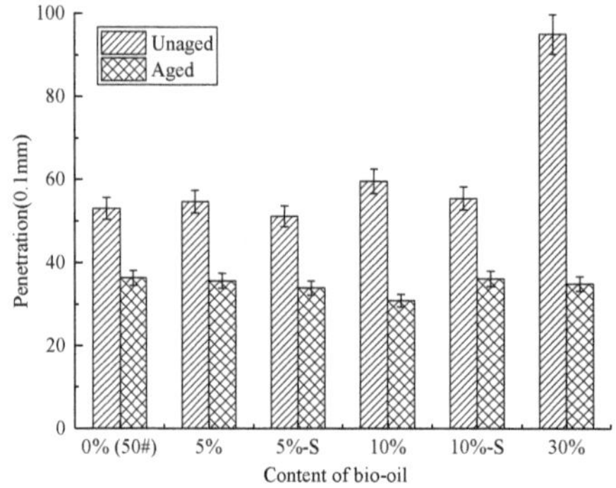


(a)

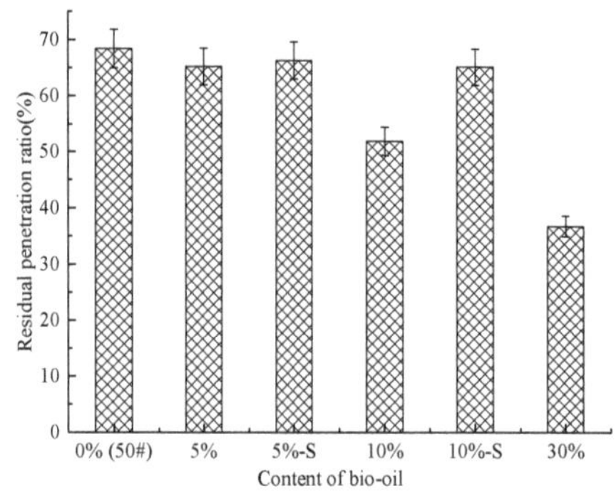


(b)

**Figure 3.** Penetration ratio and residual penetration ratio of unaged and RTFO (Rolling Thin Film Oven) aged bio-asphalt. (**a**) Penetration ratio; (**b**) Residual penetration ratio.

The softening points of unaged and RTFO-aged bio-asphalt are shown in Figure 4. The softening point of unaged bio-asphalt decreased with an increase of bio-oil content. This showed that, with the incorporation of bio-oil, the high temperature performance of bio-asphalt was reduced to some extent. Meanwhile, the addition of SBS could increase the softening point of bio-asphalt with 5% and 10% bio-oil. The softening point of RTFO-aged bio-asphalt increased with an increase of bio-oil content. This illustrated that, with the incorporation of bio-oil, the asphalt became harder, and the resistance to rutting is increased. The softening point of unaged and RTFO-aged 10%-S increased more than that of 5% bio-oil. The incorporation of SBS could improve the high temperature performance of bio-asphalt. Compared with the original asphalt, the difference of the softening point of unaged and RTFO-aged bio-asphalts with 5%, 10%, and 30% bio-oil content were 9.6 °C, 16.5 °C, and 20.6 °C, respectively. With an increase of bio-oil content, the softening point increased. This was caused by the aging of bio-asphalt, and indicated that with a greater amount of bio-oil, the aging degree of bio-asphalt increases.

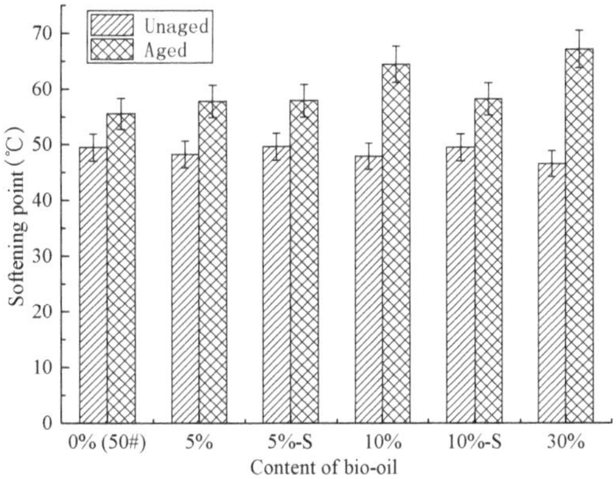


**Figure 4.** Softening point of unaged and RTFO aged bio-asphalt.

As seen in Figure 5, with the increase of bio-oil content, the ductility of unaged bio-asphalt increased. As compared with the base asphalt binder, the ductility of bio-asphalts with 5%, 10%, and 30% bio-oil content increased by 2.6%, 8.7%, and 52.8%, respectively. This showed that the addition of bio-oil could improve the low temperature performance of bio-asphalt, and improve the low temperature, anti-cracking ability of asphalt before RTFO-aging. Meanwhile, the ductilities of 5%-S and 10%-S were increased by 16.5% and 13.2%, respectively, indicating that the incorporation of SBS would improve the low temperature properties of the bio-asphalt binder. Referring to the RTFO-aged bio-asphalt, with an increase in bio-oil content, ductility decreased, indicating that the asphalt became relatively hard after short-term aging. However, for 5%-S and 10%-S, the ductility indicated that SBS could also improve the low temperature properties of RTFO-aged bio-asphalt binders.

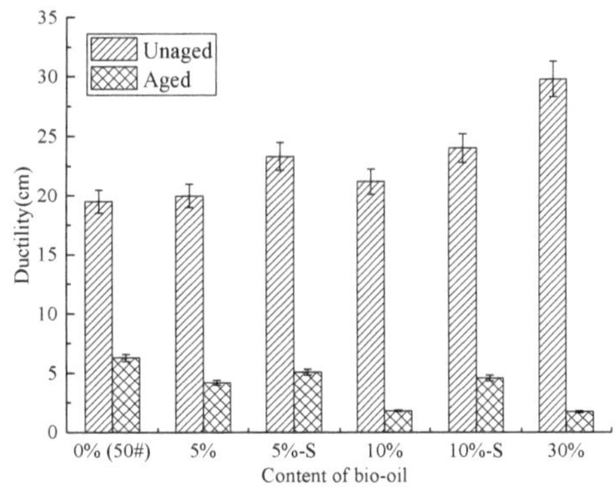


**Figure 5.** Ductility of unaged and RTFO aged bio-asphalt.

Mass losses between unaged and RTFO-aged bio-asphalt are shown in Figure 6. With the increase of bio-oil content, the mass loss of bio-asphalt increased. This showed that the addition of bio-oil made the mass of bio-asphalt decrease at 163 °C. This was due to the volatilization of light components, and the aging of the asphalt. Meanwhile, the greater the amount of bio-oil, the more aging occurred. However, for 5%-S and 10%-S, the mass loss was lower than 5% and 10% of the bio-oil content, which means that the incorporation of SBS would decrease the aging of bio-asphalt. The mass loss of bio-asphalt with 5% bio-oil content was −0.580%, which could meet the requirements described in the Technical Specifications for Construction of Highway Asphalt Pavements (JTG F40-2004). The 5%-S and 10%-S could also meet these specifications. However, the mass loss of bio-asphalt with 10% and 30% content bio-oil would not meet the requirements.

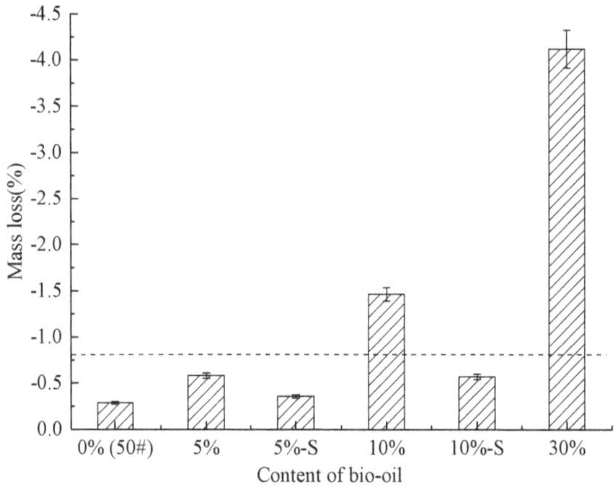


**Figure 6.** Mass loss between unaged and RTFO aged bio-asphalt.

It was found that the incorporation of bio-oil had a great influence on the mass loss of the bio-asphalt, but that the effect of different dosages was unknown. To identify the difference between the different types of asphalt, and to understand the significant changes in the mass of bio-asphalt with varying bio-oil contents, the LSD (Least Significant Difference) method was used. The LSD method is the most sensitive of the various multi-comparison methods. It is able to detect small differences between groups. The analysis through LSD of the effects of different bio-oil contents on mass loss is shown in Table 1.

**Table 1.** Analysis through LSD (Least Significant Difference) of the effects of different bio-oil contents on mass loss.

| Content of Bio-Oil | Content of Bio-Oil | Mean Difference | Standard Error | Significance | 95% Confidence Interval | |
|---|---|---|---|---|---|---|
| | | | | | Lower Limit | Upper Limit |
| 0% | 5% | −0.294000 * | 0.052726 | 0.001 | −0.41559 | −0.17241 |
| | 10% | −1.180000 * | 0.052726 | 0.000 | −1.30159 | −1.05841 |
| | 30% | −3.836000 * | 0.052726 | 0.000 | −3.95759 | −3.71441 |
| 5% | 0% | 0.294000 * | 0.052726 | 0.001 | 0.17241 | 0.41559 |
| | 10% | −0.886000 * | 0.052726 | 0.000 | −1.00759 | −0.76441 |
| | 30% | −3.542000 * | 0.052726 | 0.000 | −3.66359 | −3.42041 |
| 10% | 0% | 1.180000 * | 0.052726 | 0.000 | 1.05841 | 1.30159 |
| | 5% | 0.886000 * | 0.052726 | 0.000 | 0.76441 | 1.00759 |
| | 30% | −2.656000 * | 0.052726 | 0.000 | −2.77759 | −2.53441 |
| 30% | 0% | 3.836000 * | 0.052726 | 0.000 | 3.71441 | 3.95759 |
| | 5% | 3.542000 * | 0.052726 | 0.000 | 3.42041 | 3.66359 |
| | 10% | 2.656000 * | 0.052726 | 0.000 | 2.53441 | 2.77759 |

Note: * The significance level of the mean difference is 0.05.

According to the results of the LSD analysis, it could be seen that the significance between the various asphalts was less than 0.05, which indicates that the change of bio-oil content had a significant effect on the mass loss.

### 4.2. Temperature Sweep Test

The temperature sweep tests of unaged and RTFO-aged matrix asphalts and bio-asphalt binders were carried out, and the complex modulus G*, phase angle δ, and rutting factor G*/sin δ were analyzed.

As seen in Figure 7a,b, for unaged asphalt, the phase angle of bio-asphalts with 5% and 10% bio-oil content, and 10%-S, gradually increased with the increase of temperature; this indicated that the viscous component increased with increasing temperature. However, the phase angle of 50# matrix asphalt and 5%-S reached a maximum at 64 °C, indicating that the viscous components also reached a maximum at that temperature. The phase angle of bio-asphalt with 30% content bio-oil had a smaller downward trend; this was due to the excessive variability of the bio-oil. Compared with 50# base asphalt, the phase angles of unaged bio-asphalt with 5% and 10% bio-oil were both higher than the base asphalt, while the phase angle of bio-asphalt with 30% bio-oil was lower than that of the base asphalt. This indicated that the viscous components of bio-asphalt increased with the increase of the 5% and 10% bio-oil content, and that the high-temperature performance was affected by the bio-oil. The bio-asphalt with 30% bio-oil content reduced; this was caused by the extreme variability of the 30% bio-oil. For the RTFO-aged asphalt, the phase angle of bio-asphalt increased with the increase of temperature, and the phase angles of bio-asphalt with 5% and 10% content bio-oil were lower than those of the matrix asphalt when the temperature was lower than 70 °C, while they were higher when the temperature was greater than 70 °C. This indicated that when the temperature was greater than 70 °C, the incorporation of 5% and 10% of the bio-oil would increase the viscosity of the bio-asphalt. Meanwhile, the phase angle of the bio-asphalt with 30% bio-oil content was higher than that of the matrix asphalt and of the bio-asphalts with 5% and 10% content bio-oil, indicating that it had a larger viscous component, and that it had lower high temperature performance. The phase angles of bio-asphalt 5%-S and 10%-S were lower than those of other bio-asphalts, due to the elasticity of SBS.

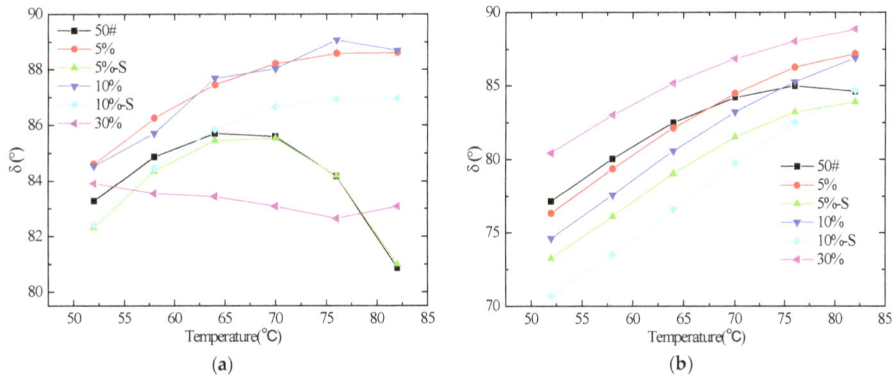


**Figure 7.** The relationship between phase angle and temperature. (**a**) Unaged; (**b**) RTFO aged.

As shown in Figure 8a,b, for the unaged and RTFO-aged asphalts, the complex modulus of bio-asphalt decreased with the increase of temperature; this indicated that, with the increase of temperature, the asphalt became soft, and its resistance to rutting declined. The sequence of complex modulus for the unaged and RTFO-aged asphalt were G* (50#) > G* (5%) = G* (10%) > G* (30%), G* (30%) > G* (10%) > G* (5%) = G* (50#), respectively. The results showed that the complex modulus of bio-asphalts without SBS decreased with an increase of bio-oil content. After short-term aging, the sequence was opposite to that of the unaged asphalt. This indicated that the bio-asphalt became softer with the addition of bio-oil for the unaged, and harder for the RTFO-aged; the results were consistent with those of the conventional tests. This was due to the aging of the bio-asphalt after the addition of significant quantities of bio-oil. The modulus of 5%- and 10%-S were improved more than those of the bio-asphalts with 5% and 10% bio-oil, but the action of SBS was different, based on the bio-oil content and RTFO aging.

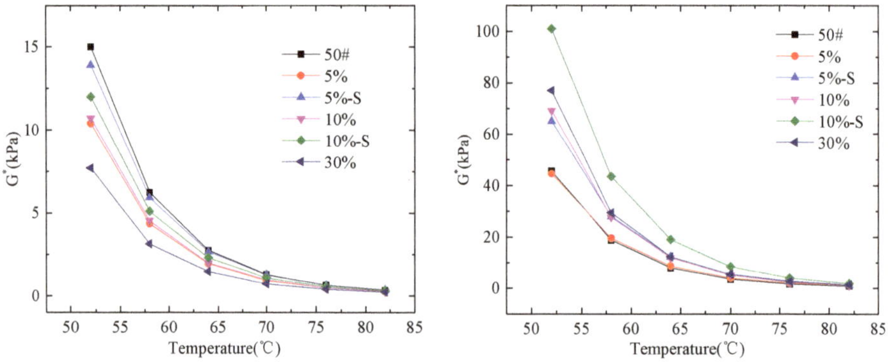


(**a**) (**b**)

**Figure 8.** The relationship between complex modulus and temperature. (**a**) Unaged; (**b**) RTFO aged.

Figure 9a,b shows the relationship between rutting factors and temperature. The changes of the rutting factors of bio-asphalt and matrix asphalt were relatively consistent with the changes of complex modulus. The sequence of rutting factors for the unaged and RTFO-aged asphalts were G*/sinδ (50#) > G*/sinδ (5%) = G*/sinδ (10%) > G*/sinδ (30%) and G*/sinδ (30%) > G*/sinδ (10%) > G*/sinδ (5%) = G*/sinδ (50#), respectively. The changes were also consistent with the changes of complex modulus. This was caused by the aging of bio-asphalt, especially for the bio-asphalt with 10% and 30% bio-oil content.

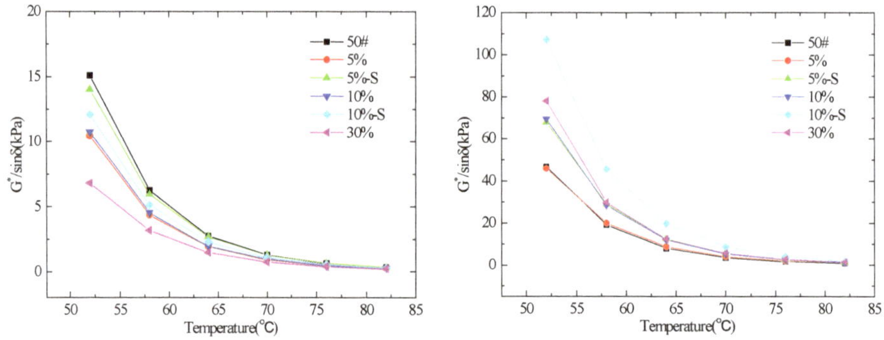


(**a**) (**b**)

**Figure 9.** The relationship between rutting factors and temperature. (**a**) Unaged; (**b**) RTFO aged.

### 4.3. *Temperature Sensitivity Analysis*

Asphalt is a viscoelastic-plastic material whose properties are affected by temperature. In detail, the rutting factors decrease with an increase in temperature. The temperature sensitivity is defined as the gradient of the rutting factors when the temperature changes. In order to analyze the temperature sensitivity of bio-asphalt, a line was utilized to describe the relationship between ln(G*/sin(δ)) and ln(T). A higher slope of the fitted line means that a higher temperature sensitivity of the asphalt is observed.

Figure 10a,b shows the linear regression between ln(G*/sin(δ)) and ln(T). As can be seen in Figure 10, ln(G*/sin(δ)) and ln(T) were linear, and the fitted formulas were obtained; they are shown in Tables 2 and 3.

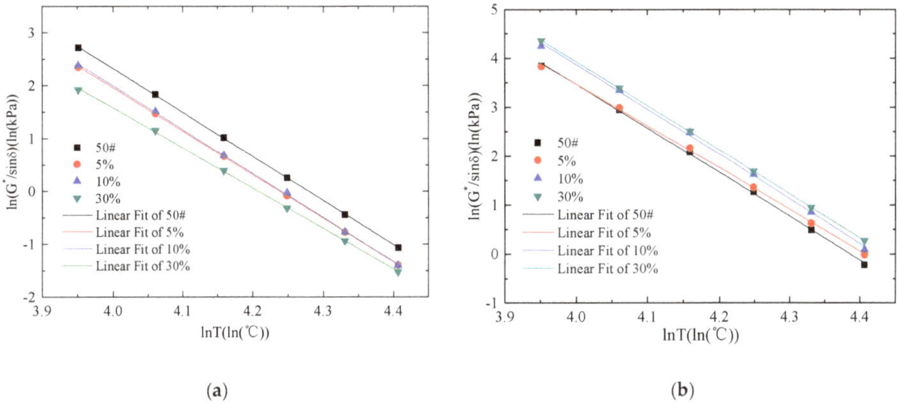


**Figure 10.** Linear regression between ln (G*/sinδ) and lnT. (**a**) Unaged; (**b**) RTFO-aged.

**Table 2.** Linear fitting results of different asphalt temperature sweep data for the unaged asphalts.

| Asphalt Type | Equation | $R^2$ |
|---|---|---|
| 50#-base asphalt | $\ln(G^*/\sin\delta) = -8.323\ln T + 35.616$ | 0.99991 |
| 5%-bio-asphalt | $\ln(G^*/\sin\delta) = -8.230\ln T + 34.876$ | 0.99994 |
| 10%-bio-asphalt | $\ln(G^*/\sin\delta) = -8.306\ln T + 35.218$ | 0.99964 |
| 30%-bio-asphalt | $\ln(G^*/\sin\delta) = -7.598\ln T + 31.966$ | 0.99965 |

**Table 3.** Linear fitting results of different asphalt temperature sweep data for the RTFO-aged asphalt.

| Asphalt Type | Equation | $R^2$ |
|---|---|---|
| 50#-base asphalt | $\ln(G^*/\sin\delta) = -8.962\ln T + 39.312$ | 0.99907 |
| 5%-bio-asphalt | $\ln(G^*/\sin\delta) = -8.526\ln T + 37.570$ | 0.99906 |
| 10%-bio-asphalt | $\ln(G^*/\sin\delta) = -9.128\ln T + 40.376$ | 0.99848 |
| 30%-bio-asphalt | $\ln(G^*/\sin\delta) = -8.989\ln T + 39.882$ | 0.99998 |

Note: RTFO is short for Rolling Thin Film Oven.

As can be seen from Tables 2 and 3, all $R^2$ values were very close to 1, indicating very good linear fitting. Thus, the rutting factor logarithm of the matrix asphalt and bio-asphalt had good linear correlation with the temperature logarithm. The linear correlation could be characterized by the Equation (2):

$$\ln (G^*/\sin\delta) = A \ln T + B \quad (2)$$

wherein, $A < 0$, $B > 0$; A and B could indicate the temperature sensitivity. As such, they were regarded as temperature sensitivity parameters, and the smaller the value of $|A|$, the lower the sensitivity of the asphalt to temperature.

Results could be drawn from Tables 2 and 3: the sequence of $|A|$ for the unaged was $|A|$ (50#) > $|A|$ (10%) > $|A|$ (5%) > $|A|$ (30%), the sequence of $|A|$ for the RTFO aged bio-asphalt was $|A|$ (10%) > $|A|$ (30%) > $|A|$ (50#) > $|A|$ (5%). For the unaged asphalt, the sensitivity did not show regularity, but the temperature sensitivity of bio-asphalt was lower than that of matrix asphalts. Among them, the temperature sensitivity of bio-asphalt with 30% content bio-oil was the lowest. For the RTFO-aged asphalt, the temperature sensitivity of bio-asphalt was not regular, but when the bio-oil content was higher than 10%, the temperature sensitivity was higher than that of the matrix asphalt. The temperature sensitivity of RTFO-aged bio-asphalt with 5% content bio-oil was the lowest.

### 4.4. Frequency Sweep Test

The bio-asphalt with 5% bio-oil and base asphalt were selected for the frequency sweep test, according to the results of conventional tests and the temperature sweep test. The changes of phase angle, storage modulus, loss modulus, and complex modulus with frequency for the unaged and RTFO-aged were analyzed.

Figure 11 shows the changes of phase angle, storage modulus, loss modulus, and complex modulus of unaged bio-asphalt with 5% bio-oil content. As can be seen in Figure 11, for the unaged bio-asphalt with 5% content bio-oil, with the increase of frequency, the phase angle reduced gradually at the same temperature, while the storage modulus, loss modulus, and complex modulus all increased gradually. This indicated that the asphalt exhibited greater elastic properties as the frequency increased, while the phase angle and the storage modulus had a certain fluctuation at 76 °C, which may be a measurement error caused by the increase in temperature. In the case of lower frequencies, the phase angle, storage modulus, loss modulus, and complex modulus varied greatly with the change of frequency. When the frequency was higher than 5 Hz, the viscoelastic curve stabilized gradually. As the temperature increased, the phase angle increased gradually, and the storage modulus, loss modulus, and complex modulus decreased gradually and the viscosity component of the bio-asphalt increased; this was consistent with the results of the temperature sweep.

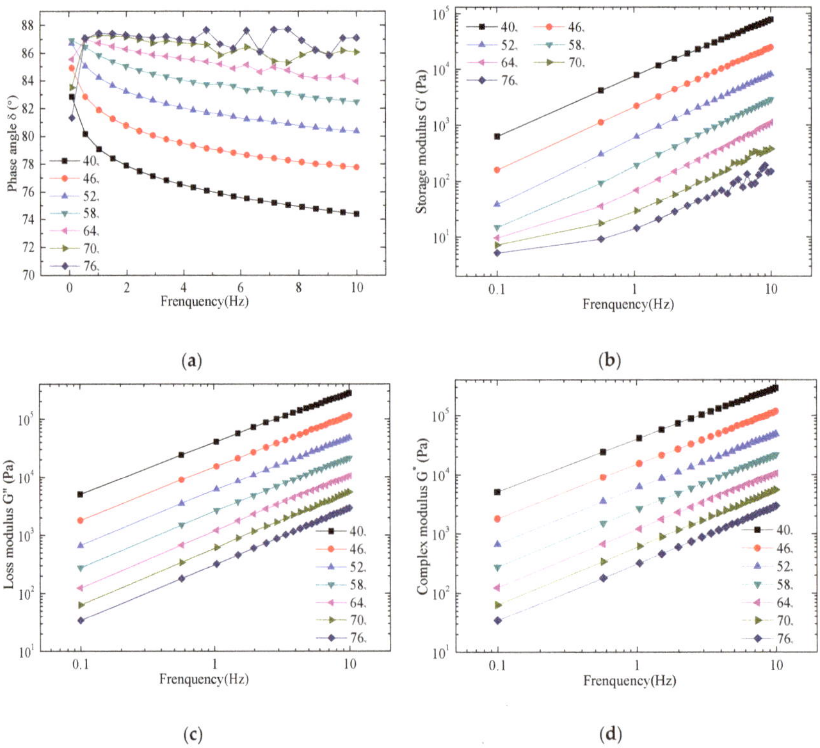


**Figure 11.** The changes of different parameters of unaged bio-asphalt with 5% content bio-oil. (**a**) Phase angle; (**b**) Storage modulus; (**c**) Loss modulus; (**d**) Complex modulus.

Figure 12 shows the changes of phase angle, storage modulus, loss modulus, and complex modulus of RTFO-aged bio-asphalt with 5% content bio-oil. It can be seen that the changes of phase angle, storage modulus, loss modulus and complex modulus of asphalt were similar to those of the unaged asphalt.

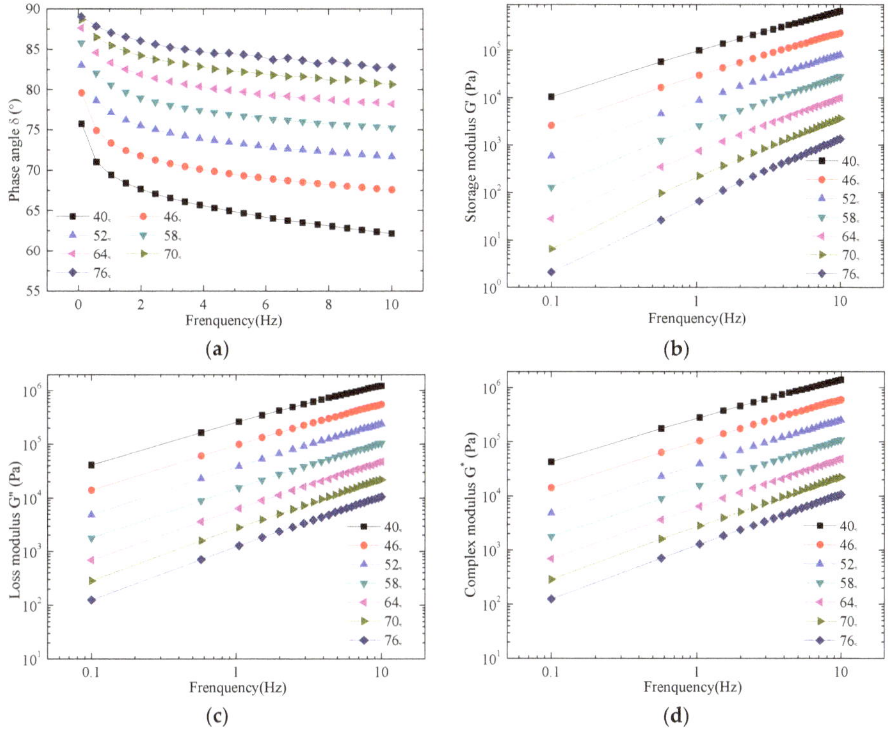


**Figure 12.** The changes of different parameters of RTFO-aged bio-asphalt with 5% bio-oil content. (**a**) Phase angle; (**b**) Storage modulus; (**c**) Loss modulus; (**d**) Complex modulus.

### *4.5. Master Curve Generation*

The master curve of the complex modulus and the rutting factors of the base asphalt and the bio-asphalt with 5% bio-oil were generated in the broadband range using the time-temperature equivalent principle. The reference temperature was 64 °C.

As seen in Figure 13a,b, unaged bio-asphalt with 5% bio-oil content, and 50# matrix asphalt, had a similar trend from a low to high-frequency range, and the complex modulus and rutting factors increased with the increase of frequency. In the lower frequency range, the complex modulus and rutting factors of unaged bio-asphalt with 5% bio-oil were consistent with those of the 50# base asphalt, indicating that the bio-asphalt with 5% bio-oil had the same resistance to rutting as the 50# base asphalt. In the higher frequency range, the complex modulus and rutting factors of the unaged bio-asphalt with 5% bio-oil reduced to some extent, compared with 50# base asphalt, which indicated that it was superior to matrix asphalt in low-temperature, anti-cracking performance in the higher frequency range.

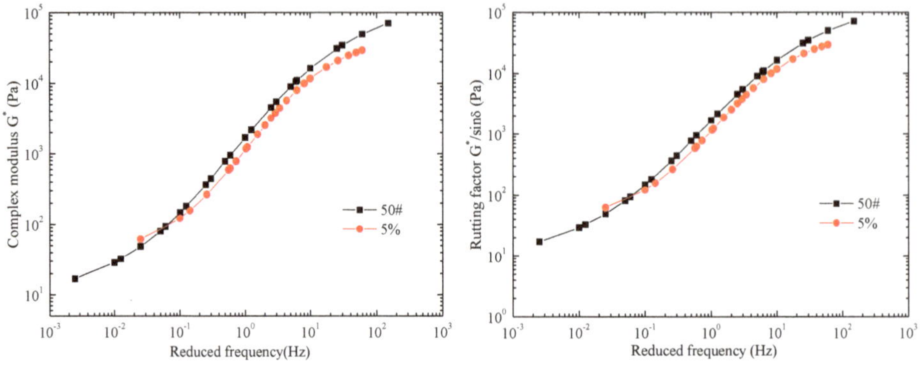


(**a**) (**b**)

**Figure 13.** The changes of different parameters with reduced frequency change for the unaged asphalt. (**a**) Complex modulus; (**b**) Rutting factor.

As shown in Figure 14a,b, RTFO-aged bio-asphalt with 5% content bio-oil and 50# matrix asphalt had a similar trend from low to high frequency range, and the complex modulus and rutting factors increased as the frequency increased. In the lower frequency range, the complex modulus and rutting factors of RTFO-aged bio-asphalt with 5% bio-oil were larger than those of the 50# base asphalt, which indicated that bio-asphalt had a better ability to resist rutting deformation. In the medium frequency range, the complex modulus and rutting factors of RTFO-aged bio-asphalt with 5% bio-oil were similar to those of the 50# matrix asphalt, and the anti-cracking performance of both was not much different. In the higher frequency range, the complex modulus and rutting factors of the RTFO-aged bio-asphalt with 5% bio-oil content were larger than those of the 50# base asphalt, and its rutting resistance increased. This was because the performances of bio-asphalt after RTFO had been affected by the aging, although the impact was not significant.

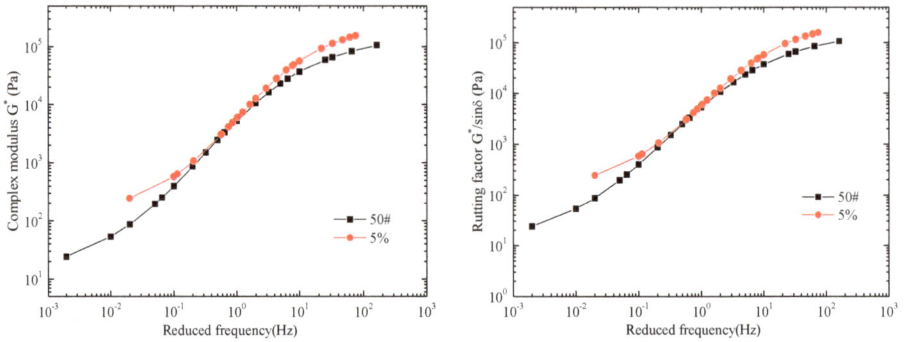


(**a**) (**b**)

**Figure 14.** The changes of different parameters with reduced frequency change for the RTFO-aged asphalt. (**a**) Complex modulus; (**b**) Rutting factor.

### *4.6. Functional Group Compositions Analysis*

In order to study the reaction mechanism and the changes of 50# base asphalt with bio-oil, the relative variations in the functional groups of 50# base asphalt, bio-asphalt with 30% bio-oil, and the bio-oil were analyzed. As shown in Figure 15, FTIR spectra were constructed, showing the changes of absorbance spectra with wavenumber.

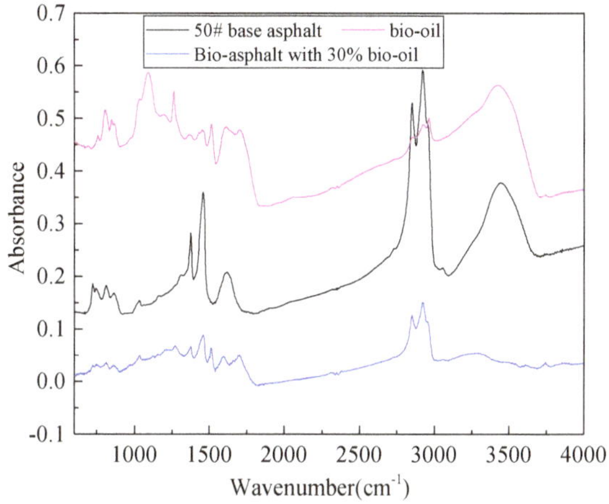


**Figure 15.** FTIR spectra of three kinds of binders.

The absorbance spectra trends of 50# base asphalt, bio-oil, and 50# base asphalt with 30% bio-oil were different. Based on the absorbance spectra peaks, the functional groups were analyzed from the three kinds of binders. The functional groups were identified and listed in Table 4.

**Table 4.** Functional groups identified of three kinds of binders.

| Binder Types | Absorption Wave Number | Functional Groups | Class of Compounds [22–24] |
|---|---|---|---|
| 50# base asphalt | 788–842 | C-H plane bending | Aromatic compounds |
| | 1024 | S=O | Sulfoxide |
| | 1379, 1452 | $CH_3$ | Aliphatic compounds |
| | 1602 | C=C ring stretch | Aromatic compounds |
| | 2852, 2925 | C-H stretching | Alkanes |
| | 3421 | O-H stretching, N-H stretching | Polymeric O-H, water, $NH_2$ |
| Bio-oil | 763–827 | C-H plane bending | Aromatic compounds |
| | 1097 | S=O | Sulfoxide |
| | 1263 | C-O stretching | Phenol, esters |
| | 1379, 1452 | $CH_3$ | Aliphatic compounds |
| | 1519 | $-NO_2$ stretching | Nitrogenous compounds |
| | 1602 | C=C ring stretch | Aromatic compounds |
| | 1706 | C=O stretching | Ketones, aldehydes, carboxylic acids |
| | 2852, 2924 | C-H stretching | Alkanes |
| | 3421 | O-H stretching, N-H stretching | Polymeric O-H, water, $NH_2$ |
| Bio-asphalt (50# base asphalt with 30% bio-oil) | 769–839 | C-H plane bending | Aromatic compounds |
| | 1024 | S=O | Sulfoxide |
| | 1379, 1452 | $CH_3$ | Aliphatic compounds |
| | 1519 | $-NO_2$ stretching | Nitrogenous compounds |
| | 1602 | C=C ring stretch | Aromatic compounds |
| | 1706 | C=O stretching | Ketones, aldehydes, carboxylic acids |
| | 2852, 2952 | C-H stretching | Alkanes |
| | 3307 | O-H stretching, N-H stretching | Polymeric O-H, water, $NH_2$ |

As shown in Figure 15 and Table 4, there was an obvious difference between 50# base asphalt and bio-oil. For the 50# base asphalt, C-H bending was dominant, while bio-oil had a large amount of different functional groups, such as O-H stretching, S=O, C=O stretching, and C-O stretching. This was mainly attributed to a great deal of oxygen in the bio-oil; it contributes to the aging process of bio-oil and bio-asphalt with bio-oil [22].

From the comparison of the three kinds of binders, it can be seen that there was an obvious peak with wavenumber 1263 $cm^{-1}$, representing the presence of esters and phenol in the bio-oil. However, this disappeared after the bio-oil was added to 50# base asphalt. The peaks at 3421 $cm^{-1}$ in bio-oil and 50# base asphalt disappeared in the bio-asphalt, and a new peak at 3307 $cm^{-1}$ appeared in the bio-asphalt. These illustrated that some chemical reactions occurred when the 50# base asphalt was mixed with bio-oil.

In addition, there are many different types of functional groups and compounds in the bio-oil and bio-asphalt, such as aldehydes, ketones, carboxylic acids, and phenol. Different compounds yielded different characteristics of the bio-asphalts with added bio-oil, as compared to conventional asphalt made from crude oil.

## 5. Conclusions

This research comprehensively investigated the high-temperature performances of bio-asphalt binders. Bio-asphalts with different contents of bio-oil extracted from wood chips were prepared. Conventional performance indexes of bio-asphalt binder were tested. The temperature sweep test and frequency sweep test of bio-asphalt with different bio-oil contents were carried out, and the temperature sensitivity of the bio-asphalt was analyzed. The master curves of matrix asphalt and bio-asphalt were constructed, and the changes of complex modulus and rutting factors of bio-asphalt under different frequencies were studied. Based on this study, the following conclusions were obtained:

(1) With an increase of bio-oil content, penetration and ductility of unaged bio-asphalt increased gradually; the anti-cracking performance increased slightly; the softening point decreased, while the anti-rutting ability of unaged bio-asphalt decreased. For the RTFO-aged asphalt, the residual penetration ratio and the softening point difference were large, indicating that the bio-asphalt underwent a certain degree of aging after heating. The content of bio-oil had a significant effect on the mass loss of the bio-asphalt.

(2) For the unaged and RTFO-aged bio-asphalts with 5% bio-oil, 5%-S, and 10%-S, the results indicated that the penetration, softening point, ductility, residual penetration ratio, mass loss, and other indicators, could meet the requirements of 50# grade asphalt.

(3) The phase angle, complex modulus, and rutting factors decreased gradually with the increase of temperature for the unaged and RTFO-aged asphalts, which indicated that the high-temperature performances of bio-asphalt decreased with increasing temperature. The sequence of the phase angle, complex modulus, and rutting factors with different contents of bio-oil showed different changes, which were caused by the aging of the bio-oil.

(4) The temperature sensitivity of both the unaged and RTFO-aged asphalts did not display regularity. The temperature sensitivity of unaged bio-asphalt was lower than that of the matrix asphalt. When the bio-oil content was higher than 10%, the temperature sensitivity of RTFO-aged bio-asphalt was higher than that of the matrix asphalt. The temperature sensitivity of RTFO-aged bio-asphalt with 5% content bio-oil was the smallest.

(5) According to the frequency sweep test, the phase angle, storage modulus, loss modulus, and complex modulus of the unaged and RTFO-aged bio-asphalts and the matrix asphalt changed greatly with the change of frequency, at the lower frequency. When the frequency was greater than 5Hz, the viscoelastic curve stabilized gradually. As the temperature increased, the phase angle increased gradually, and the storage modulus, the loss modulus, and the complex modulus decreased gradually, while the viscous component of the bio-asphalt increased. This was consistent with the results of the temperature sweep.

(6) According to the mater curve analysis, the complex modulus and rutting factors of bio-asphalt with 5% content bio-oil and 50# matrix asphalt increased with the increase of frequency, from the low frequency to the high-frequency range. Unaged bio-asphalt with 5% bio-oil content in the lower frequency range had considerable rutting resistance compared to the matrix

asphalt. In contrast, at the higher frequency range, the complex modulus and rutting factors of bio-asphalt with 5% bio-oil were lower than those of the 50# base asphalt, which was superior to matrix asphalt in low-temperature anti-cracking performance. The RTFO-aged bio-asphalt with 5% bio-oil had a higher resistance to rutting than the matrix asphalt at the low and high-frequency ranges.

(7) Chemical reactions occurred when the 50# base asphalt was mixed with bio-oil.

**Author Contributions:** J.G. and H.W. conceived and designed the experiments; J.G. performed the experiments; J.G. and H.W. analyzed the data; H.W. and Z.Y. contributed reagents/materials; J.G. wrote the paper; M.R.M.H., Y.L. and M.I. reviewed and edited the paper.

**Acknowledgments:** This research is sponsored by National Natural Science Foundation of China (No. 51578075, 51778062), the Fundamental Research Foundation of the Central Universities (No. 300102218718) and the Fundamental and Applied Research Project of the Chinese National Transportation Department (No. 2014319812180). The authors also gratefully acknowledge the financial support from China Scholarship Council (No. 201706560009).

**Conflicts of Interest:** The authors declare no conflict of interest.

## References

1. Bostancioğlu, M.; Oruç, Ş. Effect of furfural-derived thermoset furan resin on the high-temperature performance of bitumen. *Road Mater. Pavement Des.* **2014**, *16*, 227–237. [CrossRef]
2. Cavaca, L.A.S.; Afonso, C.A.M. Oleuropein: A Valuable Bio-Renewable Synthetic Building Block. *Eur. J. Org. Chem.* **2018**, *2018*, 581–589. [CrossRef]
3. Pereira, P.; Pais, J. Main flexible pavement and mix design methods in Europe and challenges for the development of an European method. *J. Traffic Transp. Eng. Engl. Ed.* **2017**, *4*, 316–346. [CrossRef]
4. Mohammad, L.; Elseifi, M.; Cooper, S.; Challa, H.; Naidoo, P. Laboratory Evaluation of Asphalt Mixtures That Contain Biobinder Technologies. *Transp. Res. Rec. J. Transp. Res. Board* **2013**, *2371*, 58–65. [CrossRef]
5. Cuadri, A.A.; García-Morales, M.; Navarro, F.J.; Partal, P. Processing of bitumens modified by a bio-oil-derived polyurethane. *Fuel* **2014**, *118*, 83–90. [CrossRef]
6. Yang, X.; You, Z.; Dai, Q. Performance evaluation of asphalt binder modified by bio-oil generated from waste wood resources. *Int. J. Pavement Res. Technol.* **2013**, *6*, 431–439.
7. Fini, E.H.; Al-Qadi, I.L.; You, Z.; Zada, B.; Mills-Beale, J. Partial replacement of asphalt binder with bio-binder: Characterisation and modification. *Int. J. Pavement Eng.* **2012**, *13*, 515–522. [CrossRef]
8. Fini, E.H.; Hosseinnezhad, S.; Oldham, D.J.; Chailleux, E.; Gaudefroy, V. Source dependency of rheological and surface characteristics of bio-modified asphalts. *Road Mater. Pavement Des.* **2016**, *18*, 408–424. [CrossRef]
9. Fini, E.H.; Kalberer, E.W.; Shahbazi, A.; Basti, M.; You, Z.; Ozer, H.; Aurangzeb, Q. Chemical Characterization of Biobinder from Swine Manure: Sustainable Modifier for Asphalt Binder. *J. Mater. Civ. Eng.* **2011**, *23*, 1506–1513. [CrossRef]
10. Fini, E.H.; Oldham, D.J.; Abu-Lebdeh, T. Synthesis and Characterization of Biomodified Rubber Asphalt: Sustainable Waste Management Solution for Scrap Tire and Swine Manure. *J. Environ. Eng.* **2013**, *139*, 1454–1461. [CrossRef]
11. Fini, E.H.; Yang, S.-H.; Xiu, S. Characterization and Application of Manure-Based Bio-binder in Asphalt Industry. *Transp. Res. Board* **2010**, *2010*, 10–2871.
12. Mills-Beale, J.; You, Z.; Fini, E.; Zada, B.; Lee, C.H.; Yap, Y.K. Aging Influence on Rheology Properties of Petroleum-Based Asphalt Modified with Biobinder. *J. Mater. Civ. Eng.* **2014**, *26*, 358–366. [CrossRef]
13. Raouf, M.; Williams, R. Temperature and Shear Susceptibility of a Nonpetroleum Binder as a Pavement Material. *Transp. Res. Rec. J. Transp. Res. Board* **2010**, *2180*, 9–18. [CrossRef]
14. Raouf, M.A.; Williams, R.C. Determination of Pre-Treatment Procedure Required for Developing Bio-Binders from Bio-Oils. In Proceedings of the 2009 Mid-Continent Transportation Research Symposium, Ames, IA, USA, 20–21 August 2009.
15. Tang, S.; Williams, R.C. Antioxidant Effect of Bio-Oil Additive ESP on Asphalt Binder. In Proceedings of the 2009 Mid-Continent Transportation Research Symposium, Ames, IA, USA, 20–21 August 2009.

16. Yang, S.H.; Suciptan, T. Rheological behavior of Japanese cedar-based biobinder as partial replacement for bituminous binder. *Constr. Build. Mater.* **2016**, *114*, 127–133. [CrossRef]
17. Wang, H.; Gao, J.; You, Z.; Tian, Y. Advances in bio-binder application on road pavement. *J. Wu Han Univ. Technol.* **2014**, *36*, 57–60. (In Chinese)
18. Gao, J.; Wang, H.; You, Z.; Yang, X. Gray relational entropy analysis of high temperature performance of bio-asphalt binder and its mixture. *Int. J. Pavement Res. Technol.* **2018**. [CrossRef]
19. Wang, H.; Gao, J.; Zhao, X.; You, Z.; Tian, Y. Rheological Properties on Bio-binder based on DSR and RV. *J. Hu Nan Univ. Nat. Sci.* **2015**, *42*, 26–33. (In Chinese)
20. He, M.; Cao, D.; Zhang, H.; Song, Z.; Wu, X. Research on Conventional Performance of Modified Bio-Asphalt. *J. Highw. Transp. Res. Dev. Engl. Ed.* **2015**, *9*, 1–6. [CrossRef]
21. Gao, J.; Wang, H.; You, Z.; Hasan, M.R.M. Research on properties of bio-asphalt binders based on time and frequency sweep test. *Constr. Build. Mater.* **2018**, *160*, 786–793. [CrossRef]
22. Zhang, R.; Wang, H.; Jiang, X.; You, Z.; Yang, X.; Ye, M. Thermal Storage Stability of Bio-Oil Modified Asphalt. *J. Mater. Civ. Eng.* **2018**, *30*, 04018054. [CrossRef]
23. Li, H.; Xia, S.; Li, Y.; Ma, P.; Zhao, C. Stability evaluation of fast pyrolysis oil from rice straw. *Chem. Eng. Sci.* **2015**, *135*, 258–265. [CrossRef]
24. Alsbou, E.; Helleur, B. Accelerated Aging of Bio-oil from Fast Pyrolysis of Hardwood. *Energy Fuels* **2014**, *28*, 3224–3235. [CrossRef]

applied sciences

MDPI

Article

# Research on Grooved Concrete Pavement Based on the Durability of Its Anti-Skid Performance

**Mulian Zheng [1,*], Yanjuan Tian [1], Xiaoping Wang [2] and Ping Peng [3]**

1 Key Laboratory for Special Area Highway Engineering of Ministry of Education, Chang'an University, South Erhuan Middle Section, Xi'an 710064, China; chdtianyj@163.com
2 Hunan Province Highway Design Limited Company, Changsha 410005, China; 2016021025@chd.edu.cn
3 Gansu Province Transportation Planning, Survey and Design Institute Limited Company, Lanzhou 730030, China; yannilb@126.com
* Correspondence: zhengml@chd.edu.cn; Tel.: +86-298-233-4846

Received: 17 April 2018; Accepted: 23 May 2018; Published: 30 May 2018

**Abstract:** The objectives of the present study are to investigate the anti-skid performance of concrete pavement and to attempt to enhance its durability by two different methods: using a longitudinally-transversely grooved (LT) form, and using a self-developed composite curing agent containing paraffin and $Na_2SiO_3$ as the main ingredients. The friction coefficient ($\mu$) was measured by self-developed equipment to evaluate the anti-skid performance of samples with three different groove forms (LT, longitudinally grooved (L), and transversely grooved (T)). Abrasion tests were then carried out to evaluate the durability of the anti-skid performance. The results indicated that anti-skid performance of LT samples was approximately 46.2% greater than that of T samples, but its durability was not as significant as that of T samples. However, the resistance to abrasion could be improved by using the aforementioned curing agent. Comparisons were carried out between samples sprayed the curing agent and control samples without any curing agent under standard conditions. It was found that the application of the curing agent increased the anti-skid durability of concrete by 35.4%~47.8%, proving it to be a useful and promising technique.

**Keywords:** concrete pavement; anti-skid performance durability; self-developed equipment; self-developed composite curing agent; longitudinally-transversely grooved form

## 1. Introduction

For several years, the anti-skid performance of cement concrete pavement has been the focus of much interest and research. Particularly on rainy days, due to a decrease in the anti-skid property of wet concrete surfaces, vehicles would skid out of control, resulting in casualties, property damage, and other serious consequences [1–3]. To enhance the anti-skid durability of concrete pavement, traditional techniques of napping, embossing, and grooving, known as the first-, second-, and third-generation anti-skid technologies, respectively, are commonly employed [4,5]. Over one or two years, napping pavements are rapidly polished and as a result the skid resistance is gradually attenuated. Embossing, with poor maneuverability, contributes little to the anti-skid performance improvement. Currently, widespread use of newer anti-skid technologies including exposed aggregate, embedded aggregate [6,7], and porous pavement is hindered by many drawbacks such as high costs, process complexity, and requirement constraints [8]. The friction performance of grooved pavement is significantly higher than that of non-grooved pavement [9]. Grooves could provide better drainage channels and improve the pavement's anti-skid performance, reducing the number of traffic accidents, especially on rainy days [10–12].

In many countries in which the grooving of concrete pavement is the most commonly used anti-skid durability technology, groove dimensions, design, and evaluation are the primary research areas [13–15]. In the United States, skid resistance force is recognized as the critical indicator of grooved concrete pavement performance by the Portland Cement Association (PCA) and American Association of State Highway and Transportation Officials (AASHTO) [16]. These organizations have adopted equally spaced rectangular grooves, of a width greater than 3 mm, a depth less than 6 mm, and spacing varying between 12 and 25 mm. In France, it was shown that increasing the groove width or reducing the groove spacing could improve the anti-skid performance of the pavement under the conditions of certain grooved surface areas [17]. It was therefore recommended that concrete pavement should be formed with transverse, equally spaced, rectangular grooves with a width between 3 and 5 mm, a depth between 5 and 6 mm, and groove spacing of 20 to 30 mm. Fwa [18] adopted grooves with a width varying from 2 to 10 mm, a depth from 1 to 10 mm, and spacing from 5 to 25 mm. Lee [19] developed an automatic instrument to measure groove dimensions in field experiments, resulting in the enhanced efficiency of grooved pavement evaluation.

The use of transverse rectangular grooves, with equal spacing of 20 mm and width and depth varying from 3 to 5 mm, has been suggested for expressways and first-class highways in China [20]. Meanwhile, municipal and rural roads in our country feature transverse rectangular grooves with smaller dimensions than higher-class roads, with widths varying from 3 to 5 mm, depths from 1 to 6 mm, and spacing from 15 to 40 mm. Li [10] proposed a simulation method using finite element software to investigate groove parameters, thus determining the optimal dimensions for longitudinally grooved (L) and transversely grooved (T) samples to be 6 mm wide, 4 mm deep, and 10 mm in spacing. To provide the required friction force, domestic concrete pavement is generally formed with transverse grooves and a wide groove spacing of approximately 20 mm.

Currently, the most commonly used concrete pavement curing agents can be divided into inorganic and organic types. Inorganic curing agents can improve the strength of concrete and have the advantage of relatively low expenses. While the inadequate surface hydration of concrete leads to early cracks, the formation of the waterproof membrane on the concrete surface is incomplete after drying. The use of paraffin emulsion-type organic curing agents results in good water retention and a smooth waterproof membrane, but does not enhance the strength of the concrete surface.

In summary, while there has been minimal progress in the research and development of concrete pavement durability, groove technology has been applied worldwide. Research on this issue in China needs to be further conducted to develop high anti-skid performance pavement. Through the evaluation of durability, the primary purpose of this study is to further investigate anti-skid technologies for grooved pavement. To enhance the anti-skid performance, a longitudinally-transversely grooved (LT) technique and new curing method are proposed. Additionally, optimal groove dimensions for better anti-skid durability are recommended for practical applications according to the laboratory tests.

## 2. Materials and Methods

Selecting the raw materials and mixture proportions for the grooved concrete pavement was the first step in this study. The LT method was proposed next, and anti-skid tests were implemented on samples of two different forms to evaluate skid resistance. Subsequently, abrasion resistance tests were designed to evaluate the anti-skid durability. Finally, we developed a new curing agent and applied it to different groove samples. The following subsections describe the step-by-step methodology that was applied in detail.

### 2.1. Raw Materials

Cement, coarse aggregate, fine aggregate, and water were used in the concrete mixtures for this investigation. Table 1 displays the mix proportions for the grooved concrete pavement used in this study, and the water-to-cement (w/c) ratio and sand ratio were maintained at 0.46% and 34%, respectively [21]. The cement used was ordinary Portland cement (type P.O.42.5, China), with the parameters listed in Table 2. The coarse aggregate was crushed limestone produced in Xianyang, China, and the fine aggregate including river sands were sourced from Bahe, China. Tables 3 and 4 display the parameters for the coarse and fine aggregates, respectively.

The self-developed concrete curing agent, adopted in the present study, has been patented (Patent no.: CN201410312792.6). It is mainly composed of $Na_2SiO_3$ (as the critical inorganic component) and paraffin wax (as the critical organic component). Its parameters are listed in Table 5.

**Table 1.** Mix proportions of grooved concrete pavement.

| Items | | Unit | Value |
|---|---|---|---|
| Concrete Strength Level | | ... | C30 |
| Water to Cement Ratio | | ... | 0.46 |
| Sand Ratio | | % | 34 |
| Mix Proportion | Cement | kg/m$^3$ | 411 |
| | Crushed Limestone | kg/m$^3$ | 1212 |
| | Sand | kg/m$^3$ | 624 |
| | Water | kg/m$^3$ | 190 |

**Table 2.** Parameters of the cement.

| Parameter | | Unit | Value |
|---|---|---|---|
| Cement Type | | ... | P.O.42.5 |
| Density | | g/cm$^3$ | 3.2 |
| Specific Surface Area | | m$^2$/kg | 355 |
| Setting Time | Initial Set | min | 215 |
| | Final Set | min | 270 |
| Compressive Strength | 3 days | MPa | 27.6 |
| | 28 days | MPa | 45.1 |
| Flexural Strength | 3 days | MPa | 5.9 |
| | 28 days | MPa | 7.8 |

**Table 3.** Parameters of the crushed limestone.

| Parameter | Unit | Value | Conclusion |
|---|---|---|---|
| Apparent Density | g/cm$^3$ | 2.69 | ... |
| Loose Density | g/cm$^3$ | 1.47 | ... |
| Crushing Value | % | 9.7 | Qualified |
| Mud Content | % | 0.827 | Qualified |
| Elongated Articles | % | 9.3 | Qualified |

**Table 4.** Parameters of the river sands.

| Parameter | Unit | Value | Conclusion |
|---|---|---|---|
| Apparent Density | g/cm$^3$ | 2.628 | ... |
| Loose Density | g/cm$^3$ | 2.621 | ... |
| Fineness Modulus | % | 2.365 | ... |
| Mud Content | % | 1.407 | Qualified |
| Moisture Content | % | 2.876 | Qualified |

**Table 5.** Parameters of the agent.

| Parameter | | Unit | Value | Conclusion |
|---|---|---|---|---|
| Effective Water Retention Rate | | % | 93.78 | $\geq$90 |
| Abrasion Loss | | kg/m$^2$ | 2.365 | $\leq$3.0 |
| Solid Content | | % | 31.04 | $\geq$20 |
| Drying Time | | h | 0.6 | $\leq$4 |
| Compressive Strength Rate | 7 days | min | 105 | $\geq$95 |
| | 28 days | min | 108 | $\geq$95 |

### 2.2. LT Method

The anti-skid force applied to vehicles mainly relates to the macro structure of the cement concrete pavement [22]. It can be increased by grooves, thus resulting in higher skid resistance in grooved compared to non-grooved pavement [23]. Grooved concrete pavements are commonly applied in two forms, L and T, and can effectively improve the skid resistance. In accordance with the aforementioned research and engineering practice, the T form is superior to the L form in terms of anti-skid performance, durability, driving comfort, direction control, and incidence of accidents [24]. Additionally, the L form is one constituent of the LT form. Thus, LT concrete pavement is proposed to provide greater braking performance to T, and to develop the advantages of both T and L pavement. Figure 1 shows a sketch of the LT form.

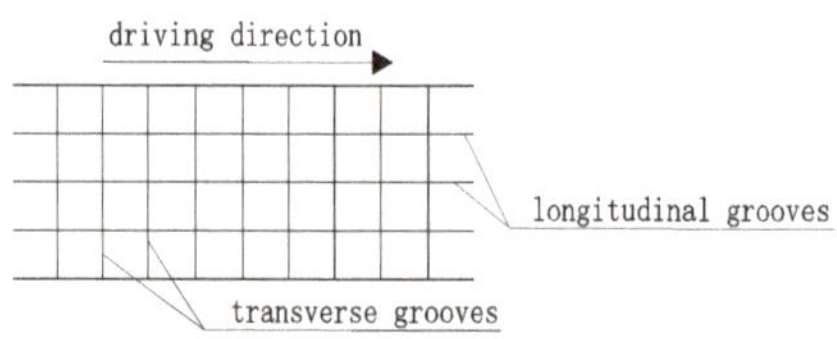


**Figure 1.** Sketch of the longitudinally-transversely grooved (LT) form.

### 2.3. Preparation of the Test Samples

Samples (300 mm × 300 mm × 50 mm) were prepared for the test. The aggregate was initially mixed with dry cement for 60 s to improve the bond between the aggregate and cement paste, before gradually mixing in the remaining water over 90 s, and then casting. After being cured for 24 h in a fog room at 20 ± 2 °C and 95% relative humidity [25], the samples were demolded. A cutting machine was used to create grooves in the concrete after it had been cured for 3 to 4 days [26]. To ensure that the grooves were straight and uniform, lines were drawn on the test samples in accordance with the design requirements. As shown in Figure 2, a straight board was placed on the test samples and used as the guide rail for the cutting machine.

(a) (b) (c)

**Figure 2.** These images show the preparation of the test samples. (**a**) The grooving technology and process; (**b**,**c**) the grooved test samples.

### 2.4. Anti-Skid Tests

The dynamic rotating friction coefficient tester developed by the research team, which has been patented (Patent no.: CN200820222395.X), was used to measure the anti-skid performance of the samples (shown in Figures 3 and 4). It was operated by installing a test slide sample under a certain load on the test disc, then placing the test disc onto the pavement. When the disc is rotated at a given speed, the torque required to drive the disc is obtained from the driving engine. The friction coefficient of the pavement surface is calculated according to Equation (1):

$$\mu = \frac{N}{dG} \tag{1}$$

where $\mu$ is the friction coefficient of the pavement surface, $N$ is the torque required to drive the disc, $d$ is the friction arm (10 cm), and $G$ is the vertical load (21.56 N).

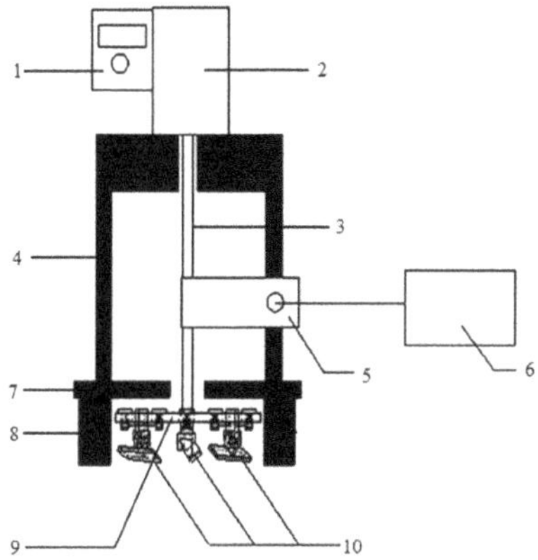


**Figure 3.** Sketch of the dynamic rotating friction coefficient tester. The tester includes the following: (1) speed control device, (2) engine, (3) axis of rotation, (4) shell, (5) torque sensor, (6) digital display instrument, (7) weighing plate, (8) bearing, (9) test disc, and (10) rubber slider.

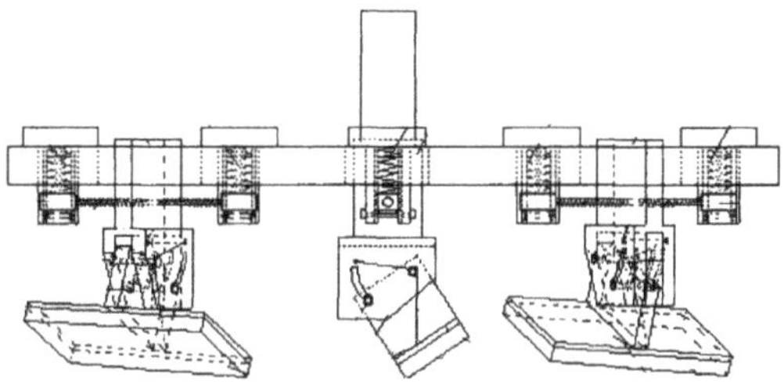

**Figure 4.** Image of the test disc of the dynamic rotating friction coefficient tester.

The texture depth (TD), the British Pendulum Number (BPN), and the friction coefficient ($\mu$) methods are all commonly used to evaluate the anti-skid performance of pavement. Table 6 shows the Chinese code performance requirements [20]. Of these methods, the TD method is the most suitable for evaluating concrete pavement. Hence, the $\mu$, adopted to evaluate the anti-skid performance in this paper, needed to be converted to TD in the laboratory tests.

**Table 6.** Anti-skid requirements for concrete pavement [1].

| Road Section | | Acceptance Value of TD (The Texture Depth) | Unit |
|---|---|---|---|
| Expressway, First-Class Highway | General Road Section | 0.7~1.10 | mm |
| | Special Road Section | 0.8~1.20 | mm |

[1] Table 6 references Table 7.2.2 in China code *Inspection and Evaluation Quality Standards for Highway Engineering Section 1 Civil Engineering (JTG F80/1-2017).*

Table 6 shows the anti-skid requirements for concrete pavement, where special road section refers to interchanges, grade crossings, and speed change lanes of expressways and first-class highways.

Contrast tests, using nine samples, were implemented between the dynamic rotating friction coefficient tester method and the sand-laying method [27]. The test results are shown in Table 7.

**Table 7.** The results of the anti-skid tests.

| Sample No. | TD/mm | $\mu$ |
|---|---|---|
| 1 | 0.91 | 0.63 |
| 2 | 1.02 | 0.64 |
| 3 | 0.89 | 0.56 |
| 4 | 0.97 | 0.59 |
| 5 | 1.09 | 0.74 |
| 6 | 0.9 | 0.60 |
| 7 | 1.24 | 0.78 |
| 8 | 1.02 | 0.80 |
| 9 | 1.01 | 0.69 |

The regression equation was obtained by the data in Table 7, as shown in Equation (2).

$$\mu = \frac{TD - 0.8904}{3.7599TD^2 - 2.9579} + 0.5427TD(R = 0.8863) \quad (2)$$

As seen in Equation (2), $\mu$ and $TD$ are positively correlated. By using the conversion relation of $\mu$ and $TD$, the results of $\mu$ should meet the specified requirements shown in Table 8.

**Table 8.** Anti-skid requirements for concrete pavement.

| Road Section | | Acceptance Value of $\mu$ |
|---|---|---|
| Expressway, First-Class Highway | General Road Section | $\geq$0.55 |
| | Special Road Section | $\geq$0.60 |

In order to compare and analyze the effects of different grooved forms on skid resistance, two types (T and LT) of grooved samples were prepared to implement comparative anti-skid tests.

2.4.1. T Schemes

In accordance with previous studies and experience, the T concrete pavement had better skid resistance when the groove width was 6 mm [10,11]. With this consistent width, samples with three different depths (2, 4, and 6 mm) and three different spacing arrangements (10, 20, and 30 mm) were prepared in this investigation, as shown in Table 9.

**Table 9.** Sample dimensions schemes.

| Sample No. | T (Transversely Grooved) Width/mm | T Depth/mm | T Spacing/mm |
|---|---|---|---|
| 1 | 6 | 2 | 10 |
| 2 | 6 | 2 | 20 |
| 3 | 6 | 2 | 30 |
| 4 | 6 | 4 | 10 |
| 5 | 6 | 4 | 20 |
| 6 | 6 | 4 | 30 |
| 7 | 6 | 6 | 10 |
| 8 | 6 | 6 | 20 |
| 9 | 6 | 6 | 30 |

2.4.2. LT Schemes

Larger groove dimensions were adopted in this study in order to select the appropriate LT sample. Table 10 shows the influence of the six factors on LT groove dimensions. In accordance with the orthogonal experimental design methodology [28], the groove dimensions were prepared in $L_{25}(5^6)$.

**Table 10.** Levels of factors (cm).

| Level | T Width (A) | T Depth (B) | T Spacing (C) | L (Longitudinally Grooved) Width (D) | L Depth (E) | L Spacing (F) |
|---|---|---|---|---|---|---|
| k1 | 2 | 2 | 15 | 2 | 2 | 10 |
| k2 | 3 | 3 | 30 | 3 | 3 | 17 |
| k3 | 4 | 4 | 45 | 4 | 4 | 24 |
| k4 | 5 | 5 | 60 | 5 | 5 | 31 |
| k5 | 6 | 6 | 75 | 6 | 6 | 37 |

2.4.3. Comparative Analysis of Different Groove Forms

To evaluate differences in skid resistance, a comparative analysis was conducted between the two best-performing T and LT grooved samples in the above anti-skid tests.

*2.5. Skid Resistance Test*

In accordance with the Chinese code, the abrasion resistance test by an abrasion tester was used to evaluate the anti-skid durability of the concrete samples [25]. The test operation process was as follows. Each sample was placed on a horizontal turntable of the abrasion tester, as shown in Figure 5, and fastened by the clamp. Samples, loaded 200 N, were ground for 30 revolutions, then removed, and weighed after brushing the grinding dust from the sample's surface. Meanwhile, the corresponding quality ($m_1$) of the sample was recorded as the initial quality. In order to promptly remove dust during the grinding process, a vacuum cleaner was aligned with the abraded surface of the samples. Each set of blades was used for one sample group test, and was replaced with a new set of blades before testing the next group.

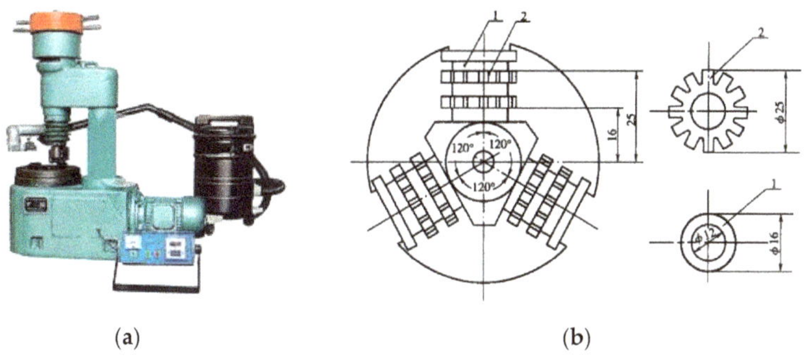


**Figure 5.** Picture of the abrasion tester. (**a**) Photo of the abrasion tester; (**b**) sketch of the grinding blades. In (b), 1 is a gasket and 2 is a blade.

Abrasion loss per unit area of each sample was calculated by Equation (3) with an accuracy of 0.001 kg/m$^2$.

$$G = \frac{m_1 - m_2}{0.0125} \tag{3}$$

where $G$ is the abrasion loss per unit area (kg/m$^2$), $m_1$ is the original quality, $m_2$ is the quality after abrasion (kg), and 0.0125 is the abrasion area (m$^2$).

### 2.5.1. Evaluation of the Grooved Samples

The samples exhibiting the best skid resistance in the aforementioned tests (the two best-performing T and LT samples) were subjected to the abrasion resistance tests to identify the differences in their skid resistance durability. In particular, each dimension sample was produced in two groups, and the mean value was adopted as the final result.

### 2.5.2. Evaluation of Samples under Different Curing Methods

The various curing methods have different effects on the abrasion resistance of concrete pavement. In this study, a composite curing agent, containing $Na_2SiO_3$ and paraffin as the primary ingredients, was self-developed to improve the concrete performance [20]. This agent has the properties of high water retention, strength, and abrasion resistance, as listed in Table 5. This paper investigated the abrasion resistance of grooved concrete under different curing methods.

The authors prepared eight accordant samples with the sample dimensions (300 mm × 300 mm × 50 mm) that offered the best skid resistance (the two best-performing T and LT samples) in the above tests, and then divided them into two groups. One group was adopted as the control group (without any curing agent), cured in standard curing box (20 °C ± 1 °C, relative humidity >90%, maintenance water 20 °C ± 1 °C) for 28 days [25]. The other group was sprayed with the curing agent at a spraying dose of 0.22 kg/m$^2$, with an inorganic to organic curing ratio of 4:6. The samples were sprayed for a second time with the same preparation 30 min later. After spraying twice, the samples were cured for 28 days under conditions identical to those of the control group. Subsequently, all samples were dried at room temperature, and surface dust was brushed away. The abrasion tester shown in Figure 5 was used to conduct abrasion resistance tests. And the abrasion loss per sample was measured and recorded; the mean value of four samples was calculated and adopted as the final result.

After the cement concrete mixture was poured, the inorganic ingredient was sprayed into the mold approximately 4 to 6 h after surface exudation. Thirty minutes later, the organic ingredient was applied by spraying between 20 and 30 °C (room temperature). The curing agent applied to the sample's surface was sprayed uniformly, and the dose was strictly controlled. Figure 6 shows the sample after spraying the curing agent.

**Figure 6.** Appearance after spraying the composite curing agent.

The samples were cured, after demolding, for 24 h under standard conditions. A curing film was applied to each surface except the grooved surface. Three or four days after grooving, the grooved surface was sprayed with the curing agent, while the other surfaces were still undergoing the film curing processing. Figure 7 shows a sample sprayed with the composite curing agent after grooving.

**Figure 7.** Photo after spraying the composite curing agent after grooving.

## 3. Results and Conclusions

This section presents the anti-skid performance and abrasion resistance results for the different grooved samples, and analyzes the impact of the curing agent on the anti-skid durability of the concrete pavement.

### 3.1. Results of Anti-Skid Tests

#### 3.1.1. Anti-Skid Results for T Schemes

Table 11 shows $\mu$ for the T samples, determined by measuring skid resistance.

**Table 11.** Results of skid-resistant performance for L samples.

| Sample No. | 1 | 2 | 3 | 4 | 5 | 6 | 7 | 8 | 9 |
|---|---|---|---|---|---|---|---|---|---|
| $\mu$ | 0.63 | 0.58 | 0.63 | 0.56 | 0.82 | 0.46 | 0.57 | 0.65 | 0.61 |

According to Table 11, $\mu$ is maintained within a certain range, as influenced by the groove dimensions. As the groove width becomes larger, the groove quality improves, but $\mu$ is reduced as the groove spacing and depth increase, with no clear correlation. This is because as the groove width increases, the area of tire embedding in the groove also increases; so, while increasing the groove depth enhances the tire plowing effect, the surface resistance of the groove is improved. When the groove spacing is small, the tire/pavement plowing effect is enhanced, thus improving pavement skid resistance.

According to the principles of skid resistance measurement, the $\mu$ of T and L samples was the same. Our test results indicated that T (L) samples 5 (6, 4, 20 mm) and 8 (6, 6, 20 mm) had the greatest skid resistance.

#### 3.1.2. Anti-Skid Results for LT Schemes

Table 12 illustrates the skid resistance performance results for LT samples, and Table 13 displays the range analysis of LT samples. $k_{ij}$ in Table 13 refers to the average value of the sum of $\mu$ in the j column under the $k_i$ level; $R_j$ refers to the difference between the maximum value and minimum value of $k_{1j}$, $k_{2j}$, $k_{3j}$, $k_{4j}$, and $k_{5j}$ in the j column, namely, range. The size of range represents the different effects of each factor on the $\mu$ value. Figure 8 shows the range analysis chart of the six factors.

**Table 12.** Skid-resistant performance of LT samples.

| **Sample No.** | 1 | 2 | 3 | 4 | 5 | 6 | 7 | 8 | 9 | 10 |
|---|---|---|---|---|---|---|---|---|---|---|
| $\mu$ | 0.758 | 0.623 | 0.573 | 0.544 | 0.528 | 0.574 | 0.919 | 0.863 | 0.505 | 0.727 |
| **Sample No.** | **11** | **12** | **13** | **14** | **15** | **16** | **17** | **18** | **19** | **20** |
| $\mu$ | 0.696 | 0.492 | 0.756 | 1.039 | 0.772 | 0.949 | 0.567 | 0.904 | 1.200 | 0.609 |
| **Sample No.** | **21** | **22** | **23** | **24** | **25** | | | | | |
| $\mu$ | 0.824 | 1.157 | 0.652 | 0.586 | 0.797 | | | | | |

**Table 13.** Range analysis of LT samples [2].

| **Level** | **A** | **B** | **C** | **D** | **E** | **F** |
|---|---|---|---|---|---|---|
| k1 | 0.605 | 0.760 | 0.917 | 0.636 | 0.709 | 0.886 |
| k2 | 0.718 | 0.752 | 0.764 | 0.728 | 0.705 | 0.792 |
| k3 | 0.751 | 0.750 | 0.677 | 0.710 | 0.763 | 0.791 |
| k4 | 0.846 | 0.775 | 0.729 | 0.793 | 0.728 | 0.637 |
| k5 | 0.605 | 0.760 | 0.917 | 0.636 | 0.709 | 0.886 |
| Range | 0.241 | 0.088 | 0.281 | 0.253 | 0.112 | 0.269 |
| Factors in Primary and Secondary Order | C > F > D > A > E > B | | | | | |
| Optimal Scheme | C1F1D5A4E5B4 | | | | | |

[2] Where A is T width, B is T depth, C is T spacing, D is L width, E is L depth, and F is L spacing.

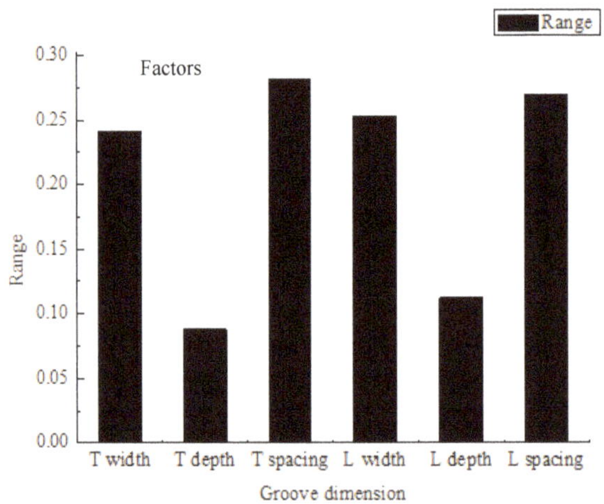


**Figure 8.** Columnar analysis diagram of the range analysis of LT samples.

The range analysis of LT samples shown in Table 13 and Figure 8 indicates that T spacing is the key factor impacting the anti-skid performance of concrete pavement, followed by L spacing, L width, T width, L depth, and T depth. The groove volume within a certain scope is the primary factor affecting $\mu$. The results of mathematical calculations and experimental findings are slightly different, but remain consistent throughout the geometric analysis. Finally, the test results show that LT samples 19 (5, 5, 30, 6, 4, 10 mm) and 22 (6, 3, 15, 6, 5, 24 mm) have the greatest skid resistance.

### 3.1.3. Comparative Analysis of Test Results for Different Groove Forms

Table 14 shows results for the two best-performing T and LT grooved samples in the above anti-skid tests.

**Table 14.** Results of skid-resistant testing for different samples (mm).

| Samples Dimensions | (6, 4, 20) | (6, 6, 20) | (5, 5, 30, 6, 4, 10) | (6, 3, 15, 6, 5, 24) |
|---|---|---|---|---|
| $\mu$ | 0.821 | 0.651 | 1.200 | 1.157 |

As can be seen in Table 14, $\mu$ of LT samples (5, 5, 30, 6, 4, 10) is approximately 46.2% greater than that of T samples (6, 4, 20). The anti-skid performance of the LT samples used in this experiment is significantly better than that of the T samples.

It can be seen that groove forms have a very important influence on anti-skid performance, and that LT schemes can effectively improve the anti-skid performance of pavement. It can be inferred that the embedded squeeze effect of the tire-road interface, the effective contact area, and the resistance to change all increase as the groove number increases; therefore, the sliding resistance increases.

### *3.2. Results of Abrasion Resistance Tests*

#### 3.2.1. Results for Grooved Samples

Table 15 shows the abrasion test results for the grooved samples.

**Table 15.** Abrasion loss of different grooved samples (kg/m$^2$).

| Sample Dimensions | (6, 4, 20) | (6, 6, 20) | (6, 3, 15, 6, 5, 24) | (5, 5, 30, 6, 4, 10) |
|---|---|---|---|---|
| $G_c$ (kg/m$^2$) | 3.17 | 3.28 | 4.56 | 3.69 |

From Table 15, it can be seen that the abrasion loss of sample (6, 4, 20) is the smallest of the four different samples; that of sample (6, 3, 15, 6, 5, 24) is the largest, but the difference between them is relatively small. This indicates that the anti-skid performance of the LT samples was improved, but the durability performance was not.

It is inferred that the improved tire-road friction is due to the grid formed by the LT form, which increases the number of prominent corners. Road surface abrasion resistance gradually decreases because of the continuous vehicle loads.

#### 3.2.2. Results of Using the Curing Agent

Table 16 shows the results of the abrasion resistance tests for the samples using different curing methods.

**Table 16.** Abrasion loss for different curing methods (kg/m$^2$).

| Samples Dimensions | Standard Method without Agent | Spraying Curing Agent |
|---|---|---|
| (6, 4, 20) | 3.17 | 1.83 |
| (6, 6, 20) | 3.28 | 2.12 |
| (6, 3, 15, 6, 5, 24) | 4.56 | 2.38 |
| (5, 5, 30, 6, 4, 10) | 3.69 | 1.97 |

As shown in Table 16, the use of the curing agent improves the abrasion resistance of grooved samples by 35.4%~47.8%. The concrete curing agent itself could improve both strength and abrasion resistance; therefore, the strength and abrasion resistance of the sprayed grooved samples were greatly enhanced. In addition, the humidity during the curing and improvement in the overall strength of the sample was ensured by covering the grooved surface and other surfaces with a plastic film.

In conclusion, the application of this curing agent results in improved abrasion resistance performance in grooved concrete, and is an advisable and promising technique to improve the anti-skid performance and durability of such surfaces.

## 4. Conclusions

This study provides an experimental investigation of the anti-skid performance and abrasion resistance of cement concrete pavements, including the LT grooved method, the curing method using a composite curing agent, anti-skid tests, and abrasion resistance tests. These tests were performed on samples with different grooved schemes, and conclusions could be drawn as follows:

(1) The two techniques used in this study, including LT grooving and a curing method using a composite concrete curing agent, could effectively enhance the anti-skid performance of the pavement.
(2) From the experimental study of the anti-skid performance of concrete pavement with different groove schemes, LT had relatively good skid resistance compared to other schemes.
(3) The improvement in durability observed for concrete cured using the sprayed composite curing agent indicates that this can be an effective method to maintain concrete pavement in practical applications.
(4) Based on the results of skid and abrasion resistance tests, the longitudinally-transversely grooved sample (5, 5, 30, 6, 4, 10) provides better skid resistance and durability and can be suggested to be adopted as the optimal dimensions for good anti-skid durability for concrete pavement. This sample's dimensions, including transverse groove width, transverse groove depth, transverse groove spacing, longitudinal groove width, longitudinal groove depth, and longitudinal groove spacing, were 5, 5, 30, 6, 4, and 10 mm, respectively.

**Author Contributions:** M.Z. conceived and designed the experiments; Y.T. performed the experiments and analyzed the data; X.W. and P.P. contributed reagents and materials; M.Z. wrote the paper.

**Acknowledgments:** This research was supported by the Fundamental Research Funds for the Central Universities in China (No. 310821163502), the Transportation Department of Hebei Province (Grant No. T-2012107 and Y-2012014), the Transportation Department of Jiangxi Province (Grant No. Ganjiaokejiao [2015], and the Transportation Department of Hubei Province of China (No. Ejiaokejiao [2012] 857). In addition, the authors would like to thank the reviewers of this paper for their ever-present support and valuable advice.

**Conflicts of Interest:** The authors declare no conflict of interest.

## References

1. Yang, X.D.; Zheng, M.L.; Zhu, H.T.; Li, Z.Z.; Wang, B.G. Numerical analysis of frictional contact condition between tire and cement concrete pavement. *J. Chang'an Univ. (Nat. Sci. Ed.)* **2010**, *30*, 13–17.
2. Liu, Y.; Tian, B.; Niu, K.M. Research on skid resistance and noise reduction properties of cement concrete pavements with different surface textures. *J. Highw. Transp. Res. Dev. (Eng. Ed.)* **2013**, *7*, 22–27. [CrossRef]
3. José, M.; Pardillo, M.; Rafael, J.P. An assessment of the skid resistance effect on traffic safety under wet-pavement conditions. *Accid. Anal. Prev.* **2009**, *41*, 881–886.
4. Wang, L.L.; Zhang, J.X.; Zhang, C.; Zhao, E.Q. Discussion about anti-skid changing-laws of cement concrete pavement. *Highw. Eng.* **2008**, *33*, 5–8.
5. Liu, Y.; Tian, B.; Niu, K.M. Skid-resistance and denoising properties of cement concrete pavement with different surface texture. *J. Highw. Transp. Res. Dev.* **2012**, *29*, 28–33.
6. Song, Y.C.; Fu, B.F.; Liang, N.X. Comprehensive evaluation of surface function of exposed-aggregate cement concrete pavement. *J. Chongqiong Jiaotong Univ. (Nat. Sci. Ed.)* **2015**, *34*, 43–47.
7. Wang, F.Z.; Yang, L.; Wang, H.; Yu, H.G. Facile preparation of photo catalytic exposed aggregate concrete with highly efficient and stable catalytic performance. *Chem. Eng. J.* **2015**, *264*, 577–586. [CrossRef]
8. Chen, Y.; Wu, X.Y. Fractal features and surface texture parameters of porous cement concrete pavement. *J. Changsha Univ. Sci. Technol. (Nat. Sci. Ed.)* **2007**, *4*, 13–17.
9. Zhu, S.Z.; Huang, X.M. Numerical simulation of tire hydroplaning speed on transverse grooved concrete pavements. *J. Southeast Univ. (Nat. Sci. Ed.)* **2016**, *46*, 1296–1300.
10. Li, B.; Han, S. Review on the selection of the parameters of the concrete pavement at home and abroad. *East China Highw.* **2010**, *2*, 81–82.

11. Huo, M. *Research on Skid Resistance Function Attenuation and Evaluation Method of Cement Concrete Pavement*; China Architecture and building Press: Beijing, China, 2011.
12. Han, S.; Li, N.; Li, B.; Huo, M.; Han, W.W. Tests study on skid resistance attenuation of cement concrete pavement. *J. Guangxi Univ. (Nat. Sci. Ed.)* **2010**, *35*, 896–900.
13. Wang, K.; Lin, L.; Li, Q.J.; Nguyen, V.; Hayhoe, G.; Larkin, A.L. Runway groove identification and evaluation using 1 mm 3D image data. In Proceedings of the Airfield and Highway Pavement Conference, Los Angeles, CA, USA, 9–12 June 2013; pp. 730–741.
14. Alexandre, L.O.; Luiz, R.P.J. Evaluation of the superficial texture of concrete pavers using digital image processing. *J. Constr. Eng. Manag.* **2015**, *141*, 04015034. [CrossRef]
15. Kelvin, C.P.W.; Luo, W.T.; Joshua, Q.L. Hydroplaning risk evaluation of highway pavements based on IMU and 1 mm 3D texture data. In *Transportation and Development Institute Congress 2nd. 2014 (T&DI 2014)*; ASCE: Reston, VA, USA, 2014; pp. 511–522. [CrossRef]
16. Lee, Y.P.K.; Liu, Y.R.; Liu, Y.; Fwa, T.F.; Choo, Y.S. Skid resistance prediction by computer simulation. *Appl. Adv. Technol. Transp. Eng.* **2004**, 465–469. [CrossRef]
17. Scott, H.R.; Joseph, R.R.; John, J.H. Criteria for predicting hydroplaning potential. *J. Transp. Eng.* **1986**, *112*, 549–553. [CrossRef]
18. Fwa, T.F.; Ong, G. Transverse pavement grooving against hydroplaning. II: Design. *J. Transp. Eng.* **2006**, *132*, 449–457. [CrossRef]
19. Lee, M.H.; Chou, C.P.; Li, K.H. Automatic measurement of runway grooving construction for pavement skid evaluation. *Autom. Constr.* **2009**, *18*, 856–863. [CrossRef]
20. MOT (Ministry of Transportation). *Technical Guideline for Construction of Highway Cement Concrete Pavements (JTG/T F30-2014)*; China Communication Press: Beijing, China, 2014.
21. MOHURD (Ministry of Housing and Urban-Rural Development). *Specification for Mix Proportion Design of Ordinary Concrete (JGJ 55-2011)*; China Architecture and Building Press: Beijing, China, 2011.
22. Wen, S.T.; Huang, P. *Principles of Tribology*, 4th ed.; Tsinghua University Press: Beijing, China, 2010; ISBN 978-7-302-30188-2.
23. MOHURD (Ministry of Housing and Urban-Rural Development). *Code for Concrete Admixture Application (GB 50119-2013)*; China Architecture and Building Press: Beijing, China, 2013.
24. MOHURD (Ministry of Housing and Urban-Rural Development). *Standard for Test Method of Mechanical Properties on Ordinary Concrete (GB/T 50081-2002)*; China Architecture and Building Press: Beijing, China, 2002.
25. MOT (Ministry of Transportation). *Test Methods of Cement and Concrete for Highway Engineering (JTG E30-2005)*; China Communication Press: Beijing, China, 2005.
26. MOT (Ministry of Transportation). *Technical Specification for Construction of Highway Concrete Pavement (JTG F30–2003)*; China Communication Press: Beijing, China, 2003.
27. MOT (Ministry of Transportation). *Field Test Methods of Subgrade and Pavement for Highway Engineering (JTG E60-2008)*; China Communication Press: Beijing, China, 2008.
28. Li, Y.Y.; Hu, C.R. *Experiment Design and Data Processing*; Chemical Industry Press: Beijing, China, 2008.

applied sciences

MDPI

*Article*

# Study on Components Determination and Performance Evaluation of LS Pre-Maintenance Agent

**Yuxiang Tian [1,2], Biao Ma [1,*], Ke Tian [3], Ning Li [1] and Xueyan Zhou [1]**

1 Key Laboratory of Special Area Highway Engineering of Ministry of Education, Chang'an University, Xi'an 710064, China; yuxiangtian0729@163.com (Y.T.); lining_sn@163.com (N.L.); 2015021028@chd.edu.cn (X.Z.)
2 Department of Civil and Environmental Engineering, Michigan Technological University, Houghton, MI 49931, USA
3 Tianjin Municipal Engineering Design & Research Institute, Tianjin 300051, China; tianking0729@126.com
* Correspondence: mb@gl.chd.edu.cn

Received: 30 April 2018; Accepted: 28 May 2018; Published: 29 May 2018

**Featured Application: Aimed at the serious aging problem of asphalt pavement in strong UV regions, developed the LS pre-maintenance agent, provided a new material and technical guidance for pre-maintenance engineering in special areas.**

**Abstract:** Adequate maintenance and taking active preventive measures can effectively prevent the early disease of asphalt pavements before significant damage occurs. By developing a light screening preventive maintenance agent (LS pre-maintenance agent) for strong ultraviolet (UV) radiation areas, based on the asphalt aging and regeneration mechanism, we analyzed the function and basic components and determined the optimum components ratio based on the best proportion of penetrant and solvent oil for solubility. The optimum ratio for quick-drying and long-term storage ability is the mass ratio of rock asphalt, reducing agent, penetrant, and solvent oil, which is 30:20:20:30. The light-shield agent is 5% of the total mass of the rock asphalt, reducing agent, penetrant, and solvent oil, and the dispersant is 0.4%. Digital image technology was used to provide an accurate measurement of the LS pre-maintenance agent penetration depth, evaluate its permeability and reasonable amount. We then used the rolling thin film oven test (RTFOT) to analyze its effect on aged asphalt and evaluated the restoration performance. Using the strong UV radiation aging test, we analyzed its anti-light aging performance. The results showed that pavement must be closed at least 2 h after the brushing LS pre-maintenance agent has been applied and this can be extended to upwards of 8 h time permitting. A dosage of 0.5 $kg/m^2$ can ensure sufficient penetration depth and curing effect. Furthermore, the agent shows excellent restorative and anti-light aging abilities, which can effectively improve the low-temperature performance of aged asphalt and meet the pre-maintenance requirements for asphalt pavement, especially in strong UV radiation areas.

**Keywords:** highway engineering; light screening preventive maintenance agent; optimum proportion; permeability; restore performance; light aging resistance

## 1. Introduction

Asphalt pavement is affected by traffic load and other environmental factors during its service period. Its performance gradually declines, as shown in Figure 1, and therefore taking active pre-maintenance before significant damage can occur effectively prevents the early disease of asphalt pavement and improves road conditions [1,2]. How to choose the targeted maintenance schedule means understanding the right time, and has been the focus of maintenance engineering research over many years. Specifically, high altitude areas, strong ultraviolet (UV) radiation, and thin asphalt surface cause serious damage and can significantly age asphalt. The UV aging range typically reaches 20–25% thickness of the surface layer, which means higher requirements for maintenance materials and technology [3]. Although some progress has been made, such as micro-surfacing and fog sealing, the application effect of existing technology is limited. Research into asphalt pavement pre-maintenance materials in strong UV radiation areas is relatively scarce [4,5]. Thus this study explores existing scarce research and develops a novel treatment schedule.

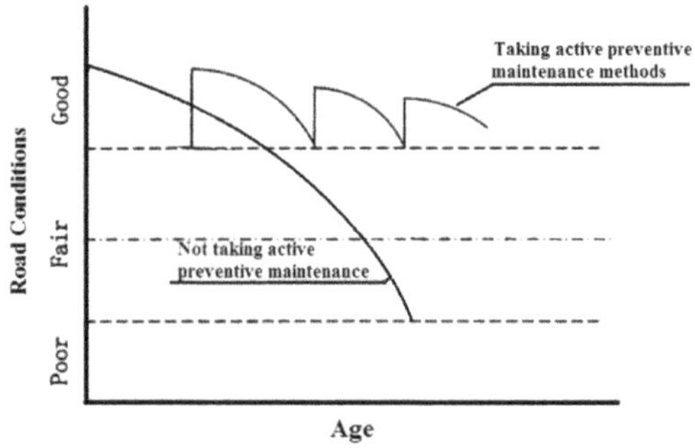


**Figure 1.** Road conditions under taking/not taking pre-maintenance work.

Existing studies have conducted research on aging and regenerating of asphalt. X.H. Lu used the thin film oven test (TFOT) and rolling thin film oven test (RTFOT) tests to simulate the aging process of asphalt. They studied the chemical and rheological properties of aging asphalt by infrared spectroscopy, chromatography, and dynamic mechanical analysis, then compared the effects of the different aging methods [6]. M. Naskarton compared the effects of the RTFOT and UV aging tests on the thermal stability and adhesion of aged asphalt, using kinetic information to predict the service life of modified asphalt [7]. H.L. Zhang used X-ray diffraction to analyze the micro-structural changes in UV aging of SBS by adding organic montmorillonite (OMMT), analyzing the light aging resistance of the OMMT/SBS asphalt [8]. W.B. Zeng used a dynamic shear rheometer (DSR) and FTIR (Fourier transform infrared spectroscopy) to analyze the temperature effect on the asphalt UV aging process, comparing the significance of temperature in an aging test [9]. Z.Q. Zhang combined component tests, combined with conventional and Superpave indicators to analyze asphalt change as a result of the RTFOT and pressure aging vessel (PAV) tests, comparing asphalt composition changes and anti-aging properties during the aging process [10]. The above research provides a multi-angle analysis on asphalt rheological properties and chemical changes in thermal oxygen aging and UV aging methods. The research outlined above of asphalt UV aging is closely related to this study.

For asphalt regeneration, the essence is to add new asphalt and an aromatics reducing agent into the aged asphalt to restore its original performance. P.L. Cong simulated a long-term light and heat aging process of an asphalt pavement, establishing a standard aging test model that analyzed the anti-aging and storage stability of antioxidants and UV absorbers [11]. Michal Varaus evaluated the restorative effects of different reducing agents on aged asphalt by bending beam rheometer (BBR) and DSR, especially on asphalt viscoelastic and low temperature performance [12]. A.Q. Chen used AFM (Atomic Force Microscopy) to study the mechanism of asphalt aging and regeneration, analyzing the change in asphalt composition from the micro-structure [13]. The above research above provides a solid theoretical basis and the experimental ideas guiding this study.

From the viewpoint of maintenance time, scientifically determining the maintenance time can effectively reduce the cost, prolong road life, and improve surface conditions; however, this requires a matched monitoring, data collecting, and analyzing system [14]. Q.Z. Wang summarized the effectiveness and applicability of common pre-maintenance measures, such as fog seal, slurry seal, thin cover and micro-surfacing. Established pavement preventive evaluation systems provided the decision-making basis [15]. C.H. Wang focused on the suitability of pre-maintenance, established an optimization model for timing and countermeasures based on the DEA (Data envelopment analysis) method, and provided a more reliable decision-making solution [16].

Asphalt pavement pre-maintenance material is mainly made up of asphalt, penetrant, and functional components, which are able to close surface micro-cracks, supply and activate aged asphalt, solidify loose aggregates, and delay disease development [17]. Among them, the Rhinophalt agent developed by the Britain's ASI Company, the ERA-C agent used by the United States, and the HAP agent developed by China have a wide range of applications [18–20]. C. Zhang established the evaluation methods and indicators for different pre-maintenance materials and provided a reasonable evaluation standard for pre-maintenance materials performance [21]. However, existing materials still have an application limitation for pavement in strong UV radiation areas with severe light-aging damage.

Therefore, this paper aims at the serious aging problem of asphalt pavement in strong UV regions by developing a LS pre-maintenance agent. We analyze and determine the raw materials and optimum composition ratio by contrast test. Using digital imaging methods, we evaluate the LS pre-maintenance agent's infiltration effect and its reasonable dosage. Using the RTFOT test, we evaluate the LS pre-maintenance agent' restorative effects on aged asphalt, contrasting the softening point of 25 °C penetration and the 15 °C ductility changes of aged asphalt after adding the LS pre-maintenance agent and Rhinophalt. Using an artificial UV aging test to accelerate the simulate natural UV aging process, we analyze the anti-light aging performance by calculating the UV aging mass loss rate of the LS pre-maintenance agent. Through the above research, we hope to provide new material and technical guidance for asphalt pavement pre-maintenance engineering in special areas.

## 2. Materials

According to asphalt pavement maintenance requirements, the LS pre-maintenance agent requires good penetration and bonding performance, which will ensure the penetration effect after spraying. By restoring and activating the aged asphalt by supplying lightweight components, this improves the asphalt's performance. Strong light screening abilities are able to inhibit the light aging damage in strong UV areas. Quick-drying and long-term storage stability reduces traffic closure time and meets the storage requirements for large-scale maintenance engineering. Finally, the construction economy, convenience, safety, and environmental protection must be guaranteed. In summary, the raw materials of the LS pre-maintenance agent include:

1. Asphalt with low viscosity and consistency, high penetration, and ductility to ensure permeability.
2. Organic reducing agent with a high light component content and low viscosity to impregnate the asphalt layer.
3. Penetrant to balance the solute and solvent molecular weight and ensure the long-term stability.
4. Solvent oil to speed up the evaporation and curing process.
5. Dispersant to reduce the asphaltene precipitation and extend material storage time.
6. Light-shield agent (LS agent) with high dispersibility, hiding power, and UV shielding capabilities.

### 2.1. Reducing Agent

We selected five commmonly used reducing agent materials (A–E). Agent A was produced by the Xi'an Xianyang Guolin Asphalt factory (Xi'an, China). B and C were produced by the Xi'an Yujian Petrochemical Co., Ltd (Xi'an, China). D and E were produced by the Shanghai Mingzhi Industrial Co., Ltd (Shanghai, China). According to the «Highway Engineering Asphalt and Asphalt Mixture Test Rules» (JTG E20-2011) [22], which measured their component proportions, including 25 °C penetration, the softening point, 5 °C ductility, 60 °C Brookfield viscosity, and other indexes. The results are shown in Table 1.

**Table 1.** The components and indexes of the reducing agent.

| Indexes | A | B | C | D | E |
|---|---|---|---|---|---|
| Saturated fragrance/% | 41.2 | 46.2 | 82.7 | 88.5 | 84 |
| Aromatics/% | 48.1 | 47.8 | 13.2 | 8.1 | 11.5 |
| Colloid/% | 3.5 | 2.1 | 0.5 | 0.2 | 0.1 |
| Asphaltene/% | 7.2 | 3.9 | 3.6 | 3.2 | 4.4 |
| Penetration/0.1 mm (25 °C,100 g, 5 s) | 140.7 | 141.3 | 165 | 185.3 | 201.3 |
| Softening point/°C | 44.8 | 44.7 | 42.2 | 40.9 | 40.3 |
| Ductility/cm (5 °C) | 71.1 | 71.9 | 83.7 | 85.5 | 96 |
| Viscosity/Pa·s (60 °C) | 1.105 | 1.085 | 0.875 | 0.755 | 0.685 |
| Mass loss/% (RTFOT) | −0.15 | −0.24 | −0.46 | −0.27 | −0.15 |
| Residual penetration ratio/% | 81 | 80.5 | 71.2 | 79.1 | 82.1 |
| Residual softening point value/°C | 1.7 | 1.9 | 4.5 | 1.3 | 1.2 |
| Residual ductility/cm (5 °C) | 46.1 | 47.5 | 46.7 | 120.5 | 106.2 |
| Residual viscosity ratio | 1.13 | 1.21 | 1.56 | 1.3 | 1.08 |

From Table 1, agents D and E had the highest penetration and ductility and smallest viscosity before and after the RTFOT test, which were more satisfied with the requirements of the reducing agent, but their price is often higher than the others. The indexes of C decreased significantly after aging; residual penetration ratio and ductility reduced by 71.2% and 44.2% compared with the data before aging, showing a poor anti-aging ability, and therefore was not suitable to use. The indexes between A and B were close, but A had a higher content of light components, which made it easier to form a stable system. It also had a stronger dissolving ability to disperse and dissolve the aging asphaltene with low viscosity and strong penetration to meet the needs of the mixing and spraying work. The mass loss and viscosity of A were relatively small after the RTFOT test.

The road spraying test was given to A, B, D, E, respectively. Agents A, B, D and E were sprayed on asphalt pavement then left for 48 h. The appearance was observed and the results are showed in Figure 2. The pavement appeared yellowing and bleeding 48 h after the D and E spray application. It turned slightly green after spraying B; however, it was still black and shiny after spraying A, which showed the overall best effect during the spraying test. In summary, A was selected as the reducing agent of the LS pre-maintenance agent.

## 2.2. Penetrant

Three penetrant agents (A, B, C) were selected. Each were colorless and transparent liquids with strong penetrating ability. Their densities are 0.89, 0.88, and 0.95 g/cm$^2$; their molecular weights are 178.3, 220.4, and 221.3 g/mol; and their boiling points are 143, 190, and 220 °C. During the heating test, A volatilized rapidly and produced a strong irritating odor after heating. B quickly condensed and became filamentous, losing its permeability. C was stable and maintained permeability after heating; its small molecular structure could penetrate the surface of the cemented material into the asphalt layer and react with the water molecules in the air to form a waterproof layer. It is characterized by stable macromolecules and as a result built a better bonding between the asphalt and aggregate. C was significantly better than A and B, so C was selected as the penetrant of the LS pre-maintenance agent.

## 2.3. Solvent Oil

Selected No.6 solvent oil, which is a colorless transparent liquid with a relative density of 0.65–0.701, a *N*-hexane content of about 30%, and 2,4-dimethylpentane and 2,3-dimethylbutane each about 20%. It is a mixture of various lower alkanes. The recent extraction process of No. 6 solvent oil can effectively reduce its harmful ingredient content, which means very low toxicity. It can be dissolved in benzene, chlorine, acetone and other organic solvents yet is insoluble in water, which is a basic organic chemical of raw materials.

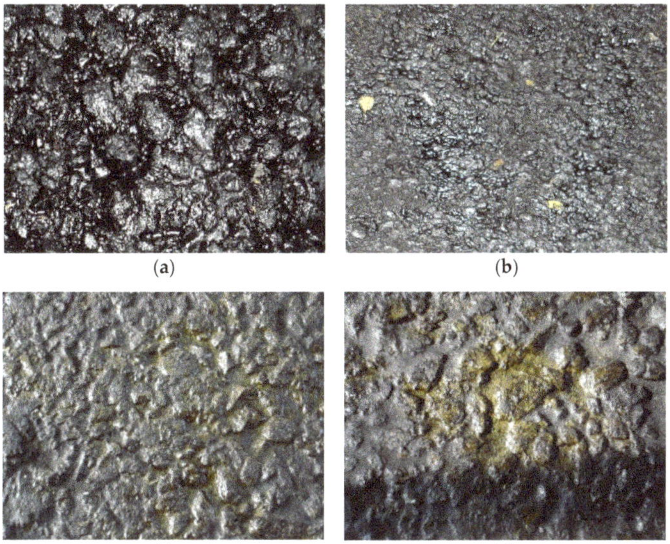

(a) (b) (c) (d)

**Figure 2.** The comparison of the road spraying test. (**a**) Agent A (black and shiny); (**b**) Agent B (green and bleeding); (**c**) Agent D (yellowing and bleeding); (**d**) Agent E (yellowing and bleeding).

### 2.4. Matrix Asphalt

110# road asphalt, SK90# asphalt, and ordinary rock asphalt were selected for the test. Among them, rock asphalt occurs in nature. Its physical properties tend to be coal while its main components are asphaltene, colloidal, and mineral asphalt matrix. The softening point can reach to 160–175 °C with good high temperature stability. Its high nitrogen content can acquire large viscosity and better anti-oxidation ability, respectively mixing them with solvent oil and penetrant can provide long-term precipitation observation. The results showed that the mixture prepared by 110# and SK90# resulted in significant precipitation, which was caused by the large molecular weight difference between the asphalt and solvent oil. This can easily break the steady state after mixing and the asphalt continued to sink until precipitation. By contrast, there was less precipitation in the rock asphalt and it showed better quick-drying performance through the pavement spraying test. The drying time was much shorter than the 110# and SK90# asphalt mixture. In summary, rock asphalt was selected as the matrix asphalt of the LS pre-maintenance agent.

### 2.5. Dispersant and LS Agent

The materials and adding dosage of the dispersant and LS agent are determined in Section 3.4, after the optimum ratio determination of the rock asphalt, reducing agent, penetrant, and solvent oil.

## 3. Optimum Composition Ratios

The dissolution degree of the rock asphalt to the penetrant and solvent oil mixture reflects the permeability and solubility, which determines the optimum range of the penetrant and solvent oil. The tratio with quick-drying properties is determined by a consolidation test that analyzes the percentage of the remaining mixture after air drying and that viscosity changes that occur after long-term storage. This determines the optimum proportion of the four main components. Finally, use consolidation and the UV aging test determine the dispersant and LS agent.

### 3.1. Optimum Ratio of Penetrant and Solvent Oil

We ground solid rock asphalt into fine powder and then mixed rock asphalt powder and penetrant (Agent C) by cutting and stirring, then added solvent oil under the premise of no precipitation, tilted the container, and observed the bottom precipitation after standing. To analyze the solubility of the mixture on rock asphalt, the asphalt dosage must not be too little; the penetrant mass must be maintained as 100 g and the limited rock asphalt minimum mass by 10% of the penetrant. The results are shown in Table 2.

**Table 2.** Proportions and solubility results.

| Components | Proportions | | | | | | |
|---|---|---|---|---|---|---|---|
| Rock asphalt/g | 50 | 30 | 20 | 20 | 10 | 10 | 10 |
| Penetrant/g | 100 | 100 | 100 | 100 | 100 | 100 | 100 |
| Solvent oil/g | <50 | 50 | 100 | >100 | 100 | 200 | >200 |
| Precipitation/YES/NO | Yes | Yes | No | Yes | No | No | Yes |

According to Table 2, when the ratio of penetrant to solvent oil was 1:1-1:2, there was no precipitation; the solution had good solubility to the rock asphalt. The results were similar to the solubility of the penetrant and solvent oil on the reducing agent, and the solubility degree was greater. After comprehensively comparing the results, the optimum ratio rage of penetrant to solvent oil should be 1:1-1:2.

### 3.2. The Composition Ratio Based on Quick-Drying Test

#### 3.2.1. The Proportion of Rock Asphalt and Reducing Agent

Considering that the penetrant is expensive and produces a slightly irritating odor when it is heated, its hydrolysis reaction with water will affect road beauty, thus its dosage must be controlled. The initial penetrant proportion must be set as 20% of the four components. The mixtures were made according to different ratios and brushed on the road and penetration condition was observed. The results showed that when the rock asphalt and reducing agent mass was above 60%, the solution viscosity was too high. As shown in Figure 3, a thin film formed on the pavement, which hindered the penetration and volatilization process. The film edge tilted and fell off after solidifying, which means a poor maintenance effect and material waste. When the mass was less than 30% the solvent oil dosage was too high, resulting in an excessive volatilization rate that had a negative effect upon maintenance. In summary, the optimum ratio of the rock asphalt and reducing agent should be 30–60%.

**Figure 3.** Thin film on the pavement.

3.2.2. Quick-Drying Test Analysis

We set 11 proportion samples and performed a quick-drying test by putting them into an empty disk and recorded how quickly their surface dried and the full consolidation time. The results are shown in Table 3.

**Table 3.** 11 proportions and quick-drying test results.

| Components | Proportions | | | | | | | | | | |
|---|---|---|---|---|---|---|---|---|---|---|---|
| Number | 1 | 2 | 3 | 4 | 5 | 6 | 7 | 8 | 9 | 10 | 11 |
| Rock asphalt/% | 40 | 40 | 30 | 30 | 30 | 25 | 20 | 15 | 20 | 35 | 40 |
| Reducing agent/% | 20 | 15 | 15 | 10 | 20 | 25 | 30 | 35 | 40 | 15 | 10 |
| Penetrant/% | 20 | 20 | 20 | 20 | 20 | 20 | 20 | 20 | 20 | 20 | 20 |
| Solvent oil/% | 20 | 25 | 35 | 40 | 30 | 30 | 30 | 30 | 20 | 20 | 20 |
| Surface dried/h | 2 | 2 | 1.5 | 1.5 | 1.5 | 2.5 | 2.5 | 3.5 | >5 | 2.5 | 1 |
| Fully consolidation/h | 4 | 5 | 2.5 | 2.5 | 2.5 | 4.5 | 4.5 | 5 | >5 | 3 | 2.5 |

From Table 3, the consolidation speed was accelerated with the increase of the rock asphalt and solvent oil, when the mass of penetrant and solvent oil varied within 40–50%. Quick-drying performance was enhanced with the increase of rock asphalt. The main reason for this is because rock asphalt is solid at normal temperatures. Once the solvent evaporates, the precipitated rock asphalt immediately condenses into solid. Quick-drying performance was also affected by the reducing agent. No. 8 and 9 showed that when the reducing agents' mass were above 30%, drying time was obviously higher than the others and not conducive to maintenance. From No. 3–5, when the rock asphalt and reducing agent mass was less than 50%, or when the reducing agent was less than 20%, the drying speeds were relatively fast, which is beneficial for maintenance. In summary, by setting the mass ratio of the penetrant as 20% when the total mass of the rock asphalt and reducing agent was 40–50% and the reducing agent was less than 20%, the material could acquire a good quick-drying ability.

*3.3. Optimum Proportion Based on Storage Stability*

3.3.1. Air-Drying Test Analysis

To compare the effective retention rate after air-drying, we prepared 5 samples within the range of the ratio determined in Section 3.2, put them into an empty disk and then tested their percentage of remaining (*POR*) after 24 h air-drying using Rhinophalt as the contrast group. The calculation method is shown in Equation (1) and the *POR* results are shown in Table 4.

$$POR = \frac{G_2 - G_P}{G_1 - G_P} \times 100\% \quad (1)$$

*POR*—Percentage of remaining after air-drying, %;
$G_P$—The weight of the plate, g;
$G_1$—The weight of plate and wet material, g;
$G_2$—The weight of plate and dry material, g.

From Table 4, the 5 samples' *POR* results were all larger than Rhinophalt. The highest *POR* is No.1, which was 1.5 times than Rhinophalt. The *POR* values of No. 1–5 were all greater than the total ratio of the rock asphalt, reducing agent and penetrant, proving that the three components were all remaining after the air-drying. It showed that the proportion based on the quick-drying analysis was reasonable; in this range the agent had good quick-drying and storage performance.

**Table 4.** *POR* test results in 5 proportions.

| Samples | Rock Asphalt/% | Reducing Agent/% | Penetrant/% | Solvent Oil/% | Plate/g | Plate & Wet Material/g | Plate & Dry Material/g | *POR*/% |
|---|---|---|---|---|---|---|---|---|
| 1 | 30 | 20 | 20 | 30 | 37.3 | 47.4 | 44.7 | 73.3 |
| 2 | 30 | 15 | 20 | 35 | 48.5 | 55.4 | 53.2 | 68.1 |
| 3 | 30 | 10 | 20 | 40 | 52.6 | 61.8 | 58.5 | 64.1 |
| 4 | 25 | 25 | 20 | 30 | 51.9 | 57.6 | 55.9 | 71.9 |
| 5 | 20 | 30 | 20 | 30 | 51.5 | 61.8 | 58.5 | 70.1 |
| Rhinophalt | \ | \ | \ | \ | 37.1 | 47.4 | 41.9 | 47.7 |

3.3.2. Storage Stability Analysis

Storage stability is mainly reflected as the viscosity changes during long-term storage. The 5 samples above were stored in a cylinder where the tube mouth was tightened. A BROOKFIELD viscometer was used to periodically measure the viscosity change at room temperature. Since the solution is very volatile, the volatilization during the sampling interferes with the accuracy of the data in the upper cylinder, so only the data in the central (C) and bottom (B) part were recorded. The results are shown in Figure 4.

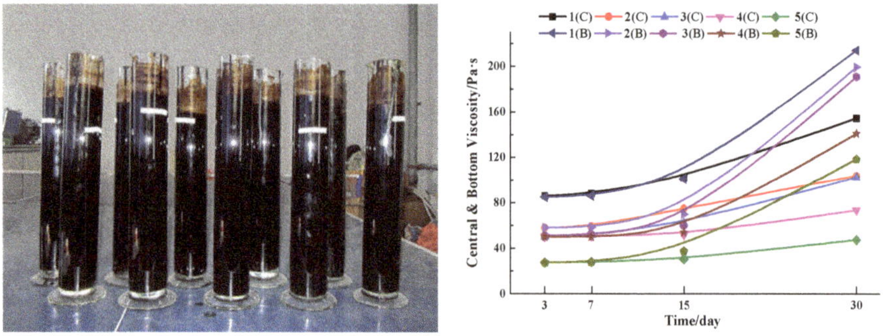


(a) (b)

**Figure 4.** Storage stability contrast test. (**a**) Storage stability samples; (**b**) Viscosity contrast results.

From Figure 4, after 3 and 7 days, the viscosity between the central and bottom were relatively small. The samples were stable, with slight changes mainly caused by temperature change. After 15 and 30 days, the viscosity showed a rapid increase because of the stratification; the viscosity gap between the central and bottom increased gradually. Comparing the 5 samples, the central and bottom viscosity difference of No. 1 and 4 were relatively small compared with 2, 3, and 5. Also No. 1 and 4 had less precipitation at the bottom, which was a controlled rock asphalt and reducing agent ratio between 30:20 and 25:25 with a penetrant and solvent oil ratio of 20:30, which resulted in better long-term storage stability and also less degradation and precipitation.

In summary, based on Sections 3.1–3.3, the optimum components ratio was determined as, rock asphalt: reducing agent: penetrant: solvent oil = 30:20:20:30.

*3.4. Dispersant and LS Agent*

3.4.1. Dispersant

We chose three dispersant materials (A, B, C). According to Figure 5, A was a white powder and commonly used as an asphaltene precipitation dispersant. B was a paraffin-like solid, used as a leveling agent with good dispersion and emulsification. C was a white ultrafine powder, used in the rubber industry with good lubricity and light stability.

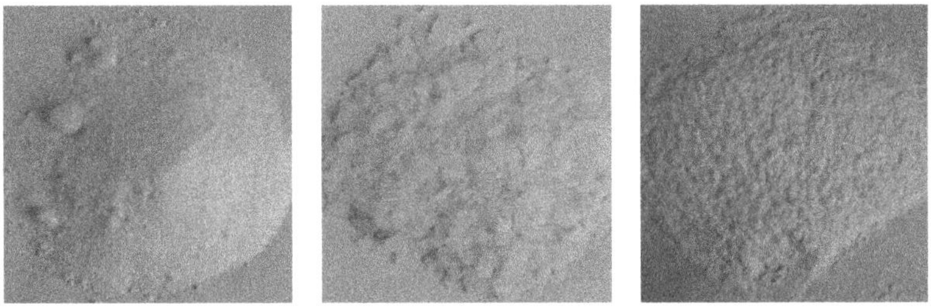

**Figure 5.** Dispersant appearance contrast. (**a**) Dispersant A; (**b**) Dispersant B; (**c**) Dispersant C.

The petroleum industry shows that a controlled dispersant dosage within 1% can significantly improve the stability of asphaltene in crude oil [23]; therefore, A, B, C were added into a pre-maintenance agent prepared at the optimum ratio. The adding amounts were 0.2, 0.4, 0.6, 0.8, and 1% of the mass of the pre-maintenance agent prepared according to the optimum ratio. The samples were sealed in glass bottles, and their consolidation condition was observed and recorded in cool, dark environment. The results are shown in Table 5.

**Table 5.** The Consolidation results of three dispersants.

| Consolidation Time/day | Dispersant Amount/% | | | | | |
|---|---|---|---|---|---|---|
| | 0 | 0.2 | 0.4 | 0.6 | 0.8 | 1 |
| A | 45 | 40 | 30 | 30 | 20 | 20 |
| B | 45 | 45 | 40 | 40 | 30 | 30 |
| C | 45 | 60 | >85 | 70 | 55 | 50 |

From Table 5, the consolidation time without any adding dispersant was 45 days. After adding A and B, the consolidation time gradually reduced with the increasing dispersant. The main reason was that adding the dispersant increases the proportion of small and medium substances, breaks the original stable state, and affects stability. The consolidation time of C was longer than A and B, and higher than 45 days in all dosages. The sample with 0.4% was not consolidated after 85 days. The storage stability of C was significantly better than A and B. From the molecular structure, A and B were short single-chain structures. C contained a positively charged and highly dispersed inorganic nuclei and two linear long hydrocarbon chains. The charged end can be wrapped in the asphaltene structure, preventing asphaltene flocculation and extending material storage time. In summary, we selected C as the dispersant agent; the dosage was 0.4% of pre-maintenance agent mass.

### 3.4.2. LS Agent

The LS agent selected was Nano-grade black powder. The main component is carbon black. We mixed it into SBRII-C modified asphalt (commonly used in high altitude strong UV regions), according to the dosage range determined by the existing studies of asphalt anti-UV aging additives [3]. The dosage was determined as 3, 5, and 7% of the asphalt mass and stirred for 10 min to ensure it dispersed evenly. The 25 °C penetration, softening point, and 5 °C ductility were measured before and after adding the LS agent. UV aging was performed for 12 h with a self-developed artificial UV aging device to measure the asphalt indexes again. The results are shown in Table 6.

**Table 6.** Asphalt indexes results.

| Indexes | 25 °C Penetration/0.1 mm | Softening Point/°C | 5 °C Ductility/cm |
|---|---|---|---|
| SBRII-C | 77.3 | 53.5 | 54.7 |
| SBR + 3% LS agent | 77.9 | 52.8 | 44.1 |
| SBR + 5% LS agent | 79.3 | 51.4 | 42.9 |
| SBR + 7% LS agent | 80.1 | 50.9 | 42.1 |
| Aged SBRII-C | 54.7 | 59.1 | 34 |
| Aged SBR + 3% LS agent | 56.3 | 56.9 | 26.1 |
| Aged SBR + 5% LS agent | 60.1 | 53.3 | 24.9 |
| Aged SBR + 7% LS agent | 60.9 | 53.2 | 23.8 |

From Table 6, the 25 °C penetration and softening point change were relatively small before UV aging. After aging, the penetration increased and the softening point decreased with the increase of the LS agent, of which the effect of 3% was limited, the effect of 5% and 7% were better and close to each other, proving that the LS agent can improve asphalt low temperature anti-cracking performance after UV aging. 5 °C ductility before and after aging were both reduced. The addition of the LS agent had a negative effect on asphalt ductility, the reason being the LS agent is a powdered solid, which will impact the plastic deformability of asphalt, thus its dosage should be controlled in order to reduce the negative effect to the asphalt while ensuring the anti-UV aging ability. Considering the significant degree of LS agent to asphalt, the effect of the LS agent in 5% and 7% were close, showing that a good anti-aging effect can be achieved in the 5–7% dosage. In order to reduce the adverse effects of the additive on asphalt rheology while controlling material costs, the dosage was determined as 5%.

## 4. Road Performance Test Plan

### *4.1. Penetration Test*

An insufficient penetration depth of the curing agent will not achieve the maintenance effect. Too much will cause material waste and affect the reasonable internal porosity of the pavement. We provide an accurate measurement of the LS pre-maintenance agent penetration depth with digital image technology following this method: we quantitatively painted a LS pre-maintenance agent on asphalt mixture specimens then used a camera to shoot the infiltration interface, adjusted the brightness and contrast by Photoshop, outlined the penetration range and fill with color, performed image scale calibration with IPP software, converted the pixel into length, separated the permeable region and measured its area and width, and then divided them to achieve the average penetration depth.

The actual dense gradation asphalt pavement porosity is about 6–7%, which gradually compacts to 3%. We prepared dense graded asphalt mixture specimens by the static pressure method, controlled the porosity under 3, 4, 5, 6, and 7%, then evenly brushed 0.5 and 1 kg/m$^2$ LS pre-maintenance agent and measured the average permeation depth after 0.5, 2, 8, and 24 h. Then we analyzed the relationship of the penetration depth, brushing amount, and penetration time, the digital image method is shown in Figure 6.

### *4.2. Restore Performance Test*

The restore effect of the LS pre-maintenance agent is determined by the recovery degree of the aged asphalt. Ordinary road petroleum asphalt was used as the matrix asphalt, based on the «Highway Engineering Asphalt and Asphalt Mixture Test Rules» (JTG E20-2011) [22], the sample was prepared and assayed according to the softening point, 25 °C penetration, and 15 °C ductility, then the asphalt was aged for 3 h by RTFOT and equal amounts of LS pre-maintenance agent and Rhinophalt agent were added. In the current maintenance project, the adding amount of the reducing agent is about 5–10% of the asphalt mass [24]. The selected reducing agent amount is 5, 7, and 9%, and the equivalent to the LS pre-maintenance agent is 25, 35, and 45% of the asphalt mass. Melted state aging asphalt is added into LS and Rhinophalt agent and for stirred 15 min. The data is compared to evaluate the restorative effects.

**Figure 6.** Infiltration interface and digital image analysis.

### 4.3. Light Aging Resistance Test

The impact of UV radiation on asphalt pavement is more serious than thermal aging, especially in high altitude areas with strong UV radiation [7,25]; however, natural light aging is a long process, which is not conducive to analyze, therefore, an indoor aging test that simulated and accelerated the natural process was used. By irradiating asphalt specimens with strong UV radiation uninterruptedly and observing the appearance change, we can calculate and compare the UV aging mass loss rate. A self-developed artificial strong UV radiation aging device is shown in Figure 7. It consists of two installed 1000 W strong UV lamps on top, two air circulation fans that are set to simulate the ventilation environment, and a ø140 mm × 9.5 mm flat disc that holds the sample.

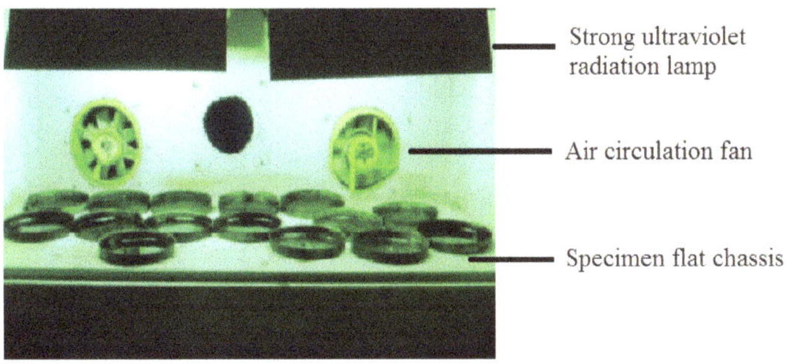


**Figure 7.** Artificial strong UV radiation aging test.

According to the relevant literature, take Lhasa, Tibet in China for example, its total annual radiation value is 195 kcal/cm$^2$ and the annual radiation time is 3006 h. The radiation intensity of the UV lamp is 1000 W × 2, the intensity on the unit surface is 390.3 kcal/cm$^2$. Thus continuous exposure over 48 h is equivalent to two years of natural radiation in Lhasa. We loaded LS and Rhinophalt agent into six plates, each 15 g, about 1 mm thick, and put them into the device and regularly rotated the position to ensure uniform radiation. We observed the appearance change after UV aging to analyze the changes in morphology and rheological properties, and calculated the average UV aging mass loss rate after 8, 12, 24, and 48 h to analyze the remaining amount of active ingredients, then contrasted them with the results of the Rhinophalt.

## 5. Road Performance Test Results Analysis

### 5.1. Penetration

The UV radiation spread with aging asphalt molecules into a 1 mm depth. Because of the existing voids, the depth of radiation ultimately reaches the maximum individual particle mix size, which is about 10 mm. Therefore, we use 10 mm as the standard penetration depth. The average penetration depth results at 0.5–24 h are shown in Figure 8.

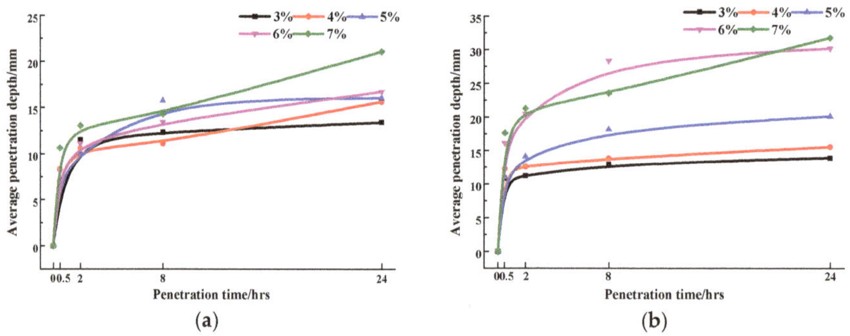


**Figure 8.** The comparison of average penetration depth. (**a**) 0.5 kg/m$^2$; (**b**) 1 kg/m$^2$.

In 0.5 kg/m$^2$, the surface groove of the specimen was fully filled, a bubble came out from a larger gap that the agent was quickly penetrating and the average penetration depth in 5–7% porosity was over 10 mm after 2 h. It kept growing and reached 15–20 mm after 24 h. In 1 kg/m$^2$, the average penetration depth reached 10 mm after 2 h, and 15–30 mm after 24 h. We can see that both dosages can completely penetrate the aged asphalt layer. The permeation depth in 2 h and 8 h accounted for about 50% and 75% of the whole 24 h results. The first 2 h was the main infiltration period. By chosing 1 kg/m$^2$ a faster and deeper penetration effect can be obtained, but it requires higher maintenance costs. In summary, the pavement should be closed at least 2 h after spraying the LS pre-maintenance agent, a conditional section can be extended to 8 h and above. 0.5 kg/m$^2$ can ensures sufficient penetration depth and effect while controlling the cost.

### 5.2. Restore Performance

The results of the matrix asphalt with adding 25, 35, and 45% asphalt mass of LS pre-maintenance agent and Rhinophalt are shown in Table 7.

**Table 7.** The asphalt indexes comparison results.

| Indexes | Original | Aging 3 hrs | 25% LS | 35% LS | 45% LS | 25% Rhinophalt | 35% Rhinophalt | 45% Rhinophalt |
|---|---|---|---|---|---|---|---|---|
| Softening point/°C | 45.9 | 53.3 | 51.5 | 36.6 | 20.3 | 43.3 | 39.8 | 22.1 |
| 25 °C Penetration/0.1 mm | 100.2 | 62.3 | 75.7 | 197 | 223 | 95.8 | 207 | 218 |
| 15 °C Ductility/cm | 158.4 | 67.1 | 33 | 29.6 | 20.9 | 31.5 | 27.9 | 20.3 |

From Table 7, the softening point declined with the increase of the LS pre-maintenance agent. 25 °C penetration increased significantly compared to 3 h aging. The softening point in adding 45% LS and Rhinophalt agent decreased by 63% and 57%. 25 °C penetration increased by 261% and 247% and it can be seen that the LS pre-maintenance agent was similar to Rhinophalt on restorative performance. The softening point and penetration of the aged asphalt varied greatly with the increase of the agent. In general, when 25 °C penetration drops below 2 mm, severe cracking of the road surface occurs.

Above 3 mm acquires good crack resistance ability, which means the LS pre-maintenance agent can significantly improve the anti-cracking ability of asphalt at low temperature. For the softening point, it is an important index to show the temperature sensitivity of the asphalt. Reducing the softening point after adding the LS pre-maintenance agent is very beneficial and improves the low temperature crack resistance of the asphalt. As for ductility, it is generally accepted that the road surface is in a crackable state if 15 °C ductility is less than 5 cm. 15 °C ductility in adding two agents were both more than 20 cm, which can meet the requirements of anti-cracking and water sealing. The effect of the two agents on the asphalt indexes were slightly different, because of the solvent oil in the LS pre-maintenance agent, which was more volatile than the naphtha in Rhinophalt, this produced the difference during the process of heating. In summary, the LS pre-maintenance agent had good restorative ability to the aged asphalt, which can effectively improve asphalt low temperature performance.

### 5.3. Light Aging Resistance

The UV aging mass loss rate (*MLR*) calculation method is shown in Equation (2), the average *MLR* results are shown in Figure 9.

$$MLR = \frac{G_1 - G_2}{G_1} \times 100\% \tag{2}$$

*MLR*—Mass loss rate, %;
$G_2$—The residual mass of the sample after light aging, g;
$G_1$—The initial mass of sample before light aging, g.

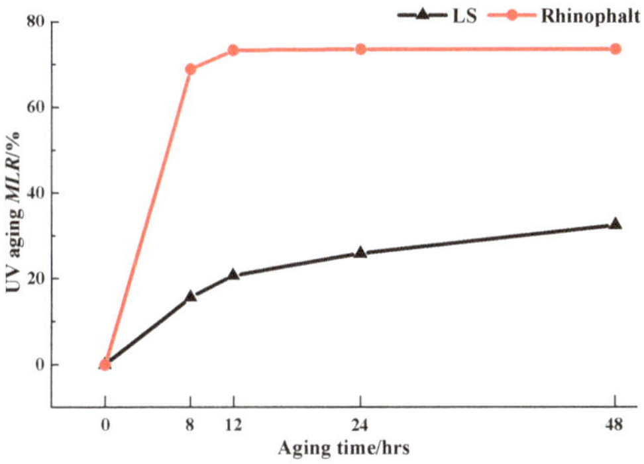


**Figure 9.** Mass loss rate (*MLR*) results versus UV aging time.

From Figure 9, the *MLR* of the LS pre-maintenance agent was obviously less than the Rhinophalt, moreover, its trend was more stable; after 48 h it reached 32.4%. The results of the Rhinophalt changed fast within 8 h and stabilized after 12 h, finally reaching 73.5%.

The appearance comparison results are shown in Figure 10.

As shown in Figure 10, the two materials were all black liquid before aging and their surfaces were smooth and consistent. For the LS pre-maintenance agent, the surface formed thin film with slight wrinkles after 8 h, the folds were further deepened with slight cracks after 12 h, the film shrunk, lower liquids became harder and brittle after 24 h, and the surface completely hardened after 48 h. For Rhinophalt, on the specimen appeared small holes and the material continuously reduced and began to harden after 8 h. The crack appeared after 12 h. The specimen was completely hardened and had lost its consistency and viscosity after 24 h.

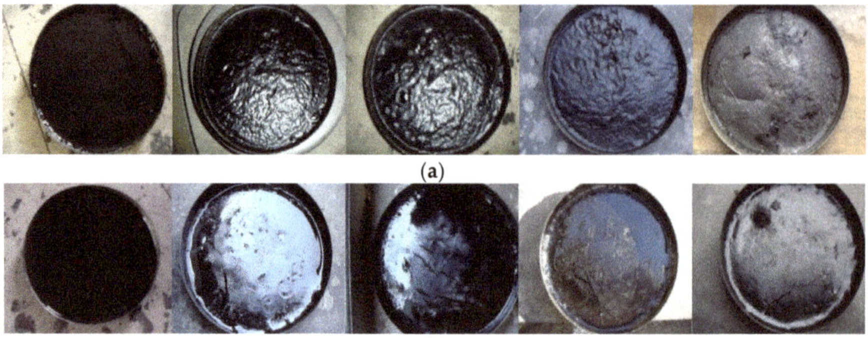

(a)

(b)

**Figure 10.** Appearance comparison after UV aging (0, 8, 12, 24, 48 h); (**a**) LS pre-maintenance agent; (**b**) Rhinophalt.

In summary, the LS pre-maintenance agent was completed aged after 48 h UV radiation, which was 2 years of natural UV radiation in Lhasa, and much longer in other lower radiation areas. The anti-light aging time of Rhinophalt was only 25% of the LS pre-maintenance agent. From the light aging resistance effect, the UV aging *MLR* of the LS pre-maintenance agent in 48 h was 32.4%. The film formed during the early aging stage can hinder the UV effect on a lower structure, thus the result of Rhinophalt was 73.5%, more than two times the LS agent. Most of the components had lost, only leaving a little rock asphalt. We can see that the LS pre-maintenance agent had better anti-light aging effects than the Rhinophalt, and can meet the requirements of pre-maintenance in strong UV areas. We can choose to spray a year and a half after the first pre-maintenance work, supply the function loss, and repair micro-cracks due to UV radiation and low temperature shrinkage, and thus prevent the disease development of the asphalt layer.

## 6. Conclusions

The study aims at the early damage of asphalt pavement in strong UV radiation areas, develops a LS pre-maintenance agent, determines its optimum composition ratio through indoor tests, and studies its effect on road performance. The main conclusions are as follows:

(1) The asphalt pavement maintenance function is determined and the main components of the LS pre-maintenance agent are rock asphalt, reducing agent, penetrant, solvent oil, dispersant, and LS agent. The raw material types are determined by contrast tests.

(2) The optimum ratio of the components is determined based on the best proportion of penetrant and solvent oil for solubility ability. The best proportion for quick-drying, long-term storage ability, and the amount of dispersant and LS agent, rock asphalt, reducing agent, penetrant and solvent oil is 30:20:20:30. The LS agent is 5% of the four components total mass. The dispersant is 0.4%.

(3) Digital image technology is used in the permeability test to provide accurate measurements of the LS pre-maintenance agent penetration depth. The results showed that the pavement should be closed at least 2 h after spraying and can extended to 8 h time permitting. The dosage of 0.5 $kg/m^2$ can ensure sufficient penetration effect.

(4) Use the RTFOT test to compare the LS pre-maintenance agent regeneration effect to the indexes of aged asphalt. The results will show that it has excellent restore and activation ability to the aged asphalt and greatly improves the low temperature anti-cracking ability of asphalt pavement.

(5) Compare the appearance and *MLR* results of the LS pre-maintenance agent after the artificial UV radiation aging test. The anti-light aging time of the LS pre-maintenance agent can last 2 years in

Lhasa. Its light resistance performance is much better than the Rhinophalt agent, which can meet the requirements of pre-maintenance work in strong UV radiation areas.

**Author Contributions:** Y.T. and B.M. conceived and designed the experiments; Y.T. and K.T. performed the experiments; N.L. and X.Z. analyzed the data; K.T. contributed reagents/materials/analysis tools; Y.T. wrote the paper.

**Acknowledgments:** The writers wish to acknowledge the financial support of this research by the Training Project of High Level Technical Personnel in Transportation Industry (No. 213021160088), National Science Foundation of China (No. 51708044), and the fundamental Research Funds for the Central Universities (No. 310821161015, 300102218408).

**Conflicts of Interest:** The authors declare no conflicts of interest.

## References

1. Li, N.Y. Highway Maintenance and Pavement Preservation Strategies in Canada. In Proceedings of the Tenth International Conference of Chinese Transportation Professionals (ICCTP), Beijing, China, 4–8 August 2010; pp. 3948–3956.
2. Wood, T.J.; Olson, R.C.; Lukanen, E.O. *Preventive Maintenance Best Management Practices of Hot Mix Asphalt Pavements*; LRRB, MN/RC, 2009-18; Minnesota Department of Transportation: Saint Paul, MN, USA, 2009.
3. Ye, F.; Sun, D.Q.; Huang, P.; Zhu, J.P.; Zhou, Z.D. Analysis of Asphalt Photooxidation Aging Property under Intensive Ultraviolet. *China J. Highw. Transp.* **2006**, *19*, 35–38.
4. Russell, G. *Hicks. Selecting a Preventive Maintenance Treatment for Flexible Pavements*; U.S. Dept. of Transportation: Washington, DC, USA, 2000.
5. Li, N.L.; Li, T.H.; Pei, J.Z.; Wang, D.W. The Study Situation of the Anti-aging Performance of Asphalt. *Mater. Rev.* **2007**, *21*, 84–86.
6. Lu, X.H. Ulf Isacsson. Effect of Ageing on Bitumen Chemistry and Rheology. *Constr. Build. Mater.* **2002**, *16*, 15–22. [CrossRef]
7. Naskar, M.; Reddy, K.S.; Chaki, T.K.; Divya, M.K.; Deshpande, A.P. Effect of Ageing on Different Modified Bituminous Binders: Comparison between RTFOT and Radiation Ageing. *Mater. Struct.* **2013**, *46*, 1227–1241. [CrossRef]
8. Zhang, H.L.; Yu, J.Y.; Wang, H.C.; Xue, L. Investigation of Microstructures and Ultraviolet Aging Properties of Organo-montmorillonite/SBS Modified Bitumen. *Mater. Chem. Phys.* **2011**, *129*, 769–776. [CrossRef]
9. Zeng, W.B.; Wu, S.P.; Wen, J.; Chen, Z. The Temperature Effects in Aging Index of Asphalt during UV Aging Process. *Constr. Build. Mater.* **2015**, *93*, 1125–1131. [CrossRef]
10. Zhang, Z.Q.; Liang, X.L.; Li, P. Evaluation Method of Asphalt Aging Properties. *J. Traffic Transp. Eng.* **2005**, *5*, 1–5.
11. Cong, P.L.; Wang, J.; Li, K.; Chen, S. Physical and Rheological Properties of Asphalt Binders Containing various Antiaging Agents. *Fuel* **2012**, *97*, 678–684. [CrossRef]
12. Varaus, M.; Koudelka, T.; Sperka, P. Rejuvenator Influence on Aged Binder Material Properties. *Key Eng. Mater.* **2017**, *730*, 380–388. [CrossRef]
13. Chen, A.Q.; Liu, G.Q.; Zhao, Y.L.; Li, J.; Pan, Y.; Zhou, J. Research on the Aging and Rejuvenation Mechanisms of Asphalt using Atomic Force Microscopy. *Constr. Build. Mater.* **2018**, *167*, 177–184. [CrossRef]
14. Peshkin, D.G.; Hoerner, T.E.; Zimmerman, K.A. *Optimal Timing of Pavement Preventive Maintenance Treatment Applications*; NCHRP Report 523; Transportation Research Board: Washington, DC, USA, 2004.
15. Wang, Q.Z.; Liu, S.Y.; Liu, J.L.; Liang, Y.G. Adaptability Analysis of Preventive Maintenance Measures for Asphalt Pavement. *Adv. Mater. Res.* **2012**, *594*, 1471–1476. [CrossRef]
16. Wang, C.H.; Wang, L.J.; Bai, J.H.; Liu, Y.D.; Wang, X.C. Research on Integration Optimization of Asphalt Pavement Preventive Maintenance Timing and Countermeasures During Period. *China J. Highw. Transp.* **2010**, *23*, 27–34.
17. Estakhri, C.K.; Agarwal, H. *Effectiveness of Fog Seals and Rejuvenators for Bituminous Pavement Surfaces*; Texas Transortation Institute: College Station, TX, USA, 1991.
18. Zhang, X.D.; Duan, Q.M.; Wu, T.; Hou, Y. Study on the Performance Analysis of Rhinophalt Pavement Material. *J. Highw. Transp. Res. Dev. (Appl. Technol. Ed.)* **2012**, *10*, 71–73.

19. Chen, R.S.; Ji, T.J.; Dou, Y.N.; Liu, B.Q.; Wang, Q.; Jiang, L. Applied Research of ERA-C on Shanghai-Nanjing Expressway. *East China Highw.* **2000**, *6*, 56–58.
20. Chen, S.L.; Wang, Y.B.; Shang, H.B. The Formation Mechanism of HAP Asphalt Pavement Protectant Strength. *J. Guangdong Commun. Polytech.* **2010**, *9*, 25–27.
21. Zhang, C.; Wang, C.Q.; Yang, S.X.; Wang, Y.X. Evaluation Indices and Methods of Performances of Preventive Maintenance Materials Spraying on Asphalt Pavement. *J. Traffic Transp. Eng.* **2009**, *9*, 1–6.
22. *Highway Engineering Asphalt and Asphalt Mixture Test Rules*; JTG E20-2011; China Communications Press: Beijing, China, 2011.
23. Zhou, Y.M.; Wang, J.Q.; Zhang, L.L.; Que, G.H. Effect of asphaltene dispersants on asphaltene concentration and asphaltene aggregation. *J. Fuel Chem. Technol.* **2008**, *36*, 65–69.
24. Li, P.L.; Zhang, Z.Q.; Wang, B.G.; Ding, Z. Evaluation on Anti-aging Performance of Asphalt Based on Changes of Viscosity and Ductility. *J. Chang'an Univ. (Nat. Sci. Ed.)* **2008**, *6*, 11–15.
25. Tan, Y.Q.; Wang, J.N.; Feng, Z.L.; Zhou, X.-Y.; Xu, H.-N. Ultraviolet aging mechanism of asphalt binder. *China J. Highw. Transp.* **2008**, *21*, 19–24.

applied sciences

MDPI

*Article*

# Characteristics of Moduli Decay for the Asphalt Mixture under Different Loading Conditions

**Songtao Lv [1,2], Xiyan Fan [1], Chengdong Xia [1,*], Jianlong Zheng [1], Dong Chen [1] and Lingyun You [2]**

[1] National Engineering Laboratory of Highway Maintenance Technology, Changsha University of Science & Technology, Changsha 410004, China; lst@csust.edu.cn (S.L.); fxy@stu.csust.edu.cn (X.F.); zjl@csust.edu.cn (J.Z.); 15878778765@163.com (D.C.)

[2] Department of Civil and Environmental Engineering, Michigan Technological University, Houghton, MI 49931, USA; liyou@mtu.edu

* Correspondence: xiachengdong@stu.csust.edu.cn; Tel.: +86-18569412272

Received: 4 May 2018; Accepted: 19 May 2018; Published: 22 May 2018

**Abstract:** In order to explore the moduli decay patterns of asphalt mixtures under different loading conditions, the nonlinear fatigue damage model was implemented in order to simulate the moduli decay patterns. Then, the direct tensile, indirect tensile, and uniaxial compression fatigue tests were employed under four kinds of stress levels with four parallel tests. The specimens of AC-13C Styrene-butadiene-styrene (SBS) modified mixtures were manufactured. Based on the test results, the decay patterns of the moduli during fatigue tests under different stress states were revealed, and the parameters of the damage model under different test conditions were obtained. By changing the values of the model parameters under a certain loading condition, fatigue curves were obtained. Then, the fatigue properties of asphalt mixtures under different stress states could be compared and analyzed directly. The result indicated that the evolution curves of fatigue damage for the direct tensile test, the indirect tensile test, and the uniaxial compression test all experienced three stages, which indicates that the fatigue damage characteristic of asphalt mixtures is non-linear. The decay patterns of the direct tensile moduli and the tensile moduli measured by the indirect tensile test are similar. The decay patterns of the uniaxial compression and the compression moduli measured by indirect tensile test are similar. The decay patterns of tensile and compressive moduli are obviously different. At the same cycle ratio state, the position of the decay curve for the compression moduli is higher than that of the tensile moduli. It indicates that the tensile failure is the main reason of the fatigue damage for asphalt mixture. The new analysis method of fatigue damage was proposed, which provides a possibility to compare the fatigue results that were obtained from different loading conditions and different specimen sizes.

**Keywords:** asphalt mixture; service life; fatigue test; moduli decay; loading condition; stress state

## 1. Introduction

Fatigue cracking is one of the most common distresses in asphalt pavement [1–3]. Fatigue cracking of asphalt pavement is an important consideration in asphalt mixture design and the structural design of flexible pavements [4,5]. Fatigue life design and maintenance strategies employ a challenging task due to the inherent nonlinear viscoelastic properties of asphalt mixtures and the complex cracking behavior observed in the field [6–10]. The damage appeared and accumulated gradually during the service life of asphalt pavement [11]. Under the action of cyclical traffic loads, the damage will increase and the pavement would suffer from the fatigue failure. Therefore, the service life of asphalt pavement would be decreased. More and more studies of fatigue performance for asphalt mixtures under the traffic and environmental conditions have been conducted [12,13]. In order to ensure the durability and usability of asphalt pavement, many researchers have been conducting the works of evaluation of fatigue characteristics via varied fatigue test methods in different specimen's size and different stress levels.

Xie Jun suggested that the direct tension strength got larger at higher strain loading rate [14]. Hyung Suk Lee conducted an indirect tensile test and determined the fundamental viscoelastic material property, which means that the strength of asphalt mixture were related to the temperature and loading time [15]. Waleed A. Zeiada employed the uniaxial compression test and observed that the fatigue endurance limit value increases by the increasing of temperature, asphalt content, and rest periods, whereas it decreased when the volume of air voids increases [16]. Different test methods produce different results. So far, three traditional test methods have been employed extensively: the direct tensile test, the indirect tensile test, and the uniaxial compression test. The direct tensile test refers to the test of a material under the action of a uniaxial tensile load. Under the assumption of homogeneity, the stress level or the stress ratio is same on each cross-section of the specimen during the test process, so the fracture position will appear at the weakest cross-section. The failure will occur for the accumulation of the micro-cracks [14]. The indirect tensile test (IDT, also called splitting test) is in a two-dimensional stress state. The lateral direction lies in the tension state, while the longitudinal direction lies in the compression one. It can simulate the stress state of asphalt pavement under traffic loads. The tensile stress near the crack area of the IDT specimen is uniform relatively [17]. The uniaxial compression test is also a one-dimensional stress test method. During the process of the uniaxial compression test, the stress level and the stress ratio are same on each cross-section of the specimen. The stress state on each cross-section is the one-dimensional compression stress state [18].

Fatigue test results of asphalt mixtures are sensitive to different test conditions. Different test methods will have the different loading conditions, which will lead to the different stress states for the specimen of asphalt mixtures. For each test method, the test results of fatigue are not correlated and are inconsistent with others. Moreover, the difference of fatigue test results is still relatively large. Therefore, even to the same material, the fatigue properties from the different test methods cannot be compared. So far, it cannot form a unified evaluation and comparison method among different fatigue test results of asphalt mixtures. Furthermore, the fatigue tests of asphalt mixtures are sensitive to the test types and size of the specimens [19,20].

The main purpose of this paper is to reveal the decay patterns of different modulus for asphalt mixture in fatigue test, which can be employed to improve the design accuracy of asphalt pavement.

## 2. Materials Preparation and Test Method

### 2.1. Materials

The main materials of the tests were the Styrene-butadiene-styrene(SBS) modified asphalt and the limestone aggregate. The AC-13 dense graded asphalt mixture was employed. The test results of asphalt are shown in Table 1. The test results of limestone aggregate are shown in Table 2. The gradation curve of dense graded asphalt mixture is shown Figure 1.

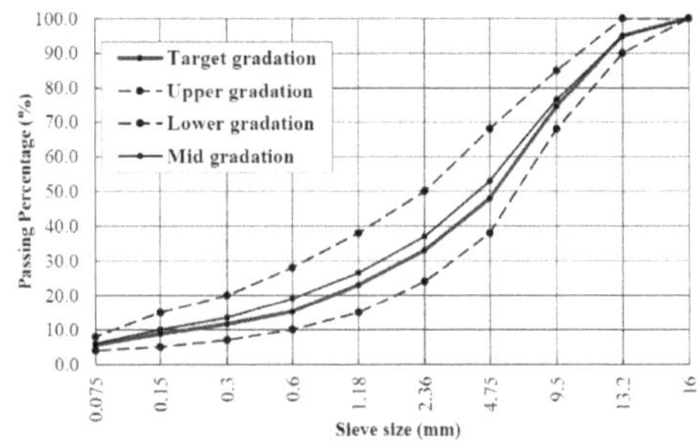


**Figure 1.** Gradation curve of dense graded asphalt mixture (AC-13).

**Table 1.** Test result of Styrene-butadiene-styrene (SBS) modified asphalt.

| Test Projects | | Test Standard: JTG F40-2004 (China) [21] | | |
|---|---|---|---|---|
| | | Technical Requirements | Test Results | Test Methods |
| Penetration (25 °C, 100 g, 5 s) (0.1 mm) | | 30~60 | 53.9 | T 0604-2000 |
| Penetration index PI | | ≥0 | 0.533 (r = 0.997) | T 0604-2000 |
| Ductility (5 cm/min, 5 °C) (cm) | | ≥20 | 36.3 | T 0605-1993 |
| Softening point (Ring ball) (°C) | | ≥60 | 72.5 | T 0606-2000 |
| Flash point (°C) | | ≥230 | 265 | T 0611-1993 |
| Solubility (%) | | ≥99 | 99.9 | T 0607-1993 |
| 135 °C kinematic viscosity (Pa.s) | | ≤3 | 2.37 | T 0620-2000 |
| Rolling Thin Film Oven Test (RTFOT) 163 °C 85 min | Mass loss (%) | ≤±1.0 | 0.23 | T 0609-1993 |
| | Residual penetration ratio (25 °C) (%) | ≥65 | 75.2 | T 0604-2000 |
| | Residual ductility (5 °C) (cm) | ≥15 | 23.4 | T 0605-1993 |

**Table 2.** Physical property of aggregates.

| Test Item | Technical Requirements [22] | Test Results | Test Methods |
|---|---|---|---|
| Crushed stone value (%) | ≤28 | 15.6 | T 0316-2005 |
| Los Angeles weared stone value (%) | ≤30 | 19.7 | T 0317-2005 |
| Apparent relative density (g/cm$^3$) | ≥2.5 | 2.6 | T 0321-2005 |
| Water absorption (%) | ≤2.0 | 1.4 | T 0304-2005 |
| Content of flat and elongated particles in coarse aggregate (%) | ≤15 | 8 | T 0312-2005 |
| <0.075 mm particle content (Washing methods) (%) | ≤1 | 0.2 | T 0310-2000 |
| Asphalt adhesion/grade | ≥4 | 6 | T 0616-1993 |
| Firmness (%) | ≤12 | 5 | T 0314-2000 |
| Content of soft stone (%) | ≤5 | 2.5 | T 0320-2000 |

The optimum asphalt-aggregate ratio was 5.2%, which was obtained by the Marshall Tests, and the test results are shown in Table 3.

**Table 3.** Marshall Test results at optimal asphalt content.

| Asphalt Aggregate Ratio (%) | Bulk Specific Gravity (g·cm$^{-3}$) | Volume of Air Voids VV (%) | Voids Filled with Asphalt VFA (%) | Voids in Mineral Aggregate VMA (%) | Marshall Stability (kN) | Flow Value (0.1 mm) |
|---|---|---|---|---|---|---|
| 5.2% | 2.44 | 4.5 | 67.2 | 16.1 | 12.7 | 27.9 |
| Specification Requirement [23] | / | 3~6 | 55~70 | >12.5 | >8 | 15~40 |

## 2.2. Specimens Preparation

According to the Chinese Standard Test Methods of Asphalt and Asphalts Mixtures for Highway Engineering (JTG E20-2011) [23], the block samples of 400 mm × 300 mm × 50 mm were fabricated through the equipment of vibrating compaction. Then, the beam specimens were cut from block samples into the size of 250 mm × 50 mm × 50 mm for the direct tensile tests, as shown in Figure 2a. During the uniaxial compression test, the Superpave Gyratory Compactor (SGC) was employed, which can control the volume of air voids of asphalt mixture more accurately. The previous studies [24,25] showed that the stress concentration would be occurred when the height is less than 50 mm. The specimens for uniaxial compression test were manufactured with 100 mm in height and 100 mm in diameter, which was shown in Figure 2b, and the indirect tensile specimens were prepared by cutting the top and the bottom surface of the specimens of uniaxial compression test to the size of 100 mm in height and 60 mm in diameter, which was shown in Figure 2c. All of the specimens were put in an environment chamber at 15 °C for 24 h before the tests. There were four parallel tests for each type of test.

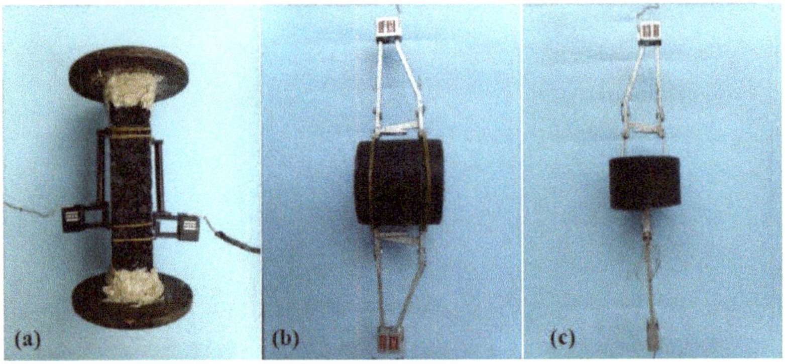


**Figure 2.** (**a**) Direct tensile test specimen; (**b**) Uniaxial compression specimen; and, (**c**) Indirect tensile specimen.

### 2.3. Test Method

In this paper, some influencing factors were considered for the fatigue test. The external factors, such as stress ratio, test temperature, loading frequency, and loading waveform were considered in the fatigue test. The internal factors, such as the fatigue test method, asphalt varieties, asphalt content, mineral type, mixture gradation, porosity, etc. were considered in the mix design. The specific test conditions are shown in Table 4.

**Table 4.** Influencing factors of fatigue test and the levels of each factor.

| Factor Type | Factor Name | Factor Level Number | Levels of Factor |
|---|---|---|---|
| External factors | Stress ratio | 1 | 0.4 |
| | Test temperature (°C) | 1 | 15 |
| | Load frequency(Hz) | 1 | 10 |
| Internal factors | Fatigue test method | 3 | Direct tensile test, Indirect tensile test, Uniaxial compression test. |
| | Asphalt types | 1 | 70 |
| | Asphalt content (%) | 1 | 5.2% |
| | Mixing gradation (AC) | 1 | AC-13 |
| | Volume of air voids (%) | 1 | 4.5% |

The process of the three kinds of fatigue tests is shown in Figure 3.

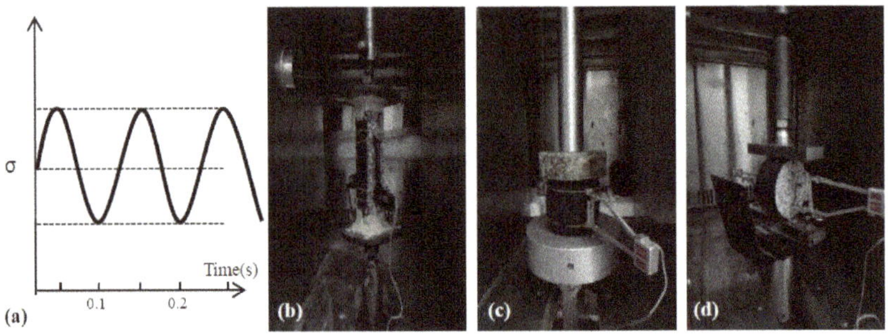


**Figure 3.** (**a**) Load waveform of fatigue test; (**b**) Direct tensile test; (**c**) Uniaxial compression test; and, (**d**) Indirect tensile test.

## 3. Test Result and Analysis

The fatigue test results of the direct tensile, the uniaxial compression, and the indirect tensile were shown in Tables 5–7, respectively.

**Table 5.** Test result of direct tensile.

| Stress Level (MPa) | Maximum Load Value (KN) | Stress Ratio | Average Fatigue Life (Times) | CV% |
|---|---|---|---|---|
| 0.25 | 0.625 | 0.223 | 398,879 | 0.176 |
| 0.5 | 1.25 | 0.446 | 41,129 | 0.203 |
| 1 | 2.5 | 0.892 | 3686 | 0.188 |
| 1.5 | 3.75 | 1.338 | 993 | 0.176 |

(*CV* is the coefficient of variation).

**Table 6.** Test result of uniaxial compression.

| Stress Level (MPa) | Maximum Load Value (KN) | Stress Ratio | Average Fatigue Life (Times) | CV% |
|---|---|---|---|---|
| 2 | 15.7 | 0.178 | 797,470 | 0.098 |
| 2.5 | 19.625 | 0.223 | 198,977 | 0.074 |
| 3 | 23.55 | 0.268 | 91,067 | 0.086 |
| 3.5 | 27.475 | 0.312 | 26,734 | 0.105 |

(*CV* is the coefficient of variation).

**Table 7.** Test result of indirect tensile.

| Stress Level (MPa) | Maximum Load Value (KN) | Stress Ratio | Average Fatigue Life (Times) | CV% |
|---|---|---|---|---|
| 0.25 | 2.386 | 0.11 | 1,283,789 | 0.021 |
| 0.5 | 4.772 | 0.22 | 50,062 | 0.166 |
| 1 | 9.544 | 0.439 | 2236 | 0.079 |
| 1.5 | 14.315 | 0.659 | 435 | 0.167 |

(*CV* is the coefficient of variation).

### 3.1. The Establishment of the Decay Model for Moduli

The conventional *S-N* fatigue Equation is widely used to analyze the fatigue performance of asphalt mixtures [26,27]. Chaboche [28] defined the traditional *S-N* fatigue Equation as:

$$N_f = k(\frac{1}{t})^n \tag{1}$$

or

$$N_f = k(\frac{1}{\sigma})^n \tag{2}$$

where, $N_f$ is fatigue life, $t$ is stress ratio, $\sigma$ is stress level. $K$, and $n$ are the material parameters of asphalt mixtures.

According to Equations (1) and (2), the fatigue curves characterized by stress ratios and stress levels were shown in Figure 4a,b, respectively.

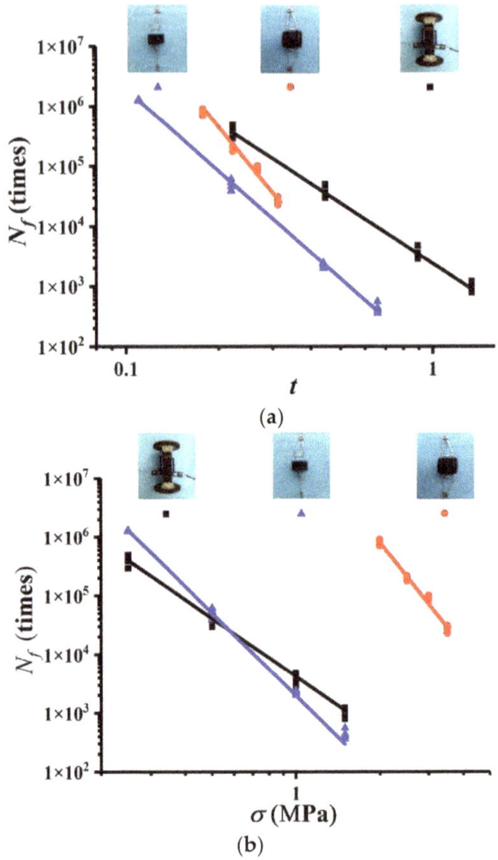


**Figure 4.** (**a**) Fatigue curves characterized by stress ratio (**b**) Fatigue curves characterized by stress level.

From Figure 4, it can be observed that the *S-N* fatigue curves of asphalt mixtures show great difference under different stress conditions. It is difficult to evaluate the fatigue performance of asphalt mixtures in different test methods [7]. Given that, in this paper, the nonlinear fatigue damage model was implemented to simulate the moduli decay patterns. Defining the moduli as the damage variable, the damage model based on the moduli decay was established.

The damage variable could be expressed, as shown in Equation (3).

$$D(N) = 1 - \frac{E(N)}{E_0} \tag{3}$$

where, $D(N)$ is damage variable, $E(N)$ is modulus in loading cycle $N_f$ the specimen, and $E_0$ is the initial value of modulus.

Chaboche [28] established another fatigue damage model. As shown in Equation (4).

$$D(N) = 1 - \left[1 - \left(\frac{N}{N_f}\right)^{\frac{1}{1-\alpha}}\right]^{\frac{1}{1+\gamma}} \tag{4}$$

where, $N_f$ is the fatigue life, $N$ is the loading cycles, and $\alpha$ and $\gamma$ are the material parameters related to the stress.

Based on Equations (3) and (4), in this paper, Equation (5) is deduced as the decay Equation of the moduli for asphalt mixtures.

$$\frac{E(N)}{E_0} = \left[1 - \left(\frac{N}{N_f}\right)^{\frac{1}{1-\alpha}}\right]^{\frac{1}{1+\gamma}} \tag{5}$$

The fitting parameters are replaced by $m$, $n$, where, $m = \frac{1}{1-\alpha}$, $n = \frac{1}{1+\gamma}$.
Equation (5) can be simplified as:

$$\frac{E(N)}{E_0} = \left[1 - \left(\frac{N}{N_f}\right)^{m}\right]^{n} \tag{6}$$

### 3.2. The Initial Values of Moduli at Different Fatigue Stress Levels

The moduli of 50th cycle is widely used as the initial moduli of fatigue test [29]. However, the fatigue lives of the same material vary at different stress levels, so the initial moduli values that were obtained by this method has a large deviation. In this paper, the initial moduli $E_0$ was defined as the average one of the 10 moduli, which were near the cycle ratio $N/N_f = 0.01$. Because the specimen of the indirect tensile test is in the state of two-dimensional stress state, the pattern of double moduli decay of the indirect tensile test was considered. The average initial moduli values of four parallel specimens under different stress state and different stress levels are summarized in Table 8:

**Table 8.** The initial moduli values in different fatigue tests.

| Stress Level/MPa | | 0.25 | 0.5 | 1 | 1.5 | 2 | 2.5 | 3 | 3.5 |
|---|---|---|---|---|---|---|---|---|---|
| Tensile moduli | $E_0$ (MPa) | 1384 | 1686 | 2121 | 2407 | / | / | / | / |
| | CV (%) | 0.054 | 0.055 | 0.058 | 0.045 | / | / | / | / |
| Compression moduli | $E_0$ (MPa) | / | / | / | / | 8852 | 10,102 | 11,167 | 11,943 |
| | CV (%) | / | / | / | / | 0.021 | 0.024 | 0.023 | 0.055 |
| Indirect tensile moduli | $E_0$ (MPa) | 3560 | 4027 | 4585 | 5057 | / | / | / | / |
| | CV (%) | 0.011 | 0.019 | 0.014 | 0.019 | / | / | / | / |
| Compression moduli measured by indirect tensile test | $E_0$ (MPa) | 7926 | 8835 | 9975 | 10,760 | / | / | / | / |
| | CV (%) | 0.015 | 0.018 | 0.018 | 0.014 | / | / | / | / |
| Tensile moduli measured by indirect tensile test | $E_0$ (MPa) | 2547 | 2935 | 3464 | 3858 | / | / | / | / |
| | CV (%) | 0.024 | 0.021 | 0.027 | 0.012 | / | / | / | / |

From Table 8, it can be observed that the initial moduli value varies in different stress levels. It also varies in the different fatigue tests, which have the different stress states. However, it shows a common characteristic that the initial values of the fatigue moduli increase with the increase of the stress levels. The initial compression moduli are larger than that of the tensile moduli.

### 3.3. The Critical Value of Moduli at Different Fatigue Stress Levels

During the fatigue tests, the moduli decrease with the increase of the load cycles until the failure of the specimens. The critical value refers to the damage value when the specimen occurs fatigue failure at the end of the fatigue test. In this paper, the average value of the moduli in the last five loading cycles in the fatigue tests were taken as the critical moduli value. The tensile, compressive, and indirect tensile moduli of asphalt mixtures, which is based on the indirect tensile test, and the calculation formula was derived on the Hooke's law in two-dimensional stress states [7]. The average critical moduli values of the four parallel specimens under different stress states and different stress levels are summarized in Table 9.

**Table 9.** The critical moduli value of fatigue test.

| Stress Level (MPa) | | 0.25 | 0.5 | 1 | 1.5 | 2 | 2.5 | 3 | 3.5 |
|---|---|---|---|---|---|---|---|---|---|
| Tensile moduli | $E_{min}$ (MPa) | 609 | 716 | 869 | 961 | / | / | / | / |
| | $CV$ (%) | 0.049 | 0.057 | 0.052 | 0.051 | / | / | / | / |
| Compression moduli | $E_{min}$ (MPa) | / | / | / | / | 7480 | 8086 | 8528 | 8787 |
| | $CV$ (%) | / | / | / | / | 0.021 | 0.022 | 0.009 | 0.010 |
| Indirect tensile moduli | $E_{min}$ (MPa) | 895 | 958 | 1040 | 1106 | / | / | / | / |
| | $CV$ (%) | 0.021 | 0.019 | 0.053 | 0.042 | / | / | / | / |
| Compression moduli measured by indirect tensile test | $E_{min}$ (MPa) | 5073 | 5593 | 6184 | 6563 | / | / | / | / |
| | $CV$ (%) | 0.027 | 0.025 | 0.018 | 0.018 | / | / | / | / |
| Tensile moduli measured by indirect tensile test | $E_{min}$ (MPa) | 560 | 616 | 693 | 733 | / | / | / | / |
| | $CV$ (%) | 0.036 | 0.024 | 0.036 | 0.021 | / | / | / | / |

From Table 9, the similar patterns of variation can be observed that the critical moduli value varies in different stress levels. It also varies in the different fatigue stress conditions. The common characteristic in the different stress states is that the critical values increase with the increase of the stress levels. The critical compression moduli are larger than that of the tensile moduli.

### 3.4. Analysis of the Fitting Results of Fatigue Tests under Different Stress Levels

In order to compare the moduli decay pattern under the same stress state, the tensile moduli and compression moduli were compared, respectively. Real-time ratio $E(N)/E_0$ for tensile moduli, as measured by direct tensile test and indirect tensile tests, was fitted with the cycle ratio by Equation (6). The fitting results of tensile moduli from the direct tensile and the indirect tensile fatigue tests in different stress levels are shown in Figures 5 and 6, respectively.

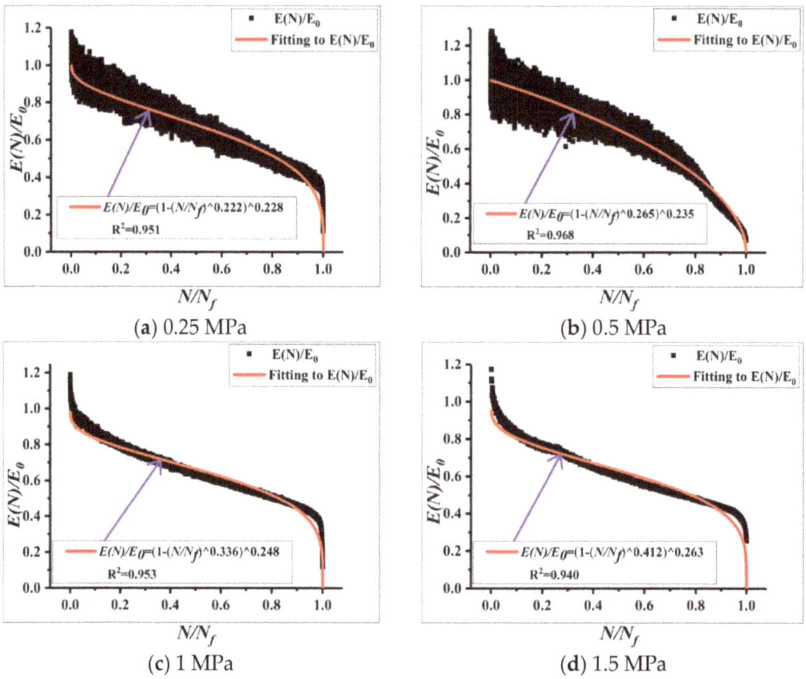


**Figure 5.** Moduli decay pattern of the direct tensile fatigue tests.

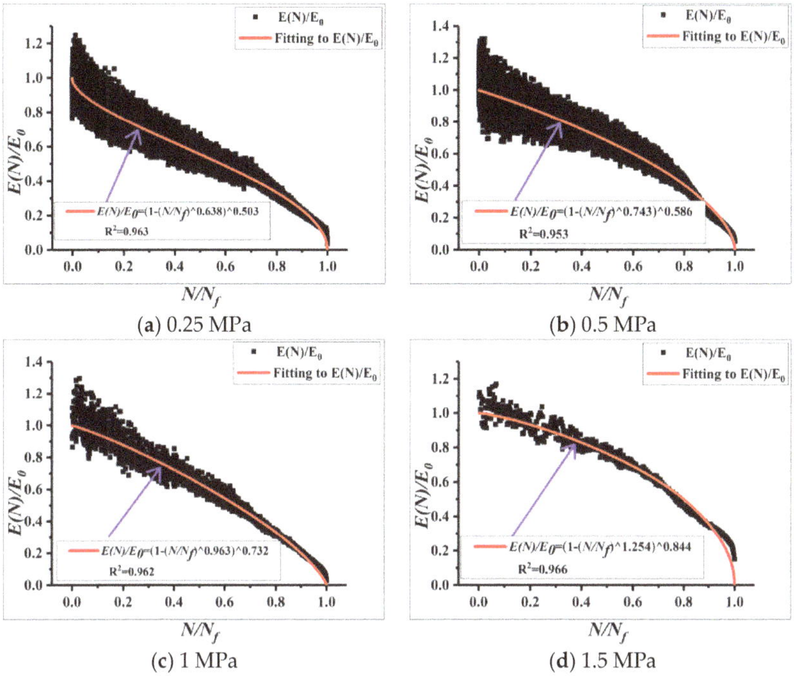


(**a**) 0.25 MPa (**b**) 0.5 MPa (**c**) 1 MPa (**d**) 1.5 MPa

**Figure 6.** Moduli decay pattern of tensile moduli measured by the indirect tensile fatigue tests.

It can be noticed from Figures 5 and 6 that the decay patterns of tensile moduli that were obtained from these two test methods are similar. There are three stages: migration stage, steady stage, and destructive stage, and the whole process is nonlinear.

However, the decay rates of different stress state are different, which are caused by different fatigue resistance of asphalt. In the direct tensile test, the specimen is within a uniform tensile condition. The main factors determining its fatigue properties are the cohesion of the asphalt mortar (the adhesion of the aggregate with the asphalt and its internal friction based on the aggregate gradation). The effect of intrusion between aggregates is relatively weak [30]. During the indirect tensile tests, the transverse tensile fatigue properties mainly depend on the adhesion (between asphalt mortar and aggregates) and the internal frictional resistance.

Figures 5 and 6 reflect the decay pattern of different stress levels under different stress states. In order to compare the decay pattern more obviously, this paper compared the decay pattern of tensile moduli under the same stress level. The parameters of the fitted curves for tensile stress were shown in the Table 10:

**Table 10.** Fitting parameters of the tensile moduli decay curves under different test conditions.

| Stress Level (MPa) | Parameters | 0.25 | 0.5 | 1 | 1.5 |
|---|---|---|---|---|---|
| Tensile moduli | $m$ | 0.222 | 0.265 | 0.337 | 0.413 |
| | $n$ | 0.229 | 0.235 | 0.248 | 0.2636 |
| Tensile moduli measured by indirect tensile test | $m$ | 0.638 | 0.743 | 0.964 | 1.255 |
| | $n$ | 0.503 | 0.586 | 0.732 | 0.844 |

The tensile moduli measured by indirect tensile test and direct tensile under the same stress level 1 MPa were compared. As shown in Figure 7:

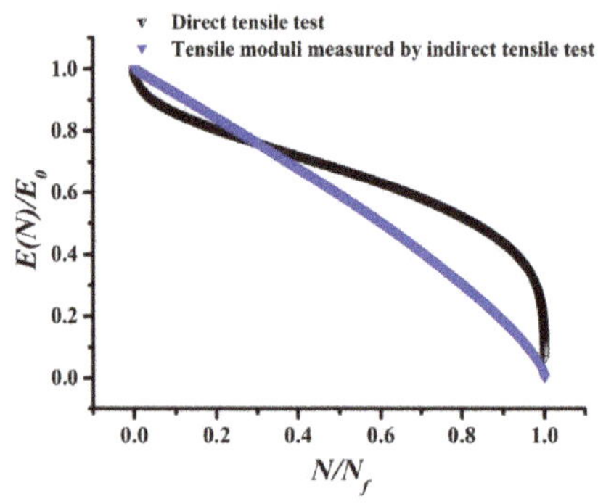


**Figure 7.** The decay pattern of tensile moduli measured by the indirect tensile test and direct tensile test.

In order to compare the decay pattern more clearly, the decay Equation is simplified. Equation (6) is simplify to $y = (1 - \chi^m)^n$, then,

$$y' = -mn\chi^{m-1}(1 - \chi^m)^{n-1} \tag{7}$$

$$y'' = m^2 n\chi^{2m-2}(n-1)(1-\chi^m)^{n-2} - mn\chi^{m-2}(m-1)(1-x^m)^{n-1} \tag{8}$$

where, $y$ is the ratio of the modulus of the material to the initial modulus of the undamaged state after loading the specimen to N, $\chi$ is the ratio of the loading cycles to the fatigue life, $m$ and $n$ are the material parameters related to the stress, $y'$ is the first derivative of $y'$, and $y''$ is the two derivative of $y'$.

While substituting parameters under the stress level of 1 MPa into Equation (8), the inflection point can be calculated, then the tangent line at the inflection point position can be adopted to compare the tangent slope. The result is as follows:

It can be noticed from Figure 7 and Table 11 that the decay patterns of tensile moduli, as measured by the indirect tensile test and the direct tensile test are nonlinear. The moduli parameters are different. The inflection point of direct tensile test is 0.381. The decay rate of direct tensile moduli decreases gradually before the inflection, while it increases after the inflection point. The inflection point of tensile moduli, as measured by the indirect tensile test is 0.114. In addition, the decay rate shows an increasing trend. There are three stages of migration stage, steady stage and destructive stage in the decay curves of the direct tensile tests and the indirect tensile tests. The decay rate of the direct tensile moduli is more quickly than that of the indirect tensile moduli during the migration state, but it is contrary during the steady stage. In addition, the decay rate of the direct tensile moduli is quicker than that of the indirect tensile moduli at the destructive stage.

**Table 11.** Inflection points and slopes of decay curves of tensile moduli under different test conditions.

| Test Type | Inflection Point | Tangency Point | Tangent Slope |
|---|---|---|---|
| Direct tensile | 0.381 | (0.381, 0.727) | −0.416 |
| Tensile moduli measured by indirect tensile test | 0.114 | (0.114, 0.908) | −0.791 |

Similarly, the decay patterns of the compression moduli were fitted as shown in Figures 8 and 9, respectively.

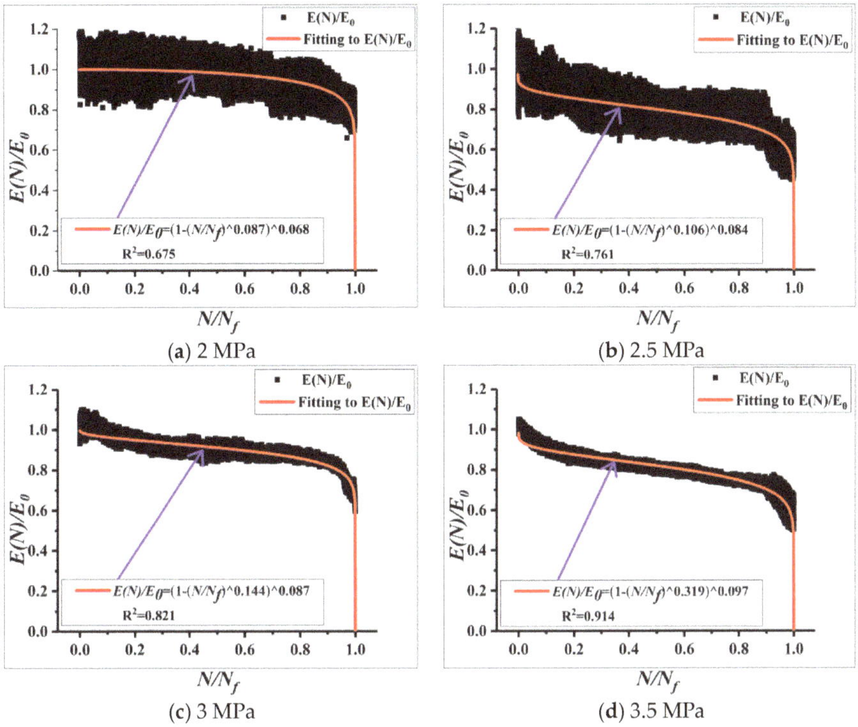


**Figure 8.** Moduli decay patterns of uniaxial compression tests.

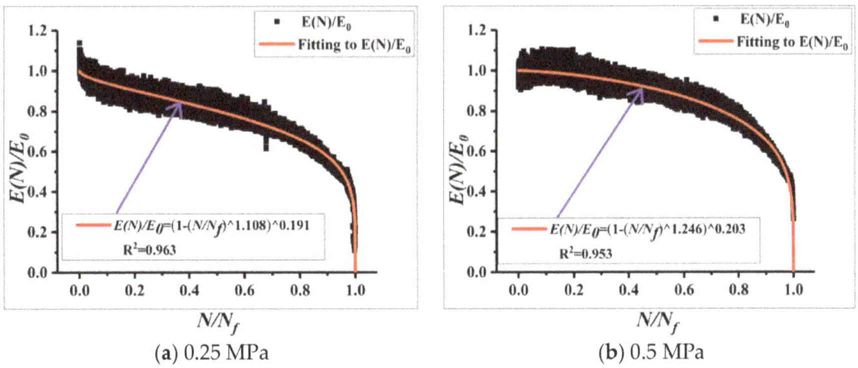


**Figure 9.** *Cont.*

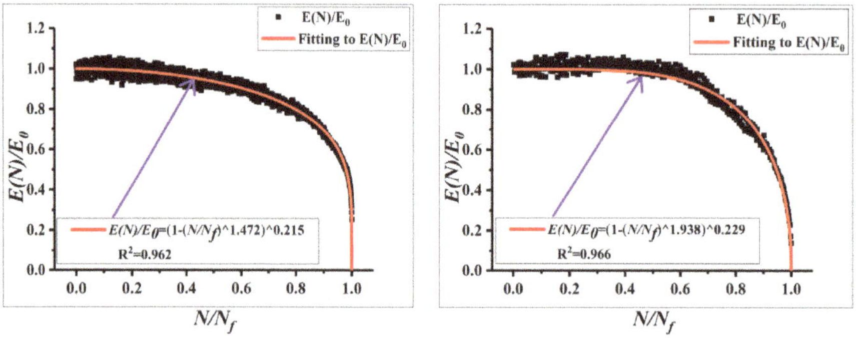


(c) 1 MPa (d) 1.5 MPa

**Figure 9.** Moduli decay patterns of compression moduli measured by the indirect tensile tests.

From Figures 8 and 9, it can be observed that the decay patterns of compression moduli also exist three stages of migration stage, steady stage, and destructive stage, and the patterns are nonlinear. But, the decay rates are different under different stress state.

For uniaxial compression test, the specimen is in a compression stress state. The skeleton structure, composition form, and friction coefficient of aggregates determine the inter-particle frictional resistance. The main factors determining its fatigue properties are the cohesion of the aggregate particles [31]. In the indirect tensile test, fatigue properties mainly depend on the interlocking effect of aggregate when the specimens are under the compression state [32]. In summary, when the materials are in a different stress state, the factors that determine the fatigue properties are different.

In order to compare the decay pattern more obviously, this paper compared the decay pattern of compression moduli under same stress level. The parameters of the fitted curves for different stress levels are shown in Table 12:

**Table 12.** Fitting parameters of fatigue moduli decay curves under different test conditions.

| Stress Level (MPa) | Parameters | 0.25 | 0.5 | 1 | 1.5 | 2 | 2.5 | 3 | 3.5 |
|---|---|---|---|---|---|---|---|---|---|
| Compression moduli | $m$ | / | / | / | / | 0.087 | 0.106 | 0.144 | 0.319 |
| | $n$ | / | / | / | / | 0.068 | 0.084 | 0.087 | 0.097 |
| Compression moduli measured by indirect tensile test | $m$ | 1.108 | 1.246 | 1.472 | 1.938 | / | / | / | / |
| | $n$ | 0.191 | 0.203 | 0.215 | 0.223 | / | / | / | / |

It is difficult to compare the moduli decay patterns. In this paper, the parameters $m$ and $n$ of compression moduli were fitted with stress levels. The fitting results were shown in Equations (9) and (10).

$$m(\sigma) = 0.086 + 0.00002e^{2.74521\sigma} \tag{9}$$

$$n(\sigma) = -23.514 + 23.549e^{0.00075\sigma} \tag{10}$$

As there are four stress level (0.25, 0.5, 1, 1.5) of compression moduli, as measured by the indirect tensile test. Taking $\sigma$ = 1 MPa into Equations (9) and (10), respectively, then the parameters $m$ and $n$ of uniaxial compression moduli under 1 MPa can be obtained, as shown in Table 13.

**Table 13.** Fitting parameters of fatigue moduli decay curve in different tests method under 1 MPa stress level.

| Test Type | $m$ | $n$ |
|---|---|---|
| Compression moduli | 0.086 | 0.053 |
| Compression moduli measured by indirect tensile test | 1.472 | 0.215 |

The compression moduli measured by indirect tensile test and uniaxial compression moduli under 1 MPa stress level were compared, as shown in Figure 10. The decay patterns of the two compression moduli are experienced in three stages: migration stage, steady stage, and destructive stage, and the whole process is nonlinear.

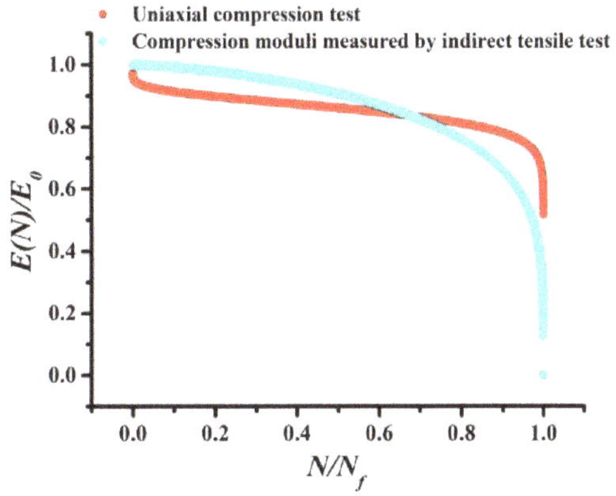


**Figure 10.** Decay pattern of compression moduli.

In order to compare the decay rate, substituting the parameters in Table 13 into Equation (8), the inflection point of the decay curve can be calculated. Then, the tangent line at the inflection point position can be implemented to compare the tangent slope. The result is as follows:

From Figure 10 and Table 14, it can be notified that the compression moduli decay pattern of uniaxial compression and indirect tensile similarly exist three stages of migration stage, steady stage, and destructive stage, and the patterns are nonlinear, too. The inflection point of uniaxial compression moduli is 0.370. The decay rate decreases before the inflection point, while it increases after the inflection point. There is no inflection point of indirect tensile moduli. In order to compare the decay rate on same position, the tangency point of indirect test is also taken as 0.37. It can be observed that the tangent slope of indirect test is larger than the uniaxial compression test on tangency point position. During the migration stage, the decay rate of uniaxial compression is faster than that of the indirect tensile. During the steady stage, the decay rate of uniaxial compression is slower than that of the indirect tensile. The decay rate of uniaxial compression is also faster than that of the indirect tensile during the destructive stage.

**Table 14.** Inflection points and slopes of decay curves of compression moduli under different test conditions.

| Test Type | Inflection Point | Tangency Point | Tangent Slope |
|---|---|---|---|
| Uniaxial compression | 0.370 | (0.370, 0.877) | −0.120 |
| Compression moduli measured by indirect tensile test | - | (0.370, 0.945) | −0.243 |

As the stress level of tensile moduli and compression moduli all is 1 MPa, the decay pattern of tensile moduli and compression moduli were compared simultaneously. The values of $m$ and $n$ in Tables 10 and 13 were substituted into Equation (6). Then, the fatigue moduli decay patterns were shown in Figure 11.

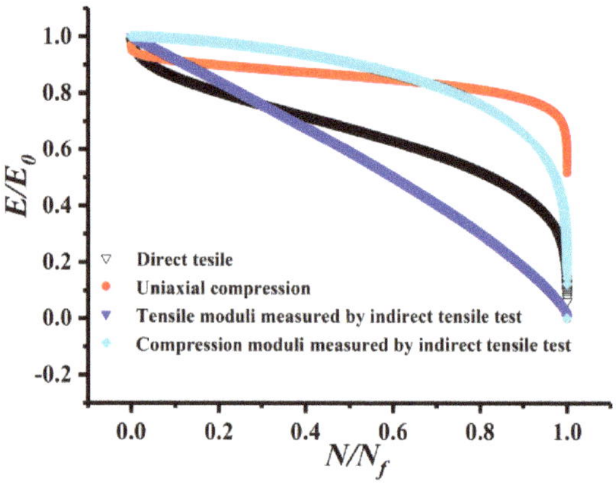


**Figure 11.** Fatigue moduli decay patterns.

From Figure 11, it can be observed that the fatigue moduli decay curves exist in the three stages of migration stage, steady stage, and destructive stage. The fatigue damage characteristics of asphalt mixtures are non-linear.

However, on one hand, the decay curves of fatigue moduli under different stress levels are quite different. In one-dimensional stress states, the tensile moduli and compression moduli are also quite different. In two-dimensional stress states, the difference is obvious between the compression moduli and tensile moduli, which were both obtained from the indirect tensile test. On the other hand, the direct tensile moduli are similar to the tensile moduli obtained from the indirect tensile test. The compression moduli of uniaxial compression test are similar to the compression moduli that were obtained from the indirect tensile test.

In addition, it can be found from Tables 11 and 14 that the inflection point of tensile moduli measured by the indirect tensile is the smallest and that of the direct tensile are the maximum. At the tangent point, the uniaxial compression has the lowest moduli decay slope and the tensile moduli, as measured by the indirect tensile test, has the largest moduli decay slope. It can be concluded that the decay rate of the tensile moduli measured by the indirect tensile test is faster than the uniaxial compression moduli at the tangent point during the course of the fatigue tests.

It also can be observed that the position of the decay curve for the compression moduli is higher than that of the tensile moduli at the same cycle ratio state. It can be concluded that the decay rate of the tensile moduli is faster than that of the compression moduli during the course of the fatigue tests, which indicates that the tensile failure is the main reason of the fatigue damage for asphalt mixture.

## 4. Summary and Conclusions

The fatigue tests and analysis of asphalt mixture under different loading conditions and stress levels were carried out. The following conclusions can be drawn from above:

(1) Under different loading conditions, the *S-N* fatigue curves for the direct tensile, indirect tensile and uniaxial compression are different. Therefore, it is difficult to compare the fatigue resistance of the same asphalt mixtures in the different stress conditions, according to the traditional method.
(2) There are three stages in the moduli decay curves under different stress levels and conditions, which are migration, stabilization, and destruction. The decay patterns are nonlinear.
(3) The decay patterns of the direct tensile moduli and the tensile moduli, as measured by the indirect tensile test, are similar. At the same time, the decay patterns of the uniaxial compression and the compression moduli, as measured by indirect tensile test, are similar. Nevertheless, the decay patterns of the tensile and the compressive moduli are obviously different.

(4) There are significant differences in the tensile and compression characteristics of asphalt mixtures. For the flexural fatigue test, the decay rate of the tensile modulus at the bottom of the specimen is greater than that of the compression modulus at the top of the specimen under the same cyclic ratio condition. So, it exhibits tensile stress failure characteristics during the flexural fatigue test of the asphalt mixture.

**Author Contributions:** Conceptualization, S.L. and X.F.; Methodology, S.L. and C.X.; Software, S.L.; Validation, S.L., X.F. and C.X.; Formal Analysis, X.F.; Investigation, S.L.; Resources, S.L.; Data Curation, C.X.; Writing-Original Draft Preparation, X.F. and C.X.; Writing-Review & Editing, S.L. and C.X.; Visualization, D.C.; Supervision, J.Z. and L.Y.; Project Administration, S.L.; Funding Acquisition, S.L. and J.Z.

**Funding:** This research was funded by National Natural Science Foundation of China (Grant number [51578081, 51608058]; The Ministry of Transport Construction Projects of Science and Technology [2015318825120]; The Guangxi Zhuang Autonomous Region Traffic and Transportation Department Transportation Projects of Science and Technology [2013-32], and The Inner Mongolia Autonomous Region Traffic and Transportation Department Transportation Projects of Science and Technology [NJ-2016-35].

**Acknowledgments:** This work is supported by Key Projects of Hunan Province-Technological Innovation Project in Industry [2016GK2096], National Engineering Laboratory Open Fund Project [kfh160102], Scientific and Technological Innovation Project of Hunan Province for University Graduate Students [CX2017B457]. The authors gratefully acknowledge their financial support.

**Conflicts of Interest:** The authors declare no conflict of interest.

## References

1. Ziari, H.; Babagoli, R.; Akbari, A. Investigation of fatigue and rutting performance of hot mix asphalt mixtures prepared by bentonite-modified bitumen. *Road Mater. Pavement Des.* **2014**, *16*, 101–118. [CrossRef]
2. Xuedong, G.; Mingzhi, S.; Wenting, D.; Kaixuan, Z.; Jilu, L. Research on the fatigue whole life of asphalt mixture based on the nonlinear damage theory. *Constr. Build. Mater.* **2017**, *156*, 546–554. [CrossRef]
3. Saboo, N.; Das, B.P.; Kumar, P. New phenomenological approach for modeling fatigue life of asphalt mixes. *Constr. Build. Mater.* **2016**, *121*, 134–142. [CrossRef]
4. Liu, G.; Jia, Y.; Yang, T.; Du, H.; Zhang, J.; Zhao, Y. Fatigue performance evaluation of asphalt mixtures based on energy-controlled loading mode. *Constr. Build. Mater.* **2017**, *157*, 348–356. [CrossRef]
5. Saha, G.; Biligiri, K.P. Stato-dynamic response analyses through semi-circular bending test: Fatigue life prediction of asphalt mixtures. *Constr. Build. Mater.* **2017**, *150*, 664–672. [CrossRef]
6. Saha, G.; Biligiri, K.P. Modulus prediction of asphalt mixtures using dynamic semicircular bending test: Estimation algorithm and nomograph development. *Int. J. Fatigue* **2018**, *109*, 137–144. [CrossRef]
7. Lv, S.; Liu, C.; Yao, H.; Zheng, J. Comparisons of synchronous measurement methods on various moduli of asphalt mixtures. *Constr. Build. Mater.* **2018**, *158*, 1035–1045. [CrossRef]
8. Bańkowski, W. Evaluation of fatigue life of asphalt concrete mixtures with reclaimed asphalt pavement. *Appl. Sci.* **2018**, *8*, 469. [CrossRef]
9. Lv, S.; Wang, X.; Liu, C.; Wang, S. Fatigue damage characteristics considering the difference of tensile-compression modulus for asphalt mixture. *J. Test. Eval.* **2018**, *46*. [CrossRef]
10. Lv, S.; Liu, C.; Zheng, J.; You, Z.; You, L. Viscoelastic fatigue damage properties of asphalt mixture with different aging degrees. *KSCE J. Civ. Eng* **2018**, *22*, 2073–2081. [CrossRef]
11. Moreno-Navarro, F.; Rubio-Gámez, M.C. A review of fatigue damage in bituminous mixtures: Understanding the phenomenon from a new perspective. *Constr. Build. Mater.* **2016**, *113*, 927–938. [CrossRef]
12. You, L.; Yan, K.; Hu, Y.; Liu, J.; Ge, D. Spectral element method for dynamic response of transversely isotropic asphalt pavement under impact load. *Road Mater. Pavement Des.* **2018**, *19*, 223–238. [CrossRef]
13. You, L.; Yan, K.; Hu, Y.; Zollinger, D.G. Spectral element solution for transversely isotropic elastic multi-layered structures subjected to axisymmetric loading. *Comput. Geotech.* **2016**, *72*, 67–73. [CrossRef]
14. Xie, J.; Xiao, Y.; Wu, S.; Huang, J. Research on fracture characteristic of gneiss prepared asphalt mixture with direct tensile test. *Constr. Build. Mater.* **2012**, *28*, 476–481. [CrossRef]
15. Lee, H.S.; Kim, J. Determination of viscoelastic poisson's ratio and creep compliance from the indirect tension Test. *J. Mater. Civ. Eng.* **2009**, *21*, 416–425. [CrossRef]
16. Zeiada, W.A.; Souliman, M.I.; Kaloush, K.E.; Mamlouk, M. Endurance limit for HMA based on healing concept using uniaxial tension-compression fatigue test. *J. Mater. Civ. Eng.* **2015**, *26*, 839–844. [CrossRef]

17. Hartman, A.M.; Gilchrist, M.D.; Walsh, G. Effect of mixture compaction on indirect tensile stiffness and fatigue. *J. Transp. Eng.* **2001**, *127*, 370–378. [CrossRef]
18. Darvell, B.W. Uniaxial compression tests and the validity of indirect tensile strength. *J. Mater. Sci.* **1990**, *25*, 757–780. [CrossRef]
19. Lv, S.; Liu, C.; Chen, D.; Zheng, J.; You, Z.; You, L. Normalization of fatigue characteristics for asphalt mixtures under different stress states. *Constr. Build. Mater.* **2018**, *177*, 33–42. [CrossRef]
20. Li, N.; Molenaar, A.A.A.; van de Ven, M.F.C.; Wu, S. Characterization of fatigue performance of asphalt mixture using a new fatigue analysis approach. *Constr. Build. Mater.* **2013**, *45*, 45–52. [CrossRef]
21. *(JTG F40-2004) Technical Spcifications for Construction of Highway Asphalt Pavements*; Renmin Communication Press: Beijing, China, 2004.
22. *(JTG E42-2005) Test Methods of Aggregate for Highway Engineering*; Renmin Communication Press: Beijing, China, 2005.
23. *(JTG E20-2011) Standard Test Methods of Asphalt and Asphalts Mixtures for Highway Engineering*; Renmin Communication Press: Beijing, China, 2011.
24. Wang, H.; Zhang, C.; Li, L.; You, Z.; Diab, A. Characterization of low temperature crack resistance of crumb rubber modified asphalt mixtures using semi-circular bending tests. *J. Test. Eval.* **2016**, *44*, 845–855. [CrossRef]
25. Mitchell, M.R.; Link, R.E.; Huang, L.; Cao, K.; Zeng, M. Evaluation of semicircular bending test for determining tensile strength and stiffness modulus of asphalt mixtures. *J. Test. Eval.* **2009**, *37*, 122–128. [CrossRef]
26. Adhikari, S.; Shen, S.; You, Z. Evaluation of fatigue models of hot-mix asphalt through laboratory testing. *Transp. Res. Rec. J. Transp. Res. Board.* **2009**, *2127*, 36–42. [CrossRef]
27. Shafabakhsh, G.; Taghipoor, M.; Sadeghnejad, M.; Tahami, S.A. Evaluating the effect of additives on improving asphalt mixtures fatigue behavior. *Constr. Build. Mater.* **2015**, *90*, 59–67. [CrossRef]
28. Chaboche, J.L. A review of some plasticity and viscoplasticity constitutive theories. *Int. J. Plast.* **2008**, *24*, 1642–1693. [CrossRef]
29. Zheng, J.; Lv, S. Nonlinear fatigue damage model for asphalt mixtures. *China J. Highw. Transpt.* **2009**, *22*, 21–28.
30. Kim, Y.K.; Woo Lee, S. Comparative study on the bond strength between direct tensile test and indirect tensile test for bonded concrete overlay. *J. Korean Soc. Civ. Eng. Mag.* **2013**, *33*, 1153–1163. [CrossRef]
31. Akçaoğlu, T.; Tokyay, M.; Çelik, T. Effect of coarse aggregate size on interfacial cracking under uniaxial compression. *Mater. Lett.* **2002**, *57*, 828–833. [CrossRef]
32. Carmona, S. Effect of specimen size and loading conditions on indirect tensile test results. *Mater. Constr.* **2009**, *59*, 7–18. [CrossRef]

applied sciences

MDPI

Article

# Investigation on Possibility of Waste Vegetable Oil Rejuvenating Aged Asphalt

**Cao Xinxin [1], Cao Xuejuan [2,*], Tang Boming [1,2], Wang Yuanyuan [2] and Li Xiaolong [2]**

1 School of Civil Engineering, Chongqing Jiaotong University, Chongqing 400074, China; caoxinxin1988@foxmail.com (C.X.); tangbm6@163.com (T.B.)
2 School of Materials Science and Engineering, Chongqing Jiaotong University, Chongqing 400074, China; wyy1005@seu.edu.cn (W.Y.); lixiaolong0715@163.com (L.X.)
* Correspondence: caoxuejuan79@foxmail.com

Received: 22 April 2018; Accepted: 8 May 2018; Published: 11 May 2018

**Abstract:** In order to guarantee the service performance of recycling asphalt mixture with reclaimed asphalt pavement (RAP), asphalt rejuvenator shall be added. In the last five years, vegetable oil-based rejuvenators have received more and more attention due to their green and regenerative advantages. The object of this paper is to investigate the feasibility of rejuvenating aged asphalt by a kind of waste vegetable oil (W-oil). The effect of W-oil on the performance of aged asphalt is characterized by a safety property test, aging property test, and pavement performance tests; the pavement performance tests included traditional tests and a rheological test. The results show that both the safety property and aging property of rejuvenated asphalt with W-oil meet the specification requirements. According to the results of traditional performance indexes (i.e., penetration, soften point, and ductility), the pavement performance of rejuvenated asphalt can be recovered to the level of virgin asphalt. According to the results of performance indices obtained from the rheological test, the optimum dosage of W-oil is determined to be 13.4 wt %. Compared with virgin asphalt, the rutting property of rejuvenated asphalt is equivalent to that of virgin asphalt, and the workability is slightly poorer; however, the fatigue property and low temperature property have been significantly enhanced. W-oil cannot only improve the pavement performance of aged asphalt, it can also guarantee good safety property and aging property. Therefore, W-oil is of great potential to serve as an asphalt rejuvenator for rejuvenating aged asphalt.

**Keywords:** asphalt; rejuvenator; vegetable oil; pavement performance; traditional indexes; rheological indexes

## 1. Introduction

Recycling reclaimed asphalt pavement (RAP) has good economic and environmental benefits. In the case of a RAP dosage between 20–50%, it can save the cost of construction by 14–34% [1], which can reduce the exploitation of non-renewable resources (e.g., stone and asphalt), and thus reduce energy consumption and pollution emission in mining and transportation [2]. However, when RAP dosage goes beyond 20%, there is a gradual increase in the deterioration of pavement's fatigue cracking and low-temperature cracking [3], and the compact of asphalt mixture will also be damaged [4]. To solve the shortcomings of rejuvenated asphalt mixture with a high dosage of RAP, it often requires the addition of a rejuvenator [5]. Rejuvenators mainly include vegetable oil and petroleum-based extracted oil [6], among which vegetable oil has attracted much attention in recent years because of its renewable advantage.

In 2012, Hallizza Asli et al. [7], on the basis of the indicators (e.g., penetration, soften point, and viscosity), pointed out that there was no clear difference in the performance between rejuvenated asphalt with fring vegetable oil and virgin asphalt. In 2014, Chen Meizhu et al. [8,9] utilized frying vegetable oil to rejuvenate aged asphalt, and the study showed that frying soybean oil significantly improved the fatigue property and low temperature anti-cracking property of aged asphalt, but the ductility was not effectively improved, and high temperature performance became poorer with the increase of frying soybean oil. In 2016, Wan et al. [10] added methyl alcohol into frying vegetable oil and made a chemical modification under alkalis catalysis, and it was found that the asphalt that was rejuvenated by the modified frying vegetable oil could achieve a better rutting resistance. In 2017, Zhang et al. [11] appraised the effects of vegetable oil with different deep-frying times on the rheological performance of aged asphalt, and the study exhibited that vegetable oil of a higher aging degree could result in a higher viscosity and a better rutting resistance. In summary, frying vegetable oil can restore the penetration of aged asphalt, as well as improve the fatigue property and low-temperature anti-cracking property of aged asphalt; however, high-temperature rutting resistance will become poorer. Vegetable oil, after modification or further aging, can reduce the damage of vegetable oil on rutting resistance. Herein, in this study, it was considered to apply highly aged waste vegetable oil of high viscosity to rejuvenate aged asphalt. The waste vegetable oil is sourced from the byproduct after the extraction of fatty acid from vegetable oil, and this byproduct is named as W-oil in this paper. W-oil is the byproduct obtained by eight-hour distillation at 300–400 °C after the acidification of vegetable oil. W-oil has a higher viscosity and deeper aging degree than that of frying vegetable oil. The output of W-oil is large in China, and the main treatment measure taken at present is combustion, which has not been effectively utilized. Therefore, it is necessary to explore the application of W-oil in rejuvenating asphalt.

## 2. Objective and Experimental Plan

This paper is aimed at observing the effects of W-oil on the performance of aged asphalt and analyzing the potential of W-oil as an asphalt rejuvenator. It can not only provide an environment-friendly way to treat W-oil, it can also become a substitute for non-renewable petroleum-based asphalt rejuvenator, which will promote the development of sustainable pavement construction.

The test plan is shown in Figure 1. First, aged asphalt is obtained by two steps: virgin asphalt is aged by rolling thin film oven test (RTFOT), then put the aged residue of RTFOT into a pressurized aging vessel (PAV) for further aging to get aged asphalt. Second, 5%, 10%, 15%, and 20% W-oil is added into the aged asphalt to prepare rejuvenated asphalt. Third, the safety property, aging property and pavement performance of virgin asphalt, aged asphalt, and rejuvenated asphalt is analyzed. Pavement performance is characterized by two indexes, namely, the traditional indexes and rheological indexes. The flash point test is to characterize the asphalt's safety, and the RTFOT test is to characterize its aging property. In the traditional indexes, the penetration test, soften point test, and ductility test are used to characterize asphalt pavement performance. Corresponding to the traditional indexes for measuring asphalt pavement performance, superpave proposed the test methods based on rheology theory, so that the rheological index can be directly correlated to the field pavement performance. The viscosity test is used to characterize the high-temperature workability, the multiple stress creep recovery (MSCR) test is used to characterize the rutting property, the time sweep (TS) test is used to characterize the medium-temperature fatigue property, and the bending beam rheometer (BBR) test is to characterize low temperature property.

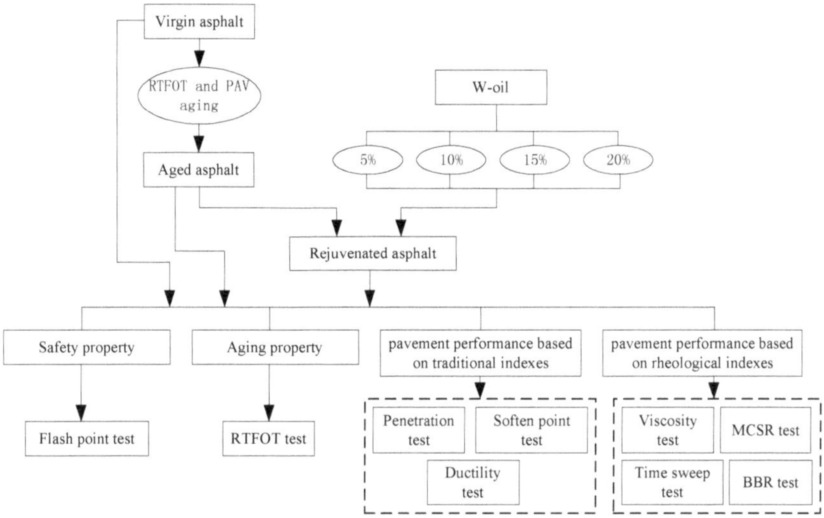


**Figure 1.** The Test Plan.

## 3. Materials and Testing Methods

### *3.1. Materials*

#### 3.1.1. W-Oil

W-oil is the byproduct after the extraction of fatty acid from vegetable oil, as shown in Figure 2. At present, the annual output of W-oil is approx. 900,000 tons, and the main treatment measure for waste oil is combustion, which produces a pungent odor and greenhouse gas, which pollutes the environment. The typical physical properties of W-oil are shown in Table 1. As shown in Table 1, the low viscosity of W-oil, as 286.7 mPa·S, means that at 60 °C, it can soften aged asphalt effectively. In general, the higher the molecular weight, the less volatile it is. The number-average molecular weight of W-oil is 1067 Daltons, so W-oil is expected to have a good anti-volatile performance in construction as a rejuvenator. Fourier transform infrared spectroscopy (FTIR) is a method of determining the chemical functional groups within a medium. The chemical functional groups are groups of atoms that are responsible for different reactions within a compound [12]. In Figure 3, a comparison of infrared spectrogram is made between virgin asphalt and W-oil. The major differences in the composition of functional groups include: (1) W-oil has a stronger absorption peak at 1150 $cm^{-1}$ and 1700 $cm^{-1}$, while virgin asphalt basically has no absorption peak, implying that W-oil contains a large number of ester bonds; (2) Virgin asphalt shows absorption peaks at 800 $cm^{-1}$ and 1580 $cm^{-1}$, where W-oil has no absorption peak, which demonstrates that W-oil does not contain benzene. Therefore, W-oil does not contain strong carcinogen-polycyclic aromatic hydrocarbon (PAH), and the use of W-oil as a rejuvenator can reduce the harm to the construction workers.

**Figure 2.** W-oil.

**Table 1.** Physical Properties of W-oil.

| Physical Properties | Density/(g/cm$^3$) | Color | Viscosity at 60 °C/mPa·S | Number-Average Molecular Weight Mn/Daltons |
|---|---|---|---|---|
| | 0.966 | Nigger-brown | 286.7 | 1067 |

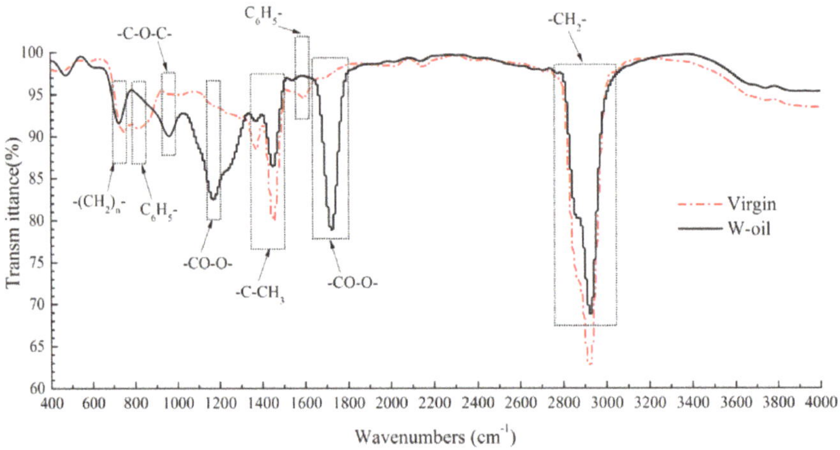


**Figure 3.** Fourier transform infrared (FTIR) Spectra of W-oil and Virgin Asphalt.

3.1.2. Asphalt

(1) Virgin asphalt and aged asphalt

PEN70 asphalt is used as the virgin asphalt. Aged asphalt is prepared by aging PEN70 asphalt in the laboratory. Asphalt aging includes short-term aging and long-term aging. Short-term aging is simulated by the rolling thin film oven test (RTFOT), long-term aging is simulated by the accelerated aging test of asphalt in a pressurized aging vessel (PAV). The PAV test is conducted at the temperature 100 °C for 20 h to simulate asphalt field aging for six to eight years. For the specific experimental method, refer to ASTM D 2872 [13] and ASTM D6521 [14]. The technical indicators for virgin asphalt and aged asphalt are as shown in Table 2.

**Table 2.** Physical Index of Virgin and Aged Asphalt.

| Index | Penetration (25 °C, 5 s)/0.1 mm | Ductility(5 °C)/cm | Soften Point/°C |
|---|---|---|---|
| Virgin asphalt | 64.6 | 11.1 | 49.2 |
| Aged asphalt | 25.2 | 0.8 | 65.6 |

(2) W-oil rejuvenated asphalt

W-oil is mixed into the aged asphalt at proportions (by weight) of 5%, 10%, 15%, and 20% at a recovering temperature of 135 °C in the stirrer at a speed of 2000 RPM for 15 min. Virgin asphalt, aged asphalt, and 5%, 10%, 15%, and 20% W-oil rejuvenated asphalt are named as Virgin, 0%WRA, 5%WRA, 10%WRA, 15%WRA, and 20%WRA, respectively.

### *3.2. Testing Methods*

#### 3.2.1. Flash Point Test

The flash point reflects the safety of asphalt in the process of mixing at a high temperature. The higher the flash point, the more safe the asphalt. The flash point test is applied to characterize the safety of virgin, aged, and rejuvenated asphalt. As shown in Figure 4, the rate of temperature rise is set between 5–6 °C/min during the last 28 °C before the flash point. The test flame is passed across the center of the test cup. For the observed flash point, the temperature is recorded at the time the test flame causes a distinct flash in the interior of the test cup. For the details of the test, refer to ASTM D 92-12b [15].

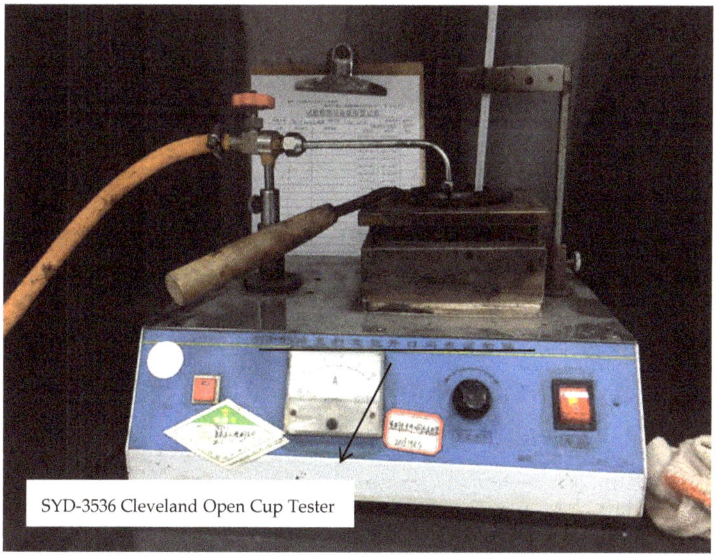


**Figure 4.** The Cleveland Open Cup Tester.

#### 3.2.2. Rolling Thin Film Oven Test (RTFOT)

RTFOT simulates the short-term aging in mixing process. The smaller mass loss that the aged asphalt has, the better the aging property is. The RTFOT test is applied to characterizing the safety of virgin, aged, and rejuvenated asphalt. As shown in Figure 5, asphalt samples are conditioned in a rolling thin film oven at 163 °C for 85 min. For the details of the test, refer to ASTM D2872 [13].

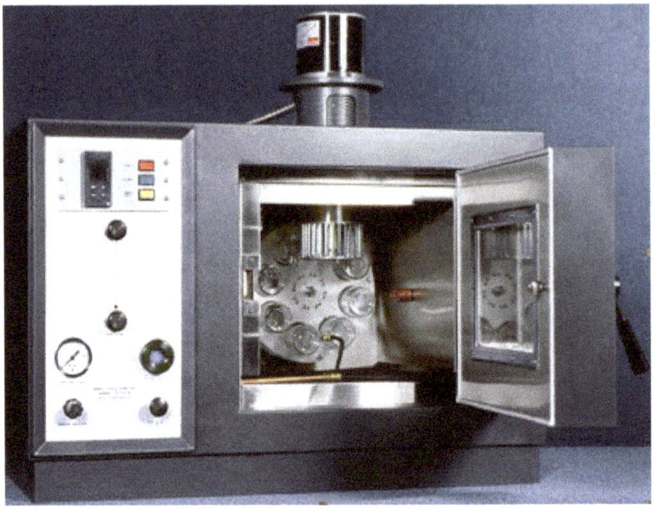

**Figure 5.** The Rolling Thin Film Oven.

### 3.2.3. Physical Tests

(1) Penetration test

Penetration is a grading index of asphalt. For example, if the penetration range is 60–80, the asphalt is defined as PEN70. As shown in Figure 6, a container filled with an asphalt sample is stored in a 25 °C water bath for 90 min, and then penetrated by a needle weighted 100 g; the penetration depth is measured as a penetration in the unit of 0.1 mm. For the details of penetration test, refer to ASTM D5 [16].

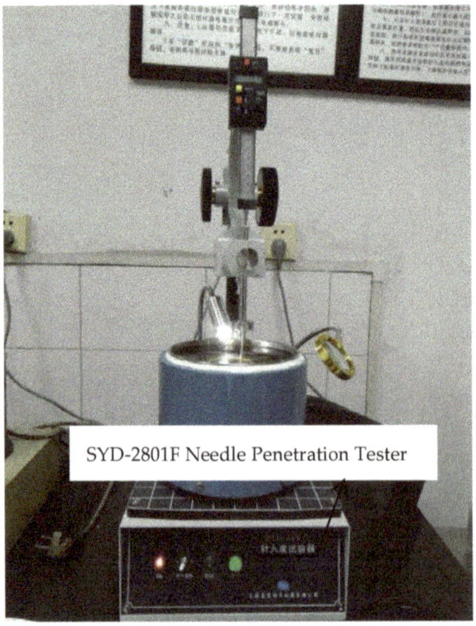


**Figure 6.** The Needle Penetration Tester.

(2) Soften point test

The soften point reflects the high-temperature stability of asphalt. The higher the soften point, the better the anti-rutting performance of the asphalt. As shown in Figure 7, two steel balls are placed on the horizontal disks of an asphalt sample contained in vertically supported metal rings. The assembly is heated in a water bath at 5 °C/min. The softening point was recorded as the average temperature at which the two disks softened enough to allow each ball, enveloped in asphalt, to fall a distance of 25 mm (1.0 in). For the details of soften point test, refer to ASTM D36 [17].

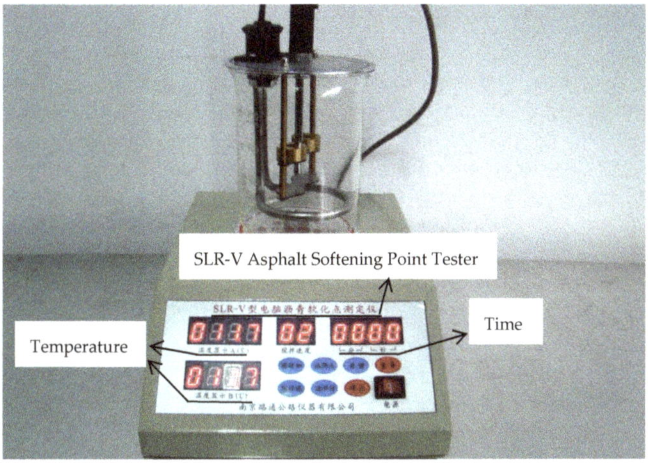


**Figure 7.** Softening Point Tester.

(3) Ductility test

Ductility reflects the ductility of asphalt. The higher the ductility, the better performance the asphalt has. The ductility test is used to measure the stretching length of a standard asphalt sample before breaking under standard testing condition (1 cm/min stretching speed at 5 °C), as shown in Figure 8. For the details of the ductility test, refer to ASTM D113 [18].

**Figure 8.** The Ductility Testing Machine.

### 3.2.4. Rheological Tests

(1) Viscosity test

In the mixing and compacting process of asphalt mixture, the viscosity of asphalt shall be kept within a certain range. Excessive viscosity will cause insufficient compacting of asphalt mixture, and too low viscosity will result in a waste of energy. The workability of asphalt can be characterized by a viscosity test. The viscosity of specimen is measured respectively at 120 °C, 135 °C, 150 °C, 165 °C, and 180 °C in this study. For the details of the test, refer to ASTM D4402 [19].

(2) Multiple Stress Creep Recovery (MSCR) test

In the Standard Specification for Performance Classification of Asphalt binder (AASHTO:M320), rutting factor ($|G^*|/\sin\delta$) is adopted as the high-temperature performance indicator of asphalt. Rutting factor can predict the high temperature performance of unmodified asphalt well, but there is still a dispute on its applicability for modified asphalt [20,21]. An MSCR test can reflect the nonlinear rheological response of modified asphalt under a large stress, and it has been confirmed that non-recoverable creep compliance ($J_{nr}$) has a good correlativity with the rutting property of modified asphalt [22]. Both the physical and chemical properties change dramatically after asphalt aging, and W-oil rejuvenated asphalt is regarded as modified asphalt. To predict the high-temperature performance of W-oil rejuvenated asphalt more accurately, a $J_{nr}$ indicator achieved by the MSCR test is adopted in this paper.

The MSCR test applies an AR 2000 dynamic shear rheometer (DSR): an 8-mm plate is applied for the specimens of 0%WRA and 5%WRA, and the gap between the parallel plates is set to be 2 mm; a 25 mm plate is applied for the specimens of Virgin, 10%WRA, 15%WRA and 20%WRA, and the gap between the rotor's parallel plates is set at 1 mm. The testing temperature for specimens is 60 °C, and the loading frequency is 10 rad/s. Specimens are loaded respectively under a stress of 0.1 kPa and 3.2 kPa; the loading process is to load for 1 s, then recover for 9 s. Each stress includes 10 cycles of loading and recovering.

(3) Time Sweep (TS) test

Superpave research proposes the fatigue factor ($|G^*| \times \sin\delta$), which is similar to the asphalt rutting factor and takes it as a control index of asphalt fatigue resistance. However, the fatigue factor cannot characterize the fatigue damage characteristics, and it has poor correlation with the asphalt mixture's fatigue property. Asphalt fatigue life, which is obtained from the time sweep test, can reflect the fatigue resistance of the corresponding asphalt mixture [23]. The number of loading cycles ($N_{f50}$) in the case of a complex modulus decreased to 50% is taken as the index for judging the asphalt fatigue [16]. The time sweep test is carried out to characterize the fatigue property of virgin, aged, and rejuvenated asphalt.

The TS test applies AR 2000 DSR, which is the same instrument as in the MSCR test. An 8-mm plate is applied for 0%WRA and 5%WRA, and the gap between the parallel plates is set at 2 mm; a 25-mm plate is applied for Virgin, 10%WRA, 15%WRA, and 20%WRA, and the gap between the parallel plates is set 1 mm. A 5% strain control mode is applied for all of the specimens, the loading frequency is 10 rad/s, and the testing temperature is 20 °C.

(4) Bending Beam Rheometer (BBR) test

The model of the testing instrument is TE-BBR. The flexural creep stiffness or flexural creep compliance, as determined from this test, describes the low-temperature stress-strain time response of the asphalt binder at the test temperature within the range of linear viscoelastic response. Creep stiffness (S) and creep rate (m) are used to assess the low temperature anti-cracking property in this paper. For the testing method, refer to ASTM D 6648 "Standard Test Method for Determining the Flexural Creep Stiffness of Asphalt Binder Using the Bending Beam Rheometer (BBR)".

## 4. Results and Discussion

### 4.1. Safety Property

Figure 9 shows the flash points of virgin asphalt, aged asphalt, and rejuvenated asphalt with different dosages of W-oil. According to Figure 9, it can be seen that the flash point increases slightly after asphalt aging, and decreases after the addition of W-oil, which demonstrates that W-oil is harmful to the safety of aged asphalt. As noted in ASTM D92-16b, the flash point can indicate the possible presence of highly volatile and flammable materials in a relatively nonvolatile or nonflammable material. The highly volatile and flammable materials are hazardous in asphalt mixing and compacting at high temperature, which may lead to fire and explosion. The flash point of aged asphalt and W-oil are 320 °C and 262 °C, respectively, which indicate that W-oil contains more highly volatile and flammable materials. As the dosage of W-oil in aged asphalt increases, the dosage of highly volatile and flammable materials in rejuvenated asphalt also increases. Therefore, W-oil is harmful to the safety of aged asphalt. Although the addition of W-oil decreases the flash point of aged asphalt, the flash point of 20%WRA, 268 °C, is still far higher than 230 °C, which is the flash point specified in the Superpave Binder Requirement (AASHTO M 320), and demonstrates that W-oil rejuvenated asphalt has good safety in construction.

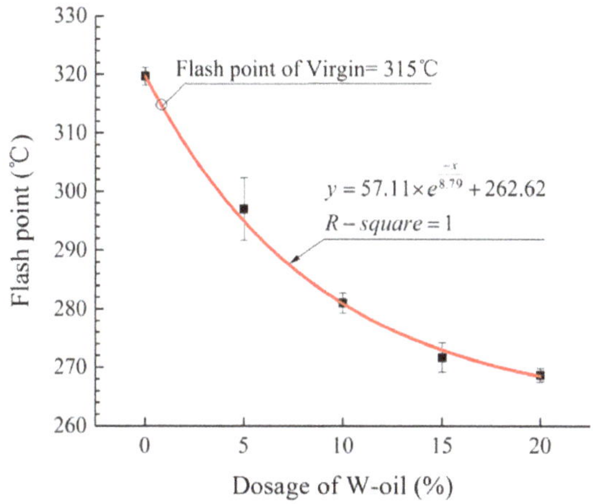


**Figure 9.** Flash Points of Virgin, Aged, and Rejuvenated Asphalt.

### 4.2. Aging Property

Figure 10 shows the mass loss of virgin asphalt, aged asphalt, and rejuvenated asphalt with different dosages of W-oil. According to Figure 10, it can be seen that the mass loss decreases after asphalt aging, which is due to the volatilization of light components in the aging process. The mass loss increases after the addition of W-oil, because W-oil contains light components with low boiling points. The mass loss of W-oil in rejuvenated asphalt is smaller than that of virgin asphalt, which implies that W-oil cannot fully replenish the light components in virgin asphalt that are lost in the aging process [24]. The mass loss presents a linear increase with the dosage of W-oil. When the dosage of W-oil is 20%, the mass loss is 0.27%, which is far smaller than the mass loss of 1% that is specified in the Superpave Binder Requirement (AASHTO M 320). The smaller mass loss the asphalt has, the better its aging resistance is, so it demonstrates that rejuvenated asphalt has a better aging property.

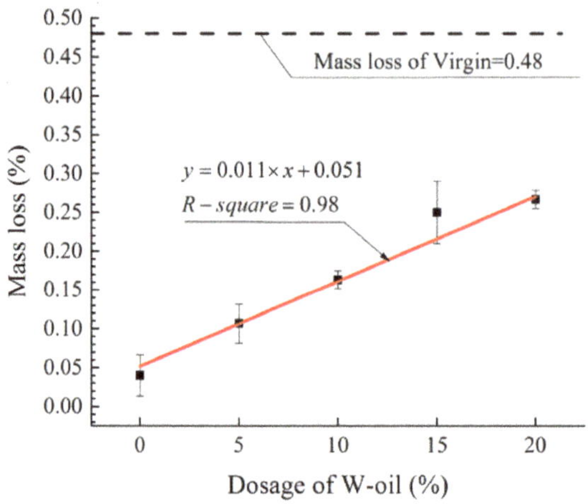


**Figure 10.** Mass Loss of Virgin, Aged, and Rejuvenated Asphalt.

### 4.3. Pavement Performance Based on Physical Tests

#### 4.3.1. Penetration

Penetration is commonly used for grading. According to Figure 11, it can be seen that the penetration decreases after asphalt aging. As the dosage of W-oil increases, the penetration of rejuvenated asphalt increases gradually, and the penetration of rejuvenated asphalt is well linear with the dosage of W-oil. When the dosage of W-oil is 9.5%, the penetration of rejuvenated asphalt is recovered to the level of virgin asphalt.

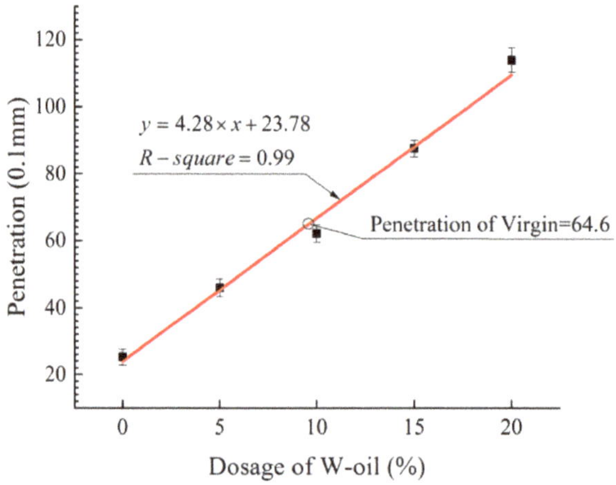


**Figure 11.** Penetration of Virgin, Aged, and Rejuvenated Asphalt.

#### 4.3.2. Soften Point

The soften point can reflect the high temperature stability of asphalt. The higher the soften point is, the better the high temperature stability. According to Figure 12, it can be seen that the soften point rises after asphalt aging, and the aging effect improves the high temperature stability of asphalt. As dosage of W-oil increases, the soften point of rejuvenated asphalt presents a linear decrease, which demonstrates that W-oil reduces the high temperature stability of aged asphalt. When the dosage of W-oil is 13%, the soften point of rejuvenated asphalt is reduced to the level of virgin asphalt.

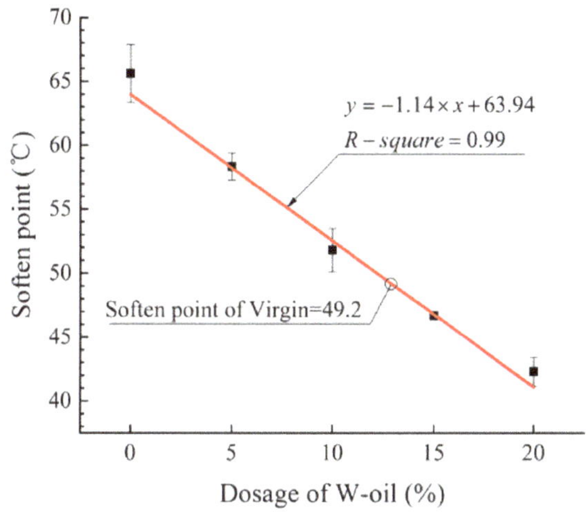


**Figure 12.** Soften Point of Virgin, Aged, and Rejuvenated Asphalt.

4.3.3. Ductility

The higher the ductility is, the better the pavement performance of the asphalt. According to Figure 13, it can be seen that the ductility decreases after asphalt aging, which demonstrates that an aging effect reduces the ductility of the asphalt. As the dosage of W-oil increases in the aged asphalt, the ductility of rejuvenated asphalt continuously increases, which implies that W-oil improves the ductility of aged asphalt. When the dosage of W-oil is 10%, the ductility of rejuvenated asphalt is restored to the level of virgin asphalt.

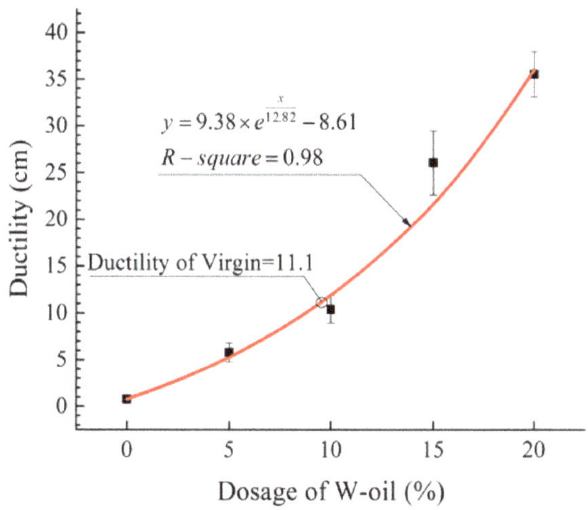


**Figure 13.** Ductility of Virgin, Aged, and Rejuvenated Asphalt.

### 4.4. Pavement Performance Based on Rheological Tests

#### 4.4.1. Workability

In order to guarantee the service performance of asphalt mixture, the viscosity of an asphalt binder shall be controlled within a certain range. In the Technical Specification for Construction of Highway Asphalt Pavement (JTG F40-2004), it is recommended that the optimum viscosity in the mixing and compacting process of asphalt binder to be 170 mPa·S and 280 mPa·S, respectively. Figure 14 shows the viscosity-temperature curve of virgin asphalt, aged asphalt, and rejuvenated asphalt with different dosages of W-oil at 120–180 °C. The line of recommending compacting viscosity and mixing viscosity are drawn at 280 mPa·S and 170 mPa·S, respectively, in which the corresponding horizontal coordinate is the compacting and mixing temperature of the related asphalt. According to Figure 14, it can be seen that an increase of temperature makes the asphalt viscosity decrease, so the control on the optimum mixing and compacting viscosity of asphalt is realized by controlling the temperature. The lower the construction temperature, the less energy is required in the heating process, and the better the workability. The viscosity of virgin asphalt increases by an aging effect, and the mixing and compacting temperature of aged asphalt under the recommended mixing and compacting viscosity are 180 °C and 168 °C, respectively. Both are increased by around 25 °C in comparison with that of virgin asphalt, so aging effect lowers the workability of virgin asphalt. As the dosage of W-oil in aged asphalt increases, the viscosity of rejuvenated asphalt reduces rapidly, which demonstrates that W-oil can improve the workability of aged asphalt. When the dosage of W-oil is 20%, rejuvenated asphalt has a viscosity-temperature curve close to that of virgin asphalt, which demonstrates that 20% W-oil shall be added to aged asphalt for restoring its workability to the virgin level. Viewing from the viscosity-temperature curves of 20%WRA and virgin, the change in the viscosity of virgin asphalt is more sensitive to temperature, which demonstrates that the addition W-oil can recover the viscosity of aged asphalt, but the temperature susceptibility cannot be recovered to the level of virgin asphalt.

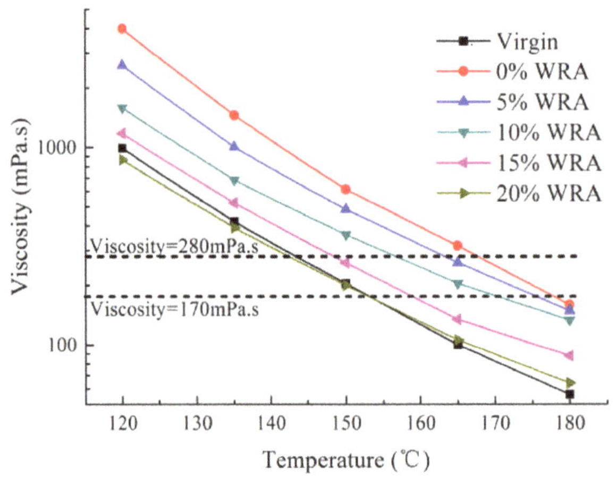


**Figure 14.** Viscosity-Temperature Curve of Virgin, Aged, and Rejuvenated Asphalt.

Figure 15 shows the mixing and compacting temperature of virgin asphalt, aged asphalt, and rejuvenated asphalt with different dosages of W-oil. It can be seen that, as the dosage of W-oil increases, the mixing and compacting temperature keep on declining. The construction temperature of rejuvenated asphalt presents a linear decrease with the addition of W-oil; it can be concluded that W-oil improves the workability of aged asphalt. When the dosage of W-oil is 20%, the viscosity of rejuvenated asphalt is basically restored to the level of virgin asphalt.

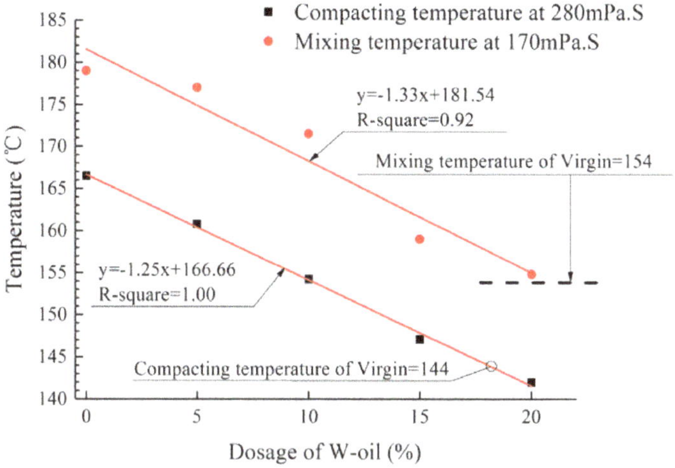


**Figure 15.** Mixing and Compacting Temperature of Virgin, Aged, and Rejuvenated Asphalt.

4.4.2. Rutting Property

Figure 16 shows the cumulative strain curve of specimens under two different stresses (0.1 kPa and 3.2 kPa). It can be seen that the cumulative deformation of 0%WRA is the minimum. The cumulative deformation of virgin asphalt is far greater than that of 0%WRA, because the aging effect improves the non-deformability of asphalt. The addition of W-oil increases the cumulative deformation of aged asphalt. The higher dosage of W-oil is, the higher the deformability of the rejuvenated asphalt. It is concluded that the deformation recovering capacity is improved by the aging effect, while the addition of W-oil lowers the deformation recovering capacity of aged asphalt. When the stress is 0.1 kPa, 15%WRA has a cumulative deformation curve close to that of virgin asphalt, while in the case of 3.2 kPa, 15%WRA has a cumulative deformation curve with a certain difference from that of virgin asphalt, which demonstrates that the trend of cumulative deformation under different stresses varies.

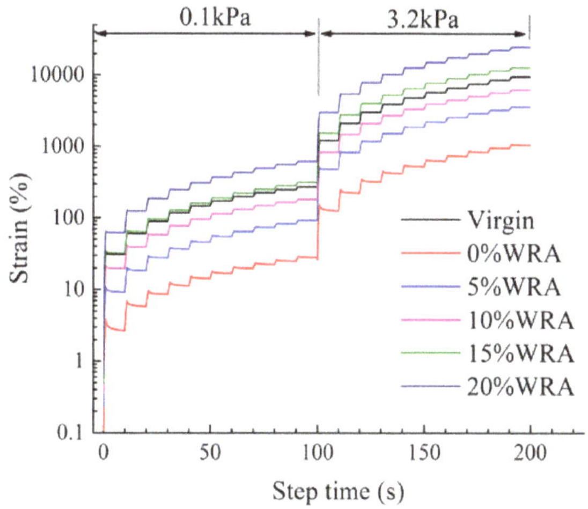


**Figure 16.** Multiple Stress Creep Recovery (MSCR) Curve of Virgin, Aged, and Rejuvenated Asphalt.

Non-recoverable creep compliance ($J_{nr}$) can characterize the anti-rutting property at high temperature well. The larger $J_{nr}$ is, the poorer the anti-rutting is; the smaller $J_{nr}$ is, the better the anti-rutting. The non-recoverable creep compliance ($J_{nr}$) of each cycle under each stress can be calculated as Equation (1):

$$J_{nr} = \frac{\gamma_{nr}}{\tau} \tag{1}$$

where $\tau$ is the loading stress for each cycle, kPa; and $\gamma_{nr}$ is the non-recoverable strain of the cycle.

According to Figure 17, it can be seen that the $J_{nr}$ of 0%WRA is far smaller than that for virgin asphalt; therefore, the aging effect improves the rutting resistance of asphalt. The larger the dosage of W-oil is, the larger $J_{nr}$ value of the rejuvenated asphalt; it is concluded that W-oil lowers the rutting property of aged asphalt. There are differences in the $J_{nr}$ values of specimens under different stresses, but no obvious rules are found.

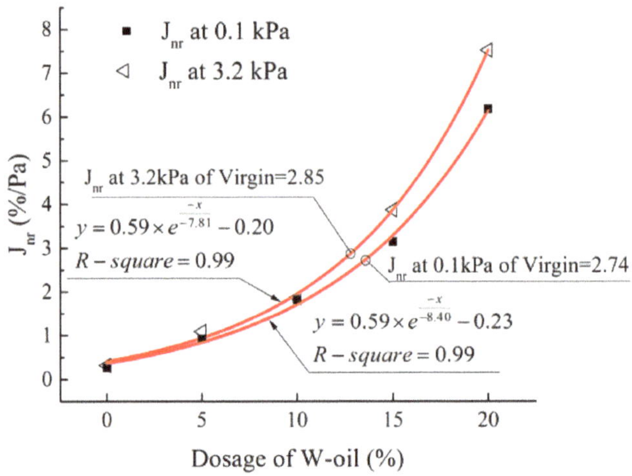


**Figure 17.** $J_{nr}$ of Virgin, Aged, and Rejuvenated Asphalt.

### 4.4.3. Fatigue Property

Figure 18 shows the impact of the number of loading cycles on the complex modulus of virgin, aged, and rejuvenated asphalt. It can be seen that as the number of loading cycles increases, the complex modulus decrease continuously, which demonstrates that loads are harmful to the structural strength of asphalt. The larger the dosage of W-oil is, the lower the initial modulus of rejuvenated asphalt. The declining trend of complex modulus becomes slower; thus, W-oil improves the fatigue property of aged asphalt on the basis of reducing the initial modulus. However, as an asphalt binder for pavement, it shall not only consider the fatigue property, but also guarantee the high initial modulus. Therefore, the optimum dosage of W-oil shall be controlled in order to ensure the high temperature and fatigue property of asphalt. Besides, the modulus curve of virgin asphalt is close to that of 10%WRA, so there is a close fatigue property between virgin asphalt and 10%WRA.

$N_{f50}$ is taken as the index for evaluating asphalt fatigue. Figure 19 shows the $N_{f50}$ values of virgin asphalt, aged asphalt, and rejuvenated asphalt with different dosages of W-oil. It can be seen that, as the dosage of W-oil increases, the fatigue life of rejuvenated asphalt increases exponentially.

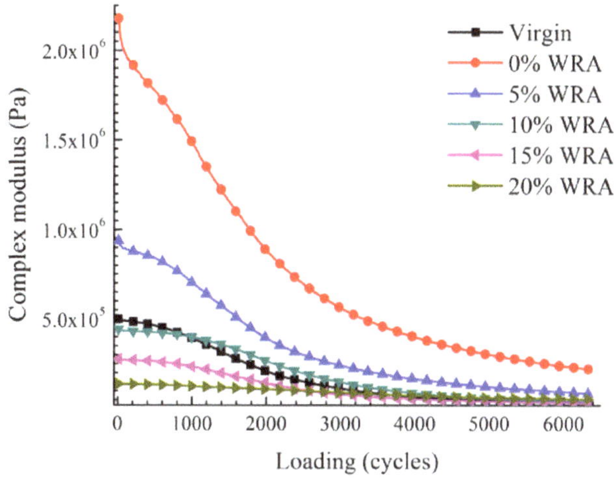


**Figure 18.** Change of Complex Modulus under Loading.

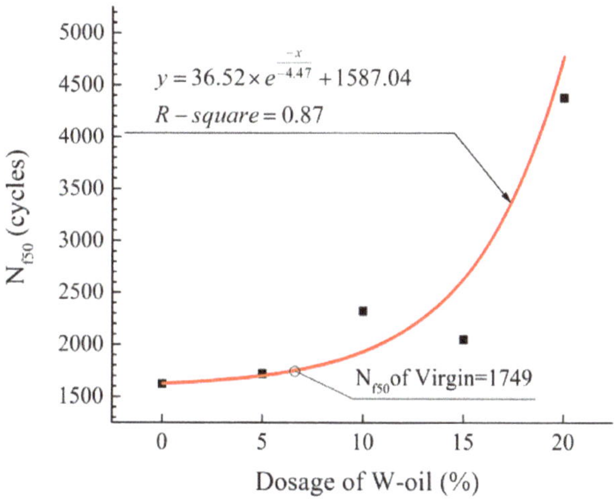


**Figure 19.** $N_{f50}$ of Virgin, Aged, and Rejuvenated Asphalt.

#### 4.4.4. Low Temperature Property

The creep stiffness (S) and creep rate (m) of virgin, aged, and rejuvenated asphalt at −12 °C are shown in Figure 20. The creep stiffness significantly increases, and the creep rate significantly declines after aging, which demonstrates that the aging effect will make asphalt harden and easily crack at low temperature. With the addition of W-oil, the creep stiffness of aged asphalt decreases, and the creep rate rises, which demonstrates that W-oil improves the low temperature anti-cracking property of aged asphalt. When the dosage of W-oil is 20%, the test fails to complete, because 20%WRA is too soft. The value of S and m in Figure 20 is set as 0 and 1, respectively. When the dosage of W-oil is 8% and 13%, the S and m values of rejuvenated asphalt are recovered respective to the level of virgin asphalt, respectively, which demonstrates that the effect of W-oil on creep stiffness is larger than its effect on the creep rate.

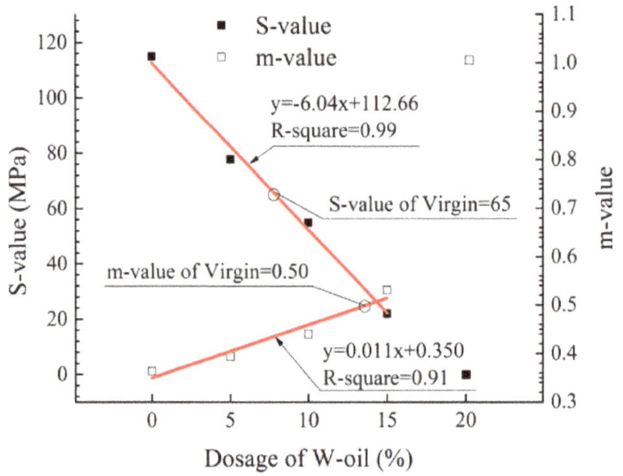


**Figure 20.** S and m Values of Virgin, Aged, and Rejuvenated Asphalt.

### 4.5. Analyzing Possibility of W-Oil as Rejuvenator

The pavement performance of asphalt becomes poor due to the aging effect, and the aim of adding rejuvenator is to restore the pavement performance of aged asphalt. First of all, the optimum dosage of W-oil shall be determined. In this paper, the optimum dosage is determined by the rheological index, because the rheological index can be directly correlated to the field pavement performance. According to the test results of performance indices obtained from the rheological test, with the increasing dosage of W-oil, the workability, fatigue property and low-temperature performance of rejuvenated asphalt are improved. However, the rutting resistance of rejuvenated asphalt is declining. The principles of determining the optimal dosage of W-oil are: under the premise of ensuring the high-temperature performance of rejuvenated asphalt, improve the workability, fatigue property, and low-temperature performance of aged asphalt to the maximum. Taking the $J_{nr}$ value of rejuvenated asphalt no lower than that of virgin asphalt as the criteria, the calculated dosage, making $J_{nr}$ in the case of 3.2 kPa and 0.1 kPa restore to the level of virgin asphalt, is respectively 13.0% and 13.7%, and the average value of 13.4% is taken as the optimum dosage of W-oil.

The feasibility of the rejuvenator is usually evaluated by comparing the properties of virgin asphalt and rejuvenated asphalt. For carrying out a comparative analysis on the performance of 13.4% W-oil rejuvenated asphalt and virgin asphalt, the safety, aging, and pavement performance indices of virgin and rejuvenated asphalt shall be processed for normalization, as shown in Figure 21. The optimum dosage of W-oil is determined as 13.4%, which is based on the same high-temperature stability as virgin asphalt. The safety of rejuvenated asphalt with 13.4% W-oil is poorer than that of virgin asphalt, while the safety can still meet the specification requirements. The aging property is superior to that of virgin asphalt. The ductility is better than that of virgin asphalt. The workability is slightly poorer than virgin asphalt, and will be further improved. The fatigue life of rejuvenated asphalt is longer than that of virgin asphalt by approximately 33%. The value m of rejuvenated asphalt is close to that of virgin asphalt, but the value S is lower by 47% than virgin asphalt; therefore, the low-temperature property of rejuvenated asphalt is better than that of virgin asphalt. In summary, W-oil can significantly improve the properties of aged asphalt, and it is of great potential to serve as an asphalt rejuvenator.

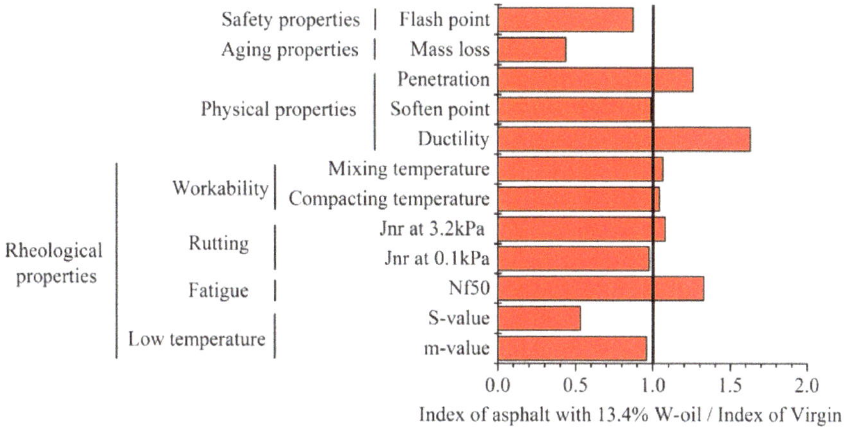


**Figure 21.** Ratio of Index of Virgin Asphalt and Rejuvenated Asphalt with 13.4% W-oil.

## 5. Conclusions

Based on the experimental results from virgin, aged, and rejuvenated asphalt in terms of safety, aging, and traditional and rheological property tests, the following conclusions can be drawn:

(1) Both the safety and aging property of W-oil rejuvenated asphalt decrease while the dosage of W-oil is increasing. When the dosage of W-oil is smaller than 20%, the flash point and mass loss of rejuvenated asphalt can still meet the specification requirements.

(2) As the dosage of W-oil increases, the penetration and ductility of rejuvenated asphalt increase, while the soften point declines. W-oil can recover the penetration, soften point, and ductility of aged asphalt to the level of virgin asphalt.

(3) As the dosage of W-oil increases, the workability of rejuvenated asphalt is improved, the fatigue life is extended, and the low-temperature property is improved, while the rutting resistance is reduced.

(4) Under the premise of the high-temperature performance of rejuvenated asphalt, the optimum dosage of W-oil is determined to be 13.4%. The safety and aging property of rejuvenated asphalt can meet the specification requirements; the pavement performance basically recovers to the level of virgin asphalt.

(5) W-oil can improve the safety, aging, and pavement performance of aged asphalt. W-oil is of great potential to serve as asphalt rejuvenator, while the workability and rutting property are to be further improved.

**Author Contributions:** C.X., C.X. and T.B. conceived and designed the experiments; C. X. and L.X. performed the experiments; C.X. and C.X. analyzed the data; T.B. and W.Y. contributed reagents/materials/analysis tools; C.X. and W.Y. wrote the paper.

**Acknowledgments:** This work is supported by Chongqing Postgraduate Scientific Innovation Project (2017B0103) and Chongqing Traffic Science and Technology Project. The authors gratefully acknowledge their financial support.

**Conflicts of Interest:** The authors declare no conflict of interest.

## References

1. Kandhal, P.S.; Mallick, R.B. *Pavement Recycling Guidelines for State and Local Governments: Participant's Reference Book*; Final Report; National Transportation Library: Washington, DC, USA, 1998.
2. Aurangzeb, Q.; Al-Qadi, I.L.; Ozer, H.; Yang, R. Hybrid life cycle assessment for asphalt mixtures with high RAP content. *Resour. Conserv. Recycl.* **2014**, *83*, 77–86. [CrossRef]

3. He, G.P.; Wong, W.G. Effects of moisture on strength and permanent deformation of foamed asphalt mix incorporating RAP materials. *Constr. Build. Mater.* **2008**, *22*, 30–40. [CrossRef]
4. Mogawer, W.; Bennert, T.; Daniel, J.S.; Bonaquist, R.; Austerman, A.; Booshehrian, A. Performance characteristics of plant produced high RAP mixtures. *Road Mater. Pavement Des.* **2012**, *13* (Suppl. 1), 183–208. [CrossRef]
5. Holden, D.A.; Strauss, J. HP-GPC characterization of rejuvenated aged CRM binders. *J. Mater. Civ. Eng.* **2007**, *19*, 515–522.
6. Zaumanis, M.; Mallick, R.B.; Frank, R. Evaluation of Rejuvenator's Effectiveness with Conventional Mix Testing for 100% Reclaimed Asphalt Pavement Mixtures. *Transp. Res. Rec. J. Transp. Res. Board.* **2013**, *28*, 17–25. [CrossRef]
7. Asli, H.; Ahmadinia, E.; Zargar, M.; Karim, M.R. Investigation on physical properties of waste cooking oil—Rejuvenated bitumen binder. *Constr. Build. Mater.* **2012**, *37*, 398–405. [CrossRef]
8. Chen, M.; Xiao, F.; Putman, B.; Leng, B.; Wu, S. High temperature properties of rejuvenating recovered binder with rejuvenator, waste cooking and cotton seed oils. *Constr. Build. Mater.* **2014**, *59*, 10–16. [CrossRef]
9. Chen, M.; Leng, B.; Wu, S.; Sang, Y. Physical, chemical and rheological properties of waste edible vegetable oil rejuvenated asphalt binders. *Constr. Build. Mater.* **2014**, *66*, 286–298. [CrossRef]
10. Azahar, W.N.A.W.; Jaya, R.P.; Hainin, M.R.; Bujang, M.; Ngadi, N. Chemical modification of waste cooking oil to improve the physical and rheological properties of asphalt binder. *Constr. Build. Mater.* **2016**, *126*, 218–226. [CrossRef]
11. Zhang, D.; Chen, M.; Wu, S.; Liu, J.; Amirkhanian, S. Analysis of the Relationships between Waste Cooking Oil Qualities and Rejuvenated Asphalt Properties. *Materials* **2017**, *10*, 508. [CrossRef] [PubMed]
12. Bowers, B.F.; Huang, B.; Shu, X.; Miller, B.C. Investigation of Reclaimed Asphalt Pavement blending efficiency through GPC and FTIR. *Constr. Build. Mater.* **2014**, *50*, 517–523. [CrossRef]
13. ASTM D2872-12e1. *Standard Test Method for Effect of Heat and Air on a Moving Film of Asphalt (Rolling Thin-Film Oven Test)*; ASTM International: West Conshohocken, PA, USA, 2012.
14. ASTM D6521-13. *Standard Practice for Accelerated Aging of Asphalt Binder Using a Pressurized Aging Vessel (PAV)*; ASTM International: West Conshohocken, PA, USA, 2013.
15. ASTM D92-16b. *Standard Test Method for Flash and Fire Points by Cleveland Open Cup Tester*; ASTM International: West Conshohocken, PA, USA, 2016.
16. ASTM D5/D5M-13. *Standard Test Method for Penetration of Bituminous Materials*; ASTM International: West Conshohocken, PA, USA, 2013.
17. ASTM D36/D36M-14e1. *Standard Test Method for Softening Point of Bitumen (Ring-and-Ball Apparatus)*; ASTM International: West Conshohocken, PA, USA, 2014.
18. ASTM D113-17. *Standard Test Method for Ductility of Asphalt Materials*; ASTM International: West Conshohocken, PA, USA, 2017.
19. ASTM D4402/D4402M-15. *Standard Test Method for Viscosity Determination of Asphalt at Elevated Temperatures Using a Rotational Viscometer*; ASTM International: West Conshohocken, PA, USA, 2015.
20. D'Angelo, J. The Relationship of the MSCR Test to Rutting. *Road Mater. Pavement Des.* **2009**, *10* (Suppl. 1), 61–80. [CrossRef]
21. D'Angelo, J.; Kluttz, R.; Dongre, R.N.; Stephens, K.; Zanzotto, L. Revision of the Superpave high temperature binder specification: The multiple stress creep recovery test (with discussion). *J. Assoc. Asphalt Paving Technol.* **2007**, *76*, 123–162.
22. Bahia, H.U.; Hanson, D.I.; Zeng, M.; Zhai, H.; Khatri, M.A.; Anderson, R.M. *Characterization of Modified Asphalt Binders in Superpave Mix Design*; The National Academies of Sciences, Engineering, and Medicine: Washington, DC, USA, 2001.
23. Nayak, P.; Sahoo, U.C. Rheological, chemical and thermal investigations on an aged binder rejuvenated with two non-edible oils. *Road Mater. Pavement Des.* **2017**, *18*, 612–629. [CrossRef]
24. Fernándezgómez, W.D.; Quintana, H.A.R.; Reyeslizcano, F. A review of asphalt and asphalt mixture aging. *Ing. Investig.* **2012**, *33*, 5–12.

applied sciences

MDPI

Article

# Interface Shear Performance between Porous Polyurethane Mixture and Asphalt Sublayer

**Jun Chen [1], Cheng Yao [1], Hao Wang [2,*], Wei Huang [2], Xie Ma [1] and Junyu Qian [3]**

1 College of Civil and Transportation Engineering, Hohai University, Nanjing 210098, China; chen_jun2728@163.com (J.C.); yc520dr9337@163.com (C.Y.); maxiehohai@163.com (X.M.)
2 Department of Civil and Environmental Engineering, Rutgers, The State University of New Jersey, Piscataway, NJ 08854, USA; huang.david@rutgers.edu
3 Department of Civil, Environmental and Ocean Engineering, Steven Institute of Technology, Hoboken, NJ 07030, USA; jqian10@stevens.edu
* Correspondence: hwang.cee@rutgers.edu

Received: 13 March 2018; Accepted: 13 April 2018; Published: 17 April 2018

**Abstract:** This paper aims to study interface shear performance between porous polyurethane mixture (PPM) and asphalt mixture with different adhesive materials. Polyurethane, epoxy resin, and SBS (styrene–butadiene–styrene) modified asphalt were selected as adhesive materials to fabricate composite specimens. The interface shear strength and shear fatigue life of composite specimen was measured using inclined shear test. The research results emphasizes that it is necessary to apply adhesive material to the interface between PPM and asphalt mixture, since the untreated interface shear strength is smaller than the ones between two asphalt mixtures. The interface shear strength is affected by the thickness of adhesive layer, temperature, and freezing–thaw condition. In general, the greatest interface shear strength was achieved by using epoxy resin followed by polyurethane and then SBS modified asphalt at 25 °C as the adhesive layer thickness is the same. However, the interface shear strengths of composite specimen with three adhesive materials are similar to each other at high and low temperatures (60 °C and −18 °C) or after freezing–thaw cycles. On the other hand, the composite specimen with epoxy resin as adhesive material has the longest fatigue life; while the SBS modified asphalt has the least fatigue life at 25 °C. The research findings can help select the appropriate adhesive materials and increase the durability and service life of pavement when PPM is used as road surface layer for safety and noise reduction.

**Keywords:** polyurethane; asphalt mixture; epoxy resin; adhesive layer; shear strength; inclined shear test

## 1. Introduction

The open graded friction course (OGFC) was mainly used as road surface layer placed on dense graded asphalt layer to improve traffic safety at rainy conditions [1,2]. With the high air void content (18%–22%), OGFC offers safety advantages in terms of high skid resistance and no splash or spray in addition to its benefits in noise reduction and water runoff quality enhancement [3–6]. Despite of the functionality benefits of OGFC, engineering practices have found that the application of OGFC faces three challenges during the service life of pavement, including clogging of air voids [7,8], raveling [9–11], and freezing-induced damage in winter [12,13]. These problems apparently reduce pavement durability and affect the functional benefits of OGFC. Due to the challenges related to OGFC, its use has been limited through the years. A survey conducted in 2016 revealed that 25 states in U.S. do not use OGFC, among which 19 states have used it before but did not continue due to freezing damaged in winter [14].

Recently, porous polyurethane mixture (PPM) has been proposed for road surface layer for noise reduction and better durability [15–18]. Although the cost of PPM is higher, PPM has three advantages over OGFC. The first is that it can be mixed at room temperature, which can significantly save heating energy and eliminate the pollution of asphalt smoke during high temperature mixing [19,20]. The second is that the PPM performs better to resist particle-related clogging due to its high air void (30%), because the pores at the greater porosity are more difficult to be clogged by small particles [18]. Lastly, the polyurethane is hard after curing and can maintain high stiffness during hot days. This makes PPM have better ability to resist deformation-related clogging and produce sufficient bonding with aggregates, which minimize the possibility of raveling [21,22]. The recent work also found that PPM can significantly retard ice-formation time and reduces the adhesive strength with ice, which shows superior anti-icing and deicing performance than OGFC [23].

Since the functional benefit and durability of PPM have been documented, the application of PPM as road surface layer is most beneficial. In this case, the interface bonding between the PPM surface layer and asphalt sublayer is important. However, previous researchers have found that early failure could be caused by the delamination of PPM with its sublayer, especially at tire braking or acceleration conditions [24]. The Japanese Public Works Research Institute (PWRI) found that adhesion of polyurethane mixture was not very durable with dense asphalt mixture but its adhesion with cement concrete was better. Therefore, they developed a semi-flexible layer (porous asphalt with the pores filled with cement) as sublayer of polyurethane mixture to solve the adhesion issue [25]. In the period 2002–2005, full-scale experiments on Japanese highways with eight different types of PPM on three locations all failed within a relative short time after construction (less than one year) owing to raveling, adhesion, and friction problems. The similar interface failure between polyurethane mixture and dense asphalt concrete layer was observed in Sweden only after four months of construction [26]. The reason causing the delamination of PPM could be insufficient bonding between PPM and underlying layer. Therefore, adhesive materials may be needed to improve the interface bonding between PPM and the traditional dense asphalt layer underneath.

Traditionally, asphalt emulsions have been used as tack coat between asphalt layers. It was found that the interface shear strength between asphalt layers was affected by temperature, tack coat type and dosage, surface roughness, and mixture type [27,28]. However, few studies have investigated the effectiveness of adhesive material on interface bonding between PPM and asphalt sublayers.

## 2. Objectives and Scope

This study aims to study interface shear performance between porous polyurethane mixture (PPM) and asphalt sublayer with different adhesive materials. Polyurethane, epoxy resin, and SBS (styrene–butadiene–styrene) modified asphalt were selected as adhesive materials to fabricate composite specimens. The interface shear strength of composite specimen was measured using inclined shear test at different temperatures and freezing–thaw conditions. The shear fatigue life of composite specimen with different adhesive materials was also investigated. The test results can be used to select the better performing adhesive material for long-term durability of PPM as surface layer.

## 3. Experimental Materials

Laboratory composite specimens were prepared for shear test with the upper layer and the bottom layer. Three types of upper layer materials were used, including one porous polyurethane mixture with the nominal maximum aggregate size of 9.5 mm (PPM-10) and two asphalt mixtures with the nominal maximum aggregate size of 13 mm (AC-13 and OGFC-13). The sublayer material is asphalt mixture with the nominal maximum aggregate size of 19 mm (AC-20).

Polyurethane was prepared by mixing two components with certain proportion followed by adding catalyst. The mass ratio of catalyst with respect to the two components is determined by the curing time of polyurethane. Based on recommendations from the manufacturer of polyurethane, 2‰ catalyst is usually used if the curing time is required within 15 min after mixing. The two components were isocyanate prepolymer (component A) and the mixture of polyether polyol and pentaerythritol (component B). The polyurethane before solidification was transparent yellow liquid under room temperature. Table 1 gives the basic properties of polyurethane. On the other hand, the SBS modified asphalt was obtained by mixing 70# base asphalt with 4% SBS polymer at high temperature. The epoxy resin was obtained by manually mixing the resin with curing agent at mass ratio of 1:1 at the temperature of 20~30 °C. It is noted that the mixing should be done unidirectional at constant speed for about one minute in order to prevent the generation of bubbles during mixing or non-uniform mass distribution after mixing.

**Table 1.** Basic properties of polyurethane binder.

| Index | Density (g/cm$^3$) | Curing Time (min) | pH Value |
|---|---|---|---|
| Value | 1.003 | 5~1200 | 6.7 |

To prepare porous polyurethane mixture (PPM), basalt aggregates with single grain size of 4.75~9.5 mm were mixed with polyurethane at the mass ratio of 94:6 under room temperature. It should be noted that the mass ratio of polyurethane to aggregate particles was determined by the inverse relationship between the dosage of polyurethane and the Cantabro mass loss of specimen measured following AASHTO T108. It was found that the scattering loss rate achieved its minimum value of 6.4% when the dosage of polyurethane was 6%. The aggregate gradation and polyurethane binder content of PPM-10 can be found in authors' previous work [23].

Three types of asphalt mixture, AC-20, AC-13 and OGFC-13, were prepared by mixing limestone aggregates with 70# base bitumen. The aggregate gradations, asphalt contents (AC), and air void (AV) contents of different asphalt mixtures are showed in Table 2. The air void contents of PPM and OGFC were determined from the measured bulk specific gravity using paraffin-coated method (AASHTO T275) and the measured theoretical maximum specific gravity (AASHTO T209); while the saturated surface dry method (AASHTO T166) was used for the bulk specific gravity of dense-graded AC specimens.

**Table 2.** Mix designs for different asphalt mixtures.

| Type | Mass Percentage (%) Passing the Below Sieve Size (mm) | | | | | | | | | | | | AC (%) | AV (%) |
|---|---|---|---|---|---|---|---|---|---|---|---|---|---|---|
| | 26.5 | 19 | 16 | 13.2 | 9.5 | 4.75 | 2.36 | 1.18 | 0.6 | 0.3 | 0.15 | 0.075 | | |
| AC-20 | 100 | 95.3 | 89.6 | 78.4 | 65.6 | 51 | 39.0 | 28.4 | 16 | 9 | 7.1 | 5.4 | 4.6 | 4.9 |
| AC-13 | 100 | 100 | 100 | 95 | 76.5 | 56.2 | 42.4 | 29 | 20 | 12.7 | 7.9 | 5.4 | 5.0 | 4.2 |
| OGFC-13 | 100 | 100 | 100 | 94.7 | 68.5 | 18.2 | 14.2 | 10.5 | 7.5 | 5.9 | 4.8 | 4 | 5.1 | 20.3 |
| PPM-10 | Single grain size of 4.75–9.0 mm | | | | | | | | | | | | 6.0 | 32.4 |

Three types of adhesive materials were used at the interface of composite specimen, including polyurethane, epoxy resin, and SBS modified asphalt. It should be noted that the desired thickness of adhesive material applied at the interface was achieved by controlling the mass of adhesive material after knowing the density and coating area. The density of polyurethane, epoxy resin, and SBS modified asphalt is 1.003 g/cm$^3$, 0.971 g/cm$^3$ and 1.030 g/cm$^3$, respectively.

## 4. Test Method

### 4.1. Specimen Preparation

A special cylindrical test mold (inner diameter of 101.6 mm and height of 150 mm) made of steel was custom-made for fabricating the composite specimen. According to the Marshall method, the sublayer (AC-20) was initially prepared in the mold after the compaction of 75 times on each side of the specimen, and cured at room temperature for 12 h without demold. After that, the adhesive material (if needed) was applied on the surface of sublayer and then the upper layer (PPM-10 or AC-13 or OGFC-13) was fabricated after the compaction of 75 times on the top of upper layer, as shown in Figure 1. The SBS modified asphalt was hot-applied at the interface at directly. When epoxy resin and polyurethane were used as adhesive materials, the upper layer should be formed within 15 min to ensure that adhesive materials were not solidified and could penetrate into the upper layer. The composite specimen was then cured for 12 h at room temperature before testing.

**Figure 1.** Preparation of composite specimen: (**a**) Schematic illustration and (**b**) picture of real specimen.

### 4.2. Inclined Shear Test

Several testing methods have been used to characterize the interface bonding performance. The modes in these testing methods can be mainly divided into shear, tension, and torsion, in which shearing is the most widely used one. Two types of shear test (direct shear test and inclined shear test) have been used to measure interface shear strength. In the direct shear test, the load direction is parallel to the interface and the confining stress is applied separately [29]. As for the inclined shear test, the load is applied in both parallel and vertical directions with respect to the interface at the same time using an inclined frame [30]. Although the setup of inclined shear test is relatively simple, the ratio of confining pressure to shear stress is defined by the inclined angle and fixed during the test. In the direct shear test, the confining pressure can be applied with varying magnitudes. Further work can be conducted to evaluate interface shear failure using different testing procedures, such as Texas overlay test.

Figure 2 presents the setup of inclined shear test in the laboratory used in this study. The inclination angle between interface axis and horizontal direction was set at 45° so that the interface was loaded with equivalent shear force and normal force. The load was applied at 5 mm/min until the layer interface failed. The shear bonding strength of interface was calculated using Equation (1).

$$\tau = \frac{F \cdot \sin 45^{o}}{S} \tag{1}$$

where, $\tau$ denotes shear bonding strength of interface; $F$ denotes the maximal vertical load; and $S$ presents the interface contact area between two layers.

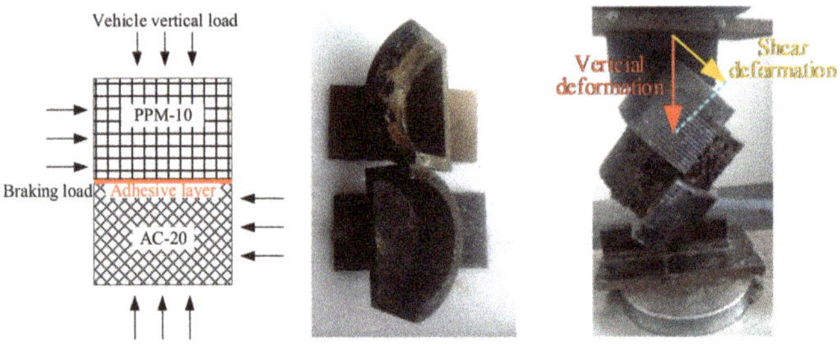


**Figure 2.** Illustration of inclined shear test: (**a**) Force on the specimen; (**b**) shear clamps; and (**c**) test setup with specimen.

### 4.3. Test Conditions

Asphalt pavement is subject to different climate conditions during its service life, including high temperature in summer, freezing–thaw in winter, and water infiltration during rainfall. At rainy days, water can drain vertically through the OGFC and laterally flow to pavement shoulder at wet weather condition due to the internal interconnected voids, which cause less moisture-related distress in OGFC pavement. However, it is documented that the open and interconnected pores in porous asphalt mixtures are easier to be damaged by the volume change due to freezing and thaw as compared to dense-graded asphalt mixtures [13]. Thus, shearing test was applied to composite specimens at different temperatures (−18 °C, 25 °C and 60 °C) and after freezing–thaw cycles. The details of test matrix were presented in Table 3. At each freezing–thaw cycle, the composite specimens were frozen under −18 °C for four hours and then immersed in water bath at 15 °C for four hours. This process was repeated for four times.

**Table 3.** Test matrix of composite specimens.

| Test | Freeze–Thaw Cycle | Thickness of Adhesive Layer (mm) | Temperature of Shearing Test (°C) |
|---|---|---|---|
| Shear test | No | 0.3, 0.5, 1.0 | −18, 25, 60 |
| Shear test | Yes | 0.5 | 25 |
| Shear fatigue test | No | 0.5 | 25 |

Although the shear strength may indicate the shear resistance of interface for the purpose of ranking, the interface between surface layer and sublayer of asphalt pavement sustains repetitive shear force due to traffic loading. Therefore, stress-controlled inclined shear fatigue test was conducted to evaluate fatigue life of the interface in the composite specimen under repeated loading cycles. The inclined shear test setup shown in Figure 2 was used for fatigue test. A sinusoidal load wave with the frequency of 10 Hz and peak stress of 0.4 MPa without rest time between cycles was applied at 25 °C for shear fatigue test. The loading frequency and magnitude was selected based on the typical shear stress magnitude experienced at near-surface under truck loading at highway speeds [24]. The cyclic loading in the fatigue test was applied until that the interface in composite specimen failed. Although currently there is no standard available for shear fatigue test, the similar fatigue test approach has been used in the previous work as the more realistic test to represent the failure process of pavement interface under repeated traffic loading [31,32].

It is noted that uniform shear stress in the interface was assumed in the calculation of shear stress. In the real case, the shear stress may have stress concentrations due to the stiffness difference between asphalt binder and aggregate. This should be investigated with microstructure-based model in future work.

## 5. Results and Discussion

### 5.1. Shear Strength of Composite Specimen without Adhesive Material

Figure 3 illustrates shear stress-deformation curves of three composite specimens without adhesive material at 25 °C. The shear deformation was the projection of total vertical deformation in the direction of interface, as shown in Figure 2c. It can be seen that the interface shear stress first increases to the peak value and then decreases with the increase of shear displacement. The composite specimen PPM-10 + AC-20 has the lowest interface shear strength among three composite specimens without adhesive material. This indicates that the bonding between PPM and asphalt mixture is insufficient compared to the interface between two asphalt layers, which may cause delamination between PPM and its underlying asphalt layer.

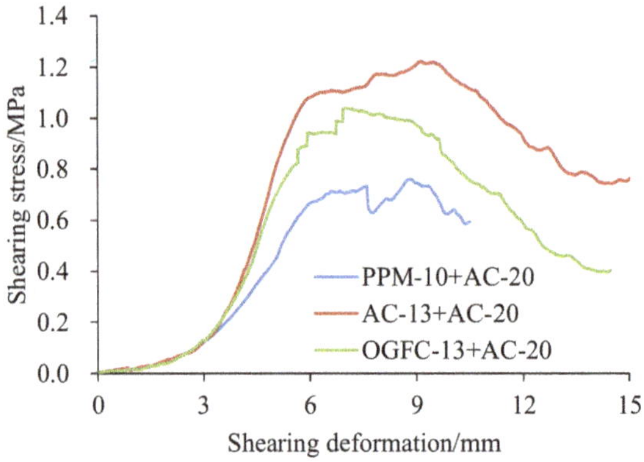


**Figure 3.** Shear stress-deformation curves of three composite specimens without adhesive materials.

The average shear strengths of three replicates of composite specimens without adhesive material are showed in Figure 4. The standard deviation values were shown as the error bar from the averages in the figure. The results show that the interface shear strength of PPM-10 + AC-20 composite specimen is the lowest at both 25 °C and 60 °C. This can be caused by two reasons. The first reason is the inter-penetration between the upper and lower layer. For AC-13 + AC-20 and OGFC-13 + AC-20 specimen, the asphalt binder of AC-20 penetrated into the upper layer when AC-13 and AC-20 specimens were compacted at high temperatures, which was different from the case of PPM that was mixed at room temperature. Another reason is that the greater air void of PPM-10 causes the interface contact area smaller than those of the other two composite specimens, resulting in the smaller shear strength. This is similar to the finding reported by previous work, in which it was found that the contact area was an important factor affecting the interface shear strength between OGFC and dense-graded asphalt mixture [32]. The effect of air void on shear strength can be proved by the observation that the interface shear strength between AC-13 and AC-20 is greater than the one between OGFC-13 and AC-20.

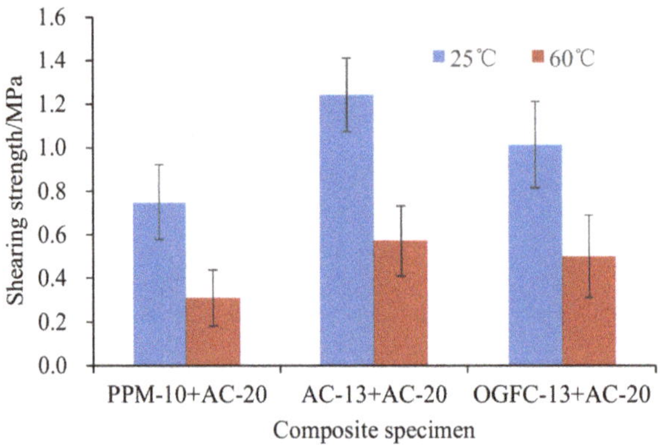


**Figure 4.** Shear strengths of composite specimen without adhesive materials.

Figure 5 shows the appearance of fractured surfaces after shear testing of composite specimen. It can be noticed that the fractured surface of PPM-10 + AC-20 composite specimen is smooth, which indicates the pure interfacial failure. However, the fractured surfaces of AC-13 + AC-20 or OGFC-13 + AC-20 composite specimen were found much rougher, which indicated that the failure was not pure interfacial sliding due to interlocking of aggregates and inter-penetration of asphalt binder across the interface. The texture profile of interface can be measured using laser sensor or microscopy if more accurate quantification is needed. The insufficient bond between PPM-10 and AC-20 may cause earlier failure of pavement due to slippage cracking or shoveling. Thus, it is necessary to apply adhesive materials to increase the bonding of PPM with the sublayer.

**Figure 5.** Appearance of fracture surface after shear test for composite specimens without adhesive materials.

### 5.2. Shear Strength of Composite Specimen with Adhesive Materials

Three types of adhesive materials were applied on PPM-10 + AC-20 composite specimens with different adhesive materials and adhesive layer thicknesses at 25 °C. The shear strengths of three replicates of each composite specimen were tested and the results are presented in Figure 6. The standard deviation values were shown as the error bar from the averages in the figure. The mass amounts of adhesive materials used for different thicknesses are shown in Table 4.

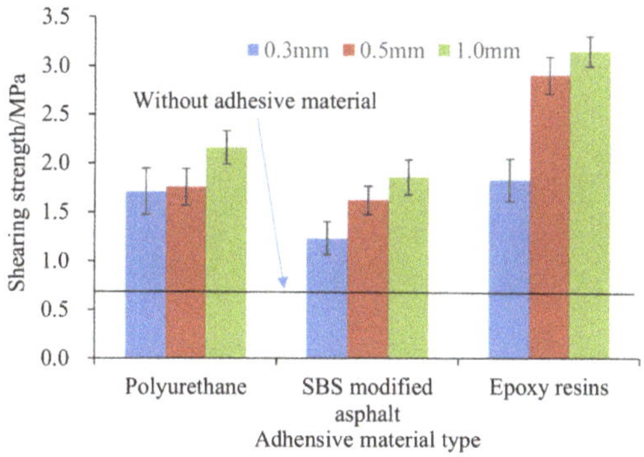


**Figure 6.** Shear strengths of PPM-10 + AC-20 composite specimen with different adhesive materials and thicknesses at 25 °C.

**Table 4.** The mass of adhesive material for different thicknesses.

| Mass of Adhesive Material at Thickness of | 0.3 mm | 0.5 mm | 1.0 mm |
|---|---|---|---|
| Polyurethane (g) | 2.4 | 4.1 | 8.1 |
| SBS modified asphalt (g) | 2.5 | 4.2 | 8.3 |
| Epoxy resins (g) | 2.4 | 3.9 | 7.9 |

According to the results in Figure 6, all the shear strengths of composite specimens are improved as the adhesive layer thickness increases. As polyurethane is used as adhesive material, the interface shear strengths with the adhesive layer thickness of 0.3 mm or 0.5 mm were very close to each other, but both increased significantly when the adhesive layer thickness increased to 1.0 mm. As SBS modified asphalt was used as adhesive material, the gradual increase of interface shear strength was notable with the increase of adhesive layer thickness, but in general the shear strength was relatively smaller. Meanwhile, the shear strength of epoxy resin increased significantly when the adhesive layer thickness increased from 0.3 mm to 0.5 mm, but increased slightly as the thickness reached 1 mm. In general, the greatest interface shear strength was achieved by using epoxy resin followed by polyurethane and then SBS modified asphalt as the adhesive layer thickness is the same.

The stress-deformation curves of interface shear test for composite specimens with 0.3 mm adhesive layer are presented in Figure 7. The results show that the stress-deformation curves are apparently different for three different adhesive materials. The differences were observed in the shear failure pattern and deformation. The interface with epoxy resin had the sudden reduction in shear stress at the maximum shear deformation, while the shear stress of the interfaces with the other two adhesive materials gradually decreased after the peak strength. Since the epoxy resin provides high adhesive strength, the glued aggregates by epoxy resin near the interface can sustain the greater shear stress. As the load increases, aggregate may get broken or be pushed out, which causes the sudden reduction of shear strength. On the other hand, the interfaces with three different adhesive materials had different shear displacements at failure.

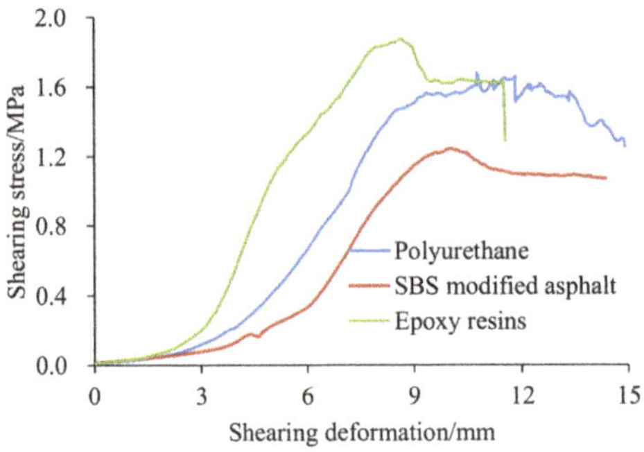


**Figure 7.** Stress-deformation curves of PPM-10 + AC-20 composite specimens during shear test with 0.3 mm adhesive layer.

### 5.3. *Influence of Temperature*

The interface shear strengths of PPM-10 + AC-20 composite specimen with 0.5 mm adhesive layer at different temperatures are shown in Figure 8. The results show that three adhesive materials offer comparative shear strengths at −18 °C and 60 °C. However, compared to SBS modified asphalt or polyurethane, epoxy resin has much greater interface shear strength at 25 °C. It is noted that although the interface shear strength between PPM and asphalt layer decreases significantly as the temperature increases, the shear strengths with adhesive materials are still greater than the ones without adhesive materials at the same temperature.

The stress-deformation curves of interface shear test for composite specimens with 0.5 mm SBS modified asphalt adhesive layer are presented in Figure 9. The stress-deformation curves are quite different in the peak (shearing strength) for three different temperatures. The shear strength increases greatly with the decrease of temperature. In addition, the failure deformation corresponding to the peak value of shear stress slightly increases as the decrease of temperature. This indicates that the interface shear failure potential of PPM-10 + AC-20 greater in summer.

The appearances of fractured surface after shear testing of PPM-10 + AC-20 composite specimens with 0.5 mm polyurethane as adhesive layer at different temperatures are shown in Figure 10. When the temperature is −18 °C or 25 °C, the major bonding failure mode is the separation of layers at the interface. However, when temperature rises to 60 °C, the damage includes not only the interfacial shear failure, but also longitudinal cracks in the upper layer and sublayer. This is due to the strength reduction of mixtures at high temperatures, combined with the stress concentration during the inclined shear test, Figure 11 illustrates the possible stress concentration after the interface slippage occurs, which may cause the cracking of mixtures.

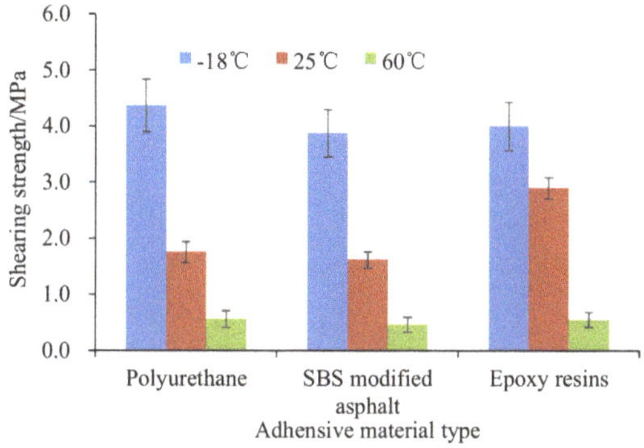


**Figure 8.** Shear strengths of PPM-10 + AC-20 composite specimen with 0.5 mm adhesive layer at different temperatures.

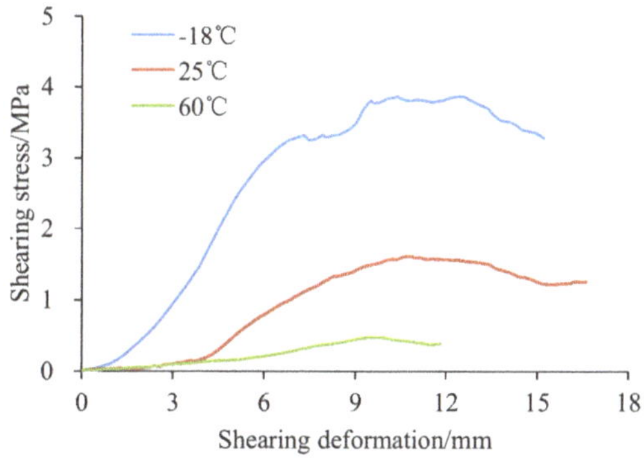


**Figure 9.** Stress-deformation curves of PPM-10 + AC-20 composite specimens with 0.5 mm SBS modified asphalt adhesive layer at different temperatures.

**Figure 10.** Appearance of PPM-10 + AC-20 composite specimen after shear test.

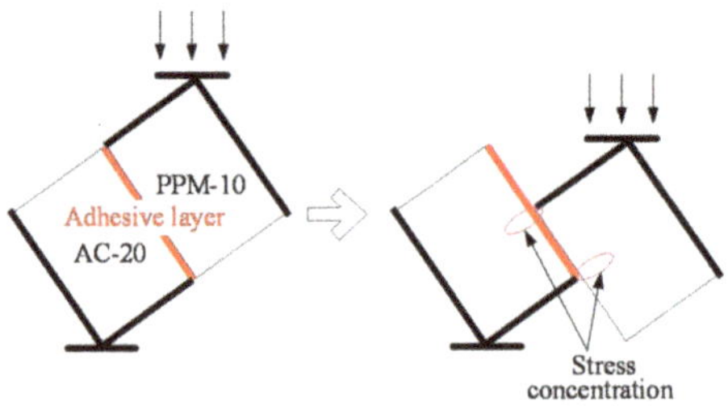


**Figure 11.** Illustration of stress concentrations during inclined shear test.

### 5.4. Influence of Freezing–Thaw Cycles

After four freezing–thaw cycles, the shear strength of PPM-10 + AC-20 composite specimen with 0.5 mm adhesive layer was tested at 25 °C. The test results are shown in Figure 12, in which the standard deviation values were shown as the error bar from the averages in the figure. The results show that all the interface shear strengths decrease after freezing–thaw cycles and the results are similar among different adhesive materials. On the other hand, the shear strength of interface with epoxy resin dropped by about 50% after freezing–thaw cycles, while the shear strength of interface with SBS modified asphalt or polyurethane had much smaller reduction. This indicates that the adhesive ability of SBS modified asphalt or polyurethane is less sensitive to freezing–thaw condition as compared to epoxy resin. Again, all the interface shear strengths after freezing–thaw cycles are greater as compared to the interface of PPM and AC-20 without any adhesive material.

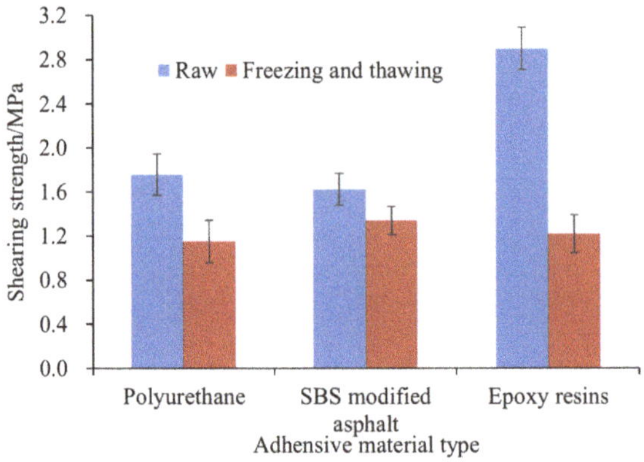


**Figure 12.** Shear strengths of PPM-10 + AC-20 composite specimen after freezing–thaw cycle with 0.5 mm adhesive layer.

### 5.5. Shear Fatigue Test

Shear fatigue tests were performed on composite specimens at 25 °C with the peak shearing stress of 0.4 MPa and the frequency of 10 Hz. Figure 13a shows the change of shear displacement throughout the test. It can be found that the displacement amplitude increases with the increase of number of loading cycles in the first half of the curve. The shear displacement of the interface with epoxy resin is stable in the back half of curve, while the shear displacements of the interfaces with SBS modified asphalt and polyurethane increases rapidly and interface shearing failure occurs at the end of the curve. The number of loading cycles corresponding to the rapid increase of shear displacement was defined as the shear fatigue life. It can be seen that the fatigue life of composite specimen with SBS modified asphalt and polyurethane is 1500 and 32,500, respectively; while the epoxy resin provides fatigue life of more than 85,000 cycles for PPM-10 + AC-20 composite specimen.

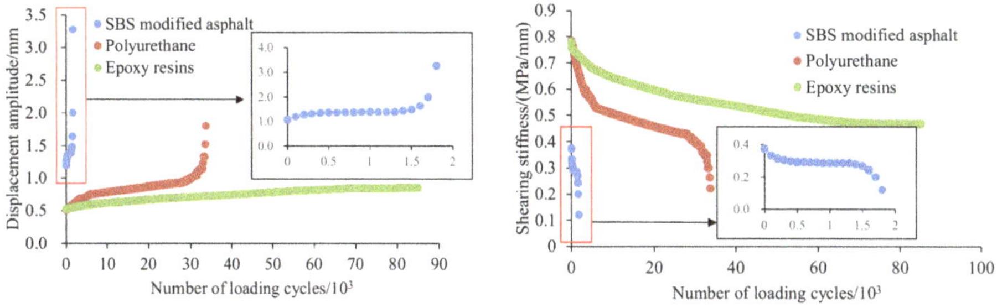


(**a**) (**b**)

**Figure 13.** Response curves of (**a**) interface displacement vs. number of loading cycles; and (**b**) shear stiffness vs. number of loading cycles.

Figure 13b shows the interface shear stiffness of composite specimen with three adhesive materials obtained from fatigue test. The interface shear stiffness was calculated using Equation (2).

$$S = \frac{\tau}{\varepsilon} \tag{2}$$

where $\tau$ is the shear stress level applied in fatigue test (MPa), which is 0.4 MPa in the paper; and $\varepsilon$ is the shear displacement (mm).

The results show that with the increase of loading cycle, the interface shear stiffness gradually decreases. As the number of loading cycle approaches fatigue life, the shear stiffness decreases rapidly. It was found that the initial stiffness of composite specimen with SBS modified asphalt was 0.37 MPa/mm, while the initial stiffness of composite specimens with polyurethane and epoxy resin reached 0.77 MPa/mm. This indicates that polyurethane and epoxy resin become stiff after curing, which provides the stronger bond than asphalt. Despite of the lowest value during three adhesive materials, the interface shear stiffness of PPM-10 + AC-20 composite specimen with SBS modified asphalt is comparable to that of composite specimen of OGFC-dense asphalt mixture with asphalt emulsion as adhesive material reported in literature [33]. The testing results also proves the feasibility of using the inclined shear fatigue test for measuring shearing fatigue life and shearing stiffness.

## 6. Conclusions

This study investigated shear performance of composite specimen with different adhesive materials using the inclined shear test. The following conclusions can be concluded:

(1) It is necessary to apply adhesive material to the interface between porous polyurethane mixture and asphalt mixture, since the untreated interface shear strength is smaller than the ones between two asphalt mixtures.

(2) The interface shear strength increases as the thickness of adhesive layer increases, although the trends vary depending on the material. In general, the greatest interface shear strength was achieved by using epoxy resin followed by polyurethane and then SBS modified asphalt at 25 °C as the adhesive layer thickness is the same.

(3) The interface shear strength decreases as the temperature increases or after freezing–thaw cycles. The interface shear strengths of composite specimen with three adhesive materials are similar to each other at high and low temperatures (60 °C and −18 °C) or after freezing–thaw cycles, which is different from the trend at 25 °C.

(4) The composite specimen with epoxy resin as adhesive material has the longest fatigue life; while the SBS modified asphalt has the least fatigue life at 25 °C. Further work will be conducted to evaluate fatigue life of interface at different temperatures and freezing–thaw conditions.

The research findings can help select the appropriate adhesive materials to enhance the interface bonding between porous polyurethane mixture and asphalt layers. Future research should be conducted to investigate the cost-effectiveness of different adhesive materials using field performance data. This will significantly increase the durability and service life of pavement and reduce the life-cycle cost when porous polyurethane mixture is used as road surface layer for safety and noise reduction.

**Acknowledgments:** The research presented herein was partially sponsored by the National Natural Science Foundation of China (No. 51208178) and the Fundamental Research Funds for the Central Universities (No. 2015B17014).

**Author Contributions:** Jun Chen and Hao Wang designed the experiment and wrote the paper; Cheng Yao and Xie Ma conducted experiments; Wei Huang and Junyu Qian helped analyzing experimental data.

**Conflicts of Interest:** The authors declare no conflicts of interest.

## References

1. Nicholls, J.C. *Review of UK Porous Asphalt Trial*; TRL Rep. 264; Transport Research Laboratory: Wokingham, UK, 1997.
2. Kandhal, P.S. *Design, Construction and Maintenance of Open-Graded Aphaltfriction Courses*; National Asphalt Pavement Association: Lanham, MD, USA, 2002; Volume 115.
3. Dell'Acqua, G.; de Luca, M.; Lamberti, R. Indirect skid resistance measurementfor porous Asphalt pavement management. *Transp. Res. Rec.* **2011**, *2205*, 147–154. [CrossRef]
4. Rungruangvirojn, P.; Kanitpong, K. Measurement of visibility loss due to splash and spray: Porous, SMA and conventional Asphalt pavements. *Int. J. Pavement Eng.* **2010**, *11*, 499–510. [CrossRef]
5. Chen, J.; Wang, H.; Zhu, H.Z. Investigation of permeability of open graded asphalt mixture considering effects of anisotropy and two-dimensional flow. *Constr. Build. Mater.* **2017**, *145*, 318–325. [CrossRef]
6. Wang, H.; Ding, Y.M.; Liao, G.Y.; Ai, C.F. Modeling and Optimization of Acoustic Absorption for Porous Asphalt Concrete. *J. Eng. Mech.* **2016**, *142*, 04016002. [CrossRef]
7. Chen, J.; Zhang, Y.Q.; Li, H.; Gao, Y. Rutting-induced permeability loss of open graded friction course mixtures. *J. Test. Eval.* **2016**, *44*, 719–724. [CrossRef]
8. Chen, J.; Li, H.; Huang, X.M.; Wu, J.T. Permeability loss of open graded friction course mixtures due to deformation-related and particle-related clogging: Understanding from a laboratory investigation. *J. Mater. Civ. Eng.* **2015**, *27*, 04015023. [CrossRef]
9. Molenaar, A.A.A.; Molenaar, J.M.M. An investigation into the contribution of the bituminous binder to the resistance to raveling of porous Asphalt. In Proceedings of the Papers Submitted for Review at 2nd Eurasphalt and Eurobitume Congress, Barcelona, Spain, 20–22 September 2000; pp. 500–508.
10. Mo, L.T.; Huurman, M.; Wu, S.; Molenaar, A.A.A. Bitumen–stone adhesive zonedamage model for the meso-mechanical mixture design of ravelling resistantporous asphalt concrete. *Int. J. Fatigue* **2011**, *33*, 1490–1503. [CrossRef]
11. Mo, L.T.; Huurman, M.; Wu, S.; Molenaar, A.A.A. Mortar fatigue model for mesomechanisticmixture design of ravelling resistant porous asphalt concrete. *Mater. Struct.* **2014**, *47*, 947–961. [CrossRef]
12. Xu, H.N.; Guo, W.; Tan, Y.Q. Internal structure evolution of asphalt mixtures during freeze–thaw cycles. *Mater. Des.* **2015**, *86*, 436–446. [CrossRef]
13. Yildirim, Y.; Dossey, T.; Fults, K.; Tahmoressi, M.; Trevino, M. *Winter Maintenance Issues Associated with New Generation Open-Graded Friction Courses*; No. FHWA/TX-08/0-4834-2; Center for Transportation Research, The University of Texas at Austin: Austin, TX, USA, 2007.
14. Hernandez-Saenz, M.A.; Caro, S.; Arámbula-Mercado, E.; Martin, A.E. Mix design, performance and maintenance of Permeable Friction Courses (PFC) in the United States: State of the Art. *Constr. Build. Mater.* **2016**, *111*, 358–367. [CrossRef]
15. Amundsen, A.H.; Klaeboe, R. *A Nordic Perspective on Noise Reduction at the Source*; Report 806/2005; Institute of Transport Economics: Oslo, Norway, 2005.

16. Goubert, L. Developing a durable and ultra-low noise poroelastic pavement. In Proceedings of the INTER-NOISE and NOISE-CON Congress and Conference, Melbourne, Australia, 16–19 November 2014.
17. Sandberg, U.; Goubert, L. PERSUADE—A European project for exceptional noise reduction by means of poroelastic road surfaces. In Proceedings of the 40th International Congress and Exposition on Noise Control Engineering 2011 (INTER-NOISE 2011), Osaka, Japan, 4–7 September 2011.
18. Yin, J.C.X.J.; Wang, H.; Ding, Y.M. Evaluation of Durability and Functional Performance of Porous Polyurethane Mixture in Porous Pavement. *J. Clean. Prod.* **2018**, *188*, 12–19.
19. Sandberg, U.; Goubert, L.; Biligiri, K.P.; Kalman, B. *State-of-the-Art Regarding Poroelastic Road Surfaces*. Deliverable D8.1 of the PERSUADE Project, VTI and BRRC. 2010. Available online: https://trid.trb.org/view/918882 (accessed on 16 April 2018).
20. Goubert, L.; Sandberg, U. The PERSUADE project: Developing the concept of poroelastic road surface into a powerful tool for abating traffic noise. In Proceedings of the INTER-NOISE 2010-39th International Congress on Noise Control Engineering 2010, Lisbon, Portugal, 15–16 June 2010.
21. Liao, G.Y.; Wang, H.; Zhu, H.Z.; Sun, P.X.; Chen, H.Q. Shear Bonding Strength between Poro-elastic Road Surface and Sublayer with Different Bonding Agents. *J. Mater. Civ. Eng.* **2018**, *30*, 04018017. [CrossRef]
22. Wang, D.; Liu, P.F.; Leng, Z.; Leng, C.; Lu, G.Y.; Bush, M.; Oeser, M. Suitability of poroelastic road surface (PERS) for urban roads in cold regions: Mechanical and functional performance assessment. *J. Clean. Prod.* **2017**, *165*, 1340–1350. [CrossRef]
23. Chen, J.; Ma, X.; Wang, H.; Xie, P.Y.; Huang, W. Experimental study on anti-icing and deicing performance of polyurethane concrete as road surface layer. *Constr. Build. Mater.* **2018**, *161*, 598–605. [CrossRef]
24. Wang, H.; Al-Qadi, I.L. Evaluation of surface-related pavement damage due to tire braking. *Road Mater. Pavement Des.* **2010**, *11*, 101–121. [CrossRef]
25. Kubo, K. Present status of porous elastic rubber surface (PERS) in Japan. In Proceedings of the INTER-NOISE and NOISE-CON Congress and Conference Proceedings 2011, Osaka, Japan, 4–7 September 2011.
26. Ejsmont, J.; Sandberg, U.; Mioduszewski, P. Tyre/road noise reduction by a poroelastic road surface. In Proceedings of the 43rd International Congress on Noise Control Engineering: Improving the World through Noise Control, INTERNOISE 2014, Melbourne, Australia, 16–19 November 2014.
27. Wang, H.; Xu, G.J.; Wang, Z.L.; Bennert, T. Flexible pavement interface bonding: Theoretical analysis and shear strength measurement. *J. Test. Eval.* **2017**, *46*, 1–9. [CrossRef]
28. Ai, C.F.; Rahman, A.; Song, J.J.; Gao, X.W.; Lu, Y. Characterization of interface bonding in asphalt pavement layers based on direct shear tests with vertical loading. *J. Mater. Civ. Eng.* **2017**, *29*, 04017102. [CrossRef]
29. Canestrari, F.; Ferrott, G.; Partl, M.N.; Santagata, E. Advanced testing and characterization of interlayer shear resistance. *Transp. Res. Rec.* **2005**, *1929*, 69–78. [CrossRef]
30. Romanoschi, S.A.; Metcalf, J.B. Characterization of asphalt concrete layer interfaces. *Transp. Res. Rec.* **2001**, *1778*, 132–139. [CrossRef]
31. Diakhaté, M.; Millien, A.; Petit, C.; Phelipot-Mardelé, A.; Pouteau, B. Experimental investigation of tack coat fatigue performance: Towards an improved lifetime assessment of pavement structure interfaces. *Constr. Build. Mater.* **2011**, *25*, 1123–1133. [CrossRef]
32. Tozzo, C.; Fiore, N.; D'Andrea, A. Dynamic shear tests for the evaluation of the effect of the normal load on the interface fatigue resistance. *Constr. Build. Mater.* **2014**, *61*, 200–205. [CrossRef]
33. Song, W.M.; Shu, X.; Huang, B.S.; Woods, M. Influence of interface characteristics on the shear performance between open-graded friction course and underlying layer. *J. Mater. Civ. Eng.* **2017**, *29*, 04017077. [CrossRef]

*applied sciences*

MDPI

*Article*

# Effect of Carbon Black Nanoparticles from the Pyrolysis of Discarded Tires on the Performance of Asphalt and its Mixtures

**Chuangmin Li [1,2], Ziran Fan [2], Shaopeng Wu [3,*], Yuanyuan Li [3,*], Youwei Gan [2] and Aoming Zhang [4]**

1 The Key Laboratory of Road Structure & Material Ministry of Transport, Beijing 100088, China; lichuangmin@126.com

2 School of Traffic and Transportation Engineering, Changsha University of Science and Technology, Changsha 410114, China; fanziran@stu.csust.edu.cn (Z.F.); ganyouwei@stu.csust.edu.cn (Y.G.)

3 State Key Laboratory of Silicate Materials for Architectures, Wuhan University of Technology, Wuhan 430070, China

4 School of Civil Engineering and Architecture, Hubei University of Arts and Science, Xiangyang 441021, China; zhangam2018@163.com

* Correspondence: wusp@whut.edu.cn (S.W.); liyuanyuan@whut.edu.cn (Y.L.); Tel.: +86-27-8716-2595 (S.W.)

Received: 15 March 2018; Accepted: 13 April 2018; Published: 17 April 2018

**Abstract:** It is of great benefit to the environment and the economy to use discarded tires pyrolysis carbon black (TPCB) nanoparticles as a modifier for asphalt binders. A base asphalt binder with 60/80 penetration (GF-70) was selected to prepare the TPCB-modified asphalt binder (TPCB/GF-70) with a 15% dosage of TPCB by the melt blending method. The test instruments, such as Fourier transform infrared spectroscopy, laser particle size analyzer, and thermogravimetric analyzer, were used to study the characteristics of TPCB. The physical performance of GF-70 and TPCB/GF-70 were tested and the rheological properties were also tested with a dynamic shear rheometer to investigate TPCB's effect on the performance of GF-70. In addition, the aromatic hydrocarbon oil (AHO) was used as the softening agent for TPCB/GF-70. The pavement performance of AC-13 and AC-20 was studied to evaluate the comprehensive effect of TPCB and AHO on the pavement performance of asphalt mixtures. Results show that a 15% dosage of TPCB can significantly improve the anti-rutting performance of GF-70, and decrease the low-temperature performance of GF-70 within one PG grade. AHO can obviously improve the low-temperature performance of TPCB/GF-70, but does not significantly decrease the high-temperature performance. With the addition of AHO and a 0.1% higher oil aggregate ratio, TPCB tends to significantly improve the anti-rutting performance and the low-temperature performance of TPCB-modified mixtures; the moisture stability of TPCB and AHO composite modified mixtures satisfies the requirement of water stability.

**Keywords:** TPCB; asphalt and asphalt mixture; physical performance; rheological properties; optimization of pavement performance

## 1. Introduction

With the rapidly increase of automobiles and trucks, more than 1.5 billion waste tires are discarded every year [1]. This will inevitably lead to environmental pollution and resource wastage if we do not appropriately deal with waste tires. At present, four main methods are used to deal with these tires, namely pyrolysis [2,3], incineration [4], landfill [5,6], and powder manufacturing [7–9]. However, three of these four methods are accompanied by other disadvantages [10]. Pyrolysis is a pioneering method that can be used to generate carbon black, tire oil, and syngas from waste tires in the absence of oxygen.

Pyrolysis has a series of advantages, such as low cost, high efficiency, little environmental pollution, and its key product being renewable carbon black (with a mass ratio of 30% and up depending on the pyrolysis temperature) [3]. However, because of the limited usage of discarded tires pyrolysis carbon black (TPCB) nanoparticles [11], TPCB is produced and immediately accumulated. Therefore, it is very important to broaden the application range of TPCB.

In recent years, plenty of studies have been conducted to use carbon black (CB) as a modifier for asphalt binders [12–20]; these works are a good start towards enlarging the application field of CB. Previous research investigated the modified effect of CB on the workability or mechanical properties of asphalt binders and their mixtures, with the results indicating that CB had good compatibility with asphalt binders and a reinforcement effect on asphalt binders, and decreased the resistivity of asphalt [12]; it could improve the complex modulus [13] and anti-aging performance (ultraviolet aging and thermo-oxidative aging) of asphalt binders [15], as well as enhance the anti-rutting performance of asphalt mixtures at high temperatures [14]. However, there was no consistent conclusion about the effect on low-temperature performance: some studies showed that carbon black could improve the low-temperature performance of asphalt, while others showed that it cannot or that it has no effect [16,17]. However, in general, micrometer and nanometer powder materials always have a stiffening effect on asphalt and can reduce the anti-crack property of asphalt at low temperatures [18–20].

In addition, previous research mostly studied the modified effect of industrial carbon black. The performance of CB depends on its structure, surface area, and particle size. TPCB differs from industrial carbon black in terms of its composition and structure form. For example, the particle size of TPCB is roughly equal to the average particle size of industrial carbon black, which indicates that TPCB is a compound material consisting of different kinds of industrial carbon black [10]. The desorption ability of TPCB is poor [19] and may not have good compatibility with rubber or may be difficult to disperse uniformly in asphalt. The modified effects of TPCB on the performance of asphalt binders or its mixtures may be different to industrial carbon blacks, so it is of great importance to investigate the TPCB effect on the pavement performance of asphalt binders and their mixtures.

In this paper, the characteristic functional groups, particle size distribution, and thermostability of TPCB are studied. A base asphalt binder of Maoming GaoFu 60/80 asphalt binder (GF-70) was designated to prepare the TPCB-modified GF-70 (TPCB/GF-70) by mixing it with a 15% dosage of TPCB; the physical performance and rheological properties were investigated to evaluate the influence of TPCB on GF-70. In order to improve the low-temperature performance of TPCB/GF-70, four levels of dosage (0.3%, 0.6%, 0.9%, and 1.2%) of aromatic hydrocarbon oil (AHO) were used as the softening agent for TPCB/GF-70. Finally, two kinds of asphalt mixtures of AC-13 and AC-20 were created to investigate the influence of TPCB on the pavement performance of asphalt mixtures.

## 2. Materials and Methods

### 2.1. Materials

#### 2.1.1. Asphalt Binder

A base asphalt binder of GF-70 was provided by Sinopec Maoming Company (Maoming, Guangdong, China). Table 1 gives the technical information of GF-70.

**Table 1.** Technical information on GF-70.

| Asphalt | Technical Information | Units | Results | Requirements | Methods |
|---|---|---|---|---|---|
| Virgin | Penetration (25 °C) | 0.1 mm | 70.3 | 60~80 | ASTM D5 [21] |
| | Softening point | °C | 47.5 | ≥46 | ASTM D36 [22] |
| | 60 °C dynamic viscosity | Pa.s | 185 | ≥180 | ASTM D4402 [23] |
| | 10 °C ductility | cm | 45 | ≥15 | ASTM D113 [24] |
| After RTFOT | Mass loss | % | −0.18 | ±0.8 | ASTM D2872 [25] |
| | 25 °C residual penetration ratio | % | 65.7 | ≥61 | ASTM D5 |
| | 10 °C Residual ductility | cm | 9.0 | ≥6 | ASTM D113 |

#### 2.1.2. Discarded Tires Pyrolysis Carbon Black (TPCB)

The TPCB used in this research was manufactured by Kimkey Environmental S&T Co., Ltd. (Shanghai, China). The appearance of TPCB is shown in Figure 1. Technical information on TPCB is listed in Table 2.

**Table 2.** Technical information on TPCB.

| Technical Information | | Units | Results |
|---|---|---|---|
| Chemical data | Iodine absorption value | g/kg | 81 |
| | PH value | - | 6.5 |
| | 45 μm mesh | ppm | 68 |
| | Heating loss | % | 2.4 |
| Tensile properties (145 °C, 30 min) | Elongation at break | % | 565.2 |
| | Tensile strength | MPa | 25.8 |
| Hardness | Shore hardness | % | 60.0 |

**Figure 1.** Appearance of TPCB.

#### 2.1.3. Softening Agent

In order to increase the low-temperature performance of TPCB modified asphalt binder, AHO was used as the softening agent to soften asphalt binder. AHO was manufactured by Hubei Guochuang Hi-tech Material Co., Ltd. (Hubei, China). AHO is a black sticky liquid, the main components of which are aromatics and colloid. It can be dissolved in asphalt to improve the component of an asphalt. At the same time, it can also increase the ductility of modified asphalt, especially at low temperatures.

#### 2.1.4. Aggregate and Filler

The crushed basalt was used as the aggregate of the asphalt mixture. In order to ensure the stability of the granular gradation, both the coarse and fine aggregate were sieved and divided into grades. The technical information on the aggregates is shown in Table 3. It should be noticed that the aggregate was first mixed according to the aggregate gradation of the AC-20 mixture (given in the following section of this paper), then divided into coarse and fine aggregates with a diameter bigger or smaller than 4.75 mm, respectively. The limestone powder was used as the filler of the asphalt mixture; there was no moisture and clumping phenomenon in limestone powder, the technical information for which is also given in Table 3.

**Table 3.** Technical information on aggregates and filler.

| Aggregate | Technical Information | | Results |
|---|---|---|---|
| Coarse aggregate | Crushed stone value | | 14.7% |
| | Los Angeles abrasion loss | | 21.8% |
| | Percentage passing 0.075 mm sieve by washed-sieve analysis | | 0.18% |
| | Soundness | | 10% |
| | Water absorption | | 1.4% |
| | Polished stone value | | 44.7% |
| | Affinity grade with asphalt | | 4 grade |
| Fine aggregate | Clay content (percentage passing 0.075 mm sieve) | | 1.1% |
| | Soundness | | 11% |
| | Sand equivalent value | | 60.9% |
| Filler | Apparent density | | 2.694 g/m$^3$ |
| | Hydrophilic coefficient | | 0.84 |
| | Plasticity index | | 3.0 |
| | Gradation by percentage of mass passing sieve size | <0.60 mm | 100% |
| | | <0.15 mm | 94.8% |
| | | <0.075 mm | 82.0% |

### 2.2. Preparation of TPCB-Modified Asphalt

The preparation procedure of TPCB-modified asphalt was as follows: first, neat asphalt was heated to a constant temperature of 155 °C; then a 15% dosage of TPCB was added to the neat asphalt (mass ratio of TPCB to neat asphalt); finally, a high-speed shearing machine was used for 40 min with a shear temperature of 155 °C and shear rate of 4000 r/min.

TPCB and AHO compound modifier modified asphalt was produced as follows: GF-70 was heated to a constant temperature of 155 °C, the designed dosages of AHO and TPCB were added to neat asphalt, and a high-speed shearing machine was used for 45 min, with a shear temperature of 155 °C and a shear rate of 4000 r/min.

### 2.3. Experiments on TPCB

#### 2.3.1. Chemical Structure Testing of TPCB

The characteristic functional groups of TPCB were tested by a Fourier transform infrared (FTIR) spectrometer. The kalium bromate (KBr) disk method was used to prepare the TPCB samples for FTIR testing, with a mass ratio of TPCB to KBr of 1 mg/100 mg. The instrument parameters of FTIR were that the scan time was set to 64 times, and scan wave number range was 4000 cm$^{-1}$~400 cm$^{-1}$.

#### 2.3.2. Laser Size Analysis Testing of TPCB

The laser particle size analyzer, produced by Jinan Micronano Particle Instrument Co., Ltd. (Jinan, Shandong, China), was used to investigate the particle size distribution of TPCB. The particle size test range of this instrument was from 1 nm to 10,000 nm; the dispersible agent was alcohol with a viscosity of 0.001096 Pa·s, the refractive index was 1.332, and the delay unit time was 50 μs.

#### 2.3.3. Thermogravimetric Testing of TPCB

The thermal properties of TPCB were investigated by thermogravimetric (TG) testing in air, with the test temperature range from 50 °C to 700 °C, the heating rate at 10 °C/min, and the mass of every sample at 10 mg.

## 2.4. Experiments on Asphalt

### 2.4.1. Physical Properties of Asphalt

Penetration testing was done according to the criterion of ASTM D5 at three temperatures, 15, 25, and 30 °C; other parameters such as PI, equivalent softening point ($T_{800}$), and equivalent brittle point ($T_{1.2}$) were calculated according to Equations (1)–(3), respectively. The softening point and ductility of GF-70 and TPCB/GF-70 were investigated depending on the criterions of ASTM D36 and ASTM D113, respectively.

$$\mathrm{PI} = \frac{20(1-25A)}{1+50A} \tag{1}$$

$$\mathrm{T}_{800} = \frac{lg800 - K}{A} = \frac{2.9031 - K}{A} \tag{2}$$

$$\mathrm{T}_{1.2} = \frac{lg1.2 - K}{A} = \frac{0.0792 - K}{A}, \tag{3}$$

where $A$ and $K$ are the slope and intercept of the linear regression equation of the logarithm of penetration at three different temperatures, 15 °C, 25 °C, and 30 °C.

### 2.4.2. Storage Stability Testing of Asphalt

First, the TPCB-modified asphalt was poured into an aluminum tube with a diameter and height of 25 mm and 140 mm, respectively. Then, we set the aluminum tube in an oven at a constant temperature of 163 °C for 48 h and cooled it in a refrigerator at 5 °C for more than 4 h. After that, the tubes were cut into three equal sections. The softening points of the top and bottom sections were tested to calculate the softening point differences.

### 2.4.3. Thermal Oxide Aging Testing

A short-term aging test was conducted, the rolling thin film oven test (RTFOT), at a temperature of 163 ± 0.5 °C for 85 min, according to the criterion of ASTM D2872. The long-term aging test was conducted by the pressure aging vessel (PAV) method at a temperature of 100 ± 0.5 °C for 20 h, according to the criterion of ASTM D6521-08.

### 2.4.4. Rheological Properties Testing

The rheological properties of both GF-70 and TPCB/GF-70, not aged by RTFOT and PAV, were investigated by a dynamic shear rheometer (DSR) in the temperature range from −10 °C to 85 °C; the frequency was 10 rad/s. In the temperature range from −10 °C to 30 °C, a rotor with a diameter of 8.0 mm was used, the thickness of the sample was 2.0 mm, and the strain was 0.05%; the rotor with a diameter of 25 mm was used in the temperature range from 30 °C to 85 °C, where the thickness of the sample was 1.0 mm and the strain was 0.5%.

### 2.4.5. BBR Testing

Before the bending beam rheometer (BBR) test, the asphalt binder was first aged by RTFOT, followed by PAV. The BBR test was conducted at a temperature of −12 °C and −18 °C, respectively, to investigate the TPCB effect on the low-temperature creep properties of GF-70. The sample size of asphalt binder beams for the BBR test was 125 mm × 12.5 mm × 6.25 mm, the binder beams were set to an absolute ethanol bath at the experiment temperature for 60 ± 5 min, then, the creep stiffness and m-values of asphalt binders were tested with a constant load of 980 ± 50 mN for 240 s.

## 2.5. Experiments on the Asphalt Mixture

Wheel tracking testing was done to study the anti-rutting performance of asphalt mixtures, where the test temperature was 60 °C and the load was 0.7 MPa; a low-temperature bending test was performed to detect the anti-crack performance and strain of asphalt mixtures at −10 °C with a loading rate of 50 mm/min. The study of water stability was mainly based on an immersion Marshall test and the freeze–thaw split test. The immersion Marshall test was undertaken in a water bath for 48 h at a constant temperature of 60 °C. Marshall stability was named "normal Marshall stability" ($MS_1$) before this test and "condition Marshall stability" ($MS_2$) after this test. The conditions for the freeze–thaw split test were freezing for 16 h at −18 °C, followed by a water bath at 60 °C for 24 h; splitting tensile strength was named "normal splitting tensile strength" ($ST_1$) before this test and "condition splitting tensile strength" ($ST_2$) after this test. The residential Marshall stability (RMS) and splitting tensile strength ratio (TSR) were calculated according to Equations (4) and (5):

$$\mathrm{RMS} = \frac{MS_2}{MS_1} \times 100 \tag{4}$$

$$\mathrm{TSR} = \frac{ST_2}{ST_1} \times 100, \tag{5}$$

where $MS_1$ (MPa) is the normal Marshall stability; $MS_2$ (MPa) is the condition Marshall stability, RMS (%) is the ratio of $MS_1$ to $MS_2$; $ST_1$ (MPa) is the normal splitting tensile strength; $ST_2$ (MPa) is the condition splitting tensile strength; and TSR (%) is the ratio of $ST_1$ to $ST_2$.

## 3. Results and Discussion

### 3.1. Characterization of TPCB

#### 3.1.1. Chemical Structure of TPCB

Figure 2 shows the FTIR spectrum of TPCB. Because of the simple graphite-like molecular structure of TPCB, most of the chemical bonds in TPCB are non-polar bonds. The infrared activity non-polar bonds are very weak [26,27], so the FTIR spectrum is simple, with four different absorption bands. The relatively wide domain absorption band at 3438 cm$^{-1}$ is caused by the stretching vibration of –OH of $H_2O$, because TPCB can absorb water from the air. The absorption bands at 1450 cm$^{-1}$ and 1604 cm$^{-1}$ [28,29] are assigned to the stretching vibration of the sp$^2$ hybrid C=C, which is caused by the carbon atoms in the graphite-like sheets of TPCB. The absorption band at 1104 cm$^{-1}$ is caused by the stretching vibration of C–O [30], which is due to the structural defects (sp$^3$ hybrid carbon atoms) in the graphite-like structures.

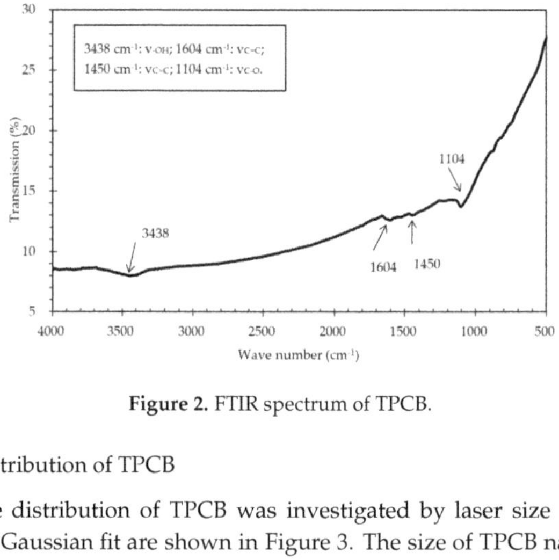


**Figure 2.** FTIR spectrum of TPCB.

#### 3.1.2. Particle Size Distribution of TPCB

The particle size distribution of TPCB was investigated by laser size analysis. The curves corresponding to the Gaussian fit are shown in Figure 3. The size of TPCB nanoparticles is mainly distributed in the diameter range from 50 nm to 5000 nm. In detail, the $D_{10}$ (accumulation of 10%), $D_{50}$ (accumulation of 50%) and $D_{90}$ (accumulation of 90%) of TPCB are 200.35 nm, 481.65 nm, and 1159.35 nm, respectively, which indicates that the particle size distribution of TPCB is not uniform. The reason is that, on the one hand, in order to satisfy the requirements of characteristics at different positions of tires, different kinds of CB are added to the tires, so that TPCB is a compound of different kinds of CB from the raw material of tires; on the other hand, the coking substances generated in the pyrolysis process can affect the particle size distribution of TPCB.

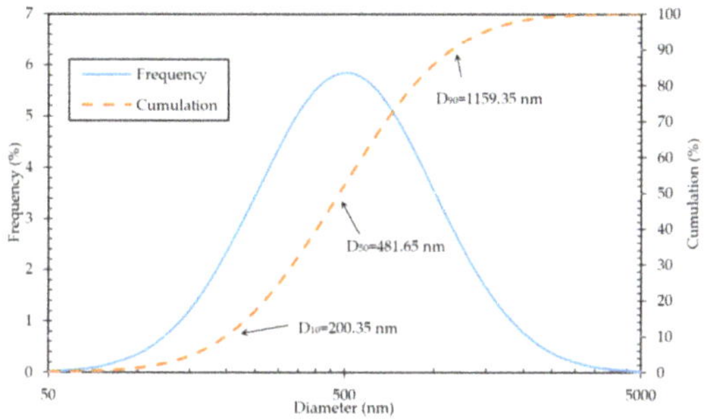


**Figure 3.** The Gaussian fit of laser size analysis of TPCB.

### 3.1.3. Thermal Properties of TPCB

The thermogravimetric (TG), differential thermogravimetric (DTG), and differential scanning calorimetric (DSC) curves are shown in Figure 4 to study the thermal properties of TPCB. There are two mass loss stages in the TG image of TPCB. In detail, the temperature range of stage one is from 50 °C to 126 °C, the mass loss ratio in this stage is 1.42%, and the fastest loss rate can be found at a temperature of 94.3 °C. The peak value of 102.9 °C in the DSC curve shows that stage one is an endothermic phase, so it corresponds to the endothermic phase of the evaporation of water by TPCB. The temperature range of stage two is from 126 °C to 700 °C. There is a peak value of 416.4 °C in the DSC curve, which indicates that stage two is a heat-releasing phase and mainly corresponds the combustion of the residual organics in TPCB. The mass loss ratio of 10.38% is relatively larger in this stage, but the heating temperature in the preparation process of TPCB/GF-70 is not higher than 160 °C and the mass loss ratio at 160 °C is only 2.76%, which shows that the thermal properties of TPCB can satisfy the requirements for preparing a TPCB-modified asphalt binder.

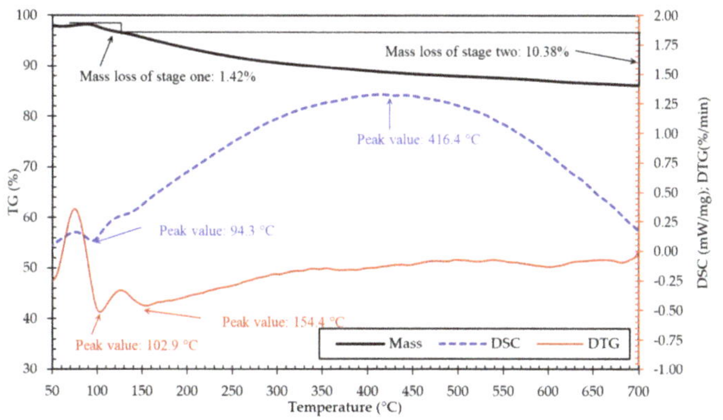


**Figure 4.** TG, DTG, and DSC curves of TPCB in the temperature range of 50 °C to 700 °C.

## 3.2. TPCB's Effect on the Performance of Asphalt

### 3.2.1. TPCB's Effect on the Physical Performance of Asphalt

The physical performance of GF-70 and TPCB/GF-70 asphalt binders is shown in Table 4. The penetration, softening point, ductility, and segregation test results of TPCB/GF-70 all meet the requirements of AH-70 B grade in China [31].

**Table 4.** Results of physical performance and segregation tests of GF-70 and TPCB/GF-70.

| Asphalt | 25 °C Penetration (0.1 mm) | Softening Point (°C) | 10°C Ductility (cm) | PI | $T_{800}$ (°C) | $T_{1.2}$ (°C) | Softening Point Difference (°C) |
|---|---|---|---|---|---|---|---|
| GF-70 | 70.3 | 47.5 | 45.0 | −0.44 | 48.9 | −21.5 | - |
| TPCB/GF-70 | 59.4 | 50.4 | 18.7 | −0.29 | 51.7 | −15.8 | 0.7 |

Penetration, softening point, and $T_{800}$ are able to reflect the high-temperature performance of an asphalt binder: the lower the penetration, and the higher the $T_{800}$ and softening point, the better the high-temperature performance [32]. The penetration of GF-70 decreases by 15.5% with the addition of a 15% dosage of TPCB, while the softening point and $T_{800}$ increase by 2.9 °C and 2.8 °C, respectively, which shows that TPCB can improve the high-temperature performance of asphalt.

The ductility and $T_{1.2}$ belong to the indexes for evaluating the low-temperature performance of an asphalt binder. After adding TPCB, the ductility decreases by 58.4% and the $T_{1.2}$ increases by 5.7 °C, which shows that the 15% dosage of TPCB can obviously decrease the low-temperature performance of asphalt.

The PI can reflect the temperature sensitivity of an asphalt binder: the higher the PI, the lower the temperature sensitivity. In general, the PI of a sol-gel asphalt binder should be in the range of −2~2. It can be seen from Table 4 that the PI of both GF-70 and TPCB/GF-70 belong to the sol-gel asphalt binders. The PI of TPCB/GF-70 is bigger than that of GF-70; therefore, TPCB can decrease the temperature sensitivity of asphalt.

The softening point difference of segregation test is 0.7 °C, which shows that a 15% dosage of TPCB-modified asphalt has good storage stability, and TPCB has good compatibility with asphalt.

### 3.2.2. TPCB's Effect on the Rheological Properties of Asphalt

(1) Complex Modulus and Phase Angle

An asphalt binder must have enough elasticity to satisfy the requirements of anti-rutting at high temperatures, while having enough plasticity to avoid cracks at low temperatures. The viscoelastic properties of the asphalt binder are important to the pavement performance of asphalt mixtures [33–35]. In order to study the TPCB effect on the rheological properties of GF-70, the complex modulus ($G^*$) and phase angle ($\delta$) of asphalt binders without and with TPCB were studied by the DSR. The parameter $G^*$ can estimate the deformation resistance of an asphalt binder under repeated shear loading, and $\delta$ can reflect the ratios of the elastic and viscous characteristics of an asphalt binder [36,37]. $G^*$ and $\delta$ of asphalt binders without and with TPCB are shown in Figure 5.

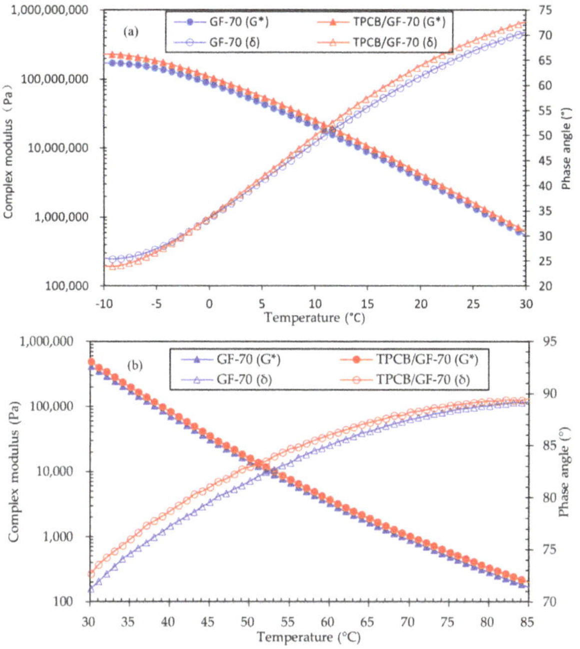


**Figure 5.** G* and δ of GF-70 and TPCB/GF-70 in the temperature range of −10~30 °C (**a**) and 30~85 °C (**b**).

It can be seen from Figure 5 that the G* of TPCB/GF-70 is bigger than that of GF-70 in the temperature range −10–30 °C, and the same tendency can be observed in the temperature range 30–85 °C, indicating that a 15% dosage of TPCB can improve the repeated shear deformation resistance of GF-70. From Figure 5a, when the temperature is lower than 0 °C, the δ of TPCB/GF-70 is smaller than that of GF-70; when the temperature is higher than 0 °C, the opposite occurs. From Figure 5b, the δ of TPCB/GF-70 is bigger than that of GF-70 in the temperature range 30–85 °C, which shows that the elastic characteristic of TPCB/GF-70 is more obvious than that of GF-70 when the temperature is lower than 0 °C; the viscous ratio of G* of TPCB/GF-70 is higher than that of GF-70 when the temperature is higher than 0 °C.

(2) Rutting Factor and Fatigue Factor

The rutting factors (G*/sinδ) of GF-70 and TPCB/GF-70 are shown in Figure 6a. The higher the G*/sinδ, the better the high-temperature anti-rutting performance of an asphalt binder. The G*/sinδ of TPCB/GF-70 is higher than that of GF-70 in the high-temperature range of 52 °C~82 °C, indicating that a 15% dosage of TPCB can improve the anti-rutting performance of GF-70. In order to ensure an asphalt binder has good rutting resistance, the G*/sinδ before aging should not be lower than 1.0 kPa. When the temperature is 69.1 °C, the G*/sinδ of GF-70 is 1.0 kPa, while the G*/sinδ of TPCB/GF-70 is 1.0 kPa at a temperature of 70.3 °C, which is 1.2 °C higher than that of GF-70. So the high-temperature performance of TPCB/GF-70 is better than that of GF-70.

The fatigue factors (G*sinδ) of GF-70 and TPCB/GF-70 are shown in Figure 6b. The lower the G*sinδ, the better the anti-fatigue cracking performance of an asphalt binder. The G*sinδ of GF-70 becomes higher after being mixed with a 15% dosage of TPCB in the middle temperature range of 19–40 °C, so that when the temperature is 22 °C, the G*sinδ of TPCB/GF-70 is 19.6% higher than that of GF-70, indicating that TPCB can decrease the anti-fatigue cracking performance of GF-70.

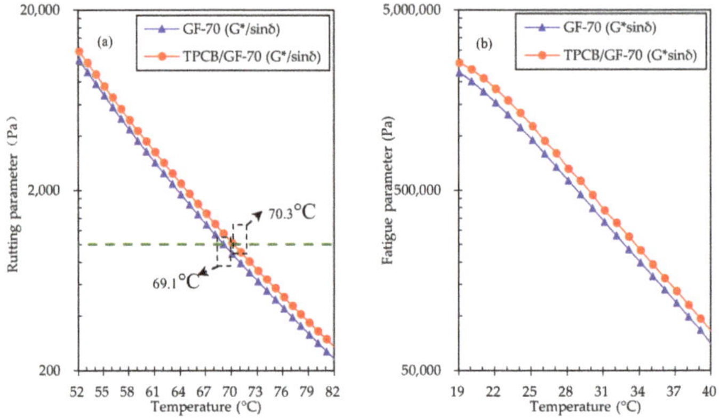


**Figure 6.** G*/sinδ (**a**) and G*sinδ (**b**) of GF-70 and TPCB/GF-70.

(3) Stiffness and m-Value

The low-temperature creep properties of GF-70 and TPCB/GF-70 were investigated by BBR testing, and the creep stiffness and m-values were recorded at 60 s. In order to ensure that asphalt binders have good low-temperature performance to anti-cracking, AASHTO M 320 requires that the stiffness not be higher than 300 MPa, and the m-value should not be lower than 0.300. The stiffness and m-value results of GF-70 and TPCB/GF-70 are shown in Figure 7; meanwhile, the temperatures correspond to the stiffness of 300 MPa and the m-value of 0.300 are shown in Table 5.

From Figure 7, the stiffness and m-values of GF-70 and TPCB/GF-70 satisfy the requirement of anti-cracking performance at a temperature of −12 °C, but do not satisfy it at a temperature of −18 °C, which shows that the low-temperature PG grade of both GF-70 and TPCB/GF-70 is −22 °C. In detail, from Table 5, the critical low temperature of GF-70 and TPCB/GF-70 based on stiffness is −24.4 °C and −22.6 °C, respectively, and the critical low temperature of GF-70 and TPCB/GF-70 based on m-value is −24.2 °C and −22.6 °C, respectively, which shows that the critical low temperature of TPCB/GF-70 is 1.6 °C higher than that of GF-70. Therefore, TPCB can decrease the anti-cracking performance of GF-70, and a 15% dosage of TPCB has an effect on the low-temperature performance of GF-70 within one PG grade.

**Table 5.** Temperature of stiffness is 300 MPa and m-value is 0.300.

| Asphalt | Unit | Critical Low Temperature Based on Stiffness | Critical Low Temperature Based on m-Value |
|---|---|---|---|
| GF-70 | °C | −24.4 | −24.2 |
| TPCB/GF-70 | °C | −22.6 | −22.6 |

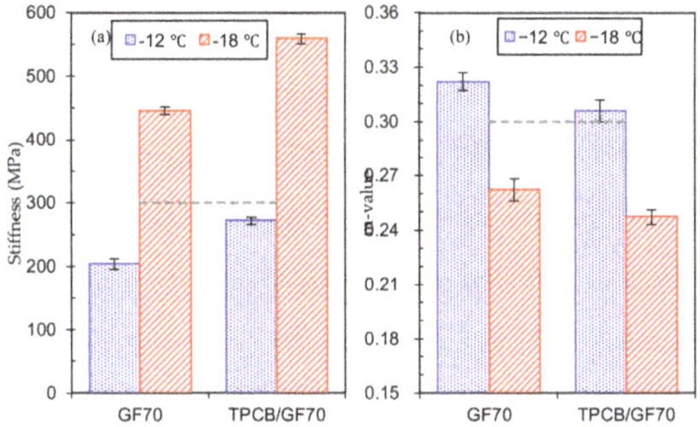


**Figure 7.** Stiffness (**a**) and m-value (**b**) of GF-70 and TPCB/GF-70.

### *3.3. Optimizing the Physical Performance of TPCB/GF-70*

The results above show that TPCB can decrease the low-temperature performance of GF-70; therefore, AHO at four different dosages (0.3%, 0.6%, 0.9%, and 1.2%) was tested to investigate whether it can improve the low-temperature performance of TPCB/GF-70. The penetration, ductility, softening point, PI, $T_{800}$, and $T_{1.2}$ results are shown in Table 6.

From Table 6, with the addition of AHO, low-temperature indictors such as ductility and $T_{1.2}$ show a good tendency. In detail, when the dosages of AHO are 0.3%, 0.6%, 0.9%, and 1.2%, the ductility of TPCB/GF-70 increases by 3.2%, 15.5%, 27.3%, and 41.2%, respectively, while the $T_{1.2}$ decreases by 0.6%, 1.3%, 2.1%, and 2.9%, respectively. Thus, AHO can significantly improve the low-temperature anti-cracking performance of TPCB-modified asphalt, and the improvement effect is more pronounced with the increase in AHO dosage. AHO can also affect the high-temperature performance of TPCB-modified asphalt: the softening point and $T_{800}$ decrease with the addition of AHO, and the penetration increases. However, the softening point and $T_{800}$ of TPCB/GF-70 only decrease by 1.5 °C and 1.3 °C with a 1.2% dosage of AHO, respectively, and the high-temperature performance of TPCB/GF-70 does not obviously decrease; the softening point and $T_{800}$ of TPCB/GF-70 are still 1.4 °C and 1.5 °C higher than that of GF-70, respectively. The PI of TPCB/GF-70 increases with the addition of AHO, which shows that the temperature sensitivity of TPCB/GF-70 decreases. In other words, AHO can obviously improve the low-temperature performance of TPCB/GF-70, but does not significantly decrease the high-temperature performance of the asphalt binder.

**Table 6.** AHO's effect on the performance of TPCB/GF-70.

| Dosage of AHO (%) | Penetration (100 g, 5 s, 0.1 mm) | | | Ductility (cm) | Softening Point (°C) | PI | $T_{800}$ (°C) | $T_{1.2}$ (°C) |
|---|---|---|---|---|---|---|---|---|
| | 30 °C | 25 °C | 15 °C | | | | | |
| 0 | 100.3 | 59.4 | 23.5 | 18.7 | 50.4 | −0.29 | 51.7 | −15.8 |
| 0.3 | 104.3 | 63.4 | 24.6 | 19.3 | 49.8 | −0.28 | 51.3 | −16.4 |
| 0.6 | 105.9 | 67.5 | 25.5 | 21.6 | 49.5 | −0.23 | 51.1 | −17.1 |
| 0.9 | 110.4 | 71.3 | 26.9 | 23.8 | 49.1 | −0.18 | 50.8 | −17.9 |
| 1.2 | 115.6 | 75.9 | 28.4 | 26.4 | 48.9 | −0.15 | 50.4 | −18.7 |

### *3.4. TPCB's Effect on the Pavement Performance of an Asphalt Mixture*

#### 3.4.1. Mix Ratio of Mixtures

Commonly used mixtures, AC-13 and AC-20, were used to study TPCB's effect on the performance of asphalt mixtures. AC-13 was selected to represent the mixture used for the surface layer, while AC-20 represented the mixture used for middle and base layers. The GF-70 asphalt with 15% TPCB and 1.2% AHO was used as the binder of asphalt mixtures and the optimal oil aggregate ratios of asphalt mixtures were determined according to the Marshall test. The aggregate gradation and optical oil aggregate ratio are given in Tables 7 and 8, respectively. The oil aggregate ratios of asphalt mixtures with TPCB-modified asphalt are 0.1% higher than those of asphalt mixtures without TPCB.

**Table 7.** Aggregate gradation of AC-13 and AC-20 mixtures.

| Mixture | Passing Ratio of Every Mesh (%) | | | | | | | | | | | |
|---|---|---|---|---|---|---|---|---|---|---|---|---|
| | 26.5 | 19 | 16 | 13.2 | 9.5 | 4.75 | 2.36 | 1.18 | 0.6 | 0.3 | 0.15 | 0.075 |
| AC-13 | 100 | 100 | 100 | 95.2 | 80.4 | 53.2 | 36.9 | 25.8 | 17.2 | 11.5 | 8.1 | 6.4 |
| AC-20 | 100 | 93.6 | 84.3 | 71.4 | 61.2 | 41.8 | 29.4 | 21.8 | 15.2 | 10.0 | 7.2 | 5.7 |

**Table 8.** Optical oil aggregate ratio of GF70 and TPCB/GF70 asphalt mixtures.

| Mixtures | Asphalt | Optical Oil Aggregate Ratio (%) |
|---|---|---|
| AC-20 | GF70 | 4.5 |
| | TPCB/GF70 | 4.6 |
| AC-13 | GF70 | 5.1 |
| | TPCB/GF70 | 5.2 |

3.4.2. High-Temperature Performance

The dynamic stability (DS) and rutting depth of GF70 and TPCB/GF70 asphalt mixtures were tested by a wheel tracking test to evaluate TPCB's and AHO's effects on the high-temperature performance of asphalt mixtures. Figure 8 illustrates the DS and rutting depths of GF70 and TPCB/GF70 asphalt mixtures.

It can be seen from Figure 8 that the DS values of TPCB/GF70 asphalt mixtures are all higher than those of GF70 asphalt mixtures, while the rutting depths of TPCB/GF70 asphalt mixtures are all smaller than those of GF70 asphalt mixtures. In detail, with the addition of 15% dosage of TPCB, the DS of the AC-13 mixture increases by 31.8% and the rutting depth decreases by 29.8%; the DS of AC-20 mixture increases by 27.2% and the rutting depth decreases by 14.8%. The higher the DS, and the smaller the rutting depth, the better the anti-rutting performance of the asphalt mixture. Therefore, TPCB can improve the anti-rutting performance of asphalt mixtures, and the improvement effect of TPCB on AC-13 mixture is more obvious than that of the AC-20 mixture. The enhanced effect of TPCB on the high-temperature performance of GF70 asphalt mixture can be explained in two ways: one reason is that TPCB is a kind of inorganic nanoparticle that has a hardening effect on GF70 and decreases the temperature sensitivity of GF-70, so GF-70 tends to be harder after being modified by TPCB; the other reason is that the TPCB has a large specific surface area, so needs more asphalt binder to cover its surface. The free asphalt content in mixtures decreases while the structural asphalt contents increase; as a result, the high-temperature performance of asphalt mixtures is enhanced.

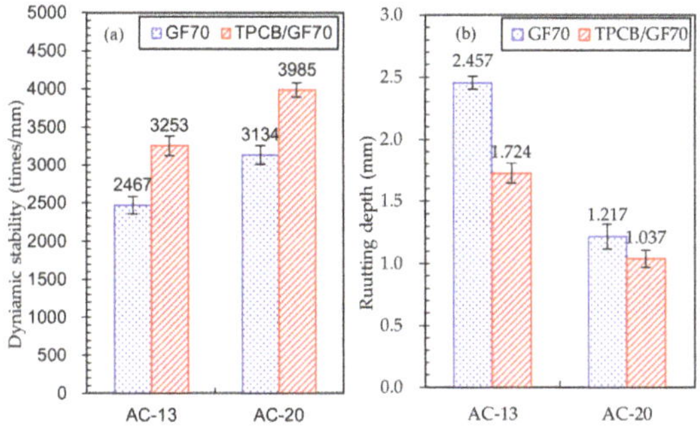


**Figure 8.** DS (**a**) and rutting deeps (**b**) of GF70 and TPCB/GF70 asphalt mixtures.

### 3.4.3. Low-Temperature Performance of Mixture

In order to investigate the TPCB and AHO effect on the low-temperature anti-cracking performance of asphalt mixtures, the flexural strength, flexural strain, and flexural modulus were tested with a low-temperature bending test. Figure 9 gives the results of a low-temperature bending test of GF70 and TPCB/GF70 asphalt mixtures.

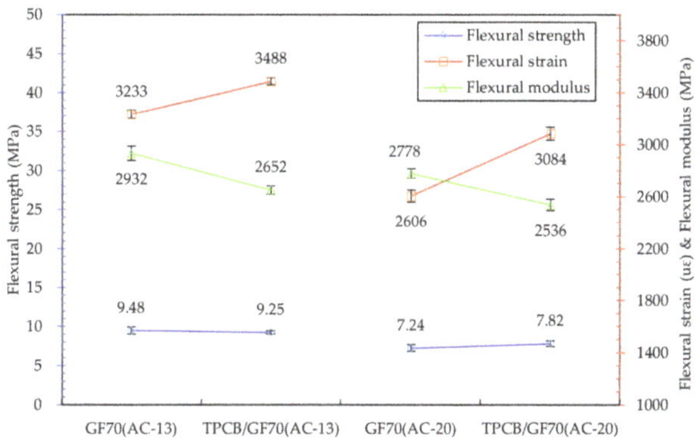


**Figure 9.** Results of low-temperature bending test of GF70 and TPCB/GF70 asphalt mixtures.

From Figure 9, we see that the AC-13 mixture has a larger flexural strain than the AC-20 mixture. Because of the softening effect of the 1.2% dosage of AHO and 0.1% higher oil aggregate ratio, the flexural strain experienced by the TPCB/GF70 asphalt mixtures is much greater than that of the GF70 asphalt mixtures, and the flexural moduli of TPCB/GF70 asphalt mixtures are lower than those of the GF70 asphalt mixtures. The flexural strain of the AC-13 mixture with TPCB is 7.9% larger than that of the AC-13 mixture without TPCB, and the flexural strain of the AC-20 mixture with TPCB is 18.3% bigger than that of the AC-13 mixture without TPCB. So it is feasible to improve the low-temperature performance of TPCB-modified mixtures by adding AHO and only a 0.1% higher oil aggregate ratio.

### 3.4.4. Water Stability of Mixture

The immersion Marshall test and freeze–thaw split test were conducted to evaluate TPCB's and AHO's effect on the water stability of asphalt mixtures; the bigger the RMS and TRS, the better the water stability of the asphalt mixtures [38]. Figure 10 gives the results of immersion Marshall tests of GF-70 and TPCB/GF-70 asphalt mixtures: $MS_1$ values of TPCB/GF-70 asphalt mixtures are all bigger than those of GF-70 asphalt mixtures; $MS_2$ values of GF-70 and TPCB/GF-70 asphalt mixtures are almost equal; RMS values decrease with the addition of TPCB—AC-13 and AC-20 mixtures are decreased by 8.2% and 5.1%—but the RMS of the AC-13 mixture with TPCB is still greater than 85%, and the AC-20 mixture is still greater than 80%.

Figure 11 gives the results of freeze–thaw split tests of GF-70 and TPCB/GF-70 asphalt mixtures. Before the freeze–thaw cycle, TPCB can improve the splitting strength of a mixture; after the freeze–thaw test, the TRS of a mixture with TPCB is significantly decreased. However, the TRS of the AC-13 mixture with TPCB is still greater than 80%, and for the AC-20 mixture it is greater than 75%. So, TPCB tends to decrease the moisture stability of mixtures, but still satisfies the requirements.

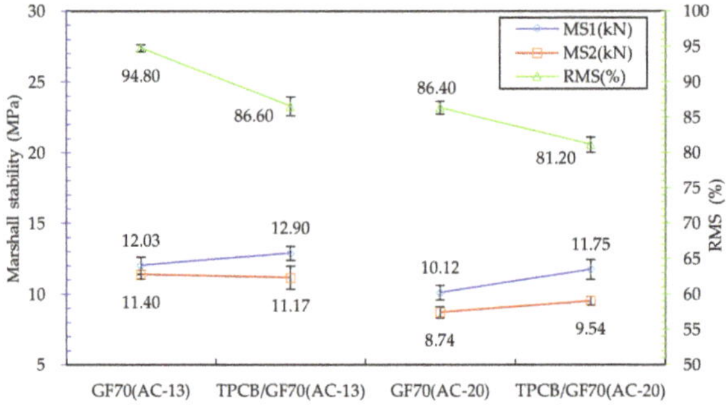


**Figure 10.** Results of immersion Marshall testing of GF70 and TPCB/GF70 asphalt mixtures.

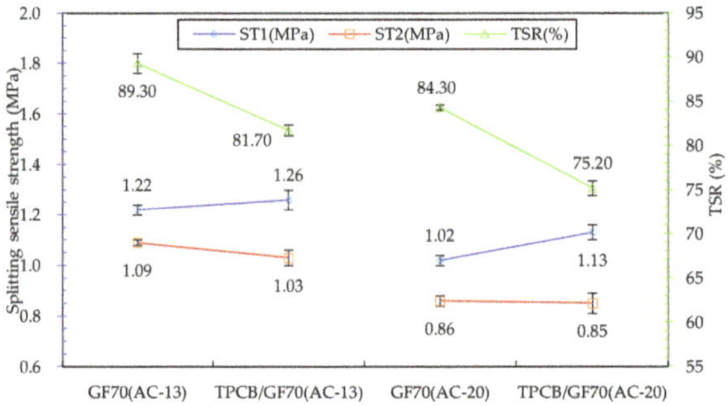


**Figure 11.** Results of freeze–thaw splitting testing of GF70 and TPCB/GF70 asphalt mixtures.

## 4. Conclusions

TPCB/GF-70 was prepared by a melt blending method. The characteristics of TPCB such as chemical structure, surface morphology, and particle size distribution were investigated by some modern testing techniques. The physical performance and rheological properties of GF-70 without and with TPCB were studied to investigate TPCB's effect on the performance of asphalt binders. AHO was used to improve the low-temperature performance of TPCB/GF-70 and its mixtures. The following conclusions can be made:

1. TPCB can decrease the temperature sensitivity and obviously improve the high-temperature performance of asphalt, while a 15% dosage decreases the low-temperature performance of GF-70 within one PG grade.
2. The softening point difference of 15% TPCB-modified asphalt in the segregation test is only 0.7 °C, which shows that TPCB has good compatibility with asphalt.
3. AHO can obviously improve the low-temperature performance and only slightly decreases the high-temperature performance of TPCB/GF-70.

4. With the addition of a 1.2% dosage of AHO and a 0.1% higher oil aggregate ratio, a 15% dosage of TPCB can significantly improve the anti-rutting performance of AC-13 and AC-20 mixtures, as well as improve the low-temperature performance.
5. Both the RMS (higher than 80%) and TSR (higher than 75%) of TPCB and AHO composite modified asphalt mixtures satisfy the requirement of water stability.

**Acknowledgments:** The authors acknowledge the funding from the Key Laboratory of Road Structure & Material Ministry of Transport, Beijing, China, 100088; and the National Key Scientific Apparatus Development Program of the Ministry of Science and Technology of China (No. 2013YQ160501).

**Author Contributions:** Shaopeng Wu, Chuangmin Li, and Yuanyuan Li conceived and designed the experiments. Ziran Fan, Yuanyuan Li, and Youwei Gan performed the experiments. Chuangmin Li, Shaopeng Wu, and Yuanyuan Li analyzed the data. Ziran Fan contributed reagents/materials/analysis tools. Ziran Fan, Yuanyuan Li, and Youwei Gan wrote the paper. Shaopeng Wu, Chuangmin Li, Ziran Fan, and Yuanyuan Li designed the software used in the analysis. Aoming Zhang edited the format of this paper. Chuangmin Li and Shaopeng Wu reviewed the paper.

**Conflicts of Interest:** The authors declare no conflict of interest.

## References

1. Uçar, S.; Karagöz, S. Co-pyrolysis of pine nut shells with scrap tires. *Fuel* **2014**, *137*, 85–93. [CrossRef]
2. González, J.F.; Encinar, J.M.; Canito, J.L.; RodríGuez, J.J. Pyrolysis of automobile tyre waste. Influence of operating variables and kinetics study. *J. Anal. Appl. Pyrolysis* **2001**, *58–59*, 667–683. [CrossRef]
3. Parthasarathy, P.; Hang, S.C.; Park, H.C.; Hwang, J.G.; Yoo, H.S.; Lee, B.K.; Upadhyay, M. Influence of process conditions on product yield of waste tyre pyrolysis—A review. *Korean J. Chem. Eng.* **2016**, *33*, 1–19. [CrossRef]
4. Unapumnuk, K.; Lu, M.; Keener, T.C. Carbon Distribution from the Pyrolysis of Tire-Derived Fuels. *Ind. Eng. Chem. Res.* **2006**, *45*, 8757–8764. [CrossRef]
5. Miranda, M.; Pinto, F.; Gulyurtlu, I.; Cabrita, I. Pyrolysis of rubber tyre wastes: A kinetic study. *Fuel* **2013**, *103*, 542–552. [CrossRef]
6. Muzenda, E. A comparative review of waste tyre pyrolysis, gasification and liquefaction (PGL) processes. In Proceedings of the International Conference on Chemical Engineering & Advanced Computational Technologies (ICCEACT'2014), Pretoria, South Africa, 24–25 November 2014.
7. Wang, H.; You, Z.; Mills-Beale, J.; Hao, P. Laboratory evaluation on high temperature viscosity and low temperature stiffness of asphalt binder with high percent scrap tire rubber. *Constr. Build. Mater.* **2012**, *26*, 583–590. [CrossRef]
8. Sienkiewicz, M.; Borzędowska-Labuda, K.; Wojtkiewicz, A.; Janik, H. Development of methods improving storage stability of bitumen modified with ground tire rubber: A review. *Fuel Process. Technol.* **2017**, *159*, 272–279. [CrossRef]
9. Peralta, J.; Silva, H.M.; Hilliou, L.; Machado, A.V.; Pais, J.; Williams, R.C. Mutual changes in bitumen and rubber related to the production of asphalt rubber binders. *Constr. Build. Mater.* **2012**, *36*, 557–565. [CrossRef]
10. Roy, C.; Chaala, A.; Darmstadt, H. The vacuum pyrolysis of used tires : End-uses for oil and carbon black products. *J. Anal. Appl. Pyrolysis* **1999**, *51*, 201–221. [CrossRef]
11. Bhadra, S.; De, P.P.; Mondal, N.; Mukhapadhyaya, R.; Gupta, S.D. Regeneration of carbon black from waste automobile tires. *J. Appl. Polym. Sci.* **2003**, *89*, 465–473. [CrossRef]
12. Wu, S.; Mo, L.; Shui, Z.; Chen, Z. Investigation of the conductivity of asphalt concrete containing conductive fillers. *Carbon* **2005**, *43*, 1358–1363. [CrossRef]
13. Wen, S.; Chung, D.D.L. Effects of carbon black on the thermal, mechanical and electrical properties of pitch-matrix composites. *Carbon* **2004**, *42*, 2393–2397. [CrossRef]
14. Cong, P.; Xu, P.; Chen, S. Effects of carbon black on the anti aging, rheological and conductive properties of SBS/asphalt/carbon black composites. *Constr. Build. Mater.* **2014**, *52*, 306–313. [CrossRef]
15. Apeagyei, A.K. Laboratory evaluation of antioxidants for asphalt binders. *Constr. Build. Mater.* **2011**, *25*, 47–53. [CrossRef]
16. Park, T.; Lovell, C.W. *Using Pyrolized Carbon Black (PCB) from Waste Tires in Asphalt Pavement (Part 1, Limestone Aggregate)*; Creep Tests: West Lafayette, IN, USA, 1996.

17. Xiao, F.; Amirkhanian, A.N.; Amirkhanian, S.N. Influence of Carbon Nanoparticles on the Rheological Characteristics of Short-Term Aged Asphalt Binders. *J. Mater. Civ. Eng.* **2011**, *23*, 423–431. [CrossRef]
18. Liu, X.; Wu, S.; Ye, Q.; Qiu, J.; Li, B. Properties evaluation of asphalt-based composites with graphite and mine powders. *Constr. Build. Mater.* **2008**, *22*, 121–126. [CrossRef]
19. Gundla, A.; Medina, J.; Gudipudi, P.; Stevens, R.; Salim, R.; Zeiada, W.; Underwood, S. Investigation of Aging in Hydrated Lime and Portland Cement Modified Asphalt Concrete at Multiple Length Scales. *J. Mater. Civ. Eng.* **2015**, *28*, 04015205. [CrossRef]
20. Kakade, V.B.; Reddy, M.A.; Reddy, K.S. Effect of aging on fatigue performance of hydrated lime modified bituminous mixes. *Constr. Build. Mater.* **2016**, *113*, 1034–1043. [CrossRef]
21. ASTM International. Standard test method for penetration of bituminous materials. In *Annual Book of ASTM Standards*; American Society for Testing and Materials Annual: Philadelphia, PA, USA, 1992.
22. ASTM International. *D36. Standard Test Method for Softening Point of Bitumen (Ring-and-Ball Apparatus)*; ASTM International: West Conshohocken, PA, USA, 2009.
23. ASTM International. *D4402. Standard Test Method for Viscosity Determination of Asphalt at Elevated Temperatures Using a Rotational Viscometer*; ASTM International: West Conshohocken, PA, USA, 2008.
24. ASTM International. Standard test method for ductility of bituminous materials. In *Annual Book of ASTM Standards*; American Society for Testing and Materials Annual: Philadelphia, PA, USA, 1979.
25. Lu, X.; Isacsson, U. Effect of ageing on bitumen chemistry and rheology. *Constr. Build. Mater.* **2002**, *16*, 15–22. [CrossRef]
26. Stewart, S.P.; Bell, S.; Mcauley, D. Determination of hydrogen peroxide concentration using a handheld Raman spectrometer: Detection of an explosives precursor. *Forensic Sci. Int.* **2012**, *216*, e5. [CrossRef] [PubMed]
27. Elbasuney, S.; El-Sherif, A.F. Complete spectroscopic picture of concealed explosives: Laser induced Raman versus infrared. *TrAC Trends Anal. Chem.* **2016**, *85*, 34–41. [CrossRef]
28. Cheng, C.H.; Lehmann, J. Ageing of black carbon along a temperature gradient. *Chemosphere* **2009**, *75*, 1021–1027. [CrossRef] [PubMed]
29. Cheng, C.H.; Lehmann, J.; Engelhard, M.H. Natural oxidation of black carbon in soils: Changes in molecular form and surface charge along a climosequence. *Geochim. Cosmochim. Acta* **2008**, *72*, 1598–1610. [CrossRef]
30. Furtado, C.A.; Kim, U.J.; Liu, X.; Gutierrez, H.R.; Chen, G.; Gupta, A.; Eklund, P.C. Raman and IR Spectroscopy of Chemically-Processed Single-Walled Carbon Nanotubes. *J. Am. Chem. Soc.* **2005**, *127*, 15437–15445.
31. China Ministry of Transport. *JTG F40-2004 Technical Specification for Construction of Highway Asphalt Pavements [S]*; China Communications Press: Beijing, China, 2004.
32. Sun, Z.; Yi, J.; Huang, Y.; Feng, D.; Guo, C. Properties of asphalt binder modified by bio-oil derived from waste cooking oil. *Constr. Build. Mater.* **2016**, *102*, 496–504. [CrossRef]
33. Liang, M.; Xin, X.; Fan, W.; Luo, H.; Wang, X.; Xing, B. Investigation of the rheological properties and storage stability of CR/SBS modified asphalt. *Constr. Build. Mater.* **2015**, *74*, 235–240. [CrossRef]
34. Rutherford, T.; Wang, Z.; Shu, X.; Huang, B.; Clarke, D. Laboratory investigation into mechanical properties of cement emulsified asphalt mortar. *Constr. Build. Mater.* **2014**, *65*, 76–83. [CrossRef]
35. Tu, L.; Wu, S.; Liu, G.; Zhou, X.; Ma, S. Effect of the Welan Gum Biopolymer on Rheological Properties and Storage Stability of Bitumens. *J. Test. Eval.* **2016**, *44*, 2211–2218. [CrossRef]
36. Zeng, W.; Wu, S.; Pang, L.; Sun, Y.; Chen, Z. The Utilization of Graphene Oxide in Traditional Construction Materials: Asphalt. *Materials* **2017**, *10*, 48. [CrossRef] [PubMed]
37. Kebria, D.Y.; Moafimadani, S.R.; Goli, Y. Laboratory investigation of the effect of crumb rubber on the characteristics and rheological behaviour of asphalt binder. *Road Mater. Pavement Des.* **2015**, *16*, 946–956. [CrossRef]
38. Cai, Y.C.; Zheng, Y.X. Experiment Study of Water Stability of Fiber-Reinforced Asphalt Mixture. *Adv. Mater. Res.* **2011**, *243–249*, 710–716. [CrossRef]

*applied sciences*

*Article*

# Design and Construction of Oblique Prestressed Concrete Pavement: A Case Study in China

**Ling Yu [1], Xu Yang [2,*], Xiaohui Yan [3,†], Xiaowei Zhang [1,†], Ting Zhao [4,†], Cong Duan [5] and Julian Mills-Beale [6]**

1 School of Transportation, Shenyang Jianzhu University, East Hunnan Road, Shenyang 110168, China; 13516094255@163.com (L.Y.); 18698277088@163.com (X.Z.)
2 Department of Civil Engineering, Monash University, Clayton VIC 3800, Australia
3 Henan Vocational College of Water Conservancy and Environment, Garden Street, Zhengzhou 450008, China; 406779136@163.com
4 Department of English for Nonmajors, Shenyang Jianzhu University, East Hunnan Road, Shenyang 110168, China; ztjoseph2018@163.com
5 National Land Resources Engineering Institute, Lianhua Campus, Kunming University of Science and Technology, Kunming 650093, China; dcworkhard@163.com
6 Department of Civil & Environmental Engineering, California Baptist University, 8432 Magnolia Ave, Riverside, CA 92504, USA; jmillsbe@calbaptist.edu
* Correspondence: xu.yang@monash.edu
† These authors contributed equally to this work.

Received: 11 February 2018; Accepted: 5 April 2018; Published: 11 April 2018

**Featured Application: This paper introduced a practical application of oblique prestressed concrete pavement in China. The design and construction experience of this study can provide useful information for the prestressed concrete pavement in other regions of China as well in as other countries.**

**Abstract:** Prestressed concrete pavement can reduce slab thickness, eliminate transverse joints and enhance durability compared to traditional concrete pavement. Traditional prestressing or precast prestressing in the longitudinal direction requires additional space for anchorage and adds more joints. This study proposed an oblique prestress concrete pavement, in which prestressed tendons were distributed with an angle to the road direction so that the prestress can be applied in both the transverse and longitudinal directions. The detailed design of the oblique prestress concrete pavement, including the selection of raw materials, design of cement concrete, anchorage area, size and distribution of prestressed tendons, stress analysis within the concrete slab, sliding layer, side reinforcement, and regular reinforcement at top and bottom are all included in this study. The slab thickness, diameter, distribution angle, and spacing of tendons were obtained based on the stress analysis to meet the requirement of fracture criteria and fatigue criteria. A demonstrative road pavement section, which has performed well after three years of traffic opening, was constructed according to the design. A step-by-step description of the construction was also presented in the study.

**Keywords:** prestressed concrete pavement; oblique; demonstrative road section; stress analysis

## 1. Introduction

The earliest application of prestressed concrete pavement was in as early as the 1940s. Prestressed concrete pavements have been constructed for highways and airfields in Europe and the United States since then [1]. Some basic geometrical and economical designs of prestressed concrete pavement have also been reported [2–4]. The benefits of prestressed concrete pavement include improved cracking resistance and reduced slab thickness; while the prestressing increases the cost of the materials and construction, the reduction in slab thickness and transverse joints can compensate some of this cost [1,5,6]. According to El-Reedy, the prestressed concrete slab with 200 mm thickness can possess equivalent design life as the traditional concrete slab with 355 mm thickness [7]. In another airfield runway study, it was determined that the thickness of the concrete slab can be reduced by half when prestressing is applied [8].

Generally, there are two types of prestressed concrete pavement: cast-in-place and precast. In the cast-in-place method, concrete slabs are constructed on-site, and the prestressing is applied either before or after construction. The most widely used prestressing approach is post-tensioning in the longitudinal direction in the early stage [9]. This technique requires some space for the anchorage area at the two ends of the slabs. Additionally, this type of prestressing limits the length of concrete slab because the prestressing loss would be high if the concrete slab is long. Due to prestressing loss, another prestressing method, known as cross tensioned, concrete pavement was introduced [1]. In the cross tensioned method, the prestressed tendons are distributed obliquely with an angle in the road direction so that the prestressing can be applied in both transverse and longitudinal directions. By adjusting the angle, the proportion of prestressing in the transverse and longitudinal directions can be changed. The prestressing in the transverse direction provides lateral constraints to the slab so that the transverse joints can be eliminated [10]. The anchorage area is on the side of the concrete slabs so that long concrete slabs can be achieved without sacrificing the prestressing [10–12].

A sliding layer between the base layer and the prestressed concrete slab is necessary to reduce the friction during the tensioning. Han et al. reported that the sliding layer can reduce the stress and deformation resulting from longer concrete slabs [11]. They proposed a design for the sliding layer with sand, cement, and polyethylene-plastic sheeting. However, the prestressed concrete pavement may be still prone to cracks if the underneath base layer is not flat [11,13]. Cement grout may be injected underneath the prestressed concrete slabs to compensate for the settlement of the ground. However, pulverization may occur due to repeated traffic loading in some cases [14].

Precast prestressed concrete pavement has been developed recently for the purpose of rapid pavement construction and rehabilitation. Compared to the cast-in-place method, the precast method can reduce construction time and improve durability [15]. The construction speed can be increased by two to three times if using the precast method instead of the case-in-place method [15]. The United States has been encouraging the application of precast prestressed concrete pavement in state highways and many demonstration sections have been constructed. While the long-term performance of prestressed concrete pavement has not been seen yet, the early-age performance indicates that it has a great potential to be used as a rapid construction or rehabilitation material [5].

In the precast prestressing concrete pavement, the prestressing is applied in the longitudinal direction, transverse direction, or both directions. Different tensioning methods have been reported in existing literatures. Transverse prestressing was applied on concrete slabs in the study by Syed and Sonparote [16]. A transverse pre-tensioning was applied during the concrete slab fabrication and a longitudinal post-tensioning was applied after the installation on the road [17]. In another study, the post-tensioning was conducted in both the transverse and longitudinal directions after the fabrication of the concrete slab [18]. The size of the precast concrete slab is relatively low. For instance, the panel size in the study by Syed and Sonparote was 3.5 × 4.5 m, and that in the study by Qu et al. was 1.5 × 1.5 m [16,18]. Special designs may be needed at the joints to help the load transfer of neighboring slabs. While the driving experience of precast prestressed concrete pavement has not been reported in existing literatures, the large amount of joints may impact driving comfort.

Finite element analysis has been applied to model the temperature and load stresses as well as mechanical responses of prestressed concrete pavement [19,20]. Naddafi and Sadeghi pointed out that a high prestressing force may impact the load transfer within concrete pavement and suggested using caution with prestress higher than 400 kN [20]. Kim et al. showed that the variation in base or subbase layers had a minimal effect on the maximum induced stresses in the precast prestressed concrete pavement via a finite element analysis [21]. Finite element analysis has also been used to find the most critical areas within the concrete slab and help the design of prestressed concrete pavement [1]. Fatigue damage is the most common distress type of prestressed concrete pavement; this distress can be detected by Fourier and wavelet analysis [10,22]. The critical stresses were found at the panel edges due to the combination of temperature stresses and traffic loading [16].

Although the cross tensioned concrete pavement has been introduced in some previous literatures, the detailed design, construction, and demonstration project have not been well reported. The influential factors that significantly affect the stress level within the concrete slab have been rarely explored. This paper aims to present the detailed design and construction of cross tensioned concrete pavement, and includes raw material selection, stress analysis, concrete design, tendon selection and distribution, tendon spacing, sliding layer design, anchorage design, and construction. This study will provide a detailed explanation of how prestressed concrete pavement is designed and constructed.

## 2. Methodology

Prestressed concrete pavement as a composite structure should be properly designed before implementation. The design should take traffic condition, or cumulative equivalent axle repetitions in the service life, into account, and the dimension of the concrete slab should be determined (the dimension should meet the requirement specified in the code of the prestressed concrete pavement in China [23]). The cement concrete should also be properly designed with enough workability and strength. The prestressed concrete tendons should be selected and the distribution (direction and spacing) of the tendons can be tentatively determined. A stress analysis may be conducted to verify the distribution of prestressed tendons. If the concrete's internal stress is not higher than the design strength of the concrete, the distribution of the prestressed tendons is acceptable. After determining if the stress is acceptable, the anchorage area and the regular reinforcement using steel bars should be designed. A stress analysis should also be carried out to verify the design. The design procedure of the prestressed concrete pavement is described in Figure 1.

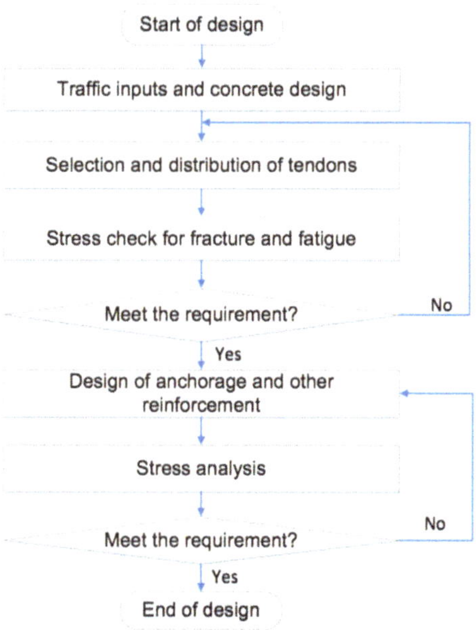


**Figure 1.** Design procedure of prestressed concrete pavement.

## 3. Stress Analysis of Oblique Prestressed Concrete Slabs

### 3.1. Prestress Loss

According to the specification of unbonded prestressed concrete structure, the effective stress of the oblique prestressed tendons can be expressed as [24]:

$$\sigma_{pe} = \sigma_{con} - \sum_{n=1}^{5} \sigma_{ln} \tag{1}$$

where,

$\sigma_{pe}$ = the effective stress;
$\sigma_{ln}$ = the value of prestress loss of item $n$ ($n$ = 1 to 5);
$\sigma_{con}$ = the tensioned control stress of prestressed reinforcement.

In general, the prestress loss include the following four aspects [24,25]:

(1) Prestress loss caused by the anchorage deformation and the shrinkage when tensioning the prestressed reinforcement, denoted as $\sigma_{l1}$. Regardless how the tendons are tensioned, the deformation of anchorage and the base plate, as well as the sliding of the prestressed tendons can cause prestress loss. The prestress loss due to the tendon shrinkage mainly occurs around the anchorage area and is non-uniform along with the tendon direction [26]. The prestress loss due to anchor deformation and tendon shrinkage can be calculated as:

$$\sigma_{l1} = \frac{a}{l} E_p \tag{2}$$

where,

$\sigma_{l1}$ = the prestress loss due to anchor deformation;
$l$ = the distance between the two anchor ends, mm;
$a$ = the shrinkage length of the tendons due to anchor deformation, for jaw vice anchorage, 6–8 mm.

(2) Prestress loss due to friction, denoted as $\sigma_{l2}$. There is friction between the prestressed tendons and the casing pipe. The closer to the anchorage area, the higher is the friction force, which can be expressed as [24]:

$$\sigma_{l2} = \left\{ \sigma_{con}\left(1 - e^{-(kx+\mu\theta)}\right) \quad kx + \mu\theta > 0.2 \; \sigma_{con}(kx + \mu\theta) \quad kx + \mu\theta \leq 0.2 \right. \tag{3}$$

where,

$\sigma_{l2}$ = the prestress loss due to friction;
$x$ = the distance from the anchorage area to the calculation cross section;
$k$ = the factor of the casing pipe;
$\mu$ = the friction coefficient between the tendons and the casing pipe;
$\theta$ = the cumulative angle.

(3) Prestress loss due to relaxation. The unrecovered deformation of steel wires at high tensioning is inevitable, which is also known as relaxation. The prestress loss due to this can be expressed as [24]:

$$\sigma_{l3} = 0.125\left(\frac{\sigma_{con}}{f_{ptk}} - 0.5\right)\sigma_{con} \tag{4}$$

where,

$$0.7 f_{ptk} < \sigma_{con} < 0.8 f_{ptk}$$

$\sigma_{l3}$ = the prestress loss due to relaxation;
$f_{ptk}$ = the tensile strength of the prestress tendons.

(4) Prestress due to concrete shrinkage and creep. Shrinkage is a common phenomenon during the curing of concrete. Creep occurs when high pressure is applied on the concrete. Both the shrinkage and creep can cause prestress loss, which can be expressed as [24]:

$$\sigma_{l3} = \frac{35 + 280\frac{\sigma_{pc}}{f'_{cu}}}{1 + 15\rho} \tag{5}$$

where,

$\rho$ = the reinforcement ratio;
$\sigma_{pc}$ = the normal compressive stress of concrete;
$f'_{cu}$ = the compressive strength of concrete under pressure.

### 3.2. Longitudinal and Transverse Stress

Figure 2 shows an example of the oblique prestressed concrete slab. Both transverse and longitudinal compressive stresses will be applied on the concrete due to the prestressing. The longitudinal and transverse stresses can be expressed as:

$$\sigma_{pL} = \frac{2(\sigma_{con} - \sigma_l)A_p cos(\alpha)}{h{\cdot}l{\cdot}tan(\alpha)} \tag{6}$$

$$\sigma_{pT} = \frac{2(\sigma_{con} - \sigma_l)A_p sin(\alpha)}{h{\cdot}l} \tag{7}$$

where,

$\sigma_{pL}$ = the longitudinal prestress;
$\sigma_{pT}$ = the transverse prestress;
$\sigma_{con}$ = the controlled design strength of tendons;
$A_p$ = the cross section area of the prestressed tendons;
$l$ = the spacing between neighboring tendons along the road direction;
$h$ = the thickness of the concrete slab.

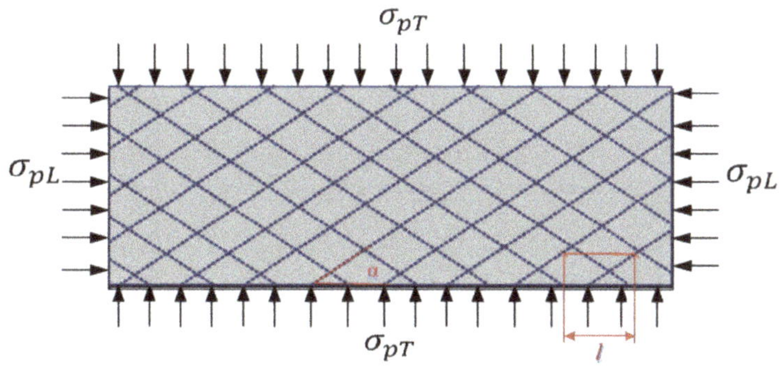


**Figure 2.** A sample concrete slab with transverse and longitudinal stress under oblique prestressed reinforcement.

According to Equations (6) and (7), the cross-sectional area of tendon ($A_p$), thickness of concrete slab ($h$), the distribution angle ($\alpha$), and the spacing ($l$) are the main influential factors affecting the stress distribution and level within the concrete slab. Thus, a parametric study was carried out to quantify the effect of these factors on the stress in the concrete slab. The once-at-a-time method was used to analyze the effect of individual factors. Some tentative values for the factors were selected based on the specification and the typical local concrete design. The slab thickness was set as 20 cm according to the typical local experience. A tendon diameter of 12.7 mm and tendon spacing of 0.8 m were tentatively used. A tentative distribution angle of 30° was used.

### *3.3. Effect of Slab Thickness*

The effect of concrete slab thickness on the stress distribution was analyzed. The diameter of the tendon was 12.7 mm, the tendon distribution angle was 30°, and tendon spacing in the longitudinal direction was 800 mm. The slab thickness varied from 140 to 240 mm with an interval of 20 mm. The stress levels in the longitudinal and transverse directions were obtained, and are shown in Figure 3. Both the longitudinal and transverse stresses decreased with an increase in slab thickness. The decreased degree of longitudinal stress was higher than that of the transverse stress, because the tendon distribution angle was smaller than 45°. This aligns with expectations—a thicker slab has an increased cross sectional area, which reduces the average stress level.

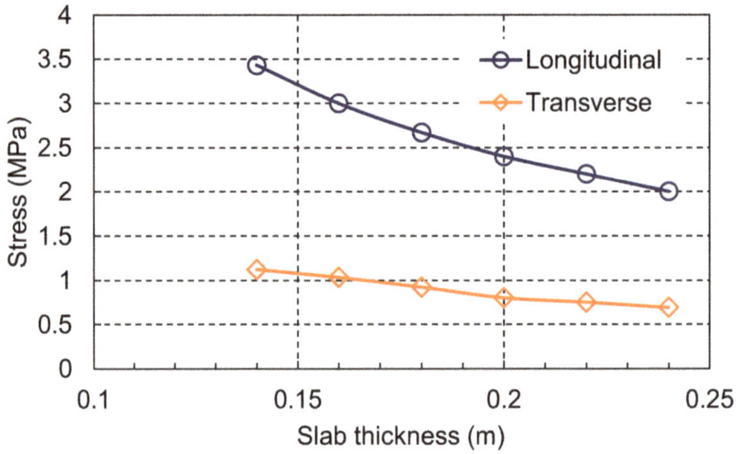


**Figure 3.** The effect of concrete slab thickness on the stress level.

### *3.4. Effect of Prestressed Tendon Diameter*

When the tendon spacing is determined, the prestress reinforcement ratio is mainly dependent on the tendon diameter. Three tentative tendon diameters were selected for the parametric study: 9.5 mm, 12.7 mm, and 15.2 mm. The concrete slab thickness, tendon angle, and tendon spacing were 20 cm, 30°, and 800 mm, respectively. Both the longitudinal and transverse stresses increased with tendon diameter, as shown in Figure 4. A bigger tendon diameter means a higher prestressed force applied on the concrete, and therefore, a higher stress level. The magnitude of increase of the longitudinal stress was higher than the transverse stress.

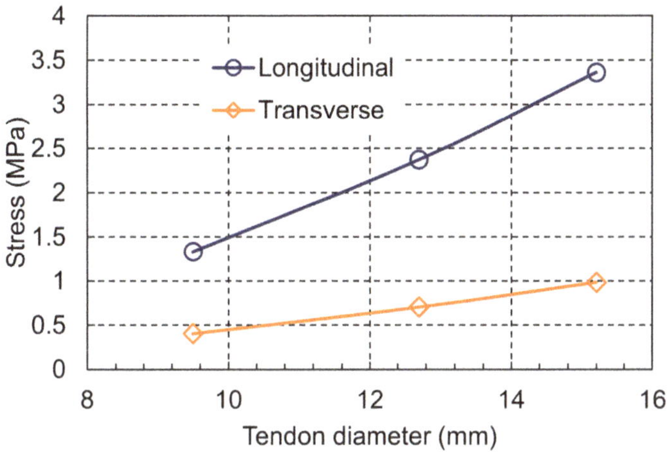


**Figure 4.** The effect of prestressed tendon diameter on the stress level in concrete slab.

### *3.5. Effect of Tendon Distribution Angle and Spacing*

The tendon distribution angle greatly affects the ratio of longitudinal stress and transverse stress. A higher tendon distribution angle is expected to bring higher longitudinal stress and lower transverse stress. Typically, a higher longitudinal stress is desirable because it demands a smaller tendon distribution angle. A low tendon distribution angle would be more challenging to design and construct the anchorage area. In addition, the transverse prestress is beneficial to resist the traffic related slab deformation; therefore, the tendon distribution angle from 25° to 45° with an interval of 5° was used for the parametric study. In terms of the spacing, three tentative values were used: 0.5 m, 0.8 m, and 1.0 m. Figure 5 displays the effect of the tendon distribution angle and tendon spacing on the stress level in the concrete slab. As expected, the longitudinal stress increased and the transverse stress decreased with an increase in tendon distribution angle. It is was observed that the transverse stress was generally linearly correlated with the tendon distribution angle, and the longitudinal stress was not; this is mainly due to the variable $\alpha$ in the Equations (6) and (7). The dependent variable for the longitudinal stress, $cos(\alpha)/tan(\alpha)$, is more complicated than that for the transverse stress, $cos(\alpha)$. In terms of the effect of tendon spacing on the stress level in the concrete slab, it can be determined from Figure 5 that an increase in space from 0.5 m to 1.0 m would result in lower longitudinal and transverse stresses. This is paralleled with expectation since a larger spacing results in a lower reinforcement rate and therefore, lower load on the concrete slab.

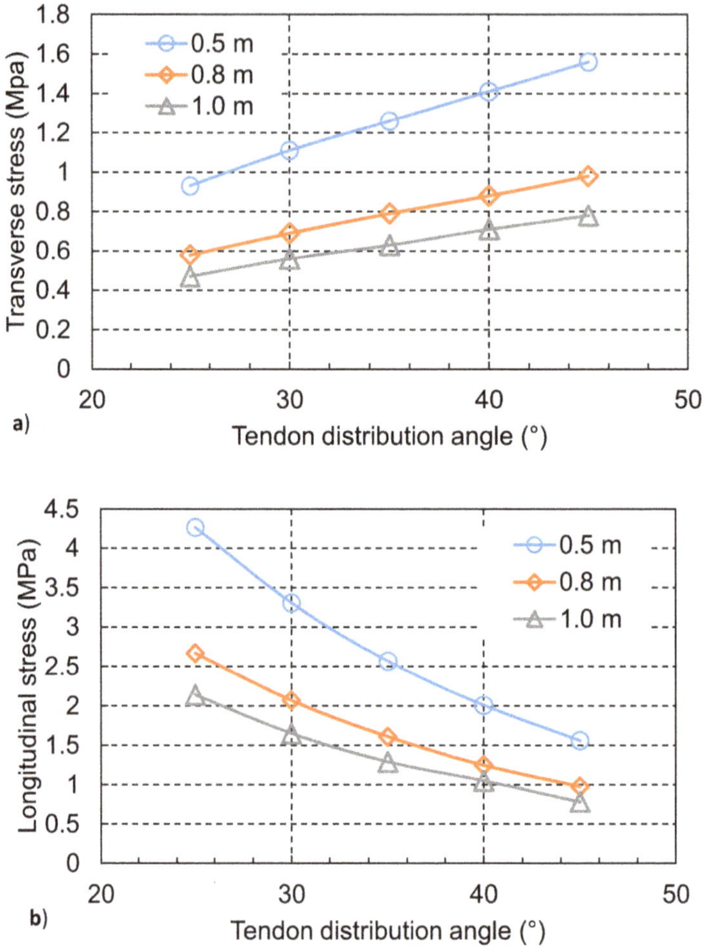


**Figure 5.** The effect of tendon distribution and spacing on the stress level in concrete slab: (**a**) longitudinal stress; and (**b**) transverse stress.

*3.6. Buckling of Concrete Slab*

Temperature differential at the top and bottom of concrete slab is very common. The maximum temperature differential which causes buckling is known as the critical temperature differential. For a concrete slab, the critical temperature can be expressed as [27]:

$$\Delta T_{cr} = \frac{H^2\pi^2}{12(1+v)\alpha}(1+\frac{1}{\lambda^2}) + \frac{K(1-v)w}{E_c H\alpha\pi^2}\frac{1}{1+\lambda^2} \tag{8}$$

where,

$H$ = the thickness to width ratio;
$E_c$ = elastic modulus of concrete;
$K$ = the coefficient of subgrade;
$w$ = the width of concrete slab;
$v$ = the Poisson's ratio of concrete;

$\lambda$ = the length to width ratio;
$\alpha$ = the linear expanding coefficient.

$$K = \frac{2\varphi(1-v^2)\pi \times 1.77}{4 \times 0.4} K_r \quad (9)$$

where,

$K_r$ = the resilient modulus of the subgrade;
$\varphi$ = the radius of the rigid plate (cm), 15 cm.

Since the prestressing applied on the concrete slab has a similar effect as the temperature differential the prestressing value can be converted to the equivalent temperature differential, as follows:

$$\Delta T_y = F_l(1+v)/(\alpha E) \quad (10)$$

where,

$\Delta T_y$ = the equivalent temperature differential caused by prestressing;
$F_l$ = the prestress level in the concrete slab.

### 3.7. Parametric Study on Buckling

In practice, some typical values for the parameters in Equations (9) and (10) have been used. Typically, the length to width ratio is between 1.0 and 3.0, the thickness to width ratio is between 1:45 and 1:30, and the compressive strength of concrete is between 15 and 40 MPa. Tentatively, we take $\lambda$, $H$, and $K$ to be 1.0, 1/35, and $1 \times 10^{-3}$ N/mm$^3$, respectively.

The effect of some individual parameters on the critical temperature differential was analyzed. Figure 6 displays the effect of various factors on the critical temperature differential in the concrete slab. All of the four investigated parameters have significant influences on the critical temperature differential. It was determined that an increase in resilient modulus of subgrade results in a lower critical temperature differential. In terms of the dimensional effect, an increase in length to width ratio results in a lower critical temperature differential, while the thickness to width ratio exhibits an opposite effect. The values of $v = 0.15$, $\alpha = 1 \times 10^{-5}$ m/°C, $H = 0.02$, $\lambda = 30$, $E = 16{,}000$ MPa were used to calculate the critical temperature causing buckling; according to Equation (8), the critical temperature differential was calculated to be 48.2 °C. This value is higher than the maximum temperature range within a day, therefore, the buckling due to natural temperature change would not occur in this case.

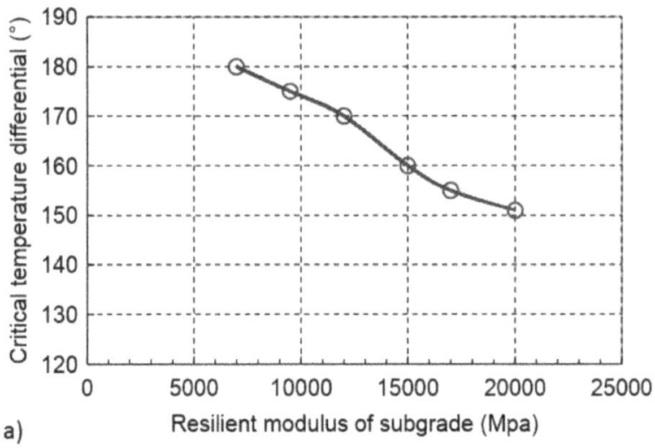


**Figure 6.** *Cont.*

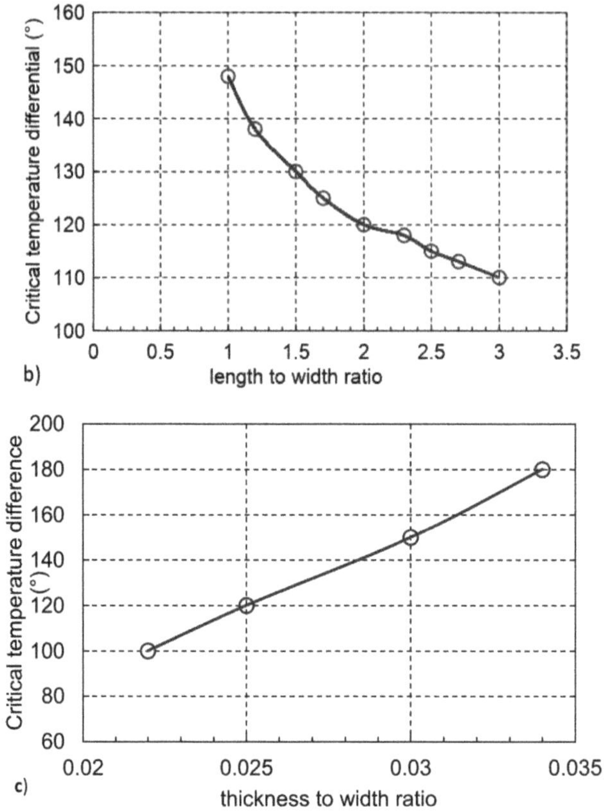


**Figure 6.** The effect on the critical temperature differential: (**a**) resilient modulus of subgrade; (**b**) length to width ratio; and (**c**) thickness to width ratio.

## 4. Design of Demonstrative Concrete Pavement Section

The trial pavement section was 150 m in length, 6 m in width, and 20 cm in thickness. The base layer was cement stabilized gravels and the cushion layer was recycled concretes. The thickness of both the base layer and the cushion layer was 20 cm, the resilient modulus was 40 MPa, and the diameter of the prestressed tendon was 12.7 mm. The design of the oblique prestressed concrete pavement is shown in Figure 7. Two layers of tendons were obliquely distributed with an angle in the longitudinal direction, as seen in Figure 7a. The prestressing was applied in the center of the concrete slab so that there is no eccentric force. In addition, spiral reinforcement was applied on the side to strengthen the anchorage area (seen in Figure 7a) and regular steel reinforcement was applied on the top and bottom of the concrete slabs (seen in Figure 7b). The design of the prestressed concrete pavement is based on the fracture criteria and fatigue criteria. The details are described below.

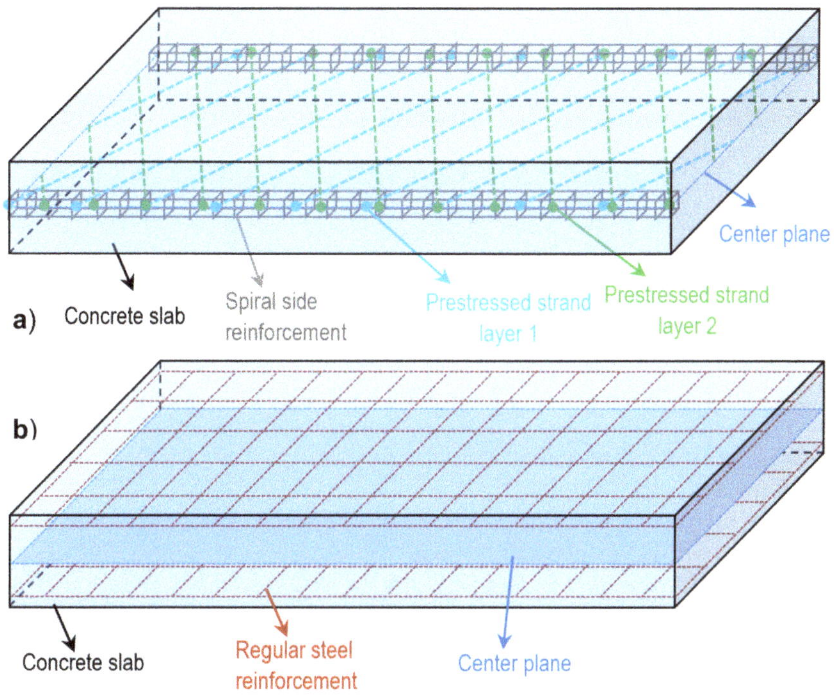


**Figure 7.** Illustration of the design of demonstrative oblique prestressed concrete pavement: (**a**) the distribution of oblique prestressed strands and side reinforcement; (**b**) the distribution of regular steel reinforcement.

## 4.1. Stress within the Concrete

### 4.1.1. Fracture Criteria

The stress within the concrete slab is determined by three components: the traffic induced stress, temperature and moisture induced stress, and the friction stress with base layer. If the combined stress is higher than the tensile strength of concrete, fracture failure will occur. Therefore, the following equation should be valid [28]:

$$\gamma(\sigma_{L,r} + \sigma_{\Delta Tr}) + \sigma_F - \sigma_{pL} \leq f_r \quad (11)$$

where,

$f_r$ = the tensile strength of concrete;
$\sigma_{\Delta Tr}$ = temperature induced stress;
$\sigma_{L,r}$ = traffic induced stress;
$\sigma_F$ = friction related stress;
$\sigma_{pL}$ = effective prestress in longitudinal direction;
$\gamma$ = coefficient of reliability.

The load induced stress level at the critical locations is expressed as:

$$\sigma_{L,r} = k_f k_c \sigma_L \quad (12)$$

where,

$\sigma_L$ = initial load induced stress;
$k_f$ = coefficient for fatigue cracking;
$k_c$ = coefficient for the impact of eccentric and dynamic loading.

The temperature induced stress level is expressed as:

$$\sigma_{\Delta Tr} = k_t \sigma_{\Delta t} \tag{13}$$

where,

$k_t$ = coefficient of cumulative fatigue stress;
$\sigma_{\Delta t}$ = the thermal stress, expressed as:

$$\sigma_{\Delta t} = \frac{E_c \alpha_c \Delta T}{2(1 - \nu_c)} \tag{14}$$

where,

$E_c$ = elastic modulus of concrete (MPa);
$\alpha_c$ = expansion coefficient of concrete;
$\nu_c$ = Poisson's ratio;
$\Delta T$ = the temperature gradient along the thickness (°C).

4.1.2. Fatigue Criteria

The fracture criteria is used to ensure the stress within the concrete slab is not higher than the tensile strength of concrete. However, fatigue crack may occur even if the stress level is lower than the tensile strength due to repeated loading. According to the specification of prestressed concrete pavement in China, the fatigue crack of concrete slab is a criterion of design. The PCA fatigue model was adopted in this study [29], expressed as:

$$\rho_\sigma = 0.972 - 0.08281 lg(N) \tag{15}$$

where,

$N$ = the fatigue life;
$\rho_\sigma$ = the ratio between the stress level to the tensile strength of concrete.

$$\rho_\sigma = \frac{\gamma(\sigma_{L,r} + \sigma_{\Delta Tr}) + \sigma_F - \sigma_{pL}}{\sigma_{pL} + f_r} \tag{16}$$

Therefore, the required prestress level in longitudinal direction should be:

$$\sigma_{pL} \geq \frac{\gamma(\sigma_{L,r} + \sigma_{\Delta Tr}) + \sigma_F}{1 + \rho_\sigma} - \frac{\rho_\sigma}{1 + \rho_\sigma} f_r \tag{17}$$

*4.2. Stress in Anchorage Area*

Anchorage area could be venerable because the force is applied in a small area. There are two types of failures for the anchorage area: one is the regional concrete failure which occurs in the small area right under the base plate, and the other is the tensile failure of concrete along with the tensioning direction which usually occurs over a much wider area [30]. The regional failure within the anchorage area was analyzed in this study, as it is a more critical concern. The regional stress level should meet the requirement below [31]:

$$F_l \leq 0.9(\beta_c \beta_l f_c + 2\alpha \rho_v \beta_{cor} f_y) A_{ln} \tag{18}$$

where,

$F_l$ = the regional stress level in the anchorage area;
$\beta_c$ = the strength coefficient of concrete;
$\beta_l$ = the increase coefficient of concrete in compression;
$f_c$ = the design compressive strength of concrete;
$\alpha$ = the reduction coefficient due to spiral reinforcement;
$\rho_v$ = the volume proportion of spiral reinforcement;
$f_y$ = the tensile strength of spiral steels;
$A_{ln}$ = the area of concrete under compression.

### 4.3. *Stress Analysis*

#### 4.3.1. Stress Analysis on Concrete

Concrete slab dimensions were selected to be $L$ = 100 m, $w$ = 4.5, $t$ = 20 cm, the tensile strength of concrete $f_r$ = 5 MPa, elastic modulus $E_c$ = 25,500 MPa, coefficient of reliability $\gamma$ = 1.08, Poisson's ratio $v_c$ = 0.15, expansion coefficient $\alpha_c = 1 \times 10^{-5}$/°C, and temperature gradient $\Delta T$ = 0.9 °C/cm.

According to the fracture criteria,

$$r = 0.537h(\frac{E_c}{E_t})^{\frac{1}{3}} = 0.537 \times 0.2 \times (\frac{30,000}{165})^{\frac{1}{3}} = 0.61$$

$$\sigma_L = 0.077r^{0.60}\,\mathrm{h}^{-2} = 0.077 \times 0.61^{0.60} \times 0.2^{-2} = 1.43\ \mathrm{MPa}$$

$$\sigma_{L,r} = k_f k_c \sigma_L = 1.69 \times 1.2 \times 1.43 = 2.90\ \mathrm{MPa}$$

$$\sigma_{\Delta T} = \frac{E_c \alpha_c \Delta T}{2(1 - v_c)} = \frac{25,500 \times 1 \times 10^{-5} \times 0.9 \times 20}{2 \times (1 - 0.15)} = 2.7\ \mathrm{MPa}$$

$$\sigma_F = \mu \rho \chi = 0.8 \times 0.024 \times 50 = 0.96\ \mathrm{MPa}$$

$$\text{Then } \sigma_{py} \geq \gamma(\sigma_{L,r} + \sigma_{\Delta Tr}) + \sigma_F - f_r = 1.08 \times (3.78 + 2.7) + 0.96 - 5 = 2.96\ \mathrm{MPa}$$

This means that the prestress applied on the concrete slab should be higher than 2.96 MPa to prevent fracture distress.

According to the fatigue criteria,

$$N = \frac{N_e}{t} = \frac{100 \times 10^4}{30} = 3.3 \times 10^4$$

$$\rho_\sigma = 0.972 - 0.08281 lgN = 0.972 - 0.08281 lg\left(3.3 \times 10^4\right) = 0.6$$

$$\text{Then, } \sigma_{py} \geq \frac{\gamma(\sigma_{L,r} + \sigma_{\Delta Tr}) + \sigma_F}{1 + \rho_d} - \frac{\rho_d}{1 + \rho_d} f_r = \frac{1.08 \times (3.78 + 2.7) + 0.96}{1 + 0.6} - \frac{0.6}{1 + 0.6} \times 5 = 3.10\ \mathrm{MPa}$$

This means that the effective prestress applied on the concrete slab should be higher than 3.1 MPa to prevent fatigue failure. Therefore, the prestress level should be higher than 3.10 MPa taking into account both the fracture and fatigue criteria.

#### 4.3.2. Determine Tendon Spacing

The tensile strength of the tendon was 1860 MPa. The total loss of prestress in the concrete slab was about 20% [32]. The longitudinal spacing can be calculated as:

$$L = \frac{2(\sigma_{con} - \sigma_1) \times A_p \times cos\alpha}{\sigma_{py} \times h \times tan\alpha} = \frac{2 \times 1116 \times 98.7 \times cos30}{3.10 \times 200 \times tan30} = 532.98\ \mathrm{mm}$$

Therefore, the spacing in longitudinal direction was selected as 500 mm.

4.3.3. Verification

According to

$$0.95\sigma_s \leq \sigma_L + \sigma_{\Delta T} \leq 1.03\sigma_s$$

$$\sigma_s = f_{cm} + \sigma_{Py} - \sigma_F = 3.33 + 3.10 - 0.96 = 5.47 \text{ MPa}$$

$$\sigma_{pr} + \sigma_{\Delta T} = 2.90 + 2.7 = 5.60 \text{ MPa}$$

$$0.95 \times 5.47 = 5.20 \text{ MPa} \leq 5.60 \text{ MPa} \leq 1.03 \times 5.47 = 5.63 \text{ MPa}$$

Therefore, the distribution angle of 30° and longitudinal spacing of 500 mm can meet the requirement according to the stress analysis. The effective prestressing in the longitudinal direction was 3.1 MPa.

### 4.4. Regional Stress Analysis in Anchorage Area

The diameter of the tendon was 12.7 mm, the dimension and thickness of the base plate were 80 mm × 80 mm and 14 mm, respectively, and the compression area was 160 mm × 200 mm. HPB235 steel was used in the spiral reinforcement, and the spacing of the steel was 40 mm. According to Equation (18), the regional stress was analyzed below:

$$F_l = 1.2\sigma_{con} A_p = 1.2 \times 0.75 \times 1860 \times 98.7 = 165.22 \text{ KN}$$

$$A_b = 160 \times 200 = 32,000 \text{ mm}^2$$

$$A_l = \frac{\pi}{4}(\alpha + 2\delta)^2 = \frac{\pi}{4}(50 + 2 \times 14)^2 = 4778 \text{ mm}^2$$

$$A_{ln} = 4778 - 98.7 = 4679.3 \text{ mm}^2$$

$$\beta_l = \sqrt{\frac{A_b}{A_l}} = 2.59 \quad \beta_c = 1 \quad f_c = 16.7$$

$$\alpha = 0.5 \text{ m} \quad f_y = 210 \text{ MPa} \quad A_{ssl} = 78.54 \text{ mm}^2 \quad d_{cor} = 100 \text{ mm}$$

$$\rho_v = \frac{4A_{ssl}}{d_{cor}S} = \frac{4 \times 78.54}{100 \times 40} = 0.079$$

$$A_{cor} = \frac{\pi}{4}{d_{cor}}^2 = \frac{\pi}{4} \times 100^2 = 7854 \text{ mm}^2 < A_b$$

$$\begin{aligned} &0.9(\beta_c\beta_l f_c + 2\alpha\rho_v\beta_{cor}f_y)A_{ln} \\ &= 0.9 \times (1 \times 2.59 \times 16.7 + 2 \times 0.5 \times 0.079 \times 1.28 \times 210) \times 4679.3 \\ &= 271.58 \text{ KN} > F_l \end{aligned}$$

Therefore, the regional stress in the anchorage area is sufficient.

## 5. Materials and Preparation

### 5.1. Cement and Aggregates

According to the standard for Unbounded Prestressed Concrete Structure in China [31], the strength grade of the concrete used in pavement should be higher than C30. Other requirements of the cement include low shrinkage and creep, rapid hardening, and high early-stage strength. The shrinkage and creep of concrete will cause prestress loss so that the effective stress within the concrete slab would be compromised. On the other hand, the rapid hardening and high early-stage strength allows for the application of the prestress soon after the construction and reduces shrinkage crack. Therefore, PO 425 silicate cement with a strength grade of C35 was used in this study to fabricate the prestressed cement concrete.

Coarse aggregates and sands are an important composition of cement concrete. Coarse aggregates with regular shape and good angularity were selected to prepare the cement concrete in this study, thus the use of flat or elongated aggregates was limited. The biggest aggregate size used was 31.5 mm and medium sand with a fineness modulus of about 2.5–3 was used as the fine aggregates. The gradation of the coarse aggregates and sands are shown in Table 1.

**Table 1.** The gradation of coarse and fine aggregates.

| Aggregates | Mesh Size (mm) | | 2.4 | 4.8 | 9.5 | 16.0 | 19.0 | 26.5 | 31.5 |
|---|---|---|---|---|---|---|---|---|---|
| Coarse | Cumulative retained (%) | Upper limit | 100 | 100 | 90 | 75 | 60 | 35 | 5 |
| | | Lower limit | 95 | 90 | 75 | 60 | 40 | 20 | 0 |
| Fine | Mesh size (mm) | | 0.15 | 0.30 | 0.60 | 1.18 | 2.36 | 4.75 | - |
| | Cumulative retained (%) | Upper limit | 100 | 92 | 70 | 50 | 25 | 10 | - |
| | | Lower limit | 90 | 70 | 41 | 10 | 0 | 0 | - |

## 5.2. Concrete Design

After the raw materials were prepared, concrete design was the next step. Concrete design mainly refers to the selection of proper fractions for each composition including water, cement, sand, coarse aggregates, and superplasticizer. Water–cement (W/C) ratio is the most important index affecting the workability and mechanical performance in concrete design. A preliminary design was carried out according to the concrete design specification in China, in which the W/C ratio was 0.39 and the sand to coarse aggregate (S/A) ratio was 0.47 and the two parameters were adjusted by plus and minus a small value [23]. Initially, three tentative W/C ratios were selected according to the specification and preliminary calculation, as seen in Table 2. The amounts of other compositions were selected correspondingly for each W/C ratio.

**Table 2.** The gradation of coarse and fine aggregates.

| Compositions | Design 1 | Design 2 | Design 3 |
|---|---|---|---|
| Water–cement ratio | 0.35 | 0.39 | 0.43 |
| Sand to coarse aggregate ratio | 0.49 | 0.47 | 0.45 |
| Cement mass percent | 17.9 | 19.7 | 20.3 |
| Sand mass percent | 24.4 | 23.3 | 22.2 |
| Coarse aggregates mass percent | 49.6 | 49.1 | 50.0 |
| Water mass percent | 7.8 | 7.8 | 7.3 |
| Superplasticizer mass percent | 0.2 | 0.2 | 0.2 |

The compressive strength and the slump of the concrete based on different designs were tested, and are shown in Figure 8. With an increase in W/C ratio, the slump of fresh concrete increased while the 7-day and 28-day compressive strength of hardened concrete decreased (this was expected). It is worth noting that the 7-day strength can arrive at about 102%, 96%, and 93% of the design strength for the three designs, respectively. A W/C ratio of 0.39 was selected to balance the workability and compressive strength.

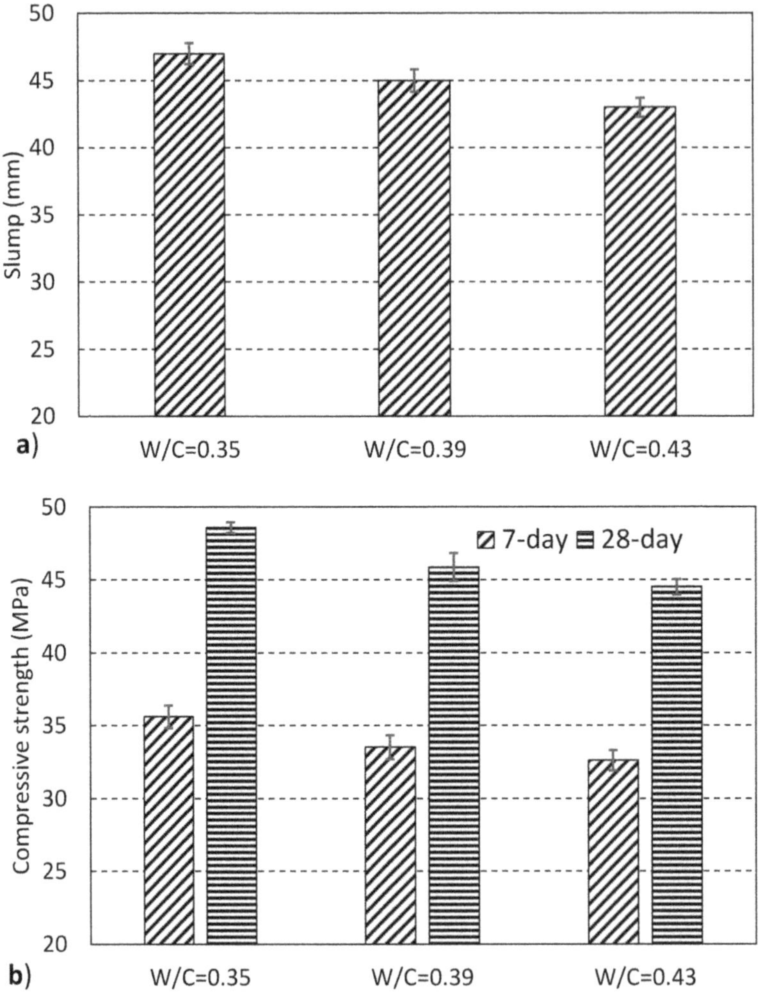


**Figure 8.** The properties of the concrete at various W/C ratios: (**a**) slump of fresh concrete; and (**b**) 7-day and 28-day compressive strength of hardened concrete.

### *5.3. Prestressed Concrete Tendons*

The prestressed tendons used in this study consisted of three parts: the prestressed wires, restrictive coating, and the paint cover. Seven high quality steel wires were included to form the tendon, with the diameter of the steel tendon being 12.7 mm. High-density polyethylene was used as the restrictive coating to prevent damage of wires during transport, storage, and placement. Asphalt or grease was used as the paint cover to reduce the friction between the tendons and concrete during prestressing. The design tensile strength of the prestressed tendons was 1860 MPa.

### *5.4. Anchorage*

The anchorage is an important part which determines the success of the prestressing. There are two anchors in an anchorage system: the fixing end anchor and the tensioning anchor. An extruding anchor and jaw vice anchorage were used for the fixing end anchor and tensioning anchor, respectively.

The diameter of the extruding anchor was 50 mm, the size and thickness of the bearing plate were 80 mm × 80 mm, 14 mm, respectively, and the size of the compressive section was 160 mm × 200 mm. The post-tension method was applied in this study, so the anchors were placed in the concrete during the casting.

## 6. Construction of Demonstrative Pavement Section

### 6.1. Side Formwork

Steel side formwork was used in the construction. Cement mortar was placed in the low-lying area to make sure the bottom of the side formwork contacted well with the base layer. After installing the formwork, the template junctions and the inside were examined, and the height difference between two templates in the joint area were measured to be no higher than 3 mm. The reserved length of the prestressed reinforcement ranged from 40 to 60 cm. Some holes were reserved on the formwork to allow the prestressed tendons to go through. It is noteworthy that the height of the tendons should be the same so that uniform stresses can be applied.

### 6.2. Sliding Layer

A sliding layer was paved on the top of the base layer before placing the concrete pavement. Common materials include asphaltic felt, geotextiles, and polyethylene layers. The material and construction method followed the specification for prestressed concrete pavement in China [33]. In this study, fine granular materials with a maximum size of 0.3 mm were initially placed with a thickness of 10 mm and then geotextiles were placed on the top of the granular materials.

### 6.3. Distribution of Tendons and Tensioning

Prestressed tendons were applied in the middle of the concrete slabs. The angle between the steel tendon and the road direction was 30°, so the angle between the tendons in two layers was 60°, as shown in Figure 9. The tensioning was applied via two steps: the first tensioning was applied to 30% of the design strength (1860 MPa) of the tendon after the 12 h of the concrete casting, and the second tensioning was applied to 105% of the design strength after 7 days of the concrete placement. The distance between neighboring tendons was 0.5 m. The grade of the anchor seal concrete was higher than that in the pavement, which means higher than C35. The spiral indirect reinforcement was applied using Φ10 smooth steel (HPB235). The spacing and radius of the screw were 30 mm and 40 mm, respectively. Four anti-splitting steel bars (Φ8) with a spacing of 60 mm were arranged.

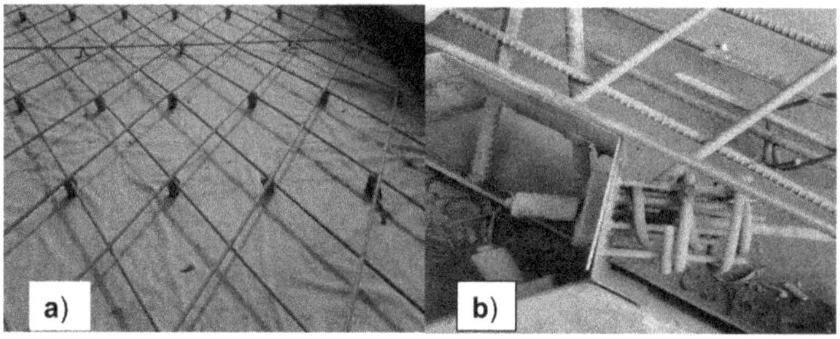


**Figure 9.** (**a**) The distribution of the prestressed tendons; and (**b**) the fixed end anchor.

### 6.4. Reinforcement at Slab Top and Bottom

Because the prestressed reinforcement was only applied to the middle part of the concrete slabs (in other words, the top and bottom were un-reinforced), an unbalanced force within the concrete slab was produced. Given this, reinforcement using Φ12 regular steel bars (with a spacing between the bars of 250 mm) was applied at the top and bottom parts of the concrete slab, as seen in Figure 10. In addition, steel reinforced sleeper beam was placed between the concrete slab and the base layer to ensure good load transfers.

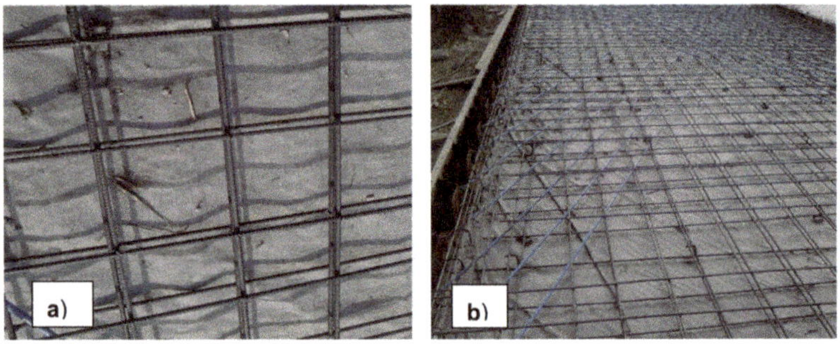


**Figure 10.** The distribution of reinforcement: (**a**) two-layer regular steel reinforcement; and (**b**) the prestressed reinforcement and the regular reinforcement.

### 6.5. Concrete Placing and Curing

The placing of concrete was completed continuously to avoid unnecessary joints. Vibration was applied during the placing to achieve high density. The side formwork and the prestressed tendons were not touched during the vibration to avoid displacement. The curing started right after the placing of concrete slabs. Geotextiles were used to cover the top of the concrete pavement and water was spread. The curing process lasted 14–21 days until the tensile strength of the concrete reached 80% of the design value.

### 6.6. Pavement Monitoring System

After the construction of the concrete pavement and the application of prestressing, the stress condition of the concrete pavement was monitored. For this purpose, a prestressed tendon anchorage dynamometer and piezoelectric sensor were placed in the base layer of the pavement. The schematic distribution of the pavement monitoring system is shown in Figure 11.

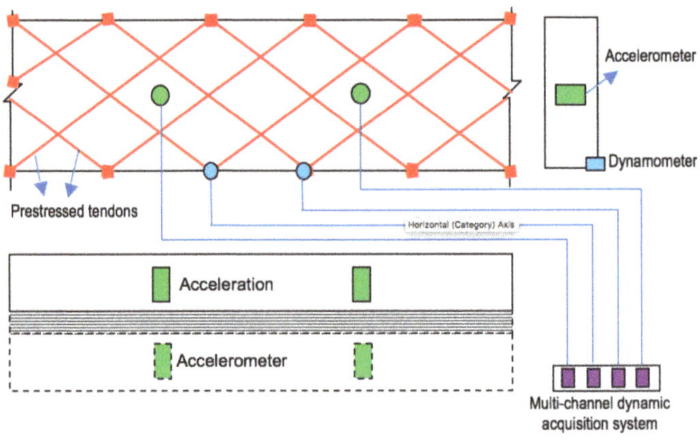


**Figure 11.** Schematic multi-views of the inclined prestressed concrete pavement monitoring system.

## 7. Summary and Conclusions

This paper presented a detailed design and construction of oblique prestressed concrete pavement. Prestressed tendons were distributed obliquely with the longitudinal direction so that prestress was applied in both the transverse and longitudinal direction. Compared to simple prestressing in the longitudinal direction or precast prestressed concrete pavement, the oblique prestressed concrete pavement allowed a much higher joint spacing. The design of prestressed concrete pavement included the selection of raw materials, design of cement concrete, anchorage area, size and distribution of prestressed tendons, stress analysis within the concrete slab, sliding layer, side reinforcement, and regular reinforcement at top and bottom. A pavement monitoring system was embedded in the concrete pavement to survey the condition of the pavement after traffic opening. Some findings of the study can be summarized as below:

1. The cement concrete was designed based on the compressive strength and workability to meet the requirement of prestressed concrete in China;
2. The possible buckling of concrete slab due to temperature change or prestressing was evaluated and it was determined that buckling would not occur under the natural temperature change in China;
3. The required prestress applied on concrete slabs was obtained based on fracture criteria and fatigue criteria and the prestressed tendon spacing was obtained based on the stress analysis;
4. The effect of slab thickness, tendon diameter, tendon distribution angle, and tendon spacing on the stress level within the concrete slab were analyzed in detail;
5. Spiral indirect reinforcement on the side was applied to strengthen the anchorage area and regular reinforcement was applied on the top and bottom of the concrete slabs to improve the stress condition;
6. The design of the anchorage area was verified based on regional stress analysis;
7. The step-by-step construction of the oblique prestressed concrete pavement was described and the demonstration road section has performed well over the three years since traffic opening.

**Author Contributions:** Ling Yu and Xu Yang conceived and designed the experiments; Xu Yang also help finalize the manuscript; Xiaohui Yan and Xiaowei Zhang performed the experiments and wrote the draft of the paper; Ting Zhao helped improve the format and English of the manuscript; Julian Mills-Beale and Cong Duan helped improve the paper in paper organization and discussion based on his professional background in pavement engineering.

**Conflicts of Interest:** The authors declare no conflict of interest.

## References

1. Hossain, M.; Hancock, J.; Wu, Z. Cross Tensioned Concrete Pavement. *J. Transp. Eng.* **2003**, *129*, 427–433. [CrossRef]
2. Sargious, M.; Wang, S.K. Design of Prestressed Concrete Airfield Pavements under Dual and Dual Tandem Wheel Loading. *J. Prestress. Concr. Inst.* **1971**, *16*, 19. [CrossRef]
3. Sargious, M.; Wang, S.K. Economical Design of Prestressed Concrete Pavements. *J. Prestress. Concr. Inst.* **1971**, *16*, 64. [CrossRef]
4. Sargious, M. Performance of Transverse Joints in Prestressed and Reinforced-Concrete Airfield Pavements. *J. Am. Concr. Inst.* **1978**, *75*, 359–366.
5. Merritt, K.D.; Tyson, S.S. Precast Prestressed Concrete Pavement Applications in the USA. In *Proceedings of the First International Conference on Recent Advances in Concrete Technology*; Amde, A.M., Sabnis, G., Tan, J.S.Y., Eds.; Destech Publications, Inc.: Lancaster, PA, USA, 2007; pp. 457–470.
6. Zhang, D.; Xu, X.; Li, N.; Han, W. Research on the Design Method of Diagonal Post-Tensioned Prestressed Jointless Cement Concrete Pavement. In Proceedings of the ICCTP 2011: Towards Sustainable Transportation Systems, Nanjing, China, 14–17 August 2011; American Society of Civil Engineers: Reston, VA, USA, 2011.
7. El-Reedy, M.A. *Advanced Materials and Techniques for Reinforced Concrete Structures*; CRC Press: Boca Raton, FL, USA, 2009.

8. Wieckowski, A.; Sznurawa, A. Structural Aspects of Airfield Runways with Concrete Pavements. In *Contemporary Challenges of Transport Systems and Traffic Engineering*; Macioszek, E., Sierpinski, G., Eds.; Springer International Publishing AG: Cham, Switzerland, 2017; pp. 39–52.
9. Li, N.; Han, S.; Yu, J.T.; Wang, H.F. Analysis of Bridge Deck Pavement Structure Interlayer Based on Different Interface Form. In Proceedings of the 10th International Conference of Chinese Transportation Professionals (ICCTP), Beijing, China, 4–8 August 2010; American Society of Civil Engineers: Reston, VA, USA, 2010.
10. Guo, C.; Zhang, M.; Wang, Z.; Wang, X. Analysis of Cross Tensioned Concrete Pavement Damage. In *Advances of Transportation: Infrastructure and Materials*; Wang, H., Liu, Y., You, Z., Eds.; Destech Publications, Inc.: Lancaster, PA, USA, 2006; Volume 2, pp. 893–899.
11. Han, S.; Chen, D.; Ling, C.; Zhang, D. Study on sliding layer of cross-tensioned concrete pavement. *Road Mater. Pavement Des.* **2015**, *16*, 518–535. [CrossRef]
12. Li, N.; Zhang, D.; Xu, X.; Han, W. Numerical Simulation on Cross-Tensioned Prestressed Concrete Pavement. In Proceedings of the Fourth International Conference on Transportation Engineering, Chengdu, China, 19–20 October 2013; American Society of Civil Engineers: Reston, VA, USA, 2013.
13. Tayabji, S. *Precast Concrete Pavement Implementation by US Highway Agencies*; The National Academies of Sciences, Engineering, and Medicine: Washington, DC, USA, 2015.
14. Kawamura, N.; Maekawa, R.; Morohashi, K.; Shiji, A.; Kamitani, K. Development of high durable grout for airport prestressed concrete pavement. In *Advances in Transportation Geotechnics II*; CRC Press, Taylor & Francis Group: Boca Raton, FL, USA, 2012; pp. 259–264.
15. Tomek, R. Advantages of Precast Concrete in Highway Infrastructure Construction. *Procedia Eng.* **2017**, *196*, 176–180. [CrossRef]
16. Syed, A.; Sonparote, R.S. Analysis of Prestressed Precast Concrete Pavement. *Mater. Today Proc.* **2017**, *4*, 9713–9717. [CrossRef]
17. Merritt, D.K.; McCullough, B.F.; Burns, N.H. Design-construction of a precast, prestressed concrete pavement for Interstate 10, El Monte, California. *PCI J.* **2005**, *50*, 18–27. [CrossRef]
18. Qu, B.; Weng, X.; Zhang, J.; Mei, J.; Guo, T.; Li, R.; An, S. Analysis on the deflection and load transfer capacity of a prefabricated airport prestressed concrete pavement. *Constr. Build. Mater.* **2017**, *157*, 449–458. [CrossRef]
19. Zhang, M.; Guo, C.; Lu, L.; Wang, X.; Wang, Z. Cross Tension Prestressed Concrete Pavements Temperature Stresses. In *Advances of Transportation: Infrastructure and Materials*; Wang, H., Liu, Y., You, Z., Eds.; Destech Publications, Inc.: Lancaster, PA, USA, 2016; Volume 2, pp. 947–954.
20. Naddafi, M.F.; Sadeghi, V. Finite Element Investigation of the Prestressed Jointed Concrete Pavements. *J. Fundam. Appl. Sci.* **2016**, *8*, 167–178. [CrossRef]
21. Kim, K.; Tia, M.; Greene, J. Evalution of Strauctural Behavior of Precast Prestressed Concrete Pavement with Finite Element Analysis. *Transp. Res. Rec.* **2016**, *2590*, 84–93. [CrossRef]
22. Melhem, H.; Kim, H.S. Damage detection in concrete by Fourier and wavelet analyses. *J. Eng. Mech.* **2003**, *129*, 571–577. [CrossRef]
23. China Academy of Building Research. *JGJ 55-2000 Specification for Mix Proportion Design of Ordinary Concrete*; China Building Industry Press: Beijing, China, 2001.
24. Yan, Y.; Liu, C.; Huang, W. Structural Design of Continuous Reinforced Concrete Pavement as Overlay. *J. China Foreign Highw.* **2005**, *5*. (In Chinese)
25. Zirakian, T.; Showkati, H. Experiments on Distortional Buckling of I-Beams. *J. Struct. Eng.* **2007**, *133*, 1009–1017. [CrossRef]
26. Wei, W.; Dong, D. Prestress Loss Calculation due to Anchor Deformation. *J. Archit. Sci. Eng.* **2007**, *24*, 86–90.
27. Wang, F. Study on the Structural Performace of Prestressed Highway Tunnel Pavement. Master's Thesis, Chang'an University, Xi'an, China, 2013. (In Chinese)
28. *GB50422, Engineering Technical Standard of Prestressed Concrete Pavement*; Ministry of Housing and Urban Construction: Beijing, China, 2007.
29. Hazaree, C.; Wang, K.; Ceylan, H.; Gopalakrishnan, K. Capillary Transport in RCC: Water-to-Cement Ratio, Strength, and Freeze-Thaw Resistance. *J. Mater. Civ. Eng.* **2011**, *23*, 1181–1191. [CrossRef]
30. Wu, J. The Design of Anchorage Area of Post-tensioning Concrete. Master's Thesis, Tongji University, Shanghai, China, 2007.

31. Zirakian, T.; Zhang, J. Elastic distortional buckling of singly symmetric I-shaped flexural members with slender webs. *Int. J. Struct. Stab. Dyn.* **2012**, *12*, 359–376. [CrossRef]
32. Yan, X. The Application of Oblique Prestressed Concrete Pavement Technology Research. Master's Thesis, Shenyang Jianzhu University, Shenyang, China, 2015.
33. Ministry of Housing and Urban-Rural Development of China. *GB 50422-2007: Technical Code for Engineerings of Prestressed Concrete Pavement*; China Plan Publishing: Beijing, China, 2007.

*applied sciences*

MDPI

*Article*

# Evaluation of Fatigue Life of Asphalt Concrete Mixtures with Reclaimed Asphalt Pavement

**Wojciech Bańkowski**

Road and Bridge Research Institute, 03-301 Warsaw, Poland; wbankowski@ibdim.edu.pl; Tel.: +48-22-390-0-403

Received: 18 January 2018; Accepted: 13 March 2018; Published: 19 March 2018

**Featured Application: The results of the work can be used in the design of bituminous mixtures with RAP, as well as may be taken into account in the change of technical regulations in Poland regarding the use of RAP.**

**Abstract:** The topic of this article is the evaluation of the fatigue life of asphalt concrete mixtures with reclaimed asphalt pavement (RAP). The evaluation was carried out in relation to asphalt concrete mixtures AC22P and high modulus asphalt concrete ACWMS16 with 50% contents of RAP, greater than currently permitted by technical regulations in Poland. The first stage consisted of the evaluation of laboratory results, which was followed by a mechanistic analysis of the designed life of pavement structures with reclaimed asphalt. The evaluation included the results of laboratory tests (i.e., the air voids content, effective asphalt content, properties of recovered asphalt (penetration, softening point), stiffness, and resistance to fatigue of bituminous mixtures). Calculations of the design life of the structure were made using the criteria according to the 2004 AASHTO specifications for fatigue life and the Asphalt Institute for subgrade deformation. In addition, calculations were carried out using the French method. The analyses allowed for a comprehensive evaluation of the asphalt concrete mixture in the analyzed scope. The evaluation of the fatigue life of AC22P and ACWMS16 mixtures with 50% content of reclaimed asphalt as well as the results of the calculations of design life of the structure indicated positive effects. The tests have been carried out within the framework of the research project dedicated to hot recycling entitled "Reclaimed asphalt pavement: Innovative technology of bituminous mixtures using material from reclaimed asphalt pavement".

**Keywords:** pavement design; fatigue life; recycling; reclaimed asphalt pavement; RAP; mechanistic method

## 1. Introduction

The hot recycling technology has been known and used worldwide for years. In many countries it is quite commonly used, and attempts can be noted for the maximum re-use of the reclaimed asphalt pavement (RAP) from the milling of asphalt courses [1–4]. This material is valuable—its composition contains mainly mineral aggregate and asphalt binder. Various methods of asphalt pavement recycling are known: both in hot and cold technology [5–8]. Application of the hot recycling technology in Poland is very limited for many reasons, including the lack of appropriate technical guidelines and recommendations, experience, equipment resources, consent from the project owners, availability of a good quality RAP, among others [9,10]. Taking the properties of a bituminous mixture with RAP into account, it must be ensured that it meets the technical requirements, and the application of RAP does not deteriorate the properties of the new mixture [11]. Certainly, these are the assumptions to ensure that the asphalt courses will be characterized by the appropriate durability. The first factor which has a significant impact is the uniformity of the reclaimed asphalt pavement [12], and in particular the contents and properties of asphalt binder. The larger the contents of RAP in the bituminous mixture, the more important is the role of old asphalt binder. Another very important factor is the ageing process covering the bitumen courses, and the asphalt binder [13,14]. Therefore, the application of reclaimed asphalt (i.e., the material after long- and short-term ageing) may raise concerns as to the characteristics of the bituminous mixture with the addition of that material. In general, a potentially increased rigidity and better resistance to permanent strains may be expected [15]. On the other hand, those properties where greater stiffness is not favorable (i.e., the resistance to cracking and fatigue resistance) may be deteriorated. In order to verify these characteristics, it is necessary to conduct performance tests. If necessary, regenerative (rejuvenator) additives should be used, the aim of which is to improve the viscoelastic properties [16–18]. The fatigue life is the primary characteristic determining the service life of the pavement [19]. A change of rheological properties of asphalt binder due to ageing can cause deterioration of the fatigue life of mixtures with RAP [20,21]. There are also studies showing an opposite effect [22]. This publication presents the results of fatigue tests of two bituminous mixtures for the base course with 50% contents of RAP; i.e., AC22P asphalt concrete and high modulus asphalt concrete ACWMS16. An analysis of the fatigue test results and calculations of the pavement design life were conducted. These studies and analyses have not been done in Poland so far. The results of fatigue tests and structure analysis are particularly important in the context of popularizing the use of hot recycling in Poland. They show the possibilities of using mixtures with RAP while maintaining an appropriate fatigue life of mixtures and structures.

## 2. Description of Test Methods

As part of laboratory tests, a number of research methods covered by European standards were applied. In the scope of binder tests, penetration (PN-EN 1426) and softening temperature (PN-EN 1427) tests were carried out. In terms of basic physical properties of mineral-bituminous mixtures, the density (PN-EN 12697-6), maximum density (PN-EN 12697-5), air void content (PN-EN 12697-8), and soluble binder content (PN-EN 12697-1) were tested.

The fatigue life was determined using the four-point bending beam test (4PB) according to PN-EN 12697-24 standard. Tests were performed in constant strain mode (several strain amplitudes), at a frequency of 10 Hz and a temperature of 10 °C. These are typical testing conditions used in Poland. Beams with size of 50 × 63 × 380 mm were cut from plates compacted by laboratory roller compactor (PN-EN 12697-33). The analysis of fatigue life most often evaluates the $\varepsilon_6$ parameter, which determines the strain under the fatigue test, during which the life of 1 million load cycles is obtained. A higher value means potentially better fatigue properties. Additional information is provided by the fatigue characteristic, which is described by the formula:

$$N = A \cdot \varepsilon^b, \tag{1}$$

where: $N$—fatigue life, $\varepsilon$—strain during fatigue test; $A$, $b$—linear regression parameters, and $b$ is the inclination of the fatigue line.

Prior to fatigue tests, the same beams were subjected to frequency sweep test, which enabled estimation of the stiffness modulus according to PN-EN 12697-26.

## 3. Production at Asphalt Plant and Specimen Preparations

The framework of the project included production of the trial batch mixtures with reclaimed asphalt at asphalt plant. The mixtures were produced in the asphalt plant of the company Budimex S.A. (Warsaw, Poland) (Figure 1). The plant is one of the four plants in Poland equipped with a black mixing drum for dispensing hot reclaimed asphalt [23]. In hot recycling technology, RAP is heated in a separate drum to a temperature of over 120–140 °C. The granulate is crushed, and then in a loose form goes to the mixer. There is no need to overheat the aggregate excessively or extend the mixing time to crush and heat the asphalt granulate, as is the case with cold dosing technology. On the basis of trial production and basic tests, mixtures for further functional tests including fatigue and stiffness were selected.

**Figure 1.** Asphalt plant with a reclaimed asphalt pavement (RAP) drum on the top and view of the RAP pile.

The produced mixes were packed in portions of about 25 kg in paper bags and delivered to the IBDiM laboratory (Warsaw, Poland). The portion of the mixture was then heated again in the laboratory oven to the compaction temperature, then it was put to the laboratory mixer. Subsequently, plates with dimensions of 50 × 18 × 10 cm were compacted using a standardized steel roller method. Thereafter, the degree of compaction was controlled, which should be in the range 98–100%. Then, beams for fatigue and stiffness tests were cut from each plate (Figure 2).

**Figure 2.** Beam for four-point bending beam test (4PB) fatigue and stiffness test.

## 4. Tests Results

### 4.1. Basic Properties of Asphalt Binders and Bituminous Mixtures

In addition to the mixtures analyzed in this publication (i.e., AC22P and ACWMS16), mixtures for the wearing course and the binder course were also produced and tested. Properties of these mixtures are necessary for the analysis of the pavement structure. In order to determine the stiffness of a bituminous mixture using the empirical method, it is necessary to know the basic properties of the asphalt binder (penetration, softening point) and the selected physical properties of the bituminous mixture (binder and aggregate content by volume). To determine these properties of asphalt binders, laboratory tests were carried out on binders recovered from mixtures containing reclaimed asphalt. Table 1 presents the mixtures, their basic properties, and the properties of the recovered binders. All mixes were designed according to the technical requirements [24] based on PN-EN 13108-1 (asphalt concrete) and PN-EN 13108-5 (SMA). There are two mixtures for base course: AC22P and AC WMS 16. The first one is conventional asphalt concrete, while the second is high modulus asphalt concrete (HMAC). The main differences are due to the composition of these mixtures and volume properties. ACWMS has a lower air voids content, finer grading, and higher binder content. The mixture obtained in this way is characterized by high resistance to fatigue and at the same time good resistance to rutting [25–27]. In order to increase stiffness, harder binders are used (e.g., 20/30 or 10/40-65). HMAC is often used in Poland for heavily trafficked roads. Mixture AC16W is typical asphalt concrete for binder course, while AC11S and SMA11 are designed for wearing course. Grading curves of all mixtures are presented in Figure 3.

### 4.2. Results of Fatigue Tests

Table 2 presents the fatigue characteristics parameters of the mixtures of asphalt concrete AC22P 35/50 and high modulus asphalt concrete ACWMS16 25/55-60 with reclaimed asphalt. Additionally, the same parameters of that type of mixture, but without the reclaimed asphalt content, were included for comparison. The results come from the research works carried out in the Road and Bridge Research Institute in recent years.

**Table 1.** Properties of recovered binders and bituminous mixtures with reclaimed asphalt.

| Mixture | AC22P 35/50 50%RAP | AC16W 25/55-60 50%RAP | ACWMS16 25/55-60 50%RAP | AC11S 45/80-55 30%RAP | SMA11 45/80-55 15%RAP |
|---|---|---|---|---|---|
| Added (fresh) asphalt binder | 35/50 | 25/55-60 | 25/55-60 | 45/80-55 | 45/80-55 |
| Contents of reclaimed asphalt, % m/m | 50 | 50 | 50 | 30 | 15 |
| Properties of recovered asphalt | | | | | |
| Penetration at 25 °C, 0.1 mm | 28 | 28 | 32 | 49 | 52 |
| Softening point according to ring and ball method, °C | 62.3 | 66.5 | 64.5 | 60.5 | 58.9 |
| Properties of bituminous mixtures | | | | | |
| Density, kg/m$^3$ | 2431 | 2417 | 2430 | 2419 | 2379 |
| Maximum density, kg/m$^3$ | 2540 | 2518 | 2500 | 2480 | 2429 |
| Contents of soluble asphalt, % *m/m* (by weight) | 3.7 | 4.4 | 5.0 | 5.2 | 6.3 |
| Contents of aggregate mixture, % *m/m* (by weight) | 96.3 | 95.6 | 95.0 | 94.8 | 93.7 |
| Air voids content, % *v/v* (by volume) | 4.3 | 4.0 | 2.8 | 2.5 | 2.1 |
| Contents of soluble asphalt, % *v/v* (by volume) | 8.7 | 10.3 | 11.8 | 12.2 | 14.6 |
| Contents of aggregate, % *v/v* (by volume) | 87.0 | 85.7 | 85.4 | 85.3 | 83.4 |

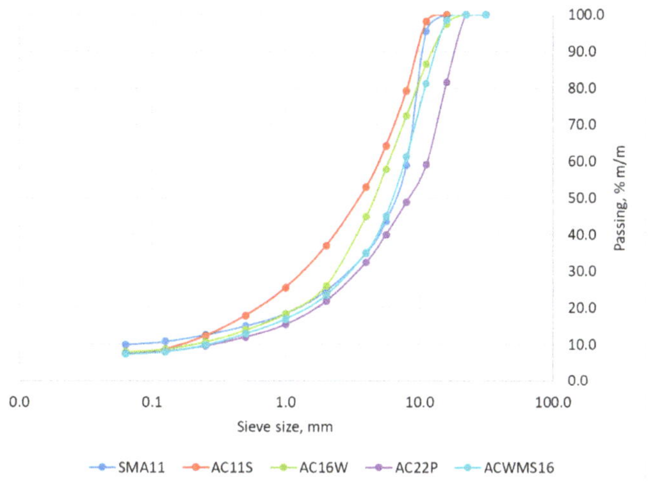


**Figure 3.** Grading curves.

**Table 2.** Results of the mixture fatigue test.

| Mixture | Fatigue Curve Parameter | | | Fatigue Resistance |
|---|---|---|---|---|
| | $A$ | $b$ | $R^2$ | $\varepsilon_6$, µm/m |
| AC22P 35/50 50%RAP | $1.7 \times 10^{19}$ | −6.2 | 0.91 | 134 |
| AC22P 35/50 ref. 1 * | $4.7 \times 10^{20}$ | −7.1 | 0.88 | 117 |
| AC22P 35/50 ref. 2 ** | $3.2 \times 10^{18}$ | −6.1 | 0.91 | 116 |
| ACWMS16 25/55-60 50%RAP | $5.3 \times 10^{18}$ | −5.7 | 0.86 | 176 |
| ACWMS16 25/55-60 ref. 1 *** | $5.9 \times 10^{22}$ | −7.4 | 0.95 | 185 |
| ACWMS16 25/55-60 ref. 2 * | $7.9 \times 10^{20}$ | −6.4 | 0.94 | 214 |

* The results of tests under the research project RID I/25, ** the results of tests under the research work described in [25], *** the results of tests under the research work described in [26].

Evaluation of the durability of AC22P and ACWMS16 mixtures with reclaimed asphalt must be carried out separately, because these are mixtures of different types and with different binders (unmodified asphalt and polymer-modified asphalt). Therefore, it is not surprising that those mixtures significantly differ in terms of fatigue resistance with an indication of the better properties of ACWMS16. It should be noted that the addition of reclaimed asphalt in a relatively large amount of 50% did not affect the scatter of results. The regression correlation coefficient $R^2$ at the level of 0.9 is the result proving small scatter of results, and does not deviate from the values obtained for mixtures without reclaimed asphalt.

The AC22P 35/50 asphalt concrete is one of the basic mixtures for base courses according to the Technical Recommendations in Poland [24], and is very often used for different traffic categories. It is designed using the empirical method and the requirements in terms of fatigue resistance were not specified. Therefore, it is not possible to determine the suitability of the AC22P 35/50 mixture with reclaimed asphalt with respect to the applicable technical requirements. However, the results obtained for this mixture may be referred to the results from other research. Hence, Table 2 and Figures 4 and 5 provide the results for two AC22P 35/50 reference mixtures. The comparison of the $\varepsilon_6$ parameter indicates higher results of the fatigue life of the mixture with reclaimed asphalt compared to the reference mixtures. This effect is also visible in diagrams of fatigue characteristics. Characteristics of AC22P mixtures have similar incline; however, the line of the mixture with RAP is shifted in the direction of longer fatigue lives. The obtained results can also be referred to slightly older results of tests of asphalt concrete BA 0/25 D50 [28], where the results of $\varepsilon_6$ from ca. 100 to 130 μm/m were obtained for the six variants of the mixture with various asphalt contents and asphalt binder origins (suppliers). On this basis, it was possible to state that the results for the AC22P 35/50 mixture and the results described above are the typical and expected results. Taking this into account, the results for the mixture with reclaimed asphalt look positive and do not allow conclusions to be drawn about the deterioration of the fatigue life.

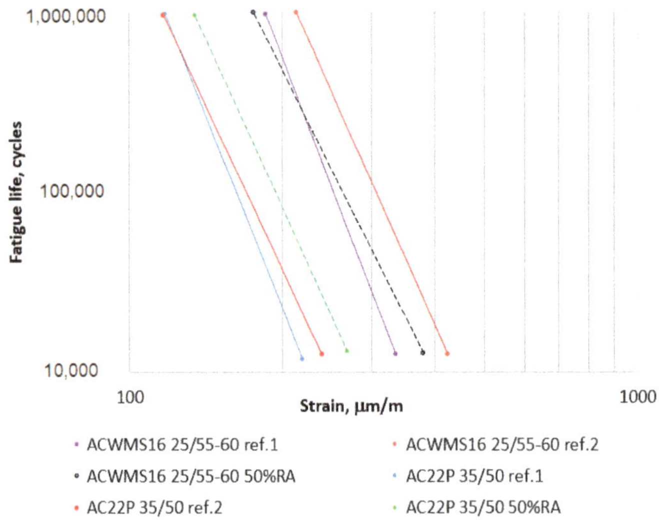


**Figure 4.** Graphical representation of the fatigue characteristics of mixtures with reclaimed asphalt and reference mixtures.

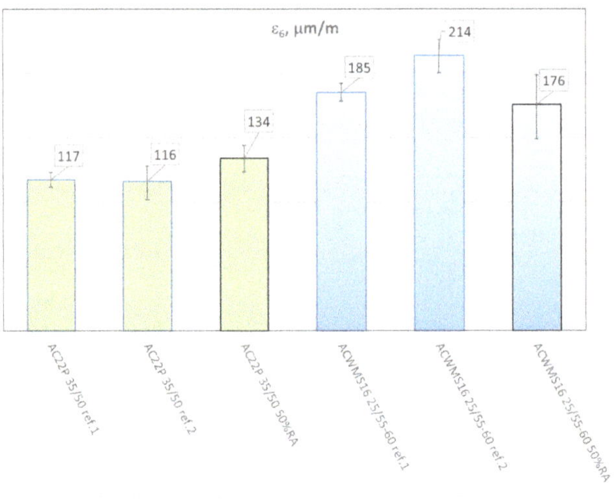


**Figure 5.** Comparison of the $\varepsilon_6$ parameter of the individual mixtures and the 95% confidence intervals.

The high modulus asphalt concrete is the only mixture in Poland for which the requirements of fatigue life are specified in the recommendations [24]. The value of the $\varepsilon_6$ parameter of the ACWMS16 25/55-60 mixture with reclaimed asphalt is much higher than the required 130 μm/m, which allows for positive evaluation of its properties in terms of fatigue life. The other two ACWMS16 25/55-60 reference mixtures without the RAP exhibit better fatigue properties in terms of the $\varepsilon_6$ parameter and the fatigue characteristics. This difference is due to the use of "fresh" polymer-modified asphalt and RAP containing unmodified asphalt binder. As we know, the use of modified binders and the extent of modification have a beneficial effect on the fatigue resistance of the mixtures. The mixture with reclaimed asphalt combines the unmodified asphalt binder and polymer-modified binder. Therefore, a lesser extent of modifications of the polymer contents was obtained than in the case of reference mixtures without RAP.

### 4.3. Determination of the Stiffness of Bituminous Mixtures

Knowledge of the stiffness modulus of bituminous mixtures is necessary for calculations of the design life of the structure using the mechanistic method. Stiffness was determined using the empirical method and the BANDS software (Shell, The Hague, The Netherlands, 1998) by using the previously specified parameters of recovered asphalts and bituminous mixtures [29]. The values were determined at a load time of 0.02 s and temperatures 10 and 13 °C. The value of stiffness at 13 °C is required in the design using the 2004 AASHTO method in accordance with the rules specified in the Polish Catalogue [30], while the result at 10 °C is necessary for the design using the French method [31]. In addition, the values of stiffness were determined at the frequency of 10 Hz and the temperature of 10 °C, which were compared with the results obtained in the laboratory tests using the four-point bending beam method. Figure 6 indicates a good correlation between the calculated values and values obtained in the laboratory, with the latter being approximately 5–10% higher (Table 3).

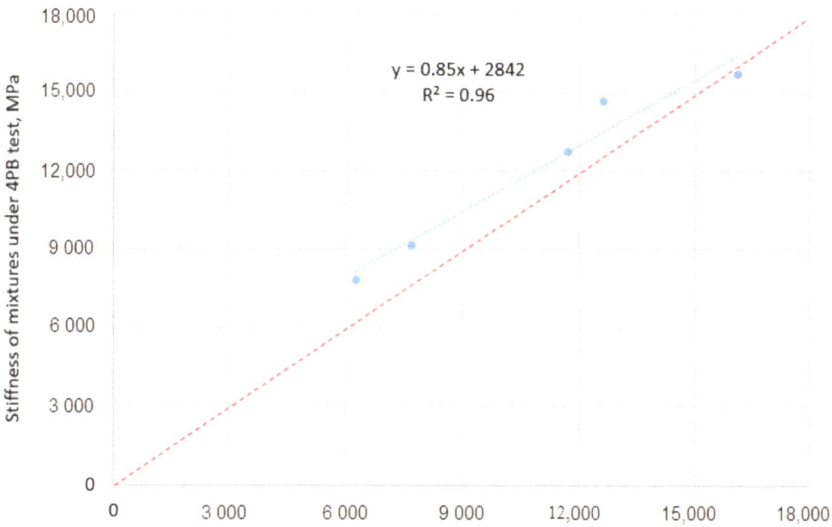


**Figure 6.** Comparison of the stiffness of trial batch mixtures obtained from laboratory tests and those calculated using the empirical method.

**Table 3.** Stiffness modulus.

| Mixture | AC22P 35/50 50%RAP | AC16W 25/55-60 50%RAP | ACWMS16 25/55-60 50%RAP | AC11S 45/80-55 30%RAP | SMA11 45/80-55 15%RAP |
|---|---|---|---|---|---|
| Stiffness (0.02 s, 13 °C) under empirical estimation, MPa | 13,900 | 10,600 | 9230 | 5900 | 4700 |
| Stiffness (0.02 s, 10 °C) under empirical estimation, MPa | 15,600 | 12,300 | 11,300 | 7260 | 5880 |
| Stiffness (10 Hz, 10 °C) under empirical estimation, MPa | 16,100 | 12,600 | 11,700 | 7700 | 6260 |
| Stiffness (10 Hz, 10 °C) under laboratory tests 4PB, MPa | 15,716 | 14,690 | 12,722 | 9189 | 7853 |

## 5. Analyses of Pavement Design Life

### 5.1. Purpose of the Analyses

The purpose of the calculations is to determine the usability of the bituminous mixtures with reclaimed asphalt for typical pavement structures in terms of calculated fatigue life. Fatigue durability is the basic parameter determining the quality of pavement structures. According to the general principle of using bituminous mixtures with RAP, they cannot reduce the durability of the structure. The analysis uses the mechanistic method, which is used in the individual design of road pavement structures in Poland. This method uses selected parameters of asphalt mixtures and recognized fatigue criteria. Therefore, the results of laboratory tests are used in the analysis. The analyses were carried out on typical flexible pavement structures according to the Catalogue [30]. Results of the analyses were compared to the required number of equivalent axle loads for given traffic load category.

### 5.2. Pavement Structures

The thicknesses of the pavement courses were left unchanged in relation to those specified in the Catalogue. Figure 7 presents upper parts of pavement structures consisting of asphalt layers and the range of equivalent axle loads for each traffic load category. The SMA11 mixture was used for the traffic load category KR3-7 wearing course, while asphalt concrete AC11S was used for KR1-2. The KR1-7 binder course and the base course adopt AC16W and AC22P mixtures, and KR5-7 structures additionally introduce variants using the ACWMS16 mixture. The lower structural layers were adopted uniformly and consisted of an unbound aggregate subbase with a thickness of 20 cm and 15 cm cement treated base course.

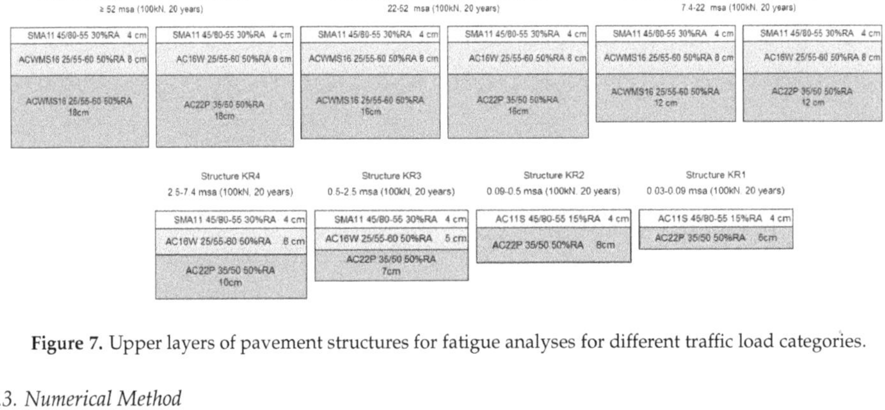


**Figure 7.** Upper layers of pavement structures for fatigue analyses for different traffic load categories.

### 5.3. Numerical Method

The mechanistic pavement design method consists of determining the fatigue life of the structure on the basis of a stress–strain state analysis. For this purpose, the surface is treated as a system of courses with specific thicknesses on the subgrade with infinite thickness. The most common model is the model of elastic layers located in the elastic half-space. The individual courses are characterized by thickness (h), the modulus of elasticity (stiffness) (E), and Poisson's ratio (ν). The asphalt layers' stiffness modulus were adopted in accordance with the results of tests and calculations stated above. The remaining parameters were selected in accordance with the instructions specified in the Catalogue [30]:

- subgrade modulus of elasticity G1 equal to 80 MPa, Poisson's ratio 0.35,
- modulus of elasticity of the cement treated sub-base 200 MPa, Poisson's ratio 0.3,
- modulus of elasticity of the unbound subbase 400 MPa, Poisson's ratio 0.3,
- Poisson's ratio of asphalt courses 0.3.

In order to make the design life calculations of the analyzed flexible pavement structure, it is necessary to determine horizontal tensile strains in the bottom of the asphalt layers and the compressive vertical strains in the upper surface of the subgrade.

Calculations of the stress–strain states in the structure were performed using the NOAH 2.0 software (Nynas NV, Stockholm, Sweden, 1996). The following assumptions were made in accordance with the provisions adopted during the development of the Catalogue [31]:

- equivalent axle load 100 kN (wheel load 50.0 kN),
- contact pressure q = 850 kPa,
- equivalent temperature 10 (French method) or 13 °C (AASHTO method).

Two criteria of pavement failure were considered; i.e., the criterion of fatigue of the asphalt layers and the criterion of the subgrade deformation.

The calculations were carried out using the criteria recommended for use in Poland [31]; i.e., the 2004 AASHTO method in relation to the fatigue life of asphalt layers and the method of the Asphalt Institute with respect to the subgrade.

The fatigue life of asphalt layers according to the 2004 AASHTO criterion is described with the following equations [32]:

$$N_{asf} = 7.3557 \cdot (10^{-6}) \cdot C \cdot k_1' \cdot (\varepsilon_t^{-3.9492}) \cdot E^{-1.281}, \tag{2}$$

where:

$$C = 10^M, \tag{3}$$

$$M = 4.84 \cdot \left( \frac{V_b}{V_a + V_b} - 0.69 \right), \tag{4}$$

$N_{asf}$—fatigue life (number of equivalent standard axles), $\varepsilon_t$—tensile strain (µm/m), $E$—stiffness (MPa), $V_b$—asphalt content by volume (% $v/v$), $V_a$—air voids (% $v/v$), $h_{ac}$—total thickness of bituminous layers (cm).

$k_1'$ is a parameter according to the formula:

$$k_1' = \frac{1}{0.000398 + \frac{0.003602}{1+e^{(11.02-1.374 \cdot h_{ac})}}}. \tag{5}$$

The criterion of subgrade strain according to the equation of the Asphalt Institute is:

$$\varepsilon_P = 0.0105 \cdot Ngr^{-0.223}, \tag{6}$$

where: $N_{gr}$—life (number of equivalent standard axles), $\varepsilon_p$—subgrade strain (µm/m).

The structure durability $N_{min}$ is the lower of the $N_{asf}$ and $N_{gr}$ values.

In addition, the fatigue calculations of the asphalt course were carried out using the so-called French method [33]. It allows for using the parameters of fatigue life of the bituminous mixture determined in the laboratory tests, unlike the 2004 AASHTO method. The equation includes the inclination of the Wöhler curve and the $\varepsilon_6$ value (value of strain under the fatigue test, during which durability is $10^6$ cycles). In France, a two-point bending trapezoidal beam method is used, which is different from the four-point bending beam method used in Poland. However, based on the experience of the Road and Bridge Research Institute from comparative fatigue tests within the framework of the RILEM committee [34,35] (which demonstrated compliance of the results obtained using those methods), it was established that the results of 4PB tests can be used in the French method.

In the French method, the relation between the elastic strain in the bottom of asphalt layers and the life of the structure is as follows:

$$\varepsilon_t = \varepsilon_6 \cdot \left(\frac{N_{asf}}{10^6}\right)^b \cdot \left(\frac{E(10\ ^\circ\mathrm{C})}{E(\theta)}\right)^{0.5} \cdot k_c \cdot k_r \cdot k_s, \tag{7}$$

where: $\varepsilon_6$—strain under the fatigue test, during which durability is $10^6$ cycles; $E$ (10 °C)—stiffness at 10 °C (MPa); $E(\theta)$—stiffness at equivalent temperature (MPa); $b$—inclination of fatigue line, see Equation (1); $k_c$—coefficient dependent on mixture type; $k_s$—coefficient dependent on the pavement subgrade modulus; $k_r$—risk level coefficient.

Parameters were taken from laboratory tests or were chosen according to French recommendations. Parameter $k_s$ was taken as 0.9 (KR1-KR4) or 1.0 (KR5-KR7). Parameter $k_c$ is 1.0 for ACWMS base course or 1.1 for typical asphalt concrete base course. Risk level coefficient was calculated with probability level 98% for KR5-KR7 structures, and 95% for other pavements.

### 5.4. Calculations of Structure Fatigue Life

Tables 4 and 5 summarize the results of calculations of strains in the asphalt base course and in the subgrade according to the individual structural variants and the results of structure durability calculations according to the 2004 AASHTO criteria (Equation (2)) and the Asphalt Institute criteria (Equation (6)). Regardless of the method used for determining the stiffness or the fatigue criterion, the results of the structure durability calculations were higher than the upper limit of the range of the design number of equivalent standard axles for the specific traffic category (red dashed lines in Figures 8 and 9). In general (with some exceptions), the results obtained according to the 2004 AASHTO criterion were higher than the results according to the French method. In the case of ACWMS structures, AASHTO results were two to four times higher than those obtained using the French method. For the corresponding constructions with typical asphalt concrete, the relation is reversed. From the designer's point of view, on this basis, it can be indicated as a safer application of the French method to the structures of ACWMS, and AASHTO methods for structures with classic asphalt concrete. The smallest differences between these methods were in the case of medium thickness structures (KR4-KR5) with AC.

**Table 4.** The results of calculations of design life of the structures (stiffness of the asphalt layers according to calculations).

| Configuration | KR7 ACWMS | KR7 | KR6 ACWMS | KR6 | KR5 ACWMS | KR5 | KR4 | KR3 | KR2 | KR1 |
|---|---|---|---|---|---|---|---|---|---|---|
| $\varepsilon_{asf}$, µm/m | 55.3 | 36.5 | 61.5 | 49.5 | 78.2 | 64.0 | 73.0 | 84.7 | 139.0 | 175.0 |
| $\varepsilon_{gr}$, µm/m | 149 | 133 | 164 | 149 | 206 | 192 | 218 | 251 | 411 | 538 |
| $N_{gr}$, MSA* (100 kN) | 193.4 | 321.9 | 125.8 | 193.4 | 45.3 | 62.1 | 35.1 | 18.7 | 2.0 | 0.6 |
| 2004 AASHTO method | | | | | | | | | | |
| $N_{asf}$, MSA (100 kN) | 153.6 | 111.0 | 101.8 | 33.6 | 40.0 | 12.3 | 7.3 | 4.4 | 0.7 | 0.4 |
| $N_{min}$, MSA (100 kN) | 153.6 | 111.0 | 101.8 | 33.6 | 40.0 | 12.3 | 7.3 | 4.4 | 0.7 | 0.4 |
| The French method | | | | | | | | | | |
| $N_{asf}$, MSA (100 kN) | 62.6 | 465.1 | 35.3 | 70.3 | 9.6 | 14.3 | 5.6 | 2.2 | 0.1 | 0.02 |
| $N_{min}$, MSA (100 kN) | 62.6 | 465.1 | 35.3 | 70.3 | 9.6 | 14.3 | 5.6 | 2.2 | 0.1 | 0.02 |

* MSA–millions of standard axle loads.

**Table 5.** The results of calculations of design life of the structures (stiffness of the asphalt layers according to laboratory tests).

| Configuration | KR7 ACWMS | KR7 | KR6 ACWMS | KR6 | KR5 ACWMS | KR5 | KR4 | KR3 | KR2 | KR1 |
|---|---|---|---|---|---|---|---|---|---|---|
| $\varepsilon_{asf}$, μm/m | 43.8 | 38.3 | 49.1 | 43.1 | 62.8 | 55.6 | 63.6 | 73.9 | 125.6 | 162 |
| $\varepsilon_{gr}$, μm/m | 133 | 117 | 141 | 131 | 179 | 168 | 192 | 223 | 404 | 502 |
| $N_{gr}$, MSA (100 kN) | 321.9 | 567.6 | 247.7 | 344.6 | 85.0 | 112.3 | 62.1 | 31.7 | 2.2 | 0.8 |
| 2004 AASHTO method | | | | | | | | | | |
| $N_{asf}$, MSA (100 kN) | 255.8 | 78.4 | 164.3 | 49.6 | 63.0 | 18.4 | 10.8 | 6.4 | 0.9 | 0.5 |
| $N_{min}$, MSA (100 kN) | 255.8 | 78.4 | 164.3 | 49.6 | 63.0 | 18.4 | 10.8 | 6.4 | 0.9 | 0.5 |
| The French method | | | | | | | | | | |
| $N_{asf}$, MSA (100 kN) | 127.6 | 241.3 | 68.9 | 116.1 | 18.2 | 23.9 | 9.2 | 3.6 | 0.1 | 0.03 |
| $N_{min}$, MSA (100 kN) | 127.6 | 241.3 | 68.9 | 116.1 | 18.2 | 23.9 | 9.2 | 3.6 | 0.1 | 0.03 |

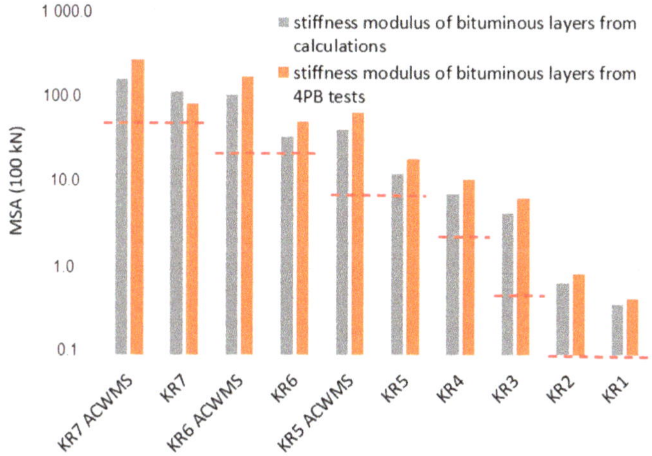


**Figure 8.** Results of calculations of pavement structure durability according to the 2004 AASHTO criterion.

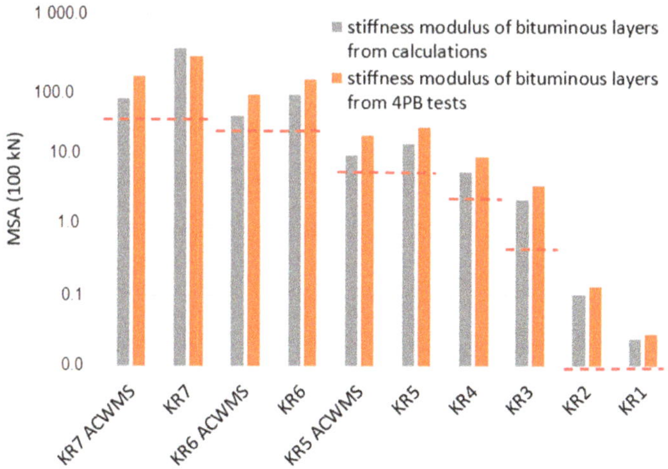


**Figure 9.** Results of calculations of pavement structure according to the French method for asphalt courses.

## 6. Summary

Hot recycling of asphalt pavement is currently a very trendy topic in Poland. This article presents the results of the research projects carried out in this field in Poland. As part of the work, tests and analyses were carried out to verify the suitability of bituminous mixtures with an increased content of RAP with regard to fatigue life. Fatigue life was evaluated in terms of mixture durability as well as pavement structure life. The results of laboratory tests of the basic properties of binders and mixtures, as well as the results of advanced tests such as fatigue life and stiffness, are presented. The test results were analyzed and the design life of the pavement structure was calculated by known calculation methods.

The evaluation of the fatigue life of AC22P and ACWMS16 mixtures with 50% content of reclaimed asphalt indicated their positive properties in this regard. Laboratory tests indicate good fatigue properties of mixtures with RAP. In the case of asphalt concrete AC22P 35/50, fatigue properties were obtained in the range typical for this type of mixture and this binder. However, in the case of ACWMS16 25/55-60, the addition of RAP slightly deteriorated the fatigue properties. The reason for this was unmodified binder in RAP, which largely replaced polymer modified binder (PMB) in the final mixture. Ultimately, the properties of the mixture in terms of fatigue were closer to the results typical for asphalt concrete with unmodified binder. Nevertheless, the mixture fulfilled the requirements of fatigue life according to Polish technical requirements. Regardless of the method used for determining the stiffness or the fatigue criterion, the results of the calculations of the structure design life were higher than the upper limit of the range of the design number of equivalent standard axles for the specific traffic category. The use of bituminous mixtures with reclaimed asphalt developed as part of the project allows the appropriate fatigue life of the pavement structure to be obtained. The results of analyses and tests also indirectly indicate the advantages of an asphalt plant with a separate drum for RAP. With this technology, it is possible to reduce RAP overheating, to limit technological aging process and proper mixing of materials. As a result, a mixture with parameters comparable to a mixture without RAP is obtained.

In the further part of the project, test sections were made and subjected to testing and observations. This will be described in subsequent publications.

**Acknowledgments:** The tests were performed within the framework of the project entitled: "Reclaimed asphalt pavement: Innovative technology of bituminous mixtures using material from reclaimed asphalt pavement" (InnGA) implemented as part of the INNOTECH programme according to agreement No. 225765/NCBR/15 in consortium with the Warsaw University of Technology and the company Budimex S.A.

**Conflicts of Interest:** The authors declare no conflicts of interest.

## References

1. Al-Qadi, I.L.; Elseifi, M.A.; Carpenter, S.H. *Reclaimed Asphalt Pavement—A Literature Review*; Research Report FHWA-ICT-07-001; Illinois Center for Transportation, University of Illinois at Urbana-Champaign: Rantoul, IL, USA, 2007.
2. Zaumanis, M.; Mallick, R.; Frank, R. 100% Hot mix asphalt recycling: Challenges and benefits. *Transp. Res. Procedia* **2016**, *14*, 3493–3502. [CrossRef]
3. Elkashef, M.; Williams, R.C.; Cochran, E. Investigation of fatigue and thermal cracking behavior of rejuvenated reclaimed asphalt pavement binders and mixtures. *Int. J. Fatigue* **2017**, *11*, 90–95. [CrossRef]
4. Visintine, B.; Khosla, P.N.; Tayebali, A. Effects of higher percentage of recycled asphalt pavement on pavement performance. *Road Mater. Pavement Des.* **2013**, *14*, 432–437. [CrossRef]
5. Buczyński, P. The Frost Resistance of Recycled Cold Mixes with Foamed Bitumen and Different Types of Road Binders. *Procedia Eng.* **2016**, *161*, 54–59. [CrossRef]
6. Iwański, M.; Buczyński, P.; Mazurek, G. Optimization of the road binder used in the base layer in the road construction. *Constr. Build. Mater.* **2015**, *125*, 1044–1054. [CrossRef]
7. Dołżycki, B.; Jaczewski, M.; Szydłowski, C. The long-term properties of mineral-cement-emulsion mixtures. *Constr. Build. Mater.* **2017**, *156*, 799–808. [CrossRef]

8. Kukiełka, J. Recykling głęboki na zimno nawierzchni asfaltowych dróg samorządowych. *Materiały Budowlane* **2014**, *12*, 11–13. (In Polish)
9. Bańkowski, W.; Sybilski, D.; Król, J.; Kowalski, K.; Radziszewski, P.; Skorek, P. Wykorzystanie destruktu asfaltowego–konieczność i innowacja. *Budownictwo i Architektura* **2016**, *15*, 157–167. (In Polish)
10. Michalski, W. Methods of batching of RAP, Polish Association of Asphalt Pavement Contractors. *Asphalt Pavements* **2014**, *2*, 16–19. (In Polish)
11. Król, J.; Włodarski, P.; Jackowski, Ł. Właściwości mieszanek mineralno-asfaltowych ze zwiększoną ilością granulatu asfaltowego. *Drogownictwo* **2014**, *11*, 373–379. (In Polish)
12. Sybilski, D. On the need to use reclaimed asphalt pavement in Poland. *Road Eng.* **2011**, *1*.
13. Radziszewski, P. *Zmiany Właściwości Lepkosprężystych Lepiszczy Modyfikowanych i Mieszanek Mineralno-Asfaltowych w Wyniku Procesu Starzenia*; Wydawnictwo Politechniki Białostockiej: Poland, Białystok, 2007.
14. Woo, W.J.; Chowdhury, A.; Glover, C. Field aging of unmodified asphalt binder in three Texas long-term performance pavements. *Transp. Res. Rec. J. Transp. Res. Board* **2008**, *2051*, 15–22. [CrossRef]
15. Mogawer, W.; Bennert, T.; Daniel, J.; Bonaquist, R.; Austerman, A.; Booshehrian, A. Performance characteristics of plant produced high RAP mixtures. *Road Mater. Pavement Des.* **2014**, *13* (Suppl. 1), 183–208. [CrossRef]
16. Mogawer, W.; Booshehrian, A.; Vahidi, S.; Austerman, A. Evaluating the effect of rejuvenators on the degree of blending and performance of high RAP, RAS, and RAP/ RAS mixtures. *Road Mater. Pavement Des.* **2014**, *14* (Suppl. 2), 193–213. [CrossRef]
17. Kowalski, K.; Król, J.; Bańkowski, W.; Radziszewski, P.; Sarnowski, M. Thermal and Fatigue Evaluation of Asphalt Mixtures Containing RAP Treated with a Bio-Agent. *Appl. Sci.* **2017**, *7*, 216. [CrossRef]
18. Tran, N.; Taylor, A.; Willis, R. *Effect of Rejuvenator on Performance Properties of HMA Mixtures with High RAP and RAS Contents*; NCAT Report 12-05; National Center for Asphalt Technology: Auburn, AL, USA, 2012.
19. Mackiewicz, P. *Fatigue Life of Bituminous Mixtures Used in Road Pavements*; Publishing House of Wrocław University of Technology: Poland, Wrocław, 2016.
20. Walubita, L.; Epps Martin, A.; Jung, S.; Glover, C.; Park, E.; Chowdhury, A.; Lytton, R. *Comparison of Fatigue Analysis Approaches for Two Hot Mix Asphalt Concrete Mixtures (FHWA/TX-05/0-4468-2)*; Texas Transportation Institute: College Station, TX, USA, 2008.
21. Daniel, J.; Gibson, N.; Tarbox, S.; Copeland, A.; Andriescu, A. Effect of long-term ageing on RAP mixtures: Laboratory evaluation of plant-produced mixtures. *Road Mater. Pavement Des.* **2013**, *14* (Suppl. 2), 173–192. [CrossRef]
22. Tapsoba, N.; Benedetto, H.; Sauzeat, C.; Baaj, H.; Ech, M. Behaviour of asphalt mixtures containing reclaimed asphalt pavement and asphalt shingle. *Road Mater. Pavement Des.* **2014**, *15*, 330–347. [CrossRef]
23. Szyller, A.; Król, J.; Bańkowski, W. Współczesne doświadczenia ze stosowania recyklingu na gorąco w wytwórni mieszanek mineralno-asfaltowych w Polsce. *Nawierzchnie Asfaltowe* **2017**, *1*, 12–19. (In Polish)
24. WT-2 2014 Część 1—Mieszanki Mineralno-Asfaltowe—Wymagania Techniczne. General Directorate for National Roads and Motorways, 2014. Available online: https://www.gddkia.gov.pl/userfiles/articles/z/zarzadzenia-generalnego-dyrektor_13901/zalacznik%20do%20zarz%2047.pdf (accessed on 31 December 2017).
25. Bańkowski, W.; Gajewski, M.; Sybilski, D. Analysis of Fatigue Damage on Test Sections Submitted to HVS Loading. *Balt. J. Road Bridge Eng.* **2013**, *8*. [CrossRef]
26. Sybilski, D.; Bańkowski, W.; Horodecka, R.; Czajkowski, P. Laboratoryjna ocena przydatności polimeroasfaltu z dodatkiem gumy do zastosowania w mieszance AC WMS. *Road Eng.* **2014**, *9*, 300–306. (In Polish)
27. Sybilski, D.; Bańkowski, W. Beton o wysokim module sztywności w Polsce. *Drogownictwo* **2011**, *66*, 232–238. (In Polish)
28. Bańkowski, W. Comparison of the conventional and energetic fatigue failure criterion for bituminous mixes. In Proceedings of the International ISAP Symposium, Zurich, Switzerland, 18–20 August 2008.
29. Hunter, R.; Self, A.; Read, J. *The Shell Bitumen Handbook*, 6th ed.; Shell: The Hague, The Netherlands, 2015.
30. Judycki, J.; Jaskuła, P.; Pszczoła, M.; Jaczewski, M.; Ryś, D.; Alenowicz, J.; Dołżycki, B.; Stienss, M. *Katalog Typowych Konstrukcij Nawierzchni Podatnych i Półsztywnych*; General Directorate for National Roads and Motorways/Gdańsk University of Technology: Gdańsk/Warsaw, Poland, 2014.
31. Judycki, J.; Jaskula, P.; Pszczoła, M.; Stienss, M. *Analiza i Projektowanie Konstrukcji Nawierzchni Podatnych i Półsztywnych*; WKiŁ: Warsaw, Poland, 2014.

32. Adhikari, S.; Shen, S. Evaluation of fatigue models of hot-mix asphalt through laboratory testing. *J. Transp. Res. Board* **2009**, *2127*, 36–42. [CrossRef]
33. Corte, J.F. *French Design Manual for Pavement Structures*; SETRA/LCPC: Bagneux, France, 1997.
34. Di Benedetto, H.; de La Roche, C.; Baaj, H.; Pronk, A.; Lundström, R. Fatigue of bituminous mixtures: Different approaches and RILEM group contribution. In *PTEBM'03 6th Rilem Symposium*; Partl, M.N., Ed.; RILEM Publications: Zurich, Switzerland, 2003; pp. 15–38.
35. Bańkowski, W.; Sybilski, D.; Soenen, H.; Gajewski, M.; Chailleux, E. *Binder Testing: Chapter 2 in Advances in Laboratory Testing and Evaluation of Bituminous Materials*; State-of-the-Art Report of the RILEM Technical Committee 206-ATB; Springer: Berlin, Germany, 2012; Volume IX, pp. 15–83.

*applied sciences*

MDPI

*Article*

# Design of SBS-Modified Bitumen Stabilizer Powder Based on the Vulcanization Mechanism

**Wengang Zhang, Longting Ding and Zhirong Jia**

1 School of Civil and Architectural Engineering, Shandong University of Technology, Zibo 255049, China
2 Highway College of Chang' an University, Xi'an 710064, China; 6675342@163.com
* Correspondence: ziwuzizwg@sdut.edu.cn (W.Z.); jiazhr@126.com (Z.J.); Tel.: +86-15288931932 (W.Z.)

Received: 10 February 2018; Accepted: 15 March 2018; Published: 16 March 2018

**Abstract:** In order to improve the problem of the bad thermal storage stability of Styrene-Butadiene-Styrene Block Copolymer Modified Bitumen (SBS-modified bitumen), the vulcanization reaction was selected as the basic mechanism for the research and development of the stabilizer. Sulfur, tetramethylthiuram disulfide (TMTD), zinc oxide, butylated hydroxytoluene (BHT), kaolin ($Al_2O_3{\cdot}2SiO_2{\cdot}2H_2O$) and carbon-white ($SiO_2$) were used as raw materials, and 32 kinds of components with different contents of raw materials were designed. The 48-h segregation softening point and other modified bitumen technical indicators, such as ductility, penetration, penetration index, viscosity, and so on, were tested and analyzed. The fluorescence microscope test was also conducted to explore the crosslinking situation based on the vulcanization reaction. The results show that the component with good performance of the SBS-modified bitumen stabilizer powder was S:TMTD:ZnO:BHT: $Al_2O_3{\cdot}2SiO_2{\cdot}2H_2O$ = 60:16:4:20:1250 or S:TMTD:ZnO:BHT:$SiO_2$ = 60:16:4:20:625. The optimum content of stabilizer powder was 0.9% (S/modified bitumen). The stabilizer was dry powder and can be directly put into bitumen. It can cross-link SBS and especially has an effect on preventing SBS segregation. The technical performance of SBS-modified bitumen with stabilizer powder, such as hot temperature performance, low temperature performance, anti-aging performance and thermal storage stability, all meet the relevant specifications.

**Keywords:** SBS-modified bitumen; stabilizer; thermal storage stability; vulcanization reaction; 48-h segregation softening point difference

## 1. Introduction

The traditional base bitumen suffers from several common defects such as temperature sensitivity, low temperature brittleness, easy flow at high temperature, poor anti-aging performance, low equivalent softening point and other shortcomings. These seriously affect the durability and service life of roads. In order to overcome the shortcomings of the bitumen material, Styrene-Butadiene-Styrene Block Copolymer (SBS), Styrene-Butadiene-Rubber (SBR), Polyethylene (PE), Ethyl-Vinyl-Acetate (EVA) and other polymer modifiers were used to modify the base bitumen. These also have also become the main ways to improve the performance of bitumen [1–5]. Practical application showed that SBS was obviously superior to other polymer modifiers in improving the low temperature toughness and the temperature sensitivity of the bitumen. As a commonly-used modifier, the SBS has outstanding performance. It can increase the softening point and low temperature ductility significantly. Besides, it can also greatly improve the temperature sensitivity. After modification, the elastic recovery rate of bitumen was particularly large. Compared to linear SBS, radial SBS has better performance in improving thermal storage stability and low temperature ductility, but the defects in thermal storage stability of SBS-modified bitumen remain. This creates serious problems for the construction technicians and the performance advantages of SBS-modified asphalt pavement [6,7].

Segregation of polymer-modified bitumen limits its application in engineering practice to a large extent. In order to fully utilize the advantages of polymer-modified bitumen in road construction and widen the application of SBS-modified bitumen, the problem of thermal storage stability must be well solved. However, at present, the thermal storage performance of SBS-modified bitumen in the market is still not satisfactory. Most contractors usually use an extra stirring device to overcome the segregation problem produced in the bitumen transportation and storage process [8,9].

In recent years, due to economic reasons, the financial support of research and development of the SBS-modified bitumen stabilizer has decreased. As a result, there have been few achievements in the research and development of stabilizers. Liu Qianju [10] from South China University of Technology pointed out that by adding the stabilizer, a stable interface adsorption layer was formed between the polymer phase and base bitumen phase. Thus, affinity between the two phases increases, which improves the thermal storage stability. Yang Yang [11] developed a new SBS-modified bitumen stabilizer by using active calcium carbonate, sulfur powder and *N*-octadecylamine polyethenoxy ether as the main raw materials. It can improve the thermal storage stability of SBS-modified bitumen by forming a network structure between bitumen and SBS. Sulfur, stearic acid, vulcanization accelerator tetramethylthiuram disulfide (TMTD) and light stabilizers were also used as the main raw materials to develop a new stabilizer that can promote the compatibility between SBS and base bitumen by Shuangrui et al. [12]. This can improve the thermal storage stability of SBS-modified bitumen. Sulfur, morpholine disulfide and other substances were used by Jinyu et al. [13] as components to prepare stabilizer. The results show that this cannot only improve the thermal storage stability of SBS-modified bitumen, but also improve the performance of the bitumen mixing material to some extent. Aguirre de Carcer et al. [14] focused on the stability of SBS-/sulfur-modified bitumen at high temperature. With the help of sulfur, the fluctuation of the softening point and viscosity became small. The rheological properties and storage stability of bitumen/SBS/montmorillonite composites were studied by Sadeghpour Galooyak et al. [15]. The presence of montmorillonite improves the storage stability of polymer-modified bitumen significantly without adverse effects on its other properties of. Sienkiewicz et al. [16] considered that the storage stability of the polymer-modified bitumen can be improved by using various types of modifiers that form the bonds between the components of these binders. Some experimental studies have been done by Amini et al. [17], and the results show that using nano-composites can improve the storage stability at high temperature. Goli et al. [18] studied the influence of carbon nanotubes on the storage stability of SBS-modified bitumen binders, and the results showed that adding carbon nanotubes to the SBS-modified binder can improve the stability. Storage stability of bitumen binder modified with recycled plastic was studied by Al-Abdul Wahhab et al. [19]. The results showed that for a recycled high-density polyethylene content below 4% and a recycled low-density polyethylene content below 6%, either SBS or Polybilt (PB) will bring good storage stability. However, the development of the aforementioned stabilizers was prohibitive due to the over complex production process, the strict preparation environment requirement and the high cost of raw materials.

The main objective of this paper was to propose a novel design of a stabilizer, which has a simple production process and lower economic cost. Based on the mechanism of the vulcanization reaction, the selection of the raw materials and corresponding content can be properly determined. The laboratory experiments were also conducted in this paper to analyze the performance of the proposed SBS-modified bitumen. "Technical Specification of Construction of Highway Asphalt Pavement (JTG F40-2004, China)" [20] was adopted in this paper as a reference.

## 2. The Mechanism of Stabilizer Components to Improve the Thermal Storage Stability of SBS-Modified Bitumen

### 2.1. Vulcanization Mechanism

Sulfur can decompose and generate sulfur free radicals at a certain temperature and in a certain time, and the sulfur free radicals have high oxidizability. Therefore, in a certain condition, hydrogen atoms in the bitumen polymer chain can be captured by the generated sulfur radicals during sulfur decomposition. Sulfur free radical and hydrogen atoms can generate hydrogen sulfide in the form of gas. Free radicals were generated at the hydrogen atom-losing site of the bitumen polymer chain. Free radicals in adjacent polymer chains can be coupled with each other in cross-link form, and free radicals can also be coupled with sulfur free radicals in bridging form, which can be coupled with $O_2$ in the form of thio-alcohol, sulfoxide, thioether or sulfone. Sulfur free radicals and the bitumen polymer chain can react with each other. Firstly, sulfur free radicals capture the benzyl carbon in bitumen and hydrogen atoms in the allyl carbon, then hydrogen atoms in the benzene ring [2,21,22]. It will be possible to further produce the following crosslinking reaction in Figure 1.

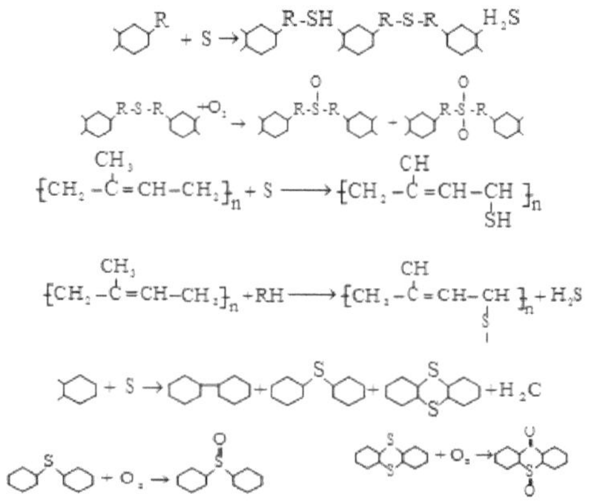

**Figure 1.** Vulcanization reaction mechanism [23].

The vulcanization reaction process includes an induction period, a hot vulcanization period, an optimum cure period and an over vulcanization period as shown in Figure 2. Vulcanization speed was different in different periods, and the torque of rubber mass will be reversed if the vulcanization time is too long.

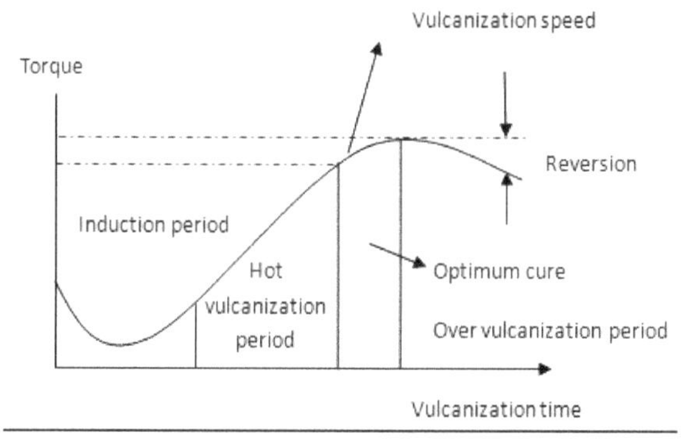


**Figure 2.** Vulcanization reaction process [24].

Rubber mass performance for different times of vulcanization reaction process was different as shown in Figure 3; tensile strength, tear strength, resilience, hardness, 300% stretching strength, elongation and permanent deformation of rubber mass changed with the vulcanization time increase. There was also an optimum vulcanization time by considering the performance above.

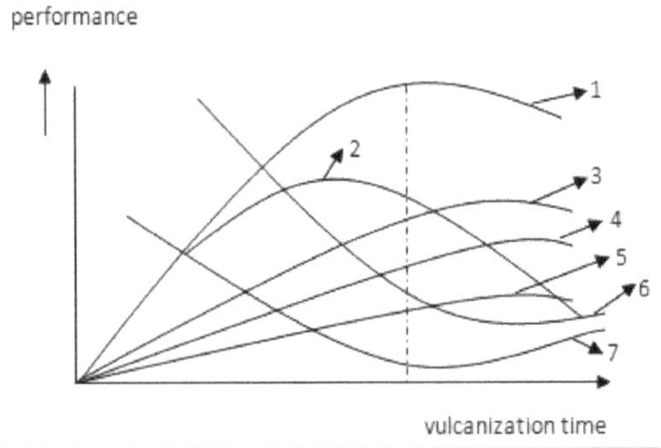


**Figure 3.** Rubber mass performance for different times of the vulcanization reaction process [25]. 1, Tensile strength; 2, tear strength; 3, resilience; 4, hardness; 5, 300% stretching strength; 6, elongation; 7, permanent deformation.

### *2.2. Other Components' Mechanism of Action*

Some other materials were beneficial to the vulcanization reaction or the improvement of the storage stability of polymer-modified bitumen, as follows.

(1) Vulcanization accelerators [26]: Vulcanization accelerators are additives that can accelerate vulcanization reactions, cut down the vulcanizing time, reduce the vulcanizing temperature, reduce the amount of vulcanizer and improve the physical or mechanical properties of the vulcanized rubber, such as tetramethylthiuram disulfide.
(2) Vulcanization activator: The vulcanization activator can increase the activity of accelerators, improve the vulcanizing speed and vulcanizing efficiency and improve the properties of vulcanized rubber [27–29], such as zinc oxide.
(3) Other active ingredients that can improve one or several modified bitumen technical performances, for example, antioxidant butylated hydroxytoluene (BHT): This can improve the oxidation resistance of bitumen. Kaolin ($Al_2O_3 \cdot 2SiO_2 \cdot 2H_2O$) can reduce the density difference between bitumen and SBS, which leads to modified bitumen to achieve stabilization [30]. In addition, there were also some reports on the same effect of carbon-white ($SiO_2$) on the thermal storage stability of bitumen [31,32].

## **3. Test Design of the Stabilizer Component**

### *3.1. Test Design*

The corresponding tests of the component of SBS-modified bitumen stabilizer were designed based on the basic principle of the vulcanization reaction and the particularity of bitumen: sulfur, vulcanization accelerator, vulcanization activator, antioxidant and other active ingredients.

For the convenience of testing, the experimental raw material samples were labeled with English letters in lower case, and the definition was as follows: a, sulfur; b, tetramethylthiuram disulfide; c, zinc oxide; d, butylated hydroxytoluene. SBS-modified bitumen was prepared by using different Component No. stabilizers (stabilizer content was 0.1% of bitumen quality), and the 48-h segregation softening point was tested and analyzed. Then, the median of each component range was calculated. The details of the experimental design are shown in Table 1.

**Table 1.** Stabilizer component test design.

| Component No. | A:B:C:D = a:b:c:d | Remarks | Component No. | A:B:C:D = a:b:c:d | Remarks |
|---|---|---|---|---|---|
| 1 | 30:16:4:20 | Determine a = the content of S | 13 | a:b:0:20 | Determine c = the content of ZnO |
| 2 | 40:16:4:20 | | 14 | a:b:3:20 | |
| 3 | 50:16:4:20 | | 15 | a:b:6:20 | |
| 4 | 60:16:4:20 | | 16 | a:b:9:20 | |
| 5 | 70:16:4:20 | | 17 | a:b:12:20 | |
| 6 | 80:16:4:20 | | 18 | a:b:15:20 | |
| 7 | a:0:4:20 | Determine b = the content of TMTD | 19 | a:b:c:0 | Determine d = the content of BHT |
| 8 | a:10:4:20 | | 20 | a:b:c:5 | |
| 9 | a:20:4:20 | | 21 | a:b:c:15 | |
| 10 | a:30:4:20 | | 22 | a:b:c:20 | |
| 11 | a:40:4:20 | | 23 | a:b:c:30 | |
| 12 | a:50:4:20 | | 24 | a:b:c:40 | |

TMTD, tetramethylthiuram disulfide; BHT, butylated hydroxytoluene.

According to the components in Table 1, firstly, Component Nos. 1–4 were tested, and the value of a (the content of S) was determined based on the test results and analysis. Secondly, Component Nos. 7–12 were tested, and the value of b (the content of TMTD) was determined based on the test results and analysis. Thirdly, Component Nos. 13–18 were tested, and the value of c (the content of ZnO) was determined based on the test results and analysis. Lastly, Component Nos. 19–20 were tested, and the value of d (the content of BHT) was determined based on the test results and analysis. a:b:c:d comprised the basic components of the stabilizer.

### 3.2. *Raw Materials*

Sulfur, tetramethylthiuram disulfide, zinc oxide, butylated hydroxytoluene, kaolin and carbon-white were used as raw materials, and the specific technical indicators are shown below.

(1) Base bitumen: The 70# base bitumen used in this paper was refined by Qilu Petrochemical (Zibo, China), and the technical indicators are shown in Table 2.

**Table 2.** The 70# bitumen technical indicators.

| Test | Unit | Measured Value | Test Items | Unit | Measured Value |
|---|---|---|---|---|---|
| 60 °C dynamic viscosity | Pa·s | 195.3 | Saturate | % | 16.3 |
| 15 °C ductility | cm | 65.8 | Aromatic | % | 39.5 |
| 10 °C ductility | cm | 45.1 | Asphaltene | % | 13.1 |
| Softening point | °C | 49.5 | Resin | % | 31.1 |
| 15 °C density | $g/cm^3$ | 1.001 | 25 °C penetration | 0.1 mm | 65.3 |
| RTFOT | | Penetration ratio | | % | 64.5 |
| | | Loss of quality | | % | 0.6 |
| | | 10 °C residue ductility | | cm | 8 |
| | | 15 °C residue ductility | | cm | 18 |

In Table 2, "Technical Specification of Construction of Highway Asphalt Pavement (JTG F40-2004, China)" [20] and "Standard Test Methods of Bitumen and Bituminous mixtures for Highway Engineering (JTG E20-2011)" [33] were adopted as the codes for the test.

(2) Other raw materials: Sulfur in this paper was produced from Tianjin Damao Chemical Reagent Factory (Tianjin, China) with an analytical reagent more than 99.5% purity. Tetramethylthiuram disulfide was produced by Shanghai QingXi Chemical Technology Co., Ltd. (Shanghai, China). Zinc oxide was produced by Tianjin FuChen Chemical Reagent Factory (Tianjin, China). Kaolin ($Al_2O_3{\cdot}2SiO_2{\cdot}2H_2O$) was produced by Guangzhou YiFeng Chemical Technology Co., Ltd. (Guangzhou, China). Carbon-white was produced by the Shandong Jiquan Biological Technology Co., Ltd. (Jinan, China). The technical indicators of the raw materials above are shown in Table 3.

**Table 3.** Technical indicators of S, TMTD, BHT, ZnO, carbon-white and kaolin.

| | S | TMTD | BHT | ZnO | Carbon-White | Kaolin |
|---|---|---|---|---|---|---|
| Molecular formula | S | $C_6H_{12}N_2S_4$ | $C_{15}H_{24}O$ | ZnO | $SiO_2$ | $Al_2O_3{\cdot}2SiO_2{\cdot}2H_2O$ |
| Density | 2.3600 g/cm$^3$ | 1.43 g/cm$^3$ | 1.048 g/cm$^3$ | 5.6060 g/cm$^3$ | 2.2–2.6 g/cm$^3$ | 2.5 g/cm$^3$ |
| Molecular weight | 32.07 g/mol | 204.14 g/mol | 220.36 g/mol | 81.39 g/mol | 60.08 g/mol | 87–88 g/mol |
| Characteristics | Can be dissolved in benzene, inflammable substance. | Can be dissolved in alcohol, benzene, with smell. | Can be dissolved in ethyl alcohol, benzene, ether. | White or pale yellow powder. | White powder. | White powder, can be slightly soluble in hydrochloric acid and acetic acid. |

### 3.3. SBS-Modified Bitumen Preparation Process

Due to the large amount of experiments, the test error caused by the different preparation processes of modified bitumen should be avoided during the experiment. Therefore, the same modification process was adopted for the modified bitumen in different samples. The preparation process of samples of the SBS-modified bitumen is shown as follows. Firstly, put some base bitumen with SBS modifier and compatibilizer together into a container. Secondly, shear the mixture at a shear rate of 4500 r/min for 30 min at a temperature of 175 ± 5 °C, and add the right amount of stabilizer during the shearing process. Lastly, keep the SBS-modified bitumen at a temperature of 175 ± 5 °C for 3 h and keep stirring constantly after shearing is completed.

## 4. Analysis of Test Results of the Basic Component of the Stabilizer

### 4.1. Basic Component Analysis of the Stabilizer

The basic components include S, TMTD, ZnO and BHT. The basic component analysis of the stabilizer is described as below.

#### 4.1.1. The S Content of Basic Components

The SBS samples with different contents of S were prepared based on the aforementioned procedure and used in the 48-h segregation softening point test (a measured quantity of polymer-modified bitumen in a sealed aluminum tube was conditioned in a vertical position for 48 h at a temperature of 163 ± 5 °C. At the end of the conditioning period, the top and bottom portions were separated and subjected to further testing to determine the degree of separation. The softening point was the test used for this purpose in this article.). Component Nos. 1, 2, 3, 4 did not show the gelation phenomenon; Component Number 5 showed an obvious gelation phenomenon after shearing for 20 min; component number 6 showed a serious gelation phenomenon after shearing for 10 min. The test results are shown in Figure 4 below.

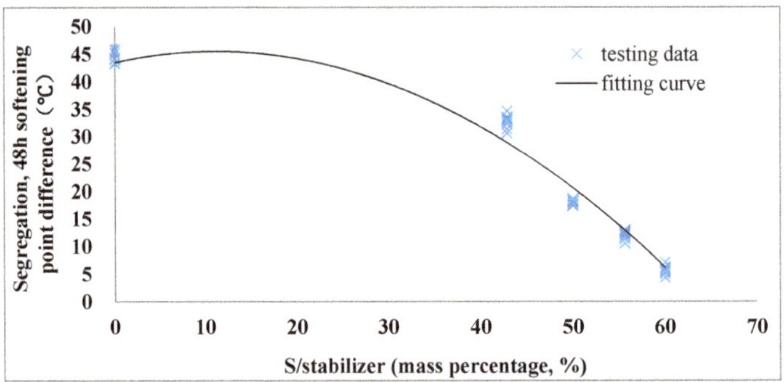


**Figure 4.** The 48-h segregation softening point with different S content.

As shown in Figure 4, the 48-h segregation softening points' difference of modified bitumen decreases gradually with the increasing content of S compared to lower content of S. With the further increase of the content of S, the gelation phenomenon occurs in the bitumen modification process. The higher the content of S is, the earlier the gelation phenomenon occurs and the more serious the gelation phenomenon. This phenomenon can be explained by the vulcanization reaction process shown in Figure 2. It can be called over vulcanization or a high degree of vulcanization. With respect to the comprehensive test phenomenon, test results and economic factors, the content of S in Component No. 4 was the optimum content, which determines a = 60.

#### 4.1.2. The TMTD Content of Basic Components

The SBS samples with different contents of TMTD were prepared based on the aforementioned procedure and used in the 48-h segregation softening point test. Component Nos. 7, 8, 9 and 10 did not show the gelation phenomenon; Component Number 11 showed the gelation phenomenon; and component number 12 showed an obvious gelation phenomenon. The test results are shown in Figure 5.

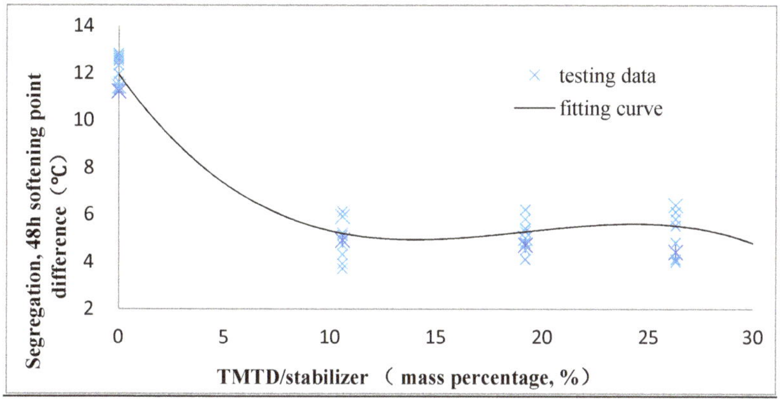


**Figure 5.** The 48-h segregation softening point with different TMTD content.

As can be seen from Figure 5, the 48-h segregation softening points' difference of the modified bitumen constantly decreased with the increasing content of TMTD. Due to the certain cross-linking effect that TMTD itself has, the gelation phenomenon appeared during the modification while the TMTD content continued increased: the higher the content was, the earlier the gelation phenomenon occurred and the more serious the degree of gelation was. The reason was that TMTD can be used as a vulcanizing agent, which can provide 13% self-quality as S, and the higher the TMTD content is, the deeper the vulcanization reaction degree is. In the content range of no cross-linking occurring, TMTD had a bit of improvement on the segregation softening point difference when the content was more than 15%. Considering the test phenomenon, test results and economic factors, it was determined that b = 16.

#### 4.1.3. The ZnO Content of Basic Components

The SBS samples with different contents of ZnO were prepared based on the aforementioned procedure and used in the 48-h segregation softening point test. There was no gelation phenomenon in Component Numbers 13, 14, 15, 16, 17 and 18, and the test results are shown in Figure 6.

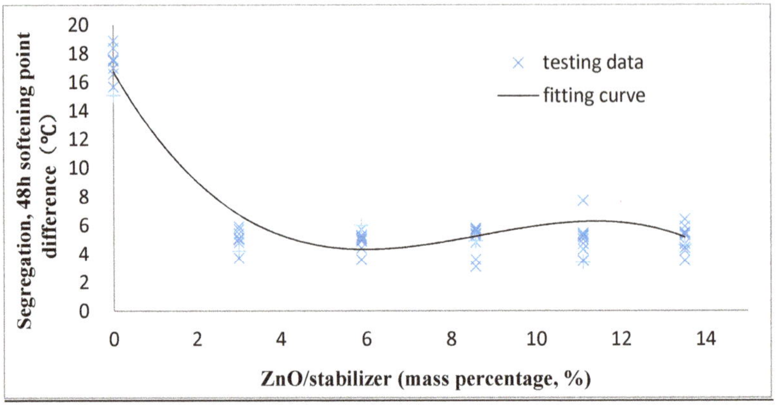


**Figure 6.** The 48-h segregation softening point difference with different ZnO content.

As can be seen from Figure 6, the 48-h segregation softening points' difference of modified bitumen decrease with the increasing content of ZnO. This can be ascribed to the activity of ZnO. However, with the continued increase of ZnO content, its activity was almost unchanged due to the effect of S and TMTD content. The reason was that with the incorporation of ZnO, the quantity of crosslinking agent in SBS-modified bitumen will increase, and TMTD will self-methyl decompose, which can promote cross-linking between macromolecules. The above phenomena can enhance the efficiency of the vulcanization reaction. Considering the test phenomenon, test results and economic factors, it was determined that c = 4.

#### 4.1.4. The BHT Content of Basic Components

According to the test design above, SBS was prepared. The content of BHT was different in each Component No. There was no gelation phenomenon observed in Component Nos. 19, 20, 21, 22, 23 and 24, and the 48-h segregation softening point test and Rolling Thin Film Oven Test (RTFOT) were carried out. The results are shown in Figure 7 and Table 4.

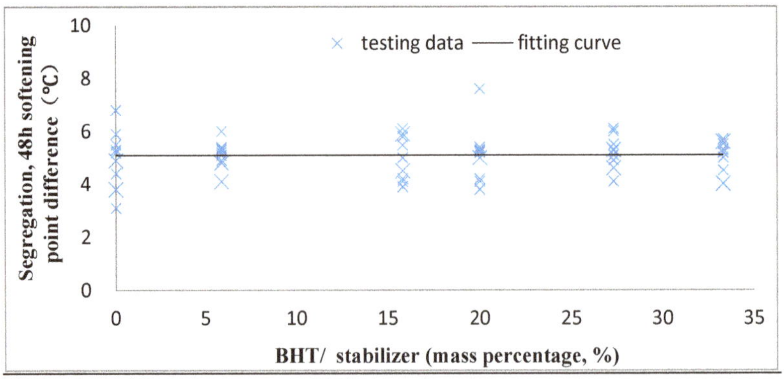


**Figure 7.** The 48-h segregation softening point difference with different BHT content.

As can be seen from Figure 7, with the increase of BHT content, the 48-h segregation softening point difference of modified bitumen was almost unchanged, which shows that BHT has little effect on the thermal storage stability of bitumen. However, BHT being an antioxidant, its ability of oxidation resistance may be useful to bitumen. Table 4 shows the anti-aging performance of SBS-modified bitumen with different BTH content.

**Table 4.** The anti-aging performance of SBS-modified bitumen with different BTH content.

| Component No. | RTFOT | | |
|---|---|---|---|
| | Loss on Heating (%) | Penetration Ratio (%) | Residue Ductility at 5 °C (cm) |
| 19 | 0.38 | 69.3 | 16.6 |
| 20 | 0.30 | 72.5 | 17.7 |
| 21 | 0.21 | 75.7 | 19.4 |
| 22 | 0.14 | 78.3 | 20.5 |
| 23 | 0.13 | 78.2 | 20.3 |
| 24 | 0.13 | 78.5 | 20.4 |

RTFOT, Rolling Thin Film Oven Test.

Table 4 shows that with the increase of BHT content, the loss on heating of SBS-modified bitumen after the rolling thin film oven test gradually decreased, the percent of residual penetration increased and the residual ductility at 5 °C slightly increased. However, the performance improvement effect was no longer obvious after the content of BHT was greater than 20%. BHT can provide H to the free radical of SBS-modified bitumen. The resistance to oxidation of SBS-modified bitumen will be increased in this way. This can also be explained by means of Chemical Reactions 1 and 2 [25] below.

$$AH + R\cdot \rightarrow A + RH \tag{1}$$

In this reaction AH is the antioxidant, R· the free radical, A· the antioxidant free radicals and RH the polymer molecules

It can also be explained by means of Chemical Reaction 2 below.

$$AH + ROO\cdot \rightarrow A\cdot + ROOH \tag{2}$$

In this reaction AH is the antioxidant, ROO· the free radical peroxidation, A· the antioxidant free radicals and ROOH the hydroperoxide.

Considering the test phenomenon, test results and economic factors, it was determined that d = 20.

From this, we can determine that the basic components of the stabilizer powder were S:TMTD:ZnO:BHT = 60:16:4:20.

### 4.2. *Analysis of Stabilizer Content in Bitumen*

In order to explore the reasonable content of stabilizer powder in SBS-modified bitumen, SBS-modified bitumen was prepared respectively with the stabilizer contents of 0%, 0.05%, 0.1%, 0.15%, 0.2%, 0.25%, 0.3% (percentage by mass of modified bitumen, and S:TMTD:ZnO:BHT = 60:16:4:20). Gelation did not occur. When the content of the stabilizers was less than 2%, slight gelation occurred at a content of 0.2%, significant gelation at a content of 0.25% and serious gelation at a content of 0.3%. The test results re shown in Table 5 and Figure 8.

In Table 5, "Technical Specification of Construction of Highway Asphalt Pavement (JTG F40-2004, China)" [20] and "Standard Test Methods of Bitumen and Bituminous mixtures for Highway Engineering (JTG E20-2011)" [33] were the referenced standards.

It can be seen from the test results shown in Table 5 that with the stabilizers' constant increase, the anti-aging properties of SBS-modified bitumen were continuously improved, the penetration of SBS modified bitumen decreased, the penetration index remained unchanged, the softening point continuously increased, Brookfield viscosity at 135 °C increased, the solubility slightly reduced (this may be related to the insolubles in the stabilizer) and elastic recovery no longer changed. When the content of stabilizer exceeds 0.2% (including 0.2%), the gelation phenomenon of bitumen occurs. With the content increasing, the gelation phenomenon becomes more and more serious.

**Table 5.** The performance of modified bitumen with different stabilizer content.

| Content, % | RTFOT | | | Penetration, 0.1 mm, 25 °C | Penetration Index PI |
|---|---|---|---|---|---|
| | Loss on Heating, % | Penetration Percentage, % | Residue Ductility at 5 °C | | |
| 0 | 1.1 | 62.6 | 11.3 | 64.6 | 0.2 |
| 0.05 | 0.41 | 74.3 | 16.1 | 62.3 | 0.2 |
| 0.1 | 0.15 | 78.3 | 20.5 | 58.5 | 0.1 |
| 0.15 | 0.08 | 79.2 | 21.0 | 55.3 | 0.1 |
| >0.2 | Gelation occurs | | | | |
| **Content, %** | **Ductility at 5 °C, cm** | **Softening Point °C** | **Brookfield viscosity at 135 °C, Pa·S** | **Solubility, %** | **Resilient at 25 °C, %** |
| 0 | 27.7 | 65.5 | 1.621 | 99.5 | 86.5 |
| 0.05 | 28.3 | 70.1 | 1.827 | 99.5 | 85.6 |
| 0.1 | 28.4 | 75.2 | 1.911 | 99.4 | 88.4 |
| 0.15 | 29.6 | 78.4 | 1.972 | 99.3 | 87.7 |
| >0.2 | Gelation occurs | | | | |

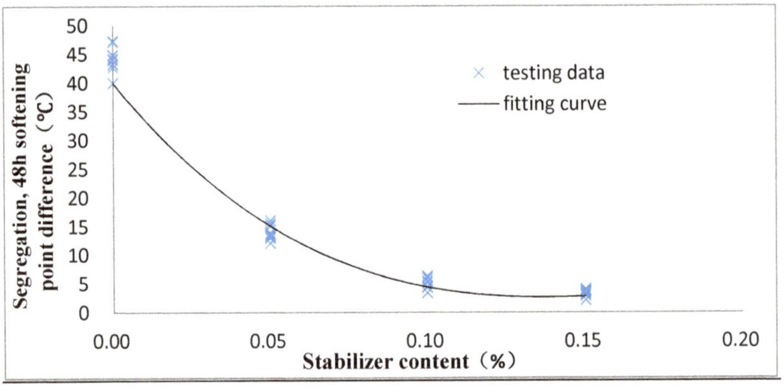


**Figure 8.** The relationship between stabilizer content and the 48-h segregation softening point difference of SBS-modified bitumen.

Figure 8 shows that the 48-h segregation softening points' difference of modified bitumen decreased with the increasing content of stabilizer content. However, even when the stabilizer content reached 0.15%, the 48-h segregation softening points' difference was no lower than 2 °C. From the analysis of the segregation softening point data and the performance of SBS-modified bitumen, the best content of the basic components of the stabilizer was 0.15%. If the content of S were taken as the reference, the optimum content of the stabilizer (expressed as the mass percentage of S content) was 0.09%.

## 5. Optimization Design of the Stabilizer Component

In this paper, kaolin ($Al_2O_3 \cdot 2SiO_2 \cdot 2H_2O$) and carbon-white ($SiO_2$) were selected as auxiliary components to improve the performance of the stabilizer, and the design component test is shown in Table 6 below.

**Table 6.** The composition with kaolin or carbon-white stabilizer.

| Component No. | S:TMTD:ZnO:BHT: $Al_2O_3{\cdot}2SiO_2{\cdot}2H_2O$ | Component No. | S:TMTD:ZnO:BHT:$SiO_2$ |
|---|---|---|---|
| 25 | 60:16:4:20:500 | 29 | 60:16:4:20:500 |
| 26 | 60:16:4:20:1000 | 30 | 60:16:4:20:1000 |
| 27 | 60:16:4:20:2000 | 31 | 60:16:4:20:2000 |
| 28 | 60:16:4:20:3000 | 32 | 60:16:4:20:3000 |

According to the above components, the different stabilizers prepared, the amount of stabilizer (in terms of the amount of S bitumen percentage) being 0.09% control stabilizer content, prepared SBS-modified bitumen and relevant tests according to the research need to be selected.

### 5.1. Kaolin Content

SBS-modified bitumen was prepared according to the test design, and the content of stabilizer (expressed as S content) was 0.09%. No gelation phenomenon occurred in the preparation process, and the 48-h segregation softening point difference was analyzed as Figure 9 shows.

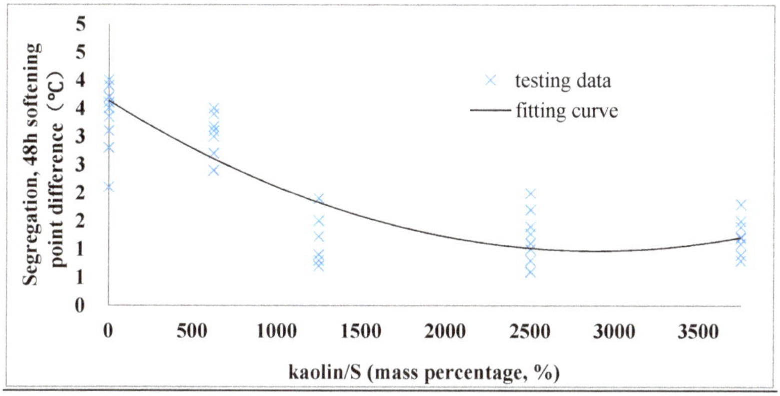


**Figure 9.** The relationship between kaolin content and the 48-h segregation softening point difference of SBS-modified bitumen.

As can be seen from Figure 9, with the increase of kaolin content, the 48-h segregation softening point difference of SBS-modified bitumen gradually decreases. When the content of kaolin (relative to S) reaches 1250%, the modified bitumen has almost reached the smallest of the 48-h segregation softening point difference; with the kaolin content continuing to increase, the 48-h segregation softening point difference seems to no longer decrease. The reason was that kaolin can reduce the density difference between bitumen and SBS, which leads to the modified bitumen achieving stabilization. Considering the result of the 48-h segregation softening point difference test, it was determined that the proper content of kaolin (relative to S) was 1250%.

It can be seen from Table 7 above that the incorporation of kaolin has a certain degree of enhancement on the softening point, the Brookfield viscosity at 135 °C and the penetration percentage after RTFOT, but the enhancement was not significant in the range of the test content. With the kaolin content increase, the ductility decreases slightly after RTFOT, and the ductility at 5 °C decreases obviously. From the viewpoint of low temperature and anti-aging performance, the content of kaolin should be controlled.

**Table 7.** Performance of SBS-modified bitumen with different kaolin content.

| Component No. | RTFOT | | | Penetration, 0.1 mm, 25 °C | Penetration Index PI |
|---|---|---|---|---|---|
| | Loss on Heating, % | Penetration Percentage, % | Residue Ductility at 5 °C | | |
| 4 | 0.08 | 79.2 | 21.0 | 55.3 | 0.1 |
| 25 | 0.11 | 77.2 | 18.9 | 54.4 | 0.1 |
| 26 | 0.11 | 77.9 | 18.8 | 54.3 | 0.1 |
| 27 | 0.11 | 78.5 | 18.4 | 54.5 | 0.1 |
| 28 | 0.10 | 78.7 | 17.5 | 54.1 | 0.1 |
| **Component No.** | **Ductility at 5 °C, cm** | **Softening Point °C** | **Brookfield viscosity at 135 °C, Pa·S** | **Solubility, %** | **Resilient at 25 °C, %** |
| 4 | 29.6 | 78.4 | 1.972 | 99.3 | 87.7 |
| 25 | 29.5 | 79.7 | 1.984 | 99.4 | 87.4 |
| 26 | 28.7 | 80.5 | 2.046 | 99.3 | 86.7 |
| 27 | 27.8 | 81.5 | 2.107 | 99.2 | 87.3 |
| 28 | 26.6 | 81.3 | 2.157 | 98.9 | 86.4 |

"Technical Specification of Construction of Highway Asphalt Pavement (JTG F40-2004, China)" [20] regulates that the 48-h segregation softening point difference should be no more than 2 °C; according to the testing data, when the content of kaolin is up to 1250%, the 48-h segregation softening point difference of SBS-modified bitumen was less than 2 °C, and considering the test phenomenon, test results and economic factors, it was determined that the proper content of kaolin (relative to S) was 1250%.

### 5.2. Carbon-White Content

SBS bitumen was prepared according to the experimental design above, and the content of the stabilizer was 0.09% (expressed as S content). No gelation phenomenon occurred during the preparation process, and the 48-h segregation softening point difference test was carried out. The test results are shown in Figure 10.

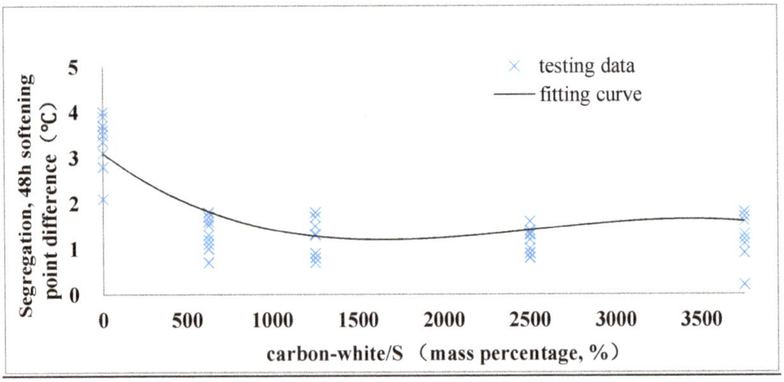


**Figure 10.** The relationship between carbon-white content and 48-h segregation softening point difference of SBS-modified bitumen.

As can be seen from Figure 10, the 48-h segregation softening point difference of SBS-modified bitumen decreased about 2 °C with the incorporation of carbon-white, but there was no significant difference after the content of carbon-white exceeded 625%, so from the perspective of improving the thermal storage stability of bitumen, the content of carbon-white (relative to S) was 625%.

Table 8 shows the performance of SBS-modified bitumen with different carbon-white content.

**Table 8.** Performance of SBS-modified bitumen with different carbon-white content.

| Component No. | RTFOT | | | Penetration, 0.1 mm, 25 °C | Penetration Index PI |
|---|---|---|---|---|---|
| | Loss on Heating, % | Penetration Percentage, % | Residue Ductility at 5 °C | | |
| 4 | 0.08 | 79.2 | 21.0 | 55.3 | 0.1 |
| 29 | 0.10 | 77.2 | 19.7 | 55.0 | 0.1 |
| 30 | 0.09 | 78.3 | 19.5 | 54.4 | 0.1 |
| 31 | 0.12 | 78.5 | 18.7 | 54.2 | 0.1 |
| 32 | 0.09 | 79.4 | 18.2 | 53.6 | 0.1 |
| **Component No.** | **Ductility at 5 °C, cm** | **Softening Point °C** | **Brookfield viscosity at 135 °C, Pa·S** | **Solubility, %** | **Resilient at 25 °C, %** |
| 4 | 29.6 | 78.3 | 1.973 | 99.3 | 87.7 |
| 29 | 29.0 | 79.1 | 1.983 | 99.1 | 87.5 |
| 30 | 28.2 | 79.3 | 1.990 | 99.1 | 87.2 |
| 31 | 27.5 | 79.8 | 1.936 | 99.1 | 87.4 |
| 32 | 26.3 | 80.2 | 1.998 | 98.8 | 87.3 |

As can be seen from Table 8 above, the incorporation of carbon-white improves the softening point, the Brookfield viscosity at 135 °C and the penetration percentage after RTFOT. However, the enhancement was not significant in the experimental range; the improvement of the modified bitumen properties above was limited; with the increase of carbon-white content, the ductility after RTFOT decreased slightly; and the ductility at 5 °C decreases obviously. From the perspective of low temperature performance, the carbon-white content should be controlled.

"Technical Specification of Construction of Highway Asphalt Pavement (JTG F40-2004, China)" regulates that the 48-h segregation softening point difference should be no more than 2 °C; according to the testing data, when the content of carbon-white was up to 625%, the 48-h segregation softening point difference of SBS-modified bitumen was less than 2 °C. Based on the economic factors and the technical performance of SBS-modified bitumen, especially the response to heat storage stability, it was suggested that the proper content of carbon-white (relative to S) be 625%.

In conclusion, the components of SBS-modified bitumen stabilizer powder were S:TMTD:ZnO:BHT: $Al_2O_3 \cdot 2SiO_2 \cdot 2H_2O$ = 60:16:4:20:1250 or S:TMTD:ZnO:BHT:$SiO_2$ = 60:16:4:20:625, and SBS-modified bitumen has good performance, especially for the thermal storage stability.

## 6. Microcosmic Test

A fluorescence microscope was used in this paper to observe the SBS bitumen from a microcosmic view after vulcanization reaction. The principle of the fluorescence microscope is described as follows. Fluorescence will occur in a short period of time after materials are excited. Most of the polymer materials can emit fluorescence light directly after high energy luminescent beam irradiation, and a red shift interval will occur between the high energy luminescent beam and fluorescence after excitation, then the wavelength of fluorescence will become longer as visible light. Bitumen cannot produce fluorescence in the blue light region, while the yellow-green fluorescence comes from SBS. The fluorescence microscope worked on the principle above.

Before the fluorescence microscope test, SBS-modified bitumen was prepared respectively with different stabilizers (S:TMTD:ZnO:BHT: $Al_2O_3 \cdot 2SiO_2 \cdot 2H_2O$ = 60:16:4:20:1250, S:TMTD:ZnO:BHT:$SiO_2$ = 60:16:4:20:625), and the stabilizer contents were both 0.09% (bitumen mass percentage, expressed as S content), while SBS content was 4% (bitumen mass percentage). SBS bitumen was put into the oven at a temperature of 163 °C for 48 h, sampling the upper 1/3 and the lower 1/3, respectively. Then, different SBS-modified bitumen was dropped on the glass, respectively, and the cover glass was pushed from one side to another lightly without air bubbles. The glass with bitumen was put under the fluorescence microscope, and it was observed with a 100 and a 400 amplified factor, as shown in the Figure 11.

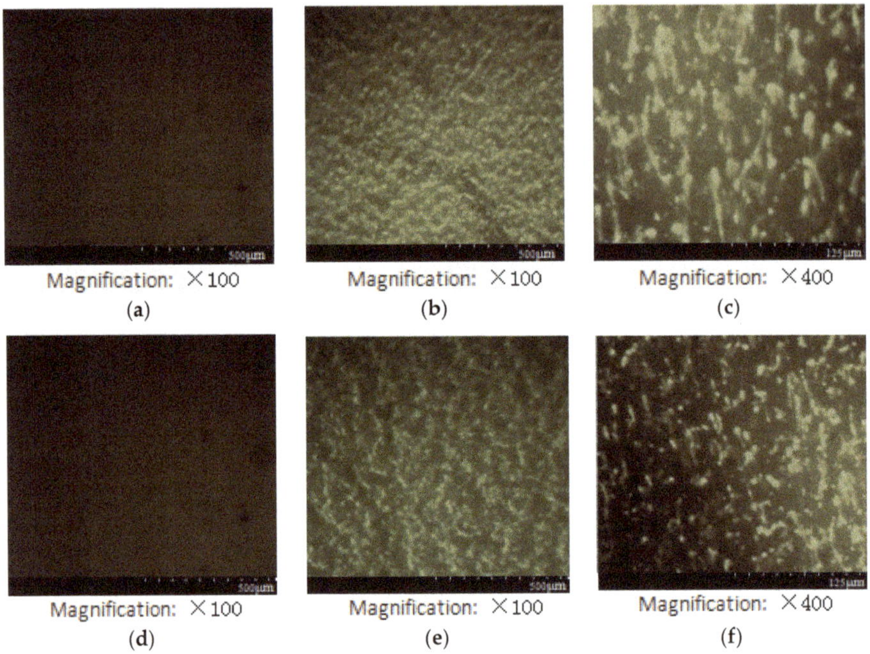


**Figure 11.** Fluorescence microscope test (stabilizer was S:TMTD:ZnO:BHT:$Al_2O_3{\cdot}2SiO_2{\cdot}2H_2O$ = 60:16:4:20:1250). (**a**) Base bitumen; (**b**) top portions (SBS bitumen); (**c**) top portions (SBS bitumen); (**d**) base bitumen; (**e**) bottom portions (SBS bitumen); (**f**) bottom portions (SBS bitumen).

In Figure 11, black represents bitumen, while white represents SBS. SBS in modified bitumen of the upper 1/3 was almost as much as the lower 1/3, which can be seen from Figure 11b,e. It shows that the stabilizer plays a good role in preventing SBS segregation. SBS was cross-linked, as shown in Figure 11c or Figure 11f, which was caused by the vulcanization reaction. The degree of vulcanization in SBS-modified bitumen was almost the same in the upper 1/3 and the lower 1/3. The same phenomenon occurred when using the stabilizer S:TMTD:ZnO:BHT:$Al_2O_3{\cdot}2SiO_2{\cdot}2H_2O$ = 60:16:4:20:1250.

According to the phenomenon that the stabilizer developed based on the vulcanization reaction in this research can allow SBS to be cross-linked, it especially has the effect of preventing SBS segregation.

## 7. Conclusions

In this paper, S, TMTD, ZnO, BHT, $Al_2O_3{\cdot}2SiO_2{\cdot}2H_2O$ and $SiO_2$ were used as raw materials to obtain SBS-modified bitumen stabilizer powder. The following conclusions can be drawn on the basis of the results and analysis.

(1) The vulcanization reaction was the main mechanism by which stabilizers improved the thermal storage stability of SBS-modified bitumen. Sulfur, vulcanization accelerator, vulcanization activator, antioxidant, other active ingredients was the basic design idea, and the basic components of the stabilizer were S:TMTD:ZnO:BHT = 60:16:4:20. The presence of the basic components of the stabilizer improves the storage stability and aging resistance of SBS-modified bitumen significantly. It can also improve the softening point and viscosity and reduce the penetration.

(2) $Al_2O_3{\cdot}2SiO_2{\cdot}2H_2O$ and $SiO_2$ can further improve the storage stability of SBS-modified bitumen significantly. However, they also reduce the low temperature property at the same time. The proper formula of SBS-modified bitumen stabilizer was S:TMTD:ZnO:BHT: $Al_2O_3{\cdot}2SiO_2{\cdot}2H_2O$ = 60:16:4:20:1250 or S:TMTD:ZnO:BHT:$SiO_2$ = 60:16:4:20:625; the optimum content was 0.9% (bitumen mass percentage, expressed as S content).

(3) The stabilizers developed in this paper were a dry powder, which can be directly put into bitumen easily, and the performance of SBS-modified bitumen with these stabilizers can meet the requirement of "Technical Specification of Construction of Highway Asphalt Pavement" (JTG F40-2004, China) [20].

**Acknowledgments:** This work was supported by the Natural Science Foundation of Shandong Province, China (Grant No. BS2015SF016) and the National Natural Science Foundation of China (Grant No. 51408044)

**Author Contributions:** W.Z. conceived of and designed the experiments. L.D. performed the experiments. W.Z. analyzed the data. Z.J. contributed reagents/materials/analysis tools. W.Z. wrote the paper.

**Conflicts of Interest:** The authors declare no conflict of interest.

## References

1. Nuria, Q.; Camila, B.; Luisa, F.C. Storage Stability of Bimodal Emulsions vs. Monomodal Emulsions. *Appl. Sci.* **2017**, *7*, 1267. [CrossRef]
2. Martínez-Estrada, A.; Chávez-Castellanos, A.E.; Herrera-Alonso, M.; Herrera-Nájera, R. Comparative study of the effect of sulfur on the morphology and rheological properties of SB- and SBS-modified asphalt. *J. Appl. Polym. Sci.* **2010**, *115*, 3409–3422. [CrossRef]
3. Ma, T.; Zhao, Y.L.; Huang, X.M.; Zhang, Y. Characteristics of desulfurized rubber asphalt and mixture. *KSCE J. Civ. Eng.* **2016**, *20*, 1347–1355. [CrossRef]
4. Yang, Z.X.; Zhang, Y.; Shi, X.M. Impact of nanoclay and carbon microfiber in combating the deterioration of asphalt concrete by non-chloride deicers. *Constr. Build. Mater.* **2018**, *160*, 514–525. [CrossRef]
5. Zhang, H.L.; Su, M.M.; Zhao, S.F.; Zhang, Y.P.; Zhang, Z.P. High and low temperature properties of nano-particles/polymer modified asphalt. *Constr. Build. Mater.* **2016**, *114*, 323–332. [CrossRef]
6. Liang, M.; Xin, X.; Fan, W.Y.; Ren, S.S.; Liang, M.; Wang, H. Effects of polymerized sulfur on rheological properties, morphology and stability of SBS modified asphalt. *Constr. Build. Mater.* **2017**, *150*, 860–871. [CrossRef]
7. Norrman, J.; Rosén, L.; Norin, M. Decision analysis for storage for reclaimed asphalt. *Environ. Eng. Sci.* **2005**, *22*, 783–801. [CrossRef]
8. Ryms, M.; Denda, H.; Jaskuła, P. Thermal stabilization and permanent deformation resistance of LWA/PCM-modified asphalt road surfaces. *Constr. Build. Mater.* **2017**, *142*, 328–341. [CrossRef]
9. Sun, D.; Ye, F.; Shi, F.Z.; Lu, W.M. Storage stability of SBS-modified road asphalt: Preparation, morphology, and rheological properties. *Pet. Sci. Technol.* **2006**, *24*, 1067–1077. [CrossRef]
10. Liu, Q.J. Study on storage stability of SBS modified asphalt and stabilizer. *Sci. Technol. Eng.* **2009**, *9*, 6579–6584. (In Chinese)
11. Yang, Y. Analysis on the structure design of rubber asphalt technology of Liao-Kai highway. *J. Liaoning Prov. Coll. Commun.* **2011**, *13*, 8–12. (In Chinese)
12. Li, S.R.; Li, Q.; Dong, S.X. Application and research on stabilizers of styrene-butadiene-styrene polymer modified asphalts. *Highway* **2007**, *9*, 153–155. (In Chinese)
13. Pang, J.Y.; Du, S.J.; Chang, R.J. The influence of inorganic clay on the performance of SBS modified asphalt. *Shanxi Sci. Technol. Commun.* **2014**, *227*, 19–21. (In Chinese)
14. Aguirre de Carcer, Í.; Masegosa, R.M.; Viñas, M.T.; Sanchez-Cabezudo, M.; Salom, C.; Prolongo, M.G.; Contreras, V.; Barceló, F.; Páez, A. Storage stability of SBS/sulfur modified bitumens at high temperature: Influence of bitumen composition and structure. *Constr. Build. Mater.* **2014**, *52*, 245–252. [CrossRef]
15. Galooyak, S.S.; Dabir, B.; Nazarbeygi, A.E.; Moeini, A. Rheological properties and storage stability of bitumen/SBS/montmorillonite composites. *Constr. Build. Mater.* **2010**, *24*, 300–307. [CrossRef]

16. Sienkiewicz, M.; Borzędowska-Labuda, K.; Wojtkiewicz, A.; Janik, H. Development of methods improving storage stability of bitumen modified with ground tire rubber: A review. *Fuel Process. Technol.* **2017**, *159*, 272–279. [CrossRef]
17. Amini, B.; Rajabbolookat, M.J.; Abdi, A.; Salehfard, R. Investigating the influence of using nano-composites on storage stability of modified bitumen and moisture damage of HMA. *Pet. Sci. Technol.* **2017**, *35*, 800–805. [CrossRef]
18. Goli, A.; Ziari, H.; Amini, A. Influence of carbon nanotubes on performance properties and storage stability of SBS modified asphalt binders. *J. Mater. Civ. Eng.* **2017**, *29*, 04017070. [CrossRef]
19. Al-Abdul Wahhab, H.I.; Dalhat, M.A.; Habib, M.A. Storage stability and high-temperature performance of asphalt binder modified with recycled plastic. *Road Mater. Pavement Des.* **2017**, *18*, 1117–1134. [CrossRef]
20. The Ministry of Communications Highway Science Institute. *Technical Specification of Construction of Highway Asphalt Pavement*; JTG F40-2004; China Communications Press: Beijing, China, 2004.
21. Polacco, G.; Filippi, S. Vulcanization accelerators as alternative to elemental sulfur to produce storage stable SBS modified asphalts. *Constr. Build. Mater.* **2014**, *58*, 94–100. [CrossRef]
22. Wen, G.; Zhang, Y.; Zhang, Y.X.; Sun, K.; Fan, Y.Z. Improved properties of SBS-modified asphalt with dynamic vulcanization. *Polym. Eng. Sci.* **2002**, *42*, 1070–1081. [CrossRef]
23. Yang, X.W.; Liu, K. Study on mechanism of sulphur modified asphalt. *J. Shijiazhuang Railw. Inst. (Nat. Sci.)* **2008**, *21*, 43–48. [CrossRef]
24. Coran, A.Y. Vulcanization. In *Science and Technology of Rubber*, 4th ed.; Mark, J.E., Erman, B., Roland, M., Eds.; Academic Press: Salt Lake City, UT, USA, 2013; pp. 337–381. [CrossRef]
25. Fu, Z. *Rubber Materials and Processing*, 1st ed.; Chemical Industry Press: Beijing, China, 2013; pp. 154–196. (In Chinese)
26. Nieuwenhuizen, P.J.; Van Duin, M.; Haasnoot, J.G.; Reedijk, J.; McGill, W.J. Limiting value of ZDMC formation: New insight into the reaction of ZnO and TMTD. *J. Appl. Polym. Sci.* **1999**, *73*, 1247–1257. [CrossRef]
27. Zhang, H.L.; Gao, Y.; Guo, G. H.; Zhao, B. J.; Yu, J.Y. Effects of ZnO particle size on properties of asphalt and asphalt mixture. *Constr. Build. Mater.* **2018**, *159*, 578–586. [CrossRef]
28. Zhao, Q.S.; Xie, H.; Ning, H.; Liu, J.L.; Zhang, H.R.; Wang, L.H.; Wang, X.B.; Zhu, Y.L.; Li, S.Y.; Wu, M.B. Intercalating petroleum asphalt into electrospun ZnO/Carbon nanofibers as enhanced free-standing anode for lithium-ion batteries. *J. Alloys Compd.* **2018**, *737*, 330–336. [CrossRef]
29. Zhu, C.Z.; Zhang, H.L.; Shi, C.J.; Li, S. Effect of nano-zinc oxide and organic expanded vermiculite on rheological properties of different bitumens before and after aging. *Constr. Build. Mater.* **2017**, *146*, 30–37. [CrossRef]
30. Ouyang, C.F.; Wang, S.F.; Zhang, Y.; Zhang, Y.X. Thermo-rheological properties and storage stability of SEBS/kaolinite clay compound modified asphalts. *Eur. Polym. J.* **2006**, *42*, 446–457. [CrossRef]
31. Wang, P.; Dong, Z.J.; Tan, Y.Q.; Liu, Z.Y. Anti-ageing properties of styrene–butadiene–styrene copolymer-modified asphalt combined with multi-walled carbon nanotubes. *Road Mater. Pavement Des.* **2017**, *18*, 533–549. [CrossRef]
32. Zhou, X.X.; Zhang, X.; Xu, S.; Wu, S.P.; Liu, Q.T.; Fan, Z.P. Evaluation of thermo-mechanical properties of graphene/carbon-nanotubes modified asphalt with molecular simulation. *Mol. Simul.* **2017**, *43*, 312–319. [CrossRef]
33. The Ministry of Communications Highway Science Institute. *Standard Test Methods of Bitumen and Bituminous mixtures for Highway Engineering*; JTG E20-2011; China Communications Press: Beijing, China, 2011.

MDPI
St. Alban-Anlage 66
4052 Basel
Switzerland
Tel. +41 61 683 77 34
Fax +41 61 302 89 18
www.mdpi.com

*Applied Sciences* Editorial Office
E-mail: applsci@mdpi.com
www.mdpi.com/journal/applsci